may the sun always
rise on your tomorrows
Roy E. Williams

Waste Production and Disposal
in Mining, Milling, and Metallurgical Industries

Other Miller Freeman Publications books for the mining and minerals processing industries:

Minerals Transportation Volume I: Proceedings of the first International Symposium on Transport and Handling of Minerals, Vancouver, British Columbia, Canada, October 1971

Minerals Transportation Volume II: Proceedings of the second International Symposium on the Transport and Handling of Minerals, Rotterdam, Netherlands, October 1973

Tailing Disposal Today: Proceedings of the first International Symposium on Tailing Disposal, Tucson, Arizona, U.S.A., October-November 1972

Copper Porphyries *by Alexander Sutulov*

World Mines Register: International directory of active mining companies, operations and key personnel

International Glossary of Mining, Processing and Geological Terms: Revised edition of over 10,000 terms and phrases. Translated in five languages: English, French, German, Spanish, and Swedish, with four separate bilingual indices

Review of the World Coal Industry to 1990 *by W. L. G. Muir*

Waste Production and Disposal *in Mining, Milling, and Metallurgical Industries*

Roy E. Williams

A WORLD MINING BOOK

Miller Freeman Publications, Inc.

Library of Congress Catalog Card Number 74-20167
International Standard Book Number 0-87930-035-3

Printed in the United States of America.

Preface

This book presents the first published, comprehensive treatment of the origin, disposal, and legal aspects of mineral resource based wastes, along with a detailed discussion of the widespread impact of the Federal Water Pollution Control Act Amendments of 1972 (Public Law 92-500).

At a time when industry, government, and society are becoming ever more concerned about the many problems created by mineral resource based wastes, the author believes that this book offers the scientific and legal background and information needed to make the changes that can lead to the reduction or elimination of these wastes.

Chapter 1 deals with Public Law 92-500 itself. The many aspects of this law pertinent to the mineral resource based industries are clearly summarized, along with interim procedures and final methods by which the law is to be implemented when it becomes effective on July 1, 1977.

Chapter 2 presents the general nature of gaseous, liquid, and solid wastes generated by mining, milling, and metallurgical industries.

Chapter 3 presents the waste treatment technologies that are available to mineral resource based industries. Readily available sources of more detailed technological information on waste water treatment are cited.

Chapter 4 treats the problem of acid mine drainage, including both the costs and the benefits of its elimination. This chapter covers the first of seven point source waste water discharge categories directly affected by Public Law 92-500.

Chapters 5 through 9 are based primarily on the government-sponsored studies from which effluent limitations guidelines for point source waste water discharges of nine major mineral resource based industries have been or will be developed. Public Law 92-500 requires that such studies be conducted to define the best practicable control technology currently available and the best control technology economically achievable for essentially all U.S. industries. These chapters also present the point source effluent limitations guidelines that have been promulgated on the basis of those studies and on the basis of reviews and revisions thereof by industry, government agencies, and the public.

Chapter 10 deals with mill wastes, which are also subject to regulation under Public Law 92-500. However, at this writing no effluent guidelines development study has been authorized.

Chapters 11 and 12 treat the subjects of mineral resource solid waste disposal and reclamation of mined areas, respectively. These subjects clearly will play an increasingly important role in the future of mineral resource based industries.

The broad scope of the book is intended to provide not only the practical perspective for much needed research and training, but also to make it appropriate as a text for several types of courses at the undergraduate and graduate levels. Chapters 5 through 9, for instance, cover subjects which make the book suitable for courses in extractive metallurgy and in sanitary engineering. The waste production and handling characteristics of several major industries are presented in parallel with the description of the processes within each industry. Chapters 4, 11, and 12 provide basic material for courses in mining engineering, hydrogeology, exploration geology, and environmental reclamation. Chapter 1 and several other portions of the book which deal with Public Law 92-500 are, in the author's opinion, especially important for university level courses treating the legal aspects of waste and resource management.

Thus, it is the author's hope that this book will be useful to both industry and education since it will facilitate the training of those many professionals who are needed to solve the crucial problems of waste disposal in the mineral resource based industries.

Roy E. Williams
Moscow, Idaho
April 1975

Acknowledgments

This book could not have been written without the help and encouragement of several dedicated individuals. Most notable among them is Dr. Alvin Van Valkenburg, Chief of the Office of University Relations, U.S. Bureau of Mines, Washington, D.C. Dr. Van Valkenburg recognized early the need for a book on mineral resource wastes that would enable universities to keep abreast of the rapidly changing demands being placed on graduates in the mineral resource field. A considerable portion of this book is based on information assembled while I was with Dr. Van Valkenburg's office under the Intergovernmental Personnel Exchange Act of 1970. Dr. Neldon Jensen and Dr. Dan Kealy, also with the Bureau of Mines, provided invaluable additional encouragement and assistance, as did Dorothy Jones, Anna Jones, and Katy Covington of the Office of University Relations. It must be noted, however, that the views and concepts presented in the book do not necessarily reflect those of the Bureau of Mines.

Thanks are also in order to Mr. Don Gilmore of the Office of Research and Development, U.S. Environmental Protection Agency, Washington, D.C. Mr. Gilmore provided much needed advice and assistance in the sometimes not so easy task of identifying, tracking down, and obtaining the latest federal documents.

At the University of Idaho, Dr. Robert Coonrod, Vice President for Academic Affairs, Dr. Rolland Reid, Dean of the College of Mines, and Dr. George Williams, Chairman of the Department of Geology, supported and granted unusual requests for sufficient leave time to complete the manuscript. Dr. Dale Ralston picked up most of the slack. For this cooperation I am most grateful. And Ms. Dianne Haynes, Administrative Assistant to the Dean of Idaho's College of Mines, must surely be one of the world's outstanding coordinators for transient authors.

On the industrial side thanks are in order to Mr. Gene Baker, Vice President, Environmental Affairs, the Bunker Hill Company, Kellogg, Idaho. Mr. Baker granted me permission to use portions of his own work in the manuscript. He also provided the opportunity for extensive study of the comprehensive Bunker Hill mining, milling, and metallurgical complex at Kellogg. Mr. William McKee, Manager of Mills at Hecla Mining Company's Star and Lucky Friday Mines, Wallace, Idaho, provided similar opportunities, as did Mr. George Deshler, Manager of ASARCO's Galena operation near Wallace. These three operations were among the first in the United States to comply with the new federal point source waste discharge permit system on the basis of the interim guidelines discussed in Chapter 1.

Last, but not least, I wish to thank Gail for putting up with a part-time husband and to thank Karen and Ed Beverly, Defense Supply Agency, and Ms. Sue Porter, Appointments Secretary for Mrs. Richard Nixon, for making my tenure in Washington, D.C. extremely pleasant.

Roy E. Williams

Contents

1. *The federal water pollution control act amendments of 1972 and related legislation*

2. *Characterization of wastes from mineral resource based industries*

3. *Methods of waste and waste water treatment available to mineral resource industries*

4. *Acid mine drainage*

5. *Bauxite refining and aluminum smelting*

6. *Phosphate fertilizer industry*

7. *Iron and steel industry and coke manufacturing*

8. *Ferroalloy industry*

9. *Recovery of copper, lead, and zinc*

10. *Mill wastes*

11. *Production and disposal of mineral resource solid wastes*

12. *Mined area reclamation*

Appendices

1. *The federal water pollution control act amendments of 1972 and related legislation*

INTRODUCTION

On October 18, 1972, the Congress of the United States passed the Federal Water Pollution Control Act Amendments of 1972. The amendments are designated Public Law 92-500. By July 1, 1977, this law will have had far-reaching effects on the mining, milling, and metallurgical industries and on the training of individuals who will work therein. Some spokesmen have predicted that the effect on industry will be severe (Driver 1973), but others feel that the controls imposed on industry will be realistic (Kostelnik 1973).

Waste water and associated solids and gases are ubiquitous throughout many mineral resource based industries. Because it is difficult to break the chain of conversion of gaseous wastes to liquid wastes to solid wastes, Public Law 92-500 is intimately associated with air pollution control, solid waste disposal, and production. Consequently, the water quality improvement aspects of the amendments will have the effect of delegating to the U.S. Environmental Protection Agency (EPA) responsibility for a large new federally sponsored research and regulatory thrust within these industries. Since the amendments delegate regulatory authority to EPA, research initiatives and regulatory power are handled by the same agency—historically a somewhat unusual situation. Planners, engineers, researchers, and administrators in the mining, milling, and metallurgical industries should find considerable value in project completion reports resulting from this dual role, particularly if cooperation among the agencies develops, as will hopefully be the case.

As an example of such potentially useful projects, on July 10, 1973, EPA let a contract to the West Virginia Surface Mining and Reclamation Association for the first phase of a \$2 million project to demonstrate the feasibility of using longwall miners and movable jacks to allow subsidence behind equipment extracting shallow underground coal seams. The objective, of course, is to demonstrate a mining method that will minimize the production of acid mine drainage after abandonment of the property, but the research may produce results applicable to a much greater variety of mining problems involving controlled caving.

Projects of this type are tabulated by EPA in publications such as Hall (1972). The most pertinent new projects are sponsored by the Office of Research and Monitoring, EPA, under the series entitled *Environmental Protection Technology*. Pertinent projects of other agencies are publicized through the well-known channels of more established branches of government, such as the U.S. Bureau of Mines.

The sections of Public Law 92-500 most pertinent to mineral resource based industries are under Title III-"Standards and Enforcement, Effluent Limitations." Sections 301, 302, and 304 are particularly pertinent. They are presented in Appendix A.

The impact of these and other sections has been discussed in some detail by Williams (1973); a summary of the broad objectives of the amendments is presented by Quarles (1973).

The amendments stipulate that the intent of Congress is to control the quality of point discharges or groups of point discharges. The definition of *point* becomes important because most waste water impoundments lose water by seepage. The definition of *point* will thus determine whether the quality of such seepage has to be controlled under the amendments. The question is important to the design and economics of disposal of mine wastes in tailing ponds (see Williams, Kealy, and Mink 1973). Cox and Walker (1973) conclude that EPA interpretation of the law extends the federal permit requirement to disposal of wastes into wells where this technique is used in combination with, or as a replacement for, facilities for discharge into navigable waters. Thus the law may be interpreted to include ground-water quality control as well as surface-water quality control.

In summary, Congress has established July 1, 1977, as the date for application of the best practicable control technology currently available and July 1, 1983, as the date for application of the best available technology economically achievable. It has further given the administrator of the Environmental Protection Agency the power to pass judgment on technology that is practicable, available, and economical, and instructed him to publish information on the quality of waste discharges which is compatible with that technology for each of several industrial categories. EPA is charged also with the task of determining what additional steps are necessary to restore and maintain the quality of the nation's waters.

Congress has also authorized under section 107 of the act, the expenditures of $30 million to assist in the control of mine water. This section is entitled "Mine Water Pollution Control Demonstrations" and is directed primarily at the problem of acid mine drainage.

THE INTERIM GUIDELINES

The Environmental Protection Agency has prepared interim guidelines for the establishment of effluent limitations for discharges. The contracts for recommending permanent guidelines have been awarded to research and consulting firms. Information that is pertinent to the completion of these studies can be obtained from the *Federal Register.* This information can also be obtained from the contractors.

The interim guidelines set forth numerical limitations based on the agency's current interpretation of "best practicable control technology currently available." These guidelines are the temporary predecessors of the permanent guidelines discussed in the amendments, but the two probably will not differ greatly. The interim guidelines for the mining and milling industry are presented here for illustrative purposes (Table 1.1). These guidelines present two categories of effluent quality limitations. Category A values reflect the agency's judgment of the effluent levels that can be achieved by the application of the highest level of control technology now considered "practicable" and "currently available." Category B values represent the agency's judgment of the minimum acceptable interim effluent levels possible for the mining and milling industries.

In the case of new plants installing pollution abatement equipment or existing plants now beginning abatement programs, EPA requires compliance with category A values. However, in some cases economic and social factors may affect the practicability of applying control techniques to achieve these values, and some modification of category A values may be allowed in particular plants.

Category B values may be applied where a discharger has, at the time the permit is issued, "commenced and made substantial progress on an abatement program that will be completed within 24 months or less from the time the discharge permit is issued. If the plant also has extensive on-going pollution abatement programs in other areas such as air pollution, the Regional Administrator may modify this 24-month period." Region X of the EPA has used these interim guidelines as a basis for the issuance of waste discharge permits to the mineral resource industries in the Coeur d'Alene mining district of North Idaho. To the knowledge of the author, these were the first such permits that were issued to the mineral resource industry under the authority granted by Public Law 92-500.

The analytical procedures acceptable for determining concentrations of pollutants in point source discharges are presented in Appendix B.*

APPLICATION OF EFFLUENT LIMITATIONS

The effluent limitations shown in categories A and B (Table 1.1) are expressed only in concentrations. Due to the wide variations in ore types processed (Table 1.2), EPA has concluded that a pound-per-day limitation is impractical. Presumably this will not be the case for all industries, and the policy might very well change for the milling industry when the final guidelines are adopted.

* These procedures are published in the *Federal Register*, Vol. 38, No. 199, October 16, 1973.

Table 1.1 *Recommended Effluent Limitations*

Criteria	*Category A*	*Category B*
Suspended solids	20 mg/l[1]	30 mg/l[1]
pH (range)	6.0 to 8.5	6.0 to 8.5
Cyanide (total)	When cyanide compounds are used in the milling process, effluent from the waste abatement system shall not exceed 0.02 mg/l.	When cyanide compounds are used in the milling process, effluent from the waste abatement system shall not exceed 0.03 mg/l.
Sulfate	The regional administrator or his designated representative shall establish an effluent limitation if beneficial downstream water uses are affected.	The regional administrator or his designated representative shall establish an effluent limitation if beneficial downstream water uses are affected.
Radioactivity	Discharge concentration shall not exceed 1/10 of Bureau of Standards Handbook 69 values, or such standards as may be developed. Ra-226 concentrations shall be as close to background as possible but in no case shall exceed the addition of 1.0 picocurie per liter.	Discharge concentration shall not exceed 1/5 of Bureau of Standards Handbook 69 values, or such standards as may be developed. Ra-226 concentrations shall be as close to background as possible but in no case shall exceed the addition of 3.0 picocurie. per liter.

Iron (dissolved)	0.5 mg/l	7.0 mg/l
Manganese	0.1 mg/l	1.0 mg/l
Heavy metals and toxic materials	The limitations on concentrations of specific metals in the discharge will be determined by the regional administrator. Additional guidance is forthcoming on those parameters which show synergistic tendencies.[2]	The limitations on concentrations of specific metals in the discharge will be determined by the regional administrator. Additional guidance is forthcoming on those parameters which show synergistic tendencies.[3]

Note: All effluent limitations apply at the point of discharge from the treatment system to the receiving water. No natural water course shall be used as a transport mechanism for untreated or partially treated waste.

1. Turbidity measurements may be utilized in lieu of suspended solids with the approval of the regional administrator or his designated representative, if the permittee can demonstrate a correlation between turbidity and suspended solids concentrations utilizing not less than two years of data generated from daily samples.

2. The maximum short-term permissible combined heavy metal discharge specified by the May 18, 1973, drafts of the waste discharge permits for the Coeur d'Alene mining district is 4.5 ppm (see Appendix D).

3. The maximum long-term permissible combined heavy metal discharge specified by the May 18, 1973, drafts of the waste discharge permits for the Coeur d'Alene mining district is 1.5 ppm (see Appendix D).

Table 1.2 *Standard Industrial Classifications Covered by Mining and Milling Industry Interim Guidelines*

SIC Number	*Classification*
1011	Iron ores
1021	Copper ores
1031	Lead and zinc ores
1042	Lode gold
1043	Placer gold
1044	Silver ores
1051	Bauxite and other aluminum ores
1062	Manganese ores
1064	Tungsten ores
1069	Ferroalloy ores, except vanadium not elsewhere classified
1092	Mercury ores
1093	Titanium ores
1094	Uranium, radium, vanadium ores
1099	Metallic minerals (ores) not elsewhere classified

Note: Other interim guidelines followed by permanent guidelines will be developed for other industries.

RATIONALE FOR DEVELOPMENT OF INTERIM EFFLUENT LIMITATIONS

The interim guidelines contain a description of the justification of the effluent limits achievable using the "best practicable control technology currently available." The rationale is:

> Wastes from the mining and milling industry commonly contain high concentrations of various toxicants, especially heavy metals, which tend to persist in receiving streams. The direct toxicity of such metals is dependent on a number of factors, such as pH, hardness, and temperature, as well as the presence of other metals which may exert a synergistic effect (U.S.D.I. 1968). It is therefore frequently necessary to treat the wastes to achieve metal removal.

EPA used several controls and treatment processes as models in developing the recommended effluent limitations for the mining and milling industry as contained in category A. The guidelines stipulate that these limitations should not be applied to all operations, but should be considered on a case-by-case basis. The essence of each model is based on standard treatment techniques which will be discussed subsequently. Currently, however, the milling industry in particular does not employ treatment to the extent suggested by the interim guidelines. This observation is based on direct observations by the author and on the following exerpts from the guidelines:

Common treatment techniques utilized by the milling industry consist of discharging the slurry into a tailings pond, with sufficient detention time to reduce the suspended-sediment concentration to approximately 20 mg/l. [See Appendix C for Metric-English units conversion table.] The clarified water is then either decanted to receiving streams or, in an increasing number of cases, recycled as mill process water. This latter procedure is preferred for environmental consideration.

The treatment considered in the development of [categories A and B] limits for discharge to streams considers primary sedimentation in a tailings pond, as is currently practiced throughout the milling industry. Following sedimentation, it is anticipated that the waste will be neutralized and treated for metals removal, as in the case of mine discharge, by sulfide precipitation. In many cases, it will be to the advantage of the industry to recycle the process water.

SAMPLING AND MONITORING REQUIREMENTS

The interim guidelines specify that the parameters required to be monitored and reported shall be those indicated in Table 1.3. The frequency and type of sampling required are indicated. Weather conditions shall be included in monitoring reports submitted to the regional administrator. However, the EPA regional administrators are given considerable flexibility in this regard. Concomitantly, waste discharge permits for the Coeur d'Alene mining district deviate somewhat from this model (see Appendix D). The regional administrators are also given considerable flexibility in requesting additional types of samples. The permit shown in Appendix D, for example, contains instructions to a mining and milling operation to install and monitor four observation wells around a tailings pond to monitor seepage.

CONSIDERATION OF EFFLUENT LIMITATIONS

Factors to be considered under the interim guidelines include the following considerations:

1. The effluent limitations shown in Table 1.1 are to be average values based on (daily) 24-hour composite samples.
2. EPA will consider it a violation of the waste discharge permit if the average of any 20 working days exceeds the average value for any parameter shown in Table 1.1.
3. EPA will consider it a violation of the waste discharge permit if any single composite exceeds the average value for any parameter by more than 50 percent.
4. Four violations in any one single year will require an action memorandum from the regional administrator containing recommendations on what action is to be taken.
5. A single violation is grounds for revocation at the discretion of the regional administrator.
6. EPA considers that the total dissolved solids (TDS) level will be acceptable in most cases, but it is believed that it should be monitored to ascertain this fact. If it interferes with beneficial downstream water uses, the regional administrator or his representative shall establish effluent limitations.

Table 1.3 *Minimum Required Sampling and Monitoring Frequency*

		Mine drainage streams	
Parameter	*Tailings pond decant*	*Before treatment*	*After treatment*
Suspended solids	Daily analysis on 24-hour composite	–	Daily analysis on 24-hour composite
pH	Hourly or continuous	Daily	Hourly or continuous
Cyanide	Daily analysis on 24-hour composite	–	Daily analysis on 24-hour composite
Sulfate	Daily analysis on 24-hour composite	Weekly analysis on 24-hour composite	Daily analysis on 24-hour composite
Radioactivity	Daily analysis on 24-hour composite, or less frequently as approved by EPA regional administrator	Quarterly	Quarterly on composite composed of weekly grabs
Iron	Daily analysis on 24-hour composite	Weekly analysis on 24-hour composite	Daily analysis on 24-hour composite
Flow	Continuous	Continuous	Continuous
Manganese	Daily analysis on 24-hour composite	Weekly analysis on 24-hour composite	Daily analysis on 24-hour composite
TDS	Weekly analysis on 24-hour composite	Weekly analysis on 24-hour composite	Daily analysis on 24-hour composite
Heavy metals	Daily analysis on 24-hour composite	Weekly analysis on 24-hour composite	Daily analysis on 24-hour composite

Table 1.4 *Contractors for Effluent Guidelines Studies in Various Industries*

Industry	*Contractor*
Cement manufacturing	Southern Research Institute
Electroplating	Battelle Columbus Laboratories
Fertilizer manufacturing and phosphate	Davy Powergas, Inc.
Iron and steel manufacturing	Cyrus W. Rice Division of NUS Corporation
Nonferrous metals manufacturing[1]	Battelle Columbus Laboratories
Ferroalloy manufacturing	Datagraphics, Inc.
Asbestos manufacturing	Sverdrup and Parcel and Associates, Inc.

1. Nonferrous metals manufacturing is further subdivided into aluminum; copper, lead, zinc; and mining and milling.

SUMMARY

The passage by Congress of Public Law 92-500, the Federal Water Pollution Control Act Amendments of 1972, established a comprehensive set of new requirements which are directed ultimately at a goal of zero discharge of water pollutants in the United States by 1985. While this goal may in reality be unattainable, the law empowers the Environmental Protection Agency with authority designed to ascertain that an all-out effort toward attainment is exerted. The law specifies:

1. July 1, 1977, as the deadline date for the implementation of the best practicable pollution control technology currently available for all point source waste water discharges (a regulatory requirement).
2. July 1, 1983, as the deadline date for implementation of the best available technology economically achievable for all point source waste water discharges (a regulatory requirement).
3. July 1, 1985, as the final date for compliance with the national goal of the complete elimination of discharge of pollutants (this is *not* a regulatory requirement).
4. The Environmental Protection Agency has established for the mining and milling industry the interim guidelines presented in Table 1.1. A typical permit issued on the basis of these interim guidelines is presented in Appendix D. Additional details of the permit system are presented by Green and Dubois (1973). Contracts have been or will be let by EPA to define permanent guidelines for several categories of industries, the most pertinent of which are listed in Table 1.4. Four of these industries–fertilizer manufacturing and phosphate, iron and steel manufacturing, nonferrous metals manufacturing, and ferroalloy manufacturing–are the bases for six of the chapters in this book. These contract completion reports have been condensed, paraphrased or heavily quoted, and combined with the author's own observations so as to summarize the origin and disposal of wastes for these industries.

Public Law 92-500 charges the administrator of the Environmental Protection Agency with the task of:

1. Establishing water quality guidelines reflecting the "best practicable pollution control technology currently available."
2. Establishing water quality guidelines reflecting the "best available technology economically achievable."
3. Defining and publishing information on these two categories of technology.
4. Identifying control measures and practices available to completely eliminate the discharge of pollutants from point sources, taking cost into account.
5. Defining and publishing information on the criteria reflecting the latest scientific knowledge of the effects of pollutants on the health of the ecosystem and on the fate of pollutants in the ecosystem.
6. Defining and publishing information on the factors necessary to restore the integrity of navigable waters, ground waters. contiguous waters, the oceans, and the biota therein.
7. Defining and regulating separately the discharge of "toxic" materials. At this writing, the elements mercury, cyanide, and cadmium have been proposed as toxic by EPA. The agency also proposes that these elements not be allowed in discharges to streams flowing less than 10 cfs.*
8. Initiating a comprehensive program to regulate ocean dumping of waste and waste waters.
9. Establishing a waste discharge permit system that will utilize the above described guidelines to regulate the discharge of wastes from point sources. These permits specify abatement programs for each discharge and specify deadlines for completion of each phase of the abatement program. (See Appendix D for an example of a typical permit and for the adjudicatory hearing procedure.)

It will be demonstrated in subsequent chapters that the Federal Water Pollution Control Act Amendments of 1972 will affect not only the disposal of waste water but also the control of air pollution and the production and disposal of solid wastes. Air pollution is controlled by converting airborne contaminants either to a liquid waste and then to a solid waste or directly to a solid waste. Liquid wastes in turn are eliminated by conversion to solid wastes for disposal. As the middle step in the sequence, the control of waste discharges automatically enhances the solid waste disposal problem and it extends the air pollution control effort.

The penalty sections of Public Law 92-500 specify that violations of the conditions of a permit or the permit system are subject to civil penalties of up to $10,000 per day, in addition to other civil limitations and administrative actions. Willful and negligent violations are subject to criminal fines of up to $25,000 per day.

The tone of Public Law 92-500, and consequently of its implementation

* See *Federal Register*, Document 73-13870, filed 7-5-73, p. 18045, and *Federal Register*, Vol. 38, No. 247, December 27, 1973, pp. 35388-35395.

mechanism, is tuned almost exclusively toward end-of-stream, primary, secondary, and tertiary waste treatment for compliance. However, on the basis of several years of observation and study, it is the opinion of the author and many of his associates that considerable process alteration and subrogation are in order if compliance is to be achieved economically and without inordinate social trauma. That is to a considerable extent what this book is about.

REFERENCES

Cox, W. E., and Walker, W. R. 1973. Ground Water Implications of Recent Federal Law. *Ground Water*, Vol. 11, No. 5, pp. 15-18.

Driver, W. J. 1973. Impact of Clean Water Act on Industry "Severe," the Chemical Industry's View. *Catalyst for Environmental Quality*, Vol. 3, No. 1, pp. 26-28.

Green, J. A., and Dubois, D. P. 1973. The Water Permit Program of EPA as applied to the Mining Industry. Paper presented at annual meeting of American Mining Congress; available from American Mining Congress, Washington, D.C.

Hall, E. P. 1972. *Projects of the Pollution Control Analysis Section.* EPA Document R2-72-134, Applied Science and Technology Branch, Office of Research and Monitoring, U.S. Environmental Protection Agency, Washington, D.C.

Kostelnik, B. M. 1973. Impact of the Federal Water Pollution Control Act Amendments of 1972 on the Mining Industry. Paper presented at annual meeting of American Mining Congress; available from American Mining Congress, Washington, D.C.

Quarles, J. R., Jr. 1973. Clean Water–National Commitment Under New Act. *Catalyst for Environmental Quality*, Vol. 3, No. 1, pp. 12-14.

USDI 1968. *Report of the Committee on Water Quality Criteria, April 1, 1968.* Federal Water Pollution Control Administration, U.S. Dept. of Interior, Washington, D.C.

Williams, R. E. 1973. Federal Water Pollution Control Act Amendments of 1972 and what the mining industry must do to comply. *World Mining*, March 1973, pp. 44-49 (U.S. edition).

Williams, R. E., Kealy, C. D., and Mink, L. L. 1973. Effects and Prevention of Leakage from Mine Tailings Ponds. *Transactions, SME, AIME,* Vol. 254, pp. 212-216.

2. Characterization of wastes from mineral resource based industries

Wastes from the mineral resource based industries—mining, milling, and metallurgy—are characterized as follows:

MINING INDUSTRY

Solid Wastes: The major solid waste disposal problem in the mining industry is the handling and relocation of overburden (stripping) and gangue. The rate of production of overburden and gangue increases annually as the depth of ore bodies mined becomes greater and grade becomes lower. Overburden is discussed here in the sense of rock which overlies the ore body regardless of whether that ore body is sought by underground, open pit, or other mining methods. Obviously, greater overburden problems are associated with open pit or strip mining methods. However, the production of waste rock by underground mining methods also constitutes a disposal problem.

Within the context of this book, the runoff from such solid waste piles is also of interest. Stream sedimentation may or may not be a problem downstream from such disposal sites, depending on the nature of the solid waste and on the nature of the climate in the mined area. The most common methods of minimizing the effect of such solid waste disposal features on stream channels are revegetative stabilization and landscaping. These operations are discussed in subsequent chapters.

Liquid Wastes: The major and most significant source of liquid waste in the mining industry is acid mine drainage. Acid mine drainage is common in areas where mine openings intersect the water table and where the rocks contain pyrite or, less commonly, certain other sulfides. Acid mine drainage is a particularly troublesome problem for the coal mining industry, because coal beds occupy reducing environments and iron sulfide is also a ubiquitous product of reducing environments. Consequently, many coal seams have pyrite associated with them. Some exceptions occur in western coals, notably in

Montana and Wyoming. However, even these are not completely pyrite free.

Acid runoff from relocated overburden is also a problem for the coal mining industry. The low pH water is produced by the sulfides associated with the overburden, which oxidize upon exposure to the atmosphere. Careful landscaping, revegetation to impede runoff, and limited chemical stabilization have been used as control measures.

Numerous techniques have been tested for the control of acid mine drainage. These techniques range from control of the ground-water flow system which impedes movement of water into mined areas, to the collection and treatment of drainage from abandoned mines by liming and sedimentation.

Gaseous Wastes: With the exception of relatively insignificant production of dust, the mining industry is blessed with few air pollution problems outside the mine. The major gaseous waste problem confronting the mining industry is the production of methane, but this problem is restricted to the atmosphere in underground openings, mostly in coal mines.

MILLING INDUSTRY

Solid Wastes: Solid wastes constitute the major disposal problem of the milling industry. The wastes consist of the host rock for the minerals which have been removed by various concentration processes. The host rock has essentially no value. The wastes are known as tailings and they are deposited in settling basins, known as tailings ponds, or mud lakes in the form of a slurry. The solids are allowed to settle in the settling basin, and the effluent is decanted from the cleanest portion. It is treated, recycled into the mill, or discarded into a stream. The solid residue piles themselves, if not properly handled, constitute sources of dust and of sediment to be distributed to streams by runoff. Vegetative stabilization is the most common method of minimizing the generation of dust and sediment from the tailings piles. In the case of underground operations, there is a trend, as areas are mined out and abandoned, toward reemplacement of greater and greater percentages of the coarse fraction of the tailings underground.

Liquid Wastes: The major liquid waste disposal problem confronting the milling industry consists of the water decanted from the tailings ponds. The decanted water contains some suspended solids and sometimes low concentrations of cyanide and other dissolved ions. Cyanide is particularly significant, since it has been proposed as a toxic element by EPA.* The cyanide is contributed to mill waste waters by reagents used in the flotation and depression of the various minerals. Several other ions present in some mill tailings pond effluents are derived from the same source. A tabulation of these reagents, their use, characteristics, and toxicity to various forms of life has been prepared by Hawley (1972). One challenging area of research for the milling industry and associated government and academic research institutions consists of the development of reagents that are either nontoxic or at least have minimum toxicity, particularly with respect to aquatic species.

* See *Federal Register*, Vol. 38, No. 247, December 27, 1973.

In addition, compliance with the suspended solids guidelines outlined in Chapter I presents a challenge to the milling industry. The removal of suspended solids from liquid wastes to the levels of 20 to 30 milligrams per liter will necessitate careful design and management of tailings ponds if downstream treatment of the effluent from the ponds is to be avoided. If the tailings pond also receives waste from metallurgical processes or acid mine drainage, which is not uncommon in the United States, downstream treatment of the effluent from the pond probably will be unavoidable if compliance with Public Law 92-500 is to be achieved.

Gaseous Wastes: With the exception of a few wet scrubbers and dust derived from the tailings piles discussed previously, the milling industry has negligible pollution problems of this type. Gaseous wastes are essentially nonexistent in the milling industry.

METALLURGICAL INDUSTRY

Solid Wastes: The term *metallurgical industry* is sufficiently broad and covers so many industries that it is difficult to define precisely what types of wastes should be considered characteristic of it. Matters are complicated further by the fact that many operations are connected both geographically and/or operationally with mines or mills or both. Consequently, the solid waste from metallurgical processes may in fact be treated identically and coincidentally with mill wastes. The solid wastes from metallurgical processes range from relatively innocuous slag from blast furnaces to unstable solids produced by the washing of gaseous wastes and metallic products. Among the most ubiquitous, but not necessarily voluminous, solid wastes produced by metallurgical industries are those that are the result of air pollution control devices such as scrubbers and cyclones used for dust control. These solid wastes are usually in the form of sludges from wet collecting devices, and they are frequently discarded in landfills if the metallurgical operation is not associated with mining or milling wherein solid waste disposal is a large-scale problem requiring a tailings pond, mud lake, or other similar device.

Liquid Wastes: Liquid wastes in the metallurgical industry range from wash water for the raw material to effluent from wet scrubbers used for air pollution control. The Federal Water Pollution Control Act Amendments of 1972 have prompted the metallurgical industry to study the concept of recycling in much greater depth than has ever been practiced before. However, it is simply not feasible to recycle all wastes for most operations because of the inevitable buildup of dissolved solids and the consistent increase or decrease in pH of the recycled water. Consequently, a bleed-off stream is inevitable for many operations; therefore, waste water treatment facilities of one type or another are unavoidable for most metallurgical operations.

Dry scrubbing constitutes one means of air pollution control which offers promise for reducing the cost of liquid waste treatment to achieve compliance with Public Law 92-500; however dry scrubbing has not been perfected for some gaseous emissions to the point where it can be used consistently to meet air pollution control standards. Even if the major constituent of the gaseous waste is simple dust, one particularly ubiquitous problem is the presence of hygroscopic substances which cause plugging or "blinding" in baghouses. Even here, however, new techniques are being developed to minimize blinding.

Two great challenges facing the metallurgical industry are the perfection of dry scrubbing techniques and the development of metallurgical processes that either minimize the use of water or eliminate it. The net result of such a transition will be a trend toward the production and disposal of solid wastes exclusively with only minimum costs and effort being expended on waste water treatment.

Gaseous Wastes: Gaseous waste is used herein to include any waste that constitutes an air pollution problem for the metallurgical industry. It is assumed to include gaseous emissions wherein particulate matter is a problem. Gaseous waste control is a continuing problem to the metallurgical industry because practically all metallurgical processes produce gaseous emissions of one kind or another. They may contain only particulate matter or they may contain only gases or they may contain a combination of both. The current practice is wet scrubbing to remove particulate matter, followed by appropriate additional treatment to remove whatever gas happens to be present. In the case of sulfur dioxide gases from copper and lead smelting operations and zinc roasters (sulfide ores), the gas emitted is predominantly sulfur dioxide. The standard technique for handling sulfur dioxide is conversion to sulfuric acid. Until recently the disposal of growing quantities of by-product sulfuric acid, which might be termed a liquid waste, was no less than difficult. However, the foreign wheat transactions, which have precipitated widespread increased activity in the agricultural sector, have stimulated the demand for sulfuric acid because of the large increase in the demand for fertilizers whose manufacture depends in large part on sulfuric acid. This increased agricultural activity has made the recovery of sulfur dioxide much more profitable and palatable to metallurgical industries. However, much of the new by-product sulfuric acid is located far from traditional markets. Several western United States copper smelters are building acid neutralization plants to get rid of surplus acid.

REFERENCES

Hawley, John R. 1972. *The Use, Characteristics and Toxicity of Mine-Mill Reagents in the Province of Ontario.* Special Projects Section, Mining Industrial Wastes Branch, Ontario Ministry of the Environment, Toronto, Ontario, Canada.

3. *Methods of waste and waste water treatment available to mineral resource industries*

INTRODUCTION

The inclusion of both "waste" and "waste water" in the title of this chapter is intended to lend emphasis to the fact that treatment techniques frequently must contend with collection and disposal of solid, liquid, and gaseous wastes. The mining, milling, and metallurgical industries, but particularly the milling industry, are somewhat unusual in this regard in that solid wastes are mixed with liquid wastes in the form of slurries wherein the liquid may be low or high in pH and contain a variety of dissolved solids and gases.

The dissolved matter may be precipitated and settled out separately from the suspended matter, or the two may be separated from the water simultaneously in the same settling basin. Consequently, most mining and milling and many metallurgical operations will have somewhere associated with them a settling basin or tailings pond to handle suspended solids. The name *tailings pond* is somewhat misleading in that, strictly speaking, the word *tailings* applies to the granulated waste rock from a mill circuit. However, tailings ponds frequently are used to remove solids from mine drainage, sandfill overflow water, and a variety of other sources. With this exception, the waste treatment procedures utilized by these three industries are not different from those utilized by most other industries producing waste water with low concentrations of organic constituents.

The treatments to which combinations of mining, milling, and sometimes metallurgical wastes are subjected usually begin with sedimentation in a settling pond or an analogy thereof. Beyond that, the treatment employed depends on the characteristics of the particular combination of wastes in question. However, the succeeding steps commonly are adjustment of pH, usually with slurried lime, combined with aeration and additional sedimentation, and thickening. Flocculating agents (also known as coagulating agents) may be added at some point in the system to aggregate small, charged, nonsettleable particles so that the neutralized, aggregated particles will settle. Most flocculating agents operate on the principle of elimination of electrical charges on the nonsettleable particles in combination with careful control of the concentration of free hydrogen ions (pH control).

A variety of treatment steps beyond sedimentation, neutralization, aeration, and flocculation are available but, on the basis of early effluent guidelines documents, they probably will not be utilized by most mineral resource operations. One exception may be those industries that produce waste waters with constituents proposed and finally adopted as toxic by EPA; for those industries concentrations in the discharge will be controlled more stringently. At this writing, EPA has proposed that the elements cyanide, mercury, and cadmium be defined as toxic. Waste waters containing suspended matter that cannot be brought to less than 30 milligrams per liter by pH control and coagulation also may have to be subjected to further treatment if economics permit.

Appendix E presents an example of a comprehensive mining, milling, and metallurgical waste water treatment scheme. It employs sedimentation, pH control, flocculation, aeration, thickening, and recycling.

SEDIMENTATION

Sedimentation utilizes gravity to remove suspended solids from waste waters. Without the addition of flocculating agents and with low flow velocities in a tailings pond, the settlement of suspended solids can be treated in terms of discrete particles as governed by Stokes' law:

$$\nu = \frac{\rho_s - \rho_l}{18\,\mu}\, gD^2 \tag{3.1}$$

where ν = settling velocity of the particle
ρ_s = specific gravity of the solid particle
D = diameter of the particle
g = gravitational acceleration
ρ_l = specific gravity of fluid
μ = viscosity of fluid

The validity of Stokes' law is based on the assumption that Reynolds' numbers are sufficiently small to reflect insignificant inertial forces and predominant viscous forces. Practically speaking, these conditions imply a low flow rate and no wave action within the pond. The role of the flow rate (assumed to be nonturbulent) can be quantified in terms of the overflow rate:

$$\nu_o = \frac{Q}{A} \tag{3.2}$$

where Q = rate of flow through the pond (i.e., inflow or overflow rate)
A = surface area of the pond

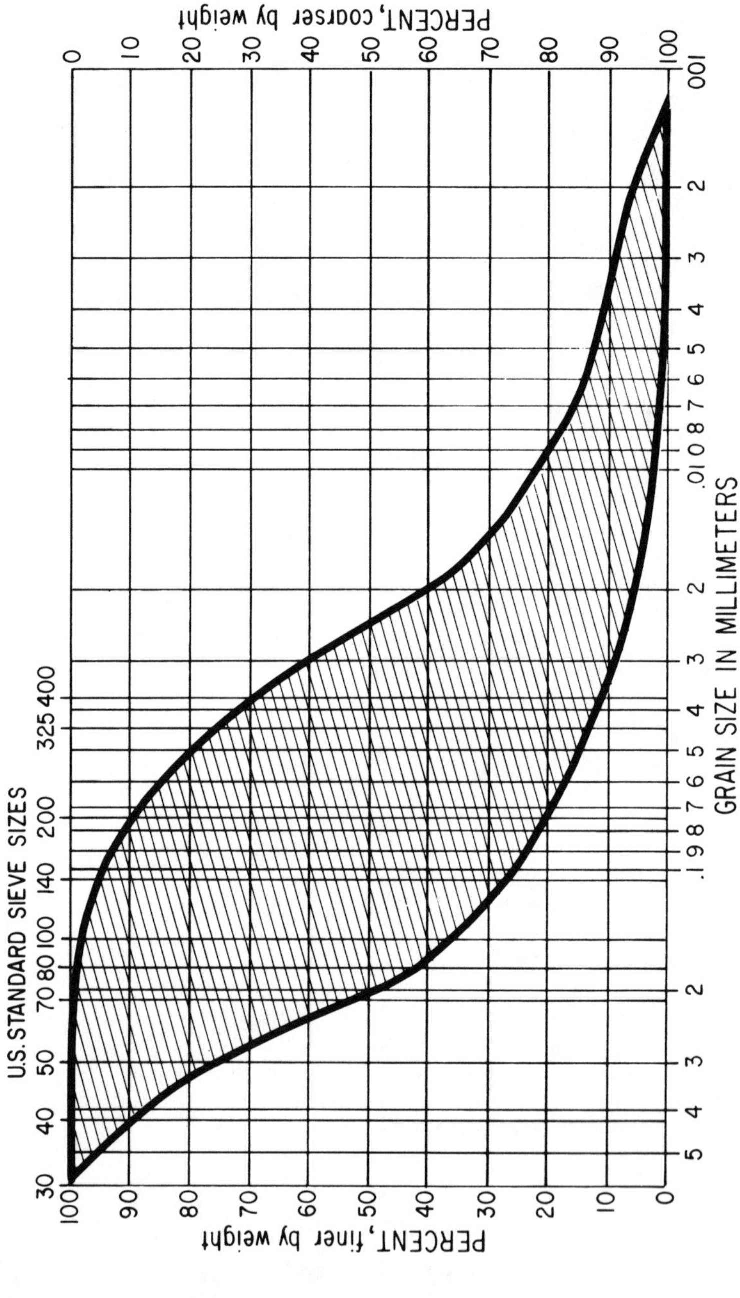

Figure 3.1. *Range of cumulative frequency distributions of the grain sizes of samples collected from a typical lead-zinc mill.*

Particles with settling velocities v greater than v_O will settle in the pond, and particles that settle at a rate less than v_O will be removed directly proportionally to the ratio v/v_O, regardless of pond depth. Total removal R_T of a suspension having a range of particle sizes can be expressed by:

$$R_T = (1 - C_o) + \frac{1}{v_O} \int_0^{C_O} v \, dC \tag{3.3}$$

where C_O = fraction of particles having a settling velocity equal to or less than v_O and similarly with C and v.

Cumulative frequency distributions such as those shown in Figure 3.1 can be used in conjunction with Equation 3.1 to calculate C_O.

Given no limitations on space, these equations facilitate the design of tailings ponds with sufficient area to settle solids that will settle as discrete particles. According to Betz (1971), the lower practical limit of grain sizes amenable to discrete particle settling for industrial wastes is 0.01 millimeter.

Tailings ponds are settling tanks with a very large A term as expressed in Equation 3.2. When A is large v_O is small and a small settling velocity (v in Equation 3.1) can be tolerated. Consequently, in the case of tailings ponds handling only discrete particles, maximizing the area is much more important than maximizing depth, subject of course to restrictions caused by channeling and wave action.

The effect of wave action, turbulence, bottom scour, and to some extent short-circuiting can be predicted on a theoretical basis. Eckenfelder (1966) provides such a treatment for conditions of turbulence and bottom scour. However, the behavior of waste water in tailings ponds is influenced by so many variables simultaneously that minimizing their undesirable effects must be done empirically on an individual-case basis. Careful control of the decant height, which should generally be located in the deepest portion of the pond, and proper manipulation of the location of inflow points can do much toward maintaining optimum conditions for sedimentation in a tailings pond. Baffles for wave action control should be movable, if possible, and careful attention should be paid to moving them whenever wave action begins to produce undesirable conditions in the discharge, especially if it cannot be recycled.

COAGULATION

Coagulation consists of the bringing together of particles that are too small to settle discretely so as to increase the term D in Equation 3.1. Increasing the diameter of a particle by the joining together of several smaller particles has an exaggerated effect on the settling velocity of the final particle because the diameter term in Equation 3.1 is squared. The equation, in fact, describes a parabola and has the associated properties thereof. Therefore, if D is assumed to increase continuously as a particle settles toward the bottom of the pond, then the settling velocity increases parabolically with D. Thus, if flocculating agents are added to a waste water, the depth of the settling basin, be it tank or tailings pond, becomes important along with its area.

The precise behavior of D as a function of time is complex, and a precise mathematical analysis is not possible. Each treatment system must be designed after the acquisition of empirical data for the waste water and flocculating agents involved. These data are obtained by jar tests to determine appropriate pH ranges and coagulating agents, followed by the use of columns with taps at varying distances from the top of the column, thereby permitting samples to be collected as a function of time and depth. Percent removal of suspended solids for each depth and time period is measured and graphed. Pond or tank depth and detention time can then be deduced from the graph. Specific examples of common problems are presented by Eckenfelder (1966) and American Water Works Association (1971).

Coagulation is accomplished in a tank or settling pond by use of coagulating agents and by pH control. The coagulating agents generally are classified as chemical or electrolytic. Chemical agents used most frequently are alum, $Al_2(SO_4)_3 \cdot 18H_2O$; ferric chloride, $FeCl_3$; ferric sulfate, $Fe_2(SO_4)_3$; copperas, $FeSO_4 \cdot 7H_2O$; magnesium oxide (MgO); and sodium aluminate ($Na_2Al_2O_4$). The electrolytic coagulants (also called coagulant aids) are high-molecular-weight, water-soluble polymers known as polyelectrolytes. These compounds contain groups that dissociate electrolytically to form charged, high-molecular-weight particles. Polyelectrolytes which dissociate to form positively charged particles are cationic and those which dissociate to form positive and negative particles are nonionic. The purpose of these polyelectrolytes is to lower the zeta potential, the difference in the electrical charge existing between the stable colloid and the dispersing medium. This observation also explains the importance of controlling pH because pH is a measure of the concentration of free $H+$ in the dispersing medium. Mink et al. (1973) report that minimum colloid stability for mill wastes occurs at pH values above 9.0 and that the minimum zeta potential they observed for mill wastes was -18.5 millivolts.

Chemical and polyelectrolyte coagulating agents probably destabilize colloids by one of three mechanisms. These are: (1) double layer compression; (2) neutralization by the adsorption of a particle of opposite charge; and (3) formation of a polymeric molecular bridge between two similarly charged colloidal particles. Details of each mechanism are presented by O'Melia (1970).

Although considerable experience has been acquired on which coagulating agents work best on given particles, it is usually determined empirically through the use of jar tests. An example of such a test is presented in Appendix F (after Wallace 1973).

Once the proper combination and rate of addition of coagulating agents has been determined experimentally, the injection is usually done at a mixing tank, or at a point of turbulence such as the inflow to a tailings pond or inflow to a settling tank or clarifier. In the case of tailings ponds, injection would likely occur at one of the inflows unless further treatment of the effluent from the tailings pond is required to comply with the waste discharge permit or if space is at a premium. If further treatment is employed, coagulant aid addition would probably occur during the neutralization or thickening stages.

CONTROL OF pH

The pH of a waste discharge is controlled for three reasons: (1) to comply with waste discharge permits; (2) to maximize the precipitation of small suspended particles; and (3) to maximize the removal of dissolved ions, particularly metals, from solution.

For the purposes of this book, the practical methods of controlling pH of waste water entering a settling pond or during downstream treatment stages are:

1. Carefully mixing wastes so that the composite is near neutral on a continuous basis.
2. Exposing acid wastes to beds of limestone.
3. Exposing acid wastes to lime, lime slurries, or dolomitic lime slurries.
4. Adding caustic soda, soda ash, or other base to the acid wastes (usually not economical).
5. Adding sulfuric acid to alkaline wastes to lower the pH to levels that will satisfy guidelines.

Only methods 2 and 3 are discussed in significant detail here because, as will be shown in Chapter 4, other methods are relatively uneconomical.

Mixing

Many mining, milling, and metallurgical processes produce waste water of a variety of qualities and many are located in close proximity. Thus, mixing wastes from different segments of an operation may be feasible, but the decision rests on several factors in addition to pH control. Frequently the volumes produced by the segments will differ more than the amount for which available storage can compensate. Thus, only partial mixing can occur and, if a reasonably desirable final product is to be realized, dual treatment facilities will be required. But perhaps the most serious problem is that mill wastes, high in pH, contain large volumes of suspended solids, whereas mine water and/or metallurgical process waters are usually high in dissolved solids and low in pH. Thus, mixing poses the threat of poorer settling characteristics of suspended solids because most fine-grained solids are more stable in suspension at pH values less than 8.0.

It may be feasible, on the other hand, to mix the waste waters after the mill wastes have been treated in a tailings pond. If space is not a deterrent, it may be desirable to subject both wastes to a sedimentation basin prior to mixing and treatment. Experimentation on an individual-case basis is the only means of determining the most efficient alternative. On the basis of the author's experience, some mixture of mining, milling, and metallurgical waste waters, where geographically feasible, is advantageous because of the desirable pH changes facilitated by the great range of pH values characteristic of these three types of effluents.

Relatively recent research suggests that mixing of mine and mill wastes may offer advantages in addition to pH alteration. Miller (1972) tested by batch tests 11 mill tailings samples from throughout the Rocky Mountain region for their effectiveness in removing dissolved contaminants from acid mine water. Miller concluded that the 11 mill tailings samples have the ability to remove dissolved contaminants from the mine drainage. The degree of removal was observed to be a function of mineralogical composition, grain-size distribution, surface area, and reactor design. The most probable reactions involved were reported to be:

1. Adsorption of metallic cations on siliceous minerals.
2. Precipitation of metal hydroxides due to the neutral pH of the tailings.
3. Precipitation of metal hydroxides due to the reaction with calcareous components of the tailings.

The average capacity for ferric iron removal was observed to be 918 mg/gm of tailings with a range from 6 mg/gm to 15 mg/gm. The maximum capacity for adsorption of iron on siliceous material was calculated to be 7.3 mg/gm with an additional 2.0 mg/gm being accounted for by pH change. A tailings sample containing high concentrations of calcareous material was observed to remove iron from mine water at a capacity in excess of 100 mg/gm, accompanied by the precipitation of ferric hydroxide.

Copper and zinc removal capacities were observed to be similar to that for iron.

Miller also attempted to test the capacity of tailings to remove iron from acid mine water on a continuous basis, but the tests were inadequate for conclusions. Clearly this line of research should be pursued. Investigations of the removal mechanism are being conducted by the author and his associates.

pH Control with Limestone

Increasing the pH of acid waste water with limestone depends on exposure of the waste water to limestone. Exposure can be accomplished in a variety of ways ranging from ordinary basins to complicated pumping systems which oscillate the solutions through beds of crushed limestone. The reaction in simplified form is:

$$CaCO_3 + H_2SO_4 \longrightarrow CaSO_4 + H_2CO_3 \tag{3.4}$$

The reaction will proceed to the right as long as excess $CaCO_3$ is available and active. Deactivation of the $CaCO_3$ occurs when the limestone becomes coated with $CaSO_4$ due to precipitation. Precipitation occurs when the acid solution approaches 5 percent or when the acid is neutralized to above 0.3 percent (Jacobs 1951). In the case of mining and metallurgical wastes, even if the solubility of $CaSO_4$ is controlled, eventual coating is likely to occur because of other constituents whose solubilities decrease as pH increases. Van Everdingen and Banner (1971), for example, report the precipitation of almost all iron, manganese, copper, lead, and zinc during utilization of limestone to neutralize acid mine drainage. Mihok et al. (1968) report that grinding of limestone improves results. But, under the best of control, considerable unused limestone will have to be discarded. Transportation and disposal of the limestone are likely to be troublesome because of the large mass involved. Perhaps the two principal advantages of limestone as a neutralization medium are its relatively low initial cost and the fact that the sludge may have better settling characteristics than that resulting from neutralization with lime (about 2.5 percent of water volume vs. 12 percent of water volume, respectively; Mihok et al. 1968; Appalachian Regional Commission 1969a). It is possible that the method of Mihok et al. can be perfected to the point of minimizing some of these problems. This technique diverts a portion of the mine drainage to a rotating (25 rpm) mill loaded with limestone chunks. The action produces a limestone slurry of approximately 400-mesh, or smaller, limestone particles, which are then mixed with the bulk of the acid water. The partly neutralized mixture is aerated and transported to a settling pond from which overflow is discharged.

pH Control with Lime

The reaction which neutralizes acid waste water by exposure to slurried lime is that of Equation 3.4, but essentially all $CaCO_3$ is consumed. The lime is physically mixed with the water, thereby allowing for tighter control of final pH. Several aspects of the process are discussed in detail in early works by Hoak (1944), Rudolfs (1943a), Rudolfs (1943b), and Smith (1943). A comprehensive bibliography on neutralization with lime slurry, as well as other methods of treating acid water, is contained in Appalachian Regional Commission (1969b). More recent research sponsored by the Environmental Protection Agency is tabulated by Hall (1972), and details of processes are presented by the American Water Works Association (1971).

Hydrated lime is a dry powder. It is produced by treating quicklime with water. It can be slurried by further mixing with water. If sufficient lime is added to an acid waste water, the pH can be raised to levels sufficient to remove most of the iron and other metals contained therein. Subsequently, however, pH level may have to be reduced to comply with pH effluent standards. Disadvantages of its use include increase in water hardness, precipitation of gypsum seals on equipment, and a high-volume, low-density sludge that is difficult to discard, even after thickening.

Control with the use of lime consists of four steps. These are:

1. Vigorous mixing of lime and acid water for 1 to 5 minutes.
2. Aeration for 15 to 30 minutes (discussed in subsequent section).
3. Sedimentation, thickening, or clarification.
4. Sludge removal and disposal.

Ordinarily the acid water is collected in basins and pumped to the treatment facilities. These facilities consist (exclusive of the holding pond) of the lime slurry system, flash mixer, aeration basin, thickener (which is optional), and sedimentation basin.

Lime is used to neutralize acidity in this waste water treatment plant in the old town of Garfield, Utah. Ferric chloride is added as a conditioner to facilitate soluble metals removal by precipitation and a coagulant, an anionic polymer, speeds the settling rate. The flow goes to two clarifiers where pollutants settle out. Sludge is pumped to a collection pond, and the clarified effluent (class C water) flows by gravity to Great Salt Lake. Kennescope, *Kennecott Copper Corporation photo*

AERATION

Aeration or oxygen transfer is normally employed along with neutralization to achieve, among other things, complete neutralization reactions and to oxidize the soluble states of iron and manganese (ferrous and manganous) to the insoluble states (ferric and manganic). Hydrogen sulfide and carbon dioxide may also be removed from solution in this fashion.

Aeration of a lagoon or tank is achieved through the dissolution of oxygen by mechanical or diffused aeration devices. According to Eckenfelder (1966), spacing of mechanical aeration units of up to 200 feet at power levels of 0.015 to 0.02 horsepower per thousand gallons of basin volume will uniformly distribute oxygen throughout a basin but will not prevent settlement of most solids. Alternatively, air may be bubbled into the tank and allowed to diffuse. Eckenfelder and Ford (1971) present an excellent review of recent concepts of oxygen transfer and the design of aeration equipment.

The mechanical method is generally used for industrial wastes. Mechanical aerators are generally classified as waterfall aerators, forced-draft aerators, and impeller systems. The simplest waterfall aerator consists of vertical risers that cascade the water onto the surface of the tank or basin. Velocity profiles and dissolved oxygen are introduced into the waste water by this method. Trays and spray nozzles may be added to increase dissolution of oxygen and mixing. Coke may be added to the trays for better iron and manganese removal. The coke imparts a catalytic effect when the iron-manganese-bearing water contacts the freshly precipitated sludge.

Forced-draft aerators utilize horizontal wood-slat trays which divide the falling water into small streams. A blower forces air across the streams in a direction opposite the flow of the streams.

Impeller units disperse compressed air by shearing and pumping action or transfer oxygen to the waste water by vortexing action. Eckenfelder (1966) presents details of operation and design of each type of unit.

THICKENING

Sludges from processes designed to remove suspended and dissolved solids frequently contain too much water for disposal as solid wastes. In some cases the excess water in the sludges creates a space problem, and in some cases the excess water is too limiting with respect to structural stability. In these cases the sludge may be partially dewatered or thickened. Thickeners are settling tanks with mechanical rakes which discharge settled solids to an underflow for disposal or drying.

According to Eckenfelder (1966), sedimentation sludges of 1 to 2 percent concentration can be gravity thickened to 6 to 10 percent. Biological and chemical sludges can be thickened less satisfactorily. However, the Phelps Dodge Corporation, for example, at its Tyrone, New Mexico, mill thickens about 45,000 tons of tailings entering at 25 percent solids per day. Tailings flow by gravity from the thickeners at about 45 percent solids (Keener 1973). Most mine tailings pond operators report thickening their wastes to 45 to 50

Five 400-foot-diameter stainless steel thickeners are being installed at a new mine south of Tucson, Arizona. The caisson thickeners will be used in CCD acid leaching of copper. The plant is expected to go on-stream in 1975.

Dorr-Oliver photo

percent. The high-density sludge process devised by the Bethlehem Steel Corporation may be even more effective (see Skelly and Loy 1973).

Thickener size required is calculated in terms of area per unit weight of solid per unit time. The surface area required is a function of percent solids entering the system and percent solids desired to leave the system. The surface area required is determined from a balance of inflow, underflow, and outflow. The unit area requirement is obtained from the equation:

$$UA = \frac{A}{Q_o C_o} = \frac{1/C_i - 1/C_u}{U_i}$$

where UA = unit area, ft^2/(lb of solid/day)
C_u = underflow concentration in lb/ft^3
C_i = solids concentration (lb/ft^3) at settling velocity U_i
U_i = interface settling velocity in ft/day

C_i is computed from: $C_i = \dfrac{C_o H_o}{H}$ where H is the depth the sludge would occupy if it were all at the interfacial concentration C_i. Eckenfelder (1966) presents details of each computation and examples of design procedures.

An indication of scale for thickeners for mill tailings is provided by Johnston (1973). Table 3.1 presents average thickener area values used by Arthur G. McKee and Company on two types of tailings.

Table 3.1 *Thickener Area Requirements for Different Types of Tailings*

Solids	*Thickener area* $ft^2/(\text{ton of solid}/24\ hr)$
Amorphous or colloidal	7 - 9
Granular	3 - 5

FLOTATION

Some operations utilize flotation instead of sedimentation as a means of removing suspended solids from waste water. The waste water frequently is pressurized with air to about 50 psi pressure. Upon rapid release of the pressure, either to atmospheric pressure or to a vacuum, small bubbles of air are released from solution. Bubble formation can also be accomplished with mechanical beaters. Suspended particles become attached to these bubbles and are floated to the surface, where they are skimmed off and discarded or recycled after draining. The effluent is drawn off the bottom of the unit and, if necessary, recycled for further clarification. Alum or silica may be added to the waste entering the tank for purposes of coagulation, and surfactants may be added to produce foam fractionation (see Horizons 1970, for an excellent discussion of the latter process). Eckenfelder (1966) presents design specifications for flotation systems, and Stander and Van Vuuren (1970) provide information on flotation system adaptability to various wastes.

On the basis of observations by the author, flotation has been used to a very limited extent in the mineral resource industries. Limited use is probably accounted for by the fact that other constituents, including pH adjustment, must be controlled also, thereby making sedimentation and neutralization more attractive to the removal of suspended solids. However, the aforementioned study by Horizons could improve the outlook for flotation combined with foam fractionation, provided recovery and reuse of surfactants can be perfected so as to reduce the cost.

REVERSE OSMOSIS

Investigations of reverse osmosis as a method of desalination of seawater or brackish water have been made by industry and the U.S. Interior department.

Osmosis has been described in the following way by H. Merten of Gulf General Atomic Corporation.

> Osmosis occurs when two solutions of different concentrations in the same solvent are separated from one another by a membrane. If the membrane is ideally semipermeable, that is, if it is permeable to the solvent and not to the solute, then a flow of solvent occurs from the more dilute into the more concentrated solution until the two solutions become equal in concentration or until the pressure in the chamber of the more concentrated solution rises to a certain well-defined value. The pressure difference at which no flow occurs is termed the osmotic pressure difference between the two solutions. If a pressure in excess of this osmotic pressure difference is applied to the more concentrated solution, then the solvent can be caused to flow into the more dilute solution.

The terms *reverse osmosis* and others are used to describe this process. At present three principal configurations of membranes are commercially available and offer reasonable success. The plate and frame type developed by Aerojet General Corporation and the spiral-wound membrane developed by Gulf General Atomic Corporation have been applied in a number of operations. A "hollow fiber" membrane is currently receiving considerable attention under the development of Havens Industries. A detailed description of the reverse osmosis process and the concomitant equipment and design parameters is presented by Goff and Gloyna (1970). Bond and Straub (1973) list all available membranes and present data on their effectiveness.

Skelly and Loy (1973) conclude that reverse osmosis is highly effective in removing nearly all of the dissolved solids in acid mine water. Recoveries of 80 to 90 percent have been demonstrated. They also point out that calcium sulfate formation on membranes can be removed from the units by operating at lower pressures for a short period. Iron fouling can be prevented by lowering the pH of the feed water to less than 3.0. Acid mine drainages containing both ferric and ferrous iron forms have been processed successfully. Mine drainage containing high concentrations of dissolved solids, including sulfates, may cause operating problems with excessive calcium sulfate formation. Biological oxidation of ferrous iron can be prevented by ultraviolet disinfection.

Product water usually contains less than 70 mg/l of dissolved solids. Quality of representative product water is presented in Table 3.2. Water of this quality is not potable due to pH, acidity, and iron and manganese content. But additional treatment by neutralization, coagulation, filtration, and disinfection would produce potable water.

The brine or reject water presents a disposal problem. Brine volume may run as much as 15 percent of the raw water feed. This results in an 8- to 12-fold concentration of the ions in the raw water. Methods of disposal of the brine would include the neutralization processes, evaporation, deep-well injection, and the neutrolosis processes.

Neutrolosis of Brine from Reverse Osmosis

The handling of brine by neutrolosis consists of treating the brine from the reverse osmosis unit by conventional lime neutralization and returning the

Table 3.2 *Typical Product Water Quality from Reverse Osmosis Systems Treating Acid Mine Drainage*

Parameter[1]	*Raw water quality*	*Product water quality*
pH	3.1 - 3.4	4.2 - 4.4
Specific conductance	1,000 - 2,000	17 - 46
Acidity	210 - 460	32 - 46
Calcium	125 - 260	0.4 - 2.2
Magnesium	92 - 170	0.3 - 1.4
Iron, total	77 - 110	0.4 - 1.2
Iron, ferrous	61 - 71	0.3 - 1.0
Aluminum	12 - 15	0.2 - 1.0
Manganese	14 - 43	0.1 - 0.5
Sulfate	660 - 1,340	0.9 - 4.6

Source: After Skelly and Loy 1973.

Note: Synopsis of tests conducted on spiral-wound, tubular, and hollow-fiber membrane systems.

1. All units are in mg/l, except specific conductance (μ mhos) and pH.

product water to the reverse osmosis unit's feed stream. The disposal problem is then reduced to landfilling the sludge. The sludge would be expected to consist mostly of ferric hydroxide and calcium sulfate.

ION EXCHANGE

As indicated by Skelly and Loy (1973), ion exchange can be defined as the reversible interchange of ions between a solid medium and an aqueous solution. To be effective, the ion exchange medium must contain exchangeable ions, be insoluble in water, and have a porous structure that allows passage of water molecules. Within the solution and the ion exchange medium, a charge balance or electroneutrality must be maintained; i.e., the number of charges, not the number of ions, must remain constant. Ion exchange materials usually possess a preference for multivalent ions, thus they give up monovalent ions. But this reaction can be reversed by increasing the concentration of monovalent ions. This reversal provides a means of regenerating the ion exchange medium once its capacity has been saturated.

The most common example of ion exchange is the softening of mineral-bearing water for domestic purposes. The calcium and magnesium content causes hardness. The ion exchange material is charged with monovalent cations, usually sodium. The hard water is passed through a bed of the material, and the divalent cations are exchanged for sodium ions. When more than one type of cation is available, the material will have an affinity for certain ones more than others. This can be an advantage or a detriment, depending on the objective.

The early ion exchange materials were zeolites, which are produced from mixtures of aluminum salts and silicates. In the 1930s, resins were developed which expanded the applications of cation exchange. In 1949, an anion exchange resin was developed which permitted the process to be used for total demineralization of water. Today, the resins available are classified as strong-acid cation, weak-acid cation, strong-base anion, and weak-base anion types. The affinity or selectivity for the various ions each type of resin will remove is given in Table 3.3, along with a summary of the status of their technologies.

OZONE OXIDATION

A study has been conducted by the Brookhaven National Laboratory of the U.S. Atomic Energy Commission on the use of ozone to oxidize ferrous iron to the ferric form (Brookhaven 1970). Ozone production was evaluated for the electrical discharge, isotopic radiation, and chemonuclear methods. Both on-site and central ozone production facilities were considered in preparing cost estimates. Following oxidation of ferrous iron to ferric iron, limestone would neutralize the acidity present in acid wastes.

Brookhaven concluded that ozone could be used to oxidize ferrous to ferric iron under acidic conditions. The process control is simpler than with aeration methods. The electric discharge method of ozone production gave the highest costs for on-site ozone production; however, this method is the only one for which production equipment is presently available. The costs are too high for adaptation to widespread use ($0.17 per thousand gallons for a one-million-gallon-per-day plant for oxidation alone).

CHEMICAL FIXATION

Chemical fixation is a process for detoxifying hazardous liquid wastes by means of a reaction of chemical additions with the waste material to form a chemically and mechanically stable solid. The process, developed and marketed under the name of Chemfix Process, can be used for the chemical fixation of polyvalent metal ions in stable and insoluble inorganic compounds. Monovalent cations and many anions are physically entrapped in the matrix structure resulting from the reaction process. Chemical fixation is costly when compared with classical treatment methods and probably would rarely be used for treating large volumes of process effluents. The process, however, might prove useful and economic for the ultimate disposal of concentrated liquor wastes which are generated from reverse osmosis and the evaporative treatment of waste water (see Chemfix 1973).

MINIMIZING WASTE WATER PRODUCTION

Many of the mineral resource operations in the United States visited by the author have not conducted detailed studies of how to minimize the production of waste water. Thus, this technique must surely be included in a discussion of alternative treatment technologies. It falls within the field of prevention rather than cure.

Table 3.3 *Typical Ion Selectivity for Modern Ion Exchange Resins*

Ion exchange resin	*Resin ion selectivity in decreasing order of preference*
Strong-acid cation (sulfonic acid type)	$Ba^{++} > Ag^{+} > Pb^{++} > Hg^{++} > Sr^{++} > Ca^{++} > Cu^{+} > Ni^{++} > Cd^{++} > Zn^{++} > Fe^{++} > Mg^{++} > Mn^{++} > K^{+} > Na^{+} > H^{+}$
Weak-acid cation (carboxylic acid type)	$H^{+} > Cu^{++} > Ca^{++} > Mg^{++} > K^{+} > Na^{+}$
Strong-base anion (quarternary ammonium, type 1)	$I^{-} > HSO_4^{-} > NO_3^{-} > Br^{-} > CN^{-} > HSO_3^{-} > NO_2^{-} > Cl^{-} > HCO_3^{-} > OH^{-} > F^{-}$
Weak-base anion	$OH^{-} > SO_4^{--} > CrO_4^{--} > NO_3^{-} > PO_4^{---} > I^{-} > Br^{-} > Cl^{-} > F^{-}$

In water treatment applications, whether for softening, demineralization, or for specific ion removal, the different ion exchange resins are generally used as follows:

Strong-acid cation resins:
Sodium form for removal of divalent cations, mainly calcium and magnesium. Hydrogen (acid) form for removal of all cations.

Weak-acid cation resins:
Hydrogen (acid) form for removal of cations associated with alkaline anions. Hardness cations associated with bicarbonate alkalinity are removed, whereas cations associated with chloride or sulfate anions are not removed.

Strong-base anion resins:
Strong-base form for removal of all anions; carbonic acid is formed in total demineralization.

Weak-base cation resins:
Weak-base form efficiently removes entire strong-acid molecules, including hydrochloric acid (HCl) or sulfuric acid (H_2SO_4).

Combinations of the available resins have been used in systems for treatment of different waters for specific purposes. The applications of these systems to the treatment of acid water has been studied mainly to produce potable water where a reduction in the total dissolved solids is required. These systems include: (1) the Desal process; (2) the Sul-biSul process; (3) the Modified Desal process; and (4) the Two Resin system. Each of these systems has been studied in pilot-sized laboratory units, and three have been evaluated in the field in larger capacity systems using acid mine drainage. From these investigations, it has been concluded that ion exchange processes can be used to demineralize mine drainage and produce water with a quality acceptable for potable use. Each of these systems is summarized by Skelly and Loy (1973). Technically they are available to the mineral resource industry, but they are too costly to use.

Source: After Skelly and Loy 1973.

Minimizing waterflows and effluent discharges is of primary importance in:

1. Reducing water costs.
2. Reducing size of treatment plants.
3. Reducing effluent treatment costs.
4. Meeting effluent discharge permit guidelines.

Methods for reducing water consumption include:

1. Control of flows by orifices in pipelines wherever deemed appropriate.
2. Conductivity controllers in rinse tanks.
3. On/off pedals at tanks with hand-operated processes.
4. Counter-current rinsing.
5. Recycle or reuse of cooling waters.
6. Fog, spray, or intermittent rinses.
7. Agitation in rinse tanks.
8. Chemical rinses after some processes.
9. Reuse of water after only partial treatment such as sedimentation.

Chalmers (1973) suggests that the factors listed in Table 3.4 should be examined to evaluate water use. According to Chalmers, all of these economy measures are in use at one plant or another in Great Britain and, where they are combined, the effect upon the use of water is dramatic. Chalmers cites three examples of unusual, but valuable, approaches to water-saving:

1. Reduction of flows to very low rates by the use of fog and spray rinses, which are recirculated and discharged at intervals (in a bleed stream). By using small tanks of 500-gallon (2,270-liter) capacity in series, and two or three rinses per process, the effluent discharge at one of the largest plating and aluminium anodic oxidation plants in the United Kingdom is reportedly reduced to 2,800 gallons per hour (12.71×10^3 per hour) from a total of six large process shops.
2. The sequential use of waste water for a number of different duties, with cross-connections between different process plants.
3. The use of vapor rinsing, using condensation on the work surface of the rinsing agent.

The data in Table 3.5 are presented by Chalmers as an example of the water requirement of a process plant before and after taking the simple economy measures shown.

The following steps might be taken as a program of investigation, recommendation, and action in the development of effluent control. Steps 1 through 10 offer maximum promise for minimizing costs at individual plants. Steps 10 through 16 are somewhat routine for all plants.

1. Establish a water balance for the plant; relate water intake to water output.

2. Establish flows of liquors at each point of use.
3. Determine contaminants discharged at each point of origin:
   a. In washwater.
   b. As strong discharge solutions.

   Construct a disposal schedule for all regular strong discharges.
4. Segregate drains from individual industrial effluent, surface water, and domestic discharges.
5. Determine the results of combining these flows from different sections of the plant.

Table 3.4 *Factors in the Examination of the Need for Water*

Water balance: Inflows should be related to outflows and to need.

Record water intake with all outputs closed or monitored. Systematic reading of strategically placed watermeters is essential.

Cooling water: Exclude cooling and uncontaminated waters from direct discharge to drain and initiate or extend reuse and recycle wherever possible. Examine bleed-off rates and minimize subject to allowable contaminant buildup in recycled water. Timing on automatic systems.

Day and night flows.

Higher night pressure in mains has been observed to produce an unnecessary increase of up to 25 percent in water use.

Stop flows when no work is being performed.

- Install on/off pedals at tanks with hand-operated processes.
- Mechanical linkages controlling water supply against plant cycles on automatic plants. Incorporate time-delay if required.
- Install conductivity controllers.

Efficiency of water use.

- Check short-circuiting in all tanks.
- Use air agitation whenever possible.
- Use counter-current rinsing.
- Maximize recycle by minimizing bleed streams.

Determine precisely the water flow needed to maintain the required quality in each rinse tank.

Incorporate water and waste economy measures at the planning stage.

Identify alternatives for air pollution control measures (wet and dry).

Source: Modified after Chalmers 1973.

6. Survey existing drainage routes including levels, capacities, and present state.
7. Determine practicability and costs of the different possible routes for the drains.
8. Determine the possibilities and effects of water or effluent recirculation or reuse.
9. Determine feasibility of dry air pollution control techniques.
10. Compare benefits and costs between local product recovery and recovery from one central plant collecting from all sources.
11. Outline flow diagram for the treatment plant(s) proposed.
12. Report recommendations, discussion, and agreement on the treatment system to be installed.
13. Discuss and negotiate with the effluent discharge permit authority.
14. Place contract including:
    a. Building works (foundations and drainage, plus building if required).
    b. Mechanical service (including pipework).
    c. Electrical services (including instrumentation and control equipment).
    d. Treatment facilities.
15. Prepare overall construction program.
16. Test installed plant.
17. Educate operating staff as required.

Other methods available for reducing the rate of production of waste water are discussed in Chapter 4.

Table 3.5 *Water Requirements of a Process Plant Before and After Simple Economy Measures*

Process	*Tank capacity (gallons)*	*Rinse flow rates* As tendered (gal/hr)	*Rinse flow rates* As modified (gal/hr)
Hot soak clean	225	105	100
Hot cathodic clean	175 }	150	50
Hot anodic clean	175 }		
Muriatic acid (HCl)	175	270	60
Cyanide copper flash and drag-out	175	270	180
Muriatic acid	175	300	60
Nickel plate and drag-out	1,250	270	60
Chrome plate and drag-out	530	360	60
Hot swill	–	45	15
Total		1,770	585

Source: After Chalmers 1973.

SOURCES OF COST INFORMATION

Each chapter of this book presents cost information if appropriate. Specific sources are cited for each subject discussed. More general sources are available also. The U.S. Environmental Protection Agency has published a two-volume report entitled: *Capital and Operating Costs of Pollution Control Equipment Modules.* It is among the more comprehensive of the sources available.

The two volumes of the EPA report are designed to aid in the estimation of purchasing and installation costs of industrial water pollution control equipment. Graphs and tables allow rapid evaluation of costs for various types of flow processing installations. The two volumes consist of the *User Guide* (Volume 1) and the *Data Manual* (Volume 2).* EPA reports that trial users have estimated costs within 10 percent of the actual constructed facility costs.

SOLID WASTE DISPOSAL

Dry air pollution control devices (baghouses and others), wet air pollution control devices, and water pollution control technologies ultimately convert air pollution control and water pollution control into a solid waste disposal problem. The most common method of discarding inorganic solid wastes is landfilling, or a modification of it such as the tailings pond. The procedures for these types of solid waste disposal, and others, are discussed in Chapter 11.

REFERENCES

American Water Works Association 1971. *Water Quality and Treatment.* 3rd ed. New York: McGraw-Hill Book Co.

Appalachian Regional Commission 1969a. *Acid Mine Drainage in Appalachia*, Vol. 1. House Document 91-180, a report pursuant to the provisions of Section 302(b) of the Appalachian Regional Development Act, Publication No. 35-4320, U.S. Govt. Printing Office, Washington, D.C.

Appalachian Regional Commission 1969b. *Acid Mine Drainage in Appalachia*, Vol. 2. House Document 91-180, Publication No. 35-4320, U.S. Govt. Printing Office, Washington, D.C.

Besselievre, E. B. 1969. *The Treatment of Industrial Wastes.* New York: McGraw-Hill Book Co.

Betz Laboratories 1971. *Betz Handbook of Industrial Water Conditioning.* 6th ed. Trevose, Pa.: Betz Laboratories Inc.

Bond, R. G., and Straub, C. P. 1973. *Handbook of Environmental Control, Vol. 3, Water Supply and Treatment.* Cleveland: Chemical Rubber Co. Press.

* The two volumes were prepared for EPA's Office of Research and Development by the ICARUS Corporation of Silver Spring, Maryland. The report (EPA Document R5-73-02a+b) is available from the National Technical Information Service of the U.S. Department of Commerce. It is also available from the Superintendent of Documents, U.S. Government Printing Office, Washington, D.C.

Brookhaven National Laboratory 1970. *Treatment of Acid Mine Drainage by Ozone Oxidation.* EPA Document 14010 FMH 12/70, U.S. Govt. Printing Office, Washington, D.C.

Chalmers, R. K. 1973. Treatment of Inorganic Liquid Wastes. In *Progress in Water Technology, Vol. 3, Water Quality; Management and Pollution Control Problems*, ed. S. H. Jenkins. Proceedings, Conference of International Association on Water Pollution Research, Jerusalem, pp. 251-264. New York: Pergamon Press.

Chemfix 1973. *Ultimate Disposal of Liquid Wastes by Chemical Fixation.* Pittsburgh: Chemfix Division, Environmental Sciences Inc.

Eckenfelder, W. W., Jr. 1966. *Industrial Water Pollution Control.* New York: McGraw-Hill Book Co.

Eckenfelder, W. W., and Ford, D. C. 1971. New Concepts of Oxygen Transfer and Aeration. In *Advances in Water Quality Improvement*, Water Resources Symposium No. 1, ed. E. F. Gloyna and W. W. Eckenfelder, Jr. Austin: Univ. of Texas Press.

Goff, D. L., and Gloyna, E. F. 1970. Reverse Osmosis Developments. In *Water Quality Improvement by Physical and Chemical Processes*, ed. E. F. Gloyna and W. W. Eckenfelder, Jr., pp. 312-326. Austin: Univ. of Texas Press.

Hall, E. P. 1972. *Projects of the Pollution Control Analysis Section.* EPA Document R2-72-134, U.S. Environmental Protection Agency, Applied Science and Technology Branch, Office of Research and Monitoring, Washington, D.C.

Hoak, R. D. 1944. Neutralization Studies on Basicity of Limestone and Lime. *Sewage Works Journal*, Vol. 16, No. 4.

Horizons, Inc. 1970. *Treatment of Acid Mine Drainage.* Document 14010 DEE 12/70, U.S. Govt. Printing Office, Washington, D.C.

Jacobs, H. L. 1951. Neutralization of Acid Wastes. *Sewage and Industrial Wastes*, Vol. 23, No. 7.

Johnston, C. E. 1973. Tailing Disposal–Its Hidden Costs. In *Tailing Disposal Today*, Proceedings, International Tailing Symposium, Tucson, Ariz., ed. C. L. Aplin and George O. Argall, Jr., pp. 762-815. San Francisco: Miller Freeman Publications.

Kealy, C. D. 1970. "Seepage Patterns in Mill Tailings Dams as Determined by Finite Element Models." M.S. thesis, College of Mines, Univ. of Idaho, Moscow. (Published in modified form as U.S. Bureau of Mines Report of investigations 7477, January 1971.)

Keener, W. H. 1973. Tailing Disposal Practices at the Tyrone Branch of Phelps Dodge Corporation. In *Tailing Disposal Today*, Proceedings, International Tailing Symposium, Tucson, Ariz., ed. C. L. Aplin and George O. Argall, Jr., pp. 224-231. San Francisco: Miller Freeman Publications.

Mihok, E. A., Duel, M., Chamberlain, D. E., and Selmeczi, J. G. 1968. *Mine Water Research–The Limestone Neutralization Process.* U.S. Bureau of Mines Report of Investigations 7191.

Miller, J. D. 1972. *Removal of Dissolved Contaminants from Mine Drainage.* EPA Document R2-72-130, U.S. Govt. Printing Office, Washington, D.C.

Mink, L. L., Williams, R. E., Wallace, A. T., and McNay, L. M. 1973. Renovation of Wastes by Mine Tailings Ponds. *Mining Engineering*, Vol. 25, No. 7, pp. 81-91.

Neal, A. W. 1971. *Industrial Waste–Its Handling, Disposal and Reuse.* London: Business Books Ltd.

O'Melia, C. R. 1970. Coagulation in Waste and Waste Water Treatment. In *Water Quality Improvement by Chemical and Physical Processes*, ed. E. F. Gloyna and W. W. Eckenfelder, Jr., pp. 219-236. Austin: Univ. of Texas Press.

Ross, R. D. 1968. *Industrial Waste Disposal.* New York: Reinhold Book Corp.

Rudolfs, W. 1943a. Neutralization with Lime. *Sewage Works Journal*, Vol. 15, No. 3.

Rudolfs, W. 1943b. Pretreatment of Acid Wastes. *Sewage Works Journal*, Vol. 15, No. 1.

Skelly and Loy Inc. 1973. *Processes, Procedures and Methods to Control Pollution from Mining Activities.* EPA Document 430/9-73-011, U.S. Environmental Protection Agency, Washington, D.C.

Smith, F. 1943. Neutralization of Pickle Liquor. *Sewage Works Journal*, Vol. 15, No. 1.

Stander, G. J., and Van Vuuren, L. R. J. 1970. Flotation of Sewage and Waste Solids. In *Water Quality Improvement by Physical and Chemical Processes*, ed. E. F. Gloyna and W. W. Eckenfelder, Jr., pp. 190-199. Austin: Univ. of Texas Press.

Van Everdingen, R. O., and Banner, J. A. 1971. *Precipitation of Heavy Metals from Natural and Synthetic Acidic Aqueous Solutions during Neutralization with Limestone.* Technical Bulletin 35, Inland Waters Branch, Dept. of Energy, Mines and Resources, Ottawa, Ontario, Canada.

Wallace, A. T. 1973. Treatment of Zinc Plant Wastes with Lime and Coagulant Aids. Unpublished report to the Bunker Hill Co., Dept. of Civil Engineering, Univ. of Idaho, Moscow.

4. *Acid mine drainage*

INTRODUCTION

The production of acid mine drainage originated on a large scale with the expansion of coal mining in the Appalachian states. Solutions to the problem are indeed difficult, not only because of economics, but because it is difficult to assign legal responsibility for the water after mining has been completed and a property abandoned. Abandonment is likely to continue to be a problem even for new mining operations. Goldberg and Power (1972) have proposed model legislation to handle this problem. Such legislation would provide for state or interstate agencies to administer legal and practical protection to a project area to insure against any activities that would cause acid mine water pollution.

Abandonment and numerous other aspects of acid mine drainage have been the subject of a voluminous literature. Some 550 articles and publications, including available bibliographies, are tabulated and discussed in a report presented to Congress by the Appalachian Regional Commission (1969). The compilation is utilized to a great degree in this chapter. The author has compiled a post-1969 bibliography, much of which is included in the 215 entries in the bibliography of Skelly and Loy (1973).

Coal mining is not unique in the production of acid mine drainage. It is produced wherever a mine of any type in permeable rock intersects the water table or a perched water body, or where surface water finds its way into a mine, and where sulfides, particularly pyrite, are present in the ore or country rock. Its occurrence is common in the metal mines of Idaho, Montana, Colorado, Washington, Arizona, Nevada, and other states.

The Appalachian Regional Commission states that in Appalachia underground mines represent 58 percent of all sources of acid mine drainage and produce 71 percent of all acid mine drainage. Active underground mines represent 5 percent of all sources and produce 19 percent of all acid mine drainage. The commission notes that surface mines represent 28 percent of all sources and produce 12 percent of all acid mine drainage. The remaining 14 percent of the sources is from combined surface and underground mines, mining waste piles, and coal preparation plants. These produce 17 percent of all acid mine drainage in Appalachia. Inactive mines contribute 78 percent of all acid mine drainage, which underscores the problem of abandonment. Acid mine drainage comprises over 50 percent of the waste water to be treated by the hard-rock mineral resource complex described in Appendix E. In Appalachia, the states of Pennsylvania and West Virginia are most plagued by acid mine drainage problems.

The commission estimates that some $4,230,000 in annual benefits (1967 dollars) would accrue if acid mine drainage were eliminated in Appalachia alone. Surprisingly, this figure excludes benefits accrued to recreation and fisheries. Some 29 percent of these benefits ($1,230,000) would be realized annually by industrial water users. The majority of the industrial savings would be realized by three industry groups. These are primary metals (65 percent), electric power generation (11 percent), and chemical products (11 percent). Thus, the mineral resource-based industry itself would appear to benefit most from cleanup. Table 4.1 presents a breakdown of savings to water users that would result from the reduction of acid mine drainage in Appalachia.

Table 4.1 *Estimated Annual Savings to Water Users from Reduction of Acid Mine Drainage in Water Supplies*

Type of user		*Savings accruing to users by the percent reduction of mine drainage pollution in water supply*		
		30%	*60%*	*90%*
Industrial and utilities	A	$ 530,000	$ 990,000	$1,230,000
	B	370,000	410,000	490,000
Navigation: barges and towboats	A & B[1]	410,000	820,000	1,370,000
facilities	A & B[1]	30,000	60,000	100,000
Municipal water supply	A & B[1]	290,000	580,000	960,000
Highway culverts and bridges	A & B[1]	170,000	340,000	570,000
Total	A	1,430,000	2,790,000	4,230,000
	B	1,270,000	2,210,000	3,490,000

Source: After Appalachian Regional Commission 1969.

Note: A = Reduction at the source by preventing the pollution formation or its introduction into surface waters. B = Reduction of acid mine drainage pollution by neutralization with lime.

1. Savings accruing to these users are estimated to be generally independent of the method of abatement.

FORMATION OF ACID MINE DRAINAGE

A comprehensive treatment of the formation of acid mine drainage is presented by Smith and Shumate (1970) and Ohio State University (1971). Much of the information in this section was obtained from these two reports.

Acid mine drainage is usually formed by the chemical reaction of oxygen (in air), water, and iron sulfide–principally pyrite (FeS_2) and/or marcasite ($Fe_{11}S_{12}$). Microorganisms may or may not (but apparently usually do) play a role. When the iron sulfide is exposed to oxygen, it is converted by oxidation to sulfuric acid and compounds of iron. The acid and iron are dissolved and the low pH water in turn increases the solubility of other metals and compounds that the water contacts. Thus, acid mine drainage may contain a variety of ions including aluminum, manganese, zinc, cadmium, and even lead, depending on the materials contacted.

The transformation of pyrite to soluble iron and sulfuric acid is characterized in simplest form by the following reactions:

(a) dry environment:

$$FeS_2 + 3O_2 \longrightarrow FeSO_4 + SO_2$$

(b) wet environment:

$$2FeS_2 + 2H_2O + 7O_2 \longrightarrow 2FeSO_4 + 2H_2SO_4$$

Acid mine drainage is generated in the presence of water; so reaction (b) is the one of interest. This produces acid mine drainage, but additional reactions may occur. The ferrous iron is converted to the ferric state according to:

$$4FeSO_4 + 2H_2SO_4 + O_2 \longrightarrow 2Fe_2(SO_4)_3 + 2H_2O$$

followed by:

$$Fe_2(SO_4)_3 + 6H_2O \longrightarrow 2Fe(OH)_3\downarrow + 3H_2SO_4$$

Schmidt and Conn (1969), Colmer and Hinkle (1947), Lorenz and Stephan (1969), Syracuse University (1971), and Baker and Wilshire (1970) present evidence that one or more of several bacteria accelerate the above reaction.

Iron is precipitated from solution when hydroxyl ions become available for combination with the sulfate radical to form ferric hydroxide which has low solubility in an acid environment. This precipitation accounts for the yellow sediment in streams receiving acid mine drainage.

The ferric ion can, and does, react with iron sulfide whereupon additional sulfide is converted to sulfate, thereby accelerating formation of acid. The reaction is:

$$7Fe_2(SO_4)_3 + FeS_2 + 8H_2O \longrightarrow 15FeSO_4 + 8H_2SO_4$$

Pyrite oxidation and subsequent transport require only a small amount of water and air. The atmosphere in a mine or over a refuse dump is adequate for

Table 4.2 *Characteristics of Seepage Water from a Tailings Pile in Elliot Lake, Ontario*

Parameter	*Concentration*[1]	*Parameter*	*Concentration*[1]
pH	2.0	Cadmium	0.05
Sulfate as $SO_4^{=}$	7,440.0	Lithium	0.07
Acidity as $CaCO_3$	14,600.0	Vanadium	20.0
Ferric iron as Fe^{+3}	1,450.0	Silver	0.05
Ferrous iron as Fe^{+3}	1,750.0	Titanium	15.0
Uranium	7.2	Magnesium	106.0
Zinc	11.4	Calcium	416.0
Nickel	3.2	Potassium	69.5
Cobalt	3.8	Sodium	920.0
Copper	3.6	Arsenic	0.74
Manganese	5.6	Phosphorus	5.0
Aluminum	588.0	Chemical oxygen demand	270.0
Lead	0.67		

Source: After Hawley and Shikaze 1971.

1. In ppm (except pH).

the basic stoichiometric reaction. Water then collects the reaction products (acid salts) and transports them out of the mine or off the refuse pile. Table 4.2 presents an analysis of seepage water collected from the down-gradient side of a high pyrite tailings pile in the Elliot Lake, Ontario, Canada, uranium district. Acid mine drainage proper is characterized in Table 4.3. The sample depicted in Table 4.3 was collected from an ore body in the Bunker Hill Mine in northern Idaho. The sulfides in the ore are approximately one-third zinc

Table 4.3 *Characteristics of Acid Mine Drainage Collected from the Bunker Hill Mine in Kellogg, Idaho*

Constituent	*Concentration*[1]	*Constituent*	*Concentration*[1]
pH	2.2	Magnesium	1,500.0
Sulfate as $SO_4^{=}$	63,000.0	Calcium	31.6
Total iron	16,250.0	Potassium	0.7
Zinc	14,560.0	Sodium	0.5
Nickel	4.8	Chromium	0.3
Copper	13.4	Chloride	38.0
Manganese	2,625.0	Nitrate as $NO_3^{=}$	77.5
Aluminum	347.0	Electrical conductivity	48,000.0
Lead	0.8	(micromhos @ 25° C)	
Cadmium	22.5		

Note: Sample collected and analyzed by Bryson Trexler, College of Mines, University of Idaho.

1. In ppm (except pH and electrical conductivity).

sulfide (sphalerite), one third lead sulfide (galena), and one-third iron sulfide (pryite). Acid mine drainage constitutes over one-half of the waste water that the Bunker Hill Company will treat in the new plant described in Appendix E.

Hill (1968) uses quality as a means of classifying acid mine drainage. His classification is presented in Table 4.4. It provides some indication of the variety in quality that can be expected of mine drainage, and it provides an approach to categorizing treatment costs in subsequent sections.

METHODS OF TREATING OR ABATING ACID MINE DRAINAGE

There are some 24 methods of treating or reducing (and perhaps eliminating) acid mine drainage. About one-half of them have been tested or applied to field or laboratory situations. The techniques can be categorized as:

1. Those preventing the formation of acid mine drainage at its source. The techniques in this category include modified mining methods such as longwall mining (see Lusk 1973), mine sealing, surface reclamation, water diversion, and the control of the ground-water flow system by well fields or other methods so as to prevent the exposure of ground water to sulfides in a mine atmosphere.
2. Those containing the polluted water. They include deep-well injection and subsurface dams and grout curtains.
3. Dilution of the acid mine drainage to achieve acceptable effluent quality.
4. Standard waste water treatment techniques which result in neutralization and removal of dissolved solids.

Tables 4.5 and 4.6 present available treatments which the Appalachian Regional Commission (1969) has categorized by degree of dependability and refinement. Category A (Table 4.5) is the preferable category. Order is not significant in Table 4.5. The less preferable methods in Table 4.6 are ranked by decreasing promise.

Table 4.4 *Mine Drainage Classes*

Constituent[1]	*Class 1* *Acid discharges*	*Class 2* *Partially oxidized and/or neutralized*	*Class 3* *Oxidized and neutralized and/or alkaline*	*Class 4* *Neutralized and not oxidized*
pH	2 - 4.5	3.5 - 6.6	6.5 - 8.5	6.5 - 8.5
Acidity ($CaCO_3$)	1,000 - 15,000	0 - 1,000	0	0
Ferrous iron	500 - 10,000	0 - 500	0	50 - 1,000
Ferric iron	0	0 - 1,000	0	0
Aluminum	0 - 2,000	0 - 20	0	0
Sulfate	1,000 - 20,000	500 - 10,000	500 - 10,000	500 - 10,000

Source: After Hill 1968.

1. In mg/l (except pH).

Table 4.5 *Control Techniques in Category A*

		Effluent quality, ppm (1)				*Cost in $/1,000 gal mine drainage* (1)
Technique	*Application*	*pH*	*TDS*	*Hardness*	*Fe*	
Neutralization	A, B, S, D, 1, 2	7 - 8	500 - 3,000	500 - 3,000	1	0.19 - 0.28
Reverse osmosis (2)	A, B, S, D, 1, 3	4 - 5	10 - 20	1	0.1	0.30 - 0.50
Streamflow regulation (3)	A, B, S, D, 1, 2, 3, 4	6 - 7	700 - 500	100 - 200	1 - 50	0.30 - 0.71
Deep-well disposal (4, 8)	A, B, S, D, 1, 2, 3, 4	–	–	–	–	0.41 - 0.55
Surface reclamation	A, B, S, 1, 2, 3, 4	(6)	(6)	(6)	(6)	–
Revegetation (8)	A, B, S, 1, 2, 3, 4	(6)	(6)	(6)	(6)	–
Pumping and drainage	A, B, S, D, 1, 2, 3, 4	–	–	–	–	0.02 - 0.05
Water diversion	A, B, S, D, 1, 2, 3, 4	(6)	(6)	(6)	(6)	–
Mine entry sealing (5)	B, D, 1, 2, 3, 4	(7)	(7)	(7)	(7)	0.04 - 0.40
Refuse treatment	A, B, S, D, 1, 2, 3, 4	(6)	(6)	(6)	(6)	–
Impoundment	A, S, 1, 2, 3, 4	(6)	(6)	(6)	(6)	–

Source: After Appalachian Regional Commission 1969.

Note: Under application, A = Active; B = Abandoned; S = Surface mine; D = Deep mine; 1, 2, 3, 4 = Mine drainage category.

(1) Factors based on moderate acidity (1,500 ppm) and 10 mgd of mine drainage.
(2) With deep-well injection.
(3) Single-purpose reservoirs system, 10:1-20:1 dilutions.
(4) 1.0 mgd of well capacity.
(5) Hydraulic sealing to abate 0.3-3.0 mgd.
(6) Uncontaminated surface-water quality.
(7) Uncontaminated ground-water quality.
(8) A supporting, not a primary, technique.

Table 4.6 *Control Techniques in Category B*

		Application		
Rank[1]	*Technique*	*Surface*	*Deep*	*Mine drainage category*
1	Desulfating	A, B	A, B	3
2	Inert gas blanket	–	A, B	1, 2, 3, 4
3	Grouting and sealing	–	A, B	1, 2, 3, 4
4	Microbiological iron and sulfur removal	A, B	A, B	1, 2, 3, 4
5	Sulfide iron removal	A, B	A, B	1, 2, 4
6	Ion exchange	A, B	A, B	3
7	Electrodialysis	A, B	A, B	3
8	Evaporation	A, B	A, B	3
9	Sterilization	–	A, B	1, 2, 3, 4
10	Microbiological control	–	B	1, 2, 3, 4
11	Internal sealing	–	A	1, 2, 3, 4
12	Permanganate iron removal	A, B	A, B	1, 2, 4

Source: After Appalachian Regional Commission 1969.

Note: A = Active mines. B = Abandoned mines.

1. According to promise of successful application.

Abatement of Acid Mine Drainage at its Source

Mine Sealing

Mine sealing is accomplished by the construction of closures which: (1) prevent the exposure of sulfides to air, thereby arresting the chemical reactions described earlier, and (2) reduce the rate of escape of water from the mine. All openings to the mine must be sealed, usually with stone, masonry, or compacted soil. Seals usually are not designed to stop all waterflow out of a mine; the bulkheads are normally traps that prevent the entry of air but permit water to escape. This is accomplished by designing the remaining opening so that it will be filled with water continually. A grout curtain is sometimes injected under the seal to prevent leakage. Drill holes also may be drilled simultaneously with sealing so as to provide pressure bleed-off after the mine openings are full. Well-designed and -sited drill holes should flow after mine openings are filled.

Halliburton (1970) used a combination of quick-setting sodium silicate cement and expansive cement to completely seal mine openings with apparent success. This technique constitutes one of the more promising approaches to mine sealing in that it demonstrates that material is available to construct a mine seal composed of rear and front bulkheads of a quick-setting slurry, separated by expansive cement filler, which seals a mine drift under a head of impounded water. Skelly and Loy (1973) also report favorable potential for single bulkhead gunite seal.

Even though the mine sealing technique may work at the point of installation, it is not without problems. The disadvantages or risks of it are: (1) water may begin to emanate from other undesirable locations because of a local increase in water-table height, and (2) roof collapse may render the method ineffective by creating new openings by which air may continue to enter the mine. In short, hydrogeologic conditions may be such that the method will not work satisfactorily. Feasibility of the use of mine seals should be determined by thorough hydrogeologic investigations at each proposed site. The resulting reports should describe hydrostatigraphic units, mine openings, fracture patterns, areas of subsidence, runoff characteristics, drill hole locations, and any other information deemed important to the method. Data are best presented on topographic maps, aerial photographs, and "stacked" two-dimensional drawings. The final decision on feasibility should be determined by teams of hydrogeologists and engineers on the basis of these data. The central question is whether or not air or water leaks will occur and whether they can be prevented or controlled. These studies will frequently be justified by the low cost of the method if it is deemed feasible.

Control of Ground-Water Recharge to the Mine

Under some hydrogeologic conditions, it is possible to minimize the recharge of ground water and the motion of ground water in the vicinity of a mine in such a way that the production of acid mine drainage can be minimized.* The method is particularly applicable to low-permeability rocks where mining has created high-permeability channels. A case study of the Bunker Hill Mine in northern Idaho by investigators at the University of Idaho in cooperation with the company illustrates what must be done if this approach is to be successful.

Acid drainage from the underground workings at the Bunker Hill Mine is discharged into the Bunker Hill tailings pond and is the major component of the effluent and a major factor in the treatment costs discussed in Appendix E. The discharge from the Bunker Hill Mine averages about 2,700 gallons per minute at an average pH of 3.3. The concentration of heavy metals is concomitantly high (see Table 4.7).

The Bunker Hill Mine of the Bunker Hill Company consists of mining properties extending along the rugged valley side of the South Fork of the Coeur d'Alene River. Mining began in 1885 with many shallow, small tunnels and stopes around the original discovery site. The Bunker Hill Mine has expanded to include over 100 miles of underground workings to depths of almost one mile.

The mine is located within a highly faulted block of the massive Revett quartzite and St. Regis formation of the Precambrian Belt rocks. These meta-

* The approach discussed here has been field-tested by Dr. Bryson Trexler, research associate at the University of Idaho, Dr. Dale Ralston of the Idaho Bureau of Mines and Geology, and the author. Publications are currently in preparation and in press. Ralston et al. (1973) is the principal document available to date.

Table 4.7 *Selected Water Quality Data for Different Sources in the Bunker Hill Mine*

Water source	*pH*	*Cadmium*	*Iron*	*Lead*	*Zinc*
Milo Creek (surface)	7.0	<0.02	<0.1	<0.1	<0.05
Diamond drill holes (country rock)	7.1	<0.02	<0.1	<0.1	<0.05
Diamond drill holes (fault)	6.9	<0.02	<0.1	<0.1	<0.05
Diamond drill holes (ore bodies)	5.3	0.05	590	0.38	545
Upper level drainage	2.6	2.8	838	1.9	1,095
Lower level drainage	6.1	0.03	3.1	1.1	15.4
Kellogg Tunnel drainage[1]	3.3	0.3	59	2.3	120

Note: In ppm (except pH). < implies below detection limit of laboratory instrument.

1. Total mine drainage.

morphic rocks are relatively impermeable to water except where faulted or mined. Ore deposits are tabular veins along or adjacent to steeply dipping fault zones (approximately 45°). The four vein shapes observable in the mine are: (1) wide brecciated zones of great vertical height, (2) narrow brecciated zones of great vertical height, (3) wide fissure-type ore veins, and (4) combinations of the first three shapes.

Mining methods used in the Bunker Hill Mine have varied with the shape of the ore body and with time (technological advances). The methods have included: (1) vertical slice, (2) cut and fill with square sets, (3) horizontal cut and fill with square sets, and (4) block caving. Mining method can be shown to have a significant effect on the production of acid mine water.

Sandfilling was implemented in 1961. Waste-fill from the development headings was used for backfilling some stopes until about 1961. Some of the waste-fill, or gob, in the upper levels of the mine contains galena, sphalerite, and pyrite. The filled stopes in the more recently mined areas do not contain sulfide-rich waste because the sandfill technique was used.

Distribution of Acid Water: Water samples have been collected from more than 400 sites throughout the mine by the investigators previously named. Locations of these sites are shown on a generalized two-dimensional diagram of the mine in Figure 4.1. Acid water is a prominent feature of the upper levels of the mine as can be seen from the pH data also shown on the figure. Low-pH water occurs from level 4 to level 19. It has been determined that the vertical pattern of acid water is the result of drainage downward through a series of interconnecting stopes between these levels (Figure 4.2). The interconnecting stopes follow two major ore bodies—the March and the Flood-Stanly ore bodies. The Flood-Stanly zinc ore body was mined from level 4 to level 12. Water from workings in the Flood-Stanly zinc ore body drains into the workings of the March ore body on level 12 and continues down to level 19.

Major downward movement of water is facilitated by the interconnection of man-made workings. As water drains down the man-made openings, some of it moves through sulfide-rich waste-fill in the stopes. This water becomes more acidic and increases in concentration of heavy metals as it flows over the sulfides. The downward movement of water is effectively halted where the levels are not well interconnected with old workings. The data show that the decrease in pH values is dependent upon the type of ore body and the time of contact with the rock. Water collected from the Flood-Stanly zinc ore body

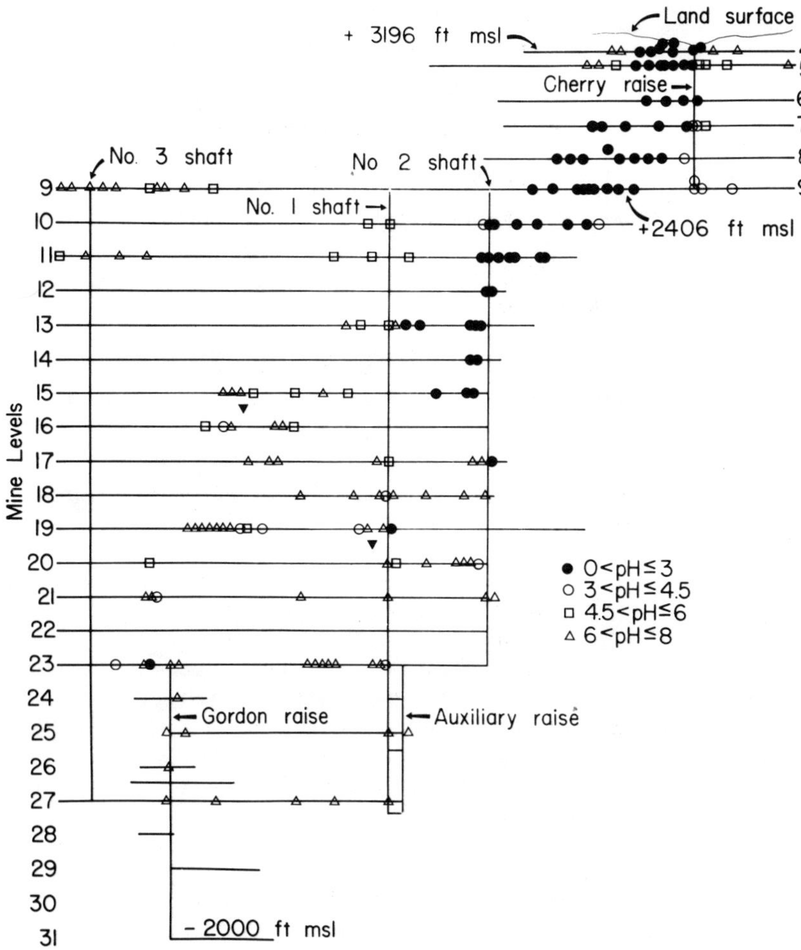

Figure 4.1. *Generalized diagram of the Bunker Hill Mine, showing location of sampling stations and the respective pH values.*

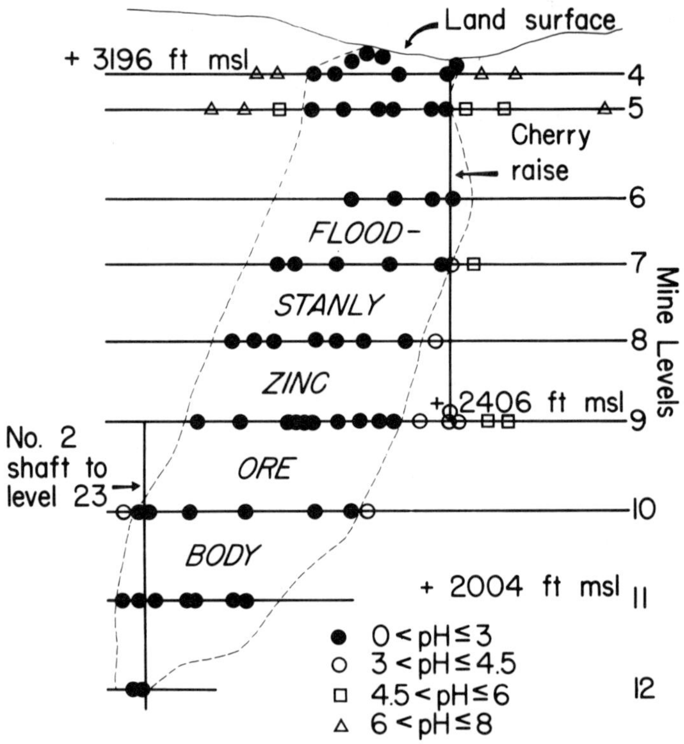

Figure 4.2. *Interconnecting stopes of the Flood-Stanly zinc ore body on generalized diagram of levels 4 through 12, Bunker Hill Mine.*

shows a decrease in pH from level 4 to level 11. The water in some stope areas has pH values as low as 0.1 where stagnant pools have been formed during low-flow periods. Zinc content of these waters runs as high as 14,000 parts per million. In comparison, the March ore body, in which mineralization is primarily a galena-ore type, shows a much higher range of pH values (5.0 to 6.5) on the same upper levels of the mine. It is only on level 12, where the Flood-Stanly zinc ore body stoping joins the March ore body by a common stope, that low-pH water occurs in the March ore body. Both ore bodies are recharged with water from the same drainage basin above the mine. These surface waters have a pH of approximately 7.0.

The quality of the water described here may be compared with water collected from drill holes that intercept: (1) the country rock, (2) faulted areas, and (3) other ore bodies. In general, the pH of the water from the faults and the country rock is higher than 6.5. Samples from drill holes that penetrate ore bodies at lower levels in the mine indicate an average pH of about 5. Water from the older levels with the gob-filled stopes has a lower pH, generally less than 3.5. Drainage from the lower levels has a much higher pH, around 5 to 6.

Data on concentrations of cadmium, iron, lead, and zinc from the various locations within the mine are presented in Table 4.7. Similar distribution patterns are reflected by the data on heavy metals. Most of the water with high concentrations of heavy metals is associated with drainage through the older workings of the Flood-Stanly zinc ore body in the upper levels of the mine. Water samples from freshly opened areas in the lower part of the mine show good-quality water.

The difference in the distribution of low-pH values between the upper, older levels and the lower levels might appear initially to be due to the presence of the ore-rich waste-fill. However, the higher pH values of the March ore body imply that the presence of sulfide-rich waste-fill, which is common to stopes in both of these ore bodies, is not the only factor affecting water quality. This observation prompted a comparison of pH values with minerals present in the ore bodies.

The Bunker Hill Mine contains three major ore types. They have been named (1) Bluebird ore, (2) Bunker Hill ore, and (3) Jersey ore. The Flood-Stanly body is a Bluebird ore and the March body is a Bunker Hill ore.

The Bluebird ore contains considerable pyrite and galena which usually exceed or equal the sphalerite content in a siderite-quartz gangue. Bunker Hill ore consists mainly of galena in a siderite-quartz gangue. The Jersey ore consists of galena with considerable sphalerite in a quartz gangue. Table 4.8 lists the minerals in the different ore types and their relative abundance.

Table 4.8 *Mineralogical Composition of the Three Ore Types in the Bunker Hill Mine*

	Ore types		
Minerals	*Bluebird*	*Bunker Hill*	*Jersey*
Galena, PbS	A	A	A
Sphalerite, ZnS	A	R-P	A
Pyrite, FeS_2	A	R-P	R
Arsenopyrite, $FeAsS$	P	N	N
Chalcopyrite, $CuFeS_2$	P	P	N
Tetrahedrite, $(CuFeZnAg)_{12}Sb_4S_{13}$	P	P	P-A
Siderite, $FeCO_3$	A	A	P-A
Ferrodolomite, $Ca(Mg, Fe)(CO_2)_2$	A	A	P-A
Quartz, SiO_2	A	P	A
Bournonite, $PbCuSbS_3$	R	N	N
Boulangerite, $Pb_5Sb_4S_{11}$	R	N	N
Stibnite, Sb_2S_3	R	N	N
Sericite, $KAl_2(AlSi_3O_{10})(OH)_2$	P	P	P
Chlorite, $Mg_3(Si_4O_{10})OH_2\text{-}Mg_3(OH_{10})_6$	P	P	P
Limonite, $FeO(OH) \cdot nH_2O$	A	N	N
Cerussite, $PbCO_3$	R	N	N

Note: A = Very abundant. P = Present. R = Rare. N = None noted.

Table 4.9 *Comparison of Water Quality from Stopes of the Three Ore Types in the Bunker Hill Mine*

pH	*Cd*	*Ca*	*Cu*	*Fe*	*K*	*Mg*	*Mn*	*Na*	*Pb*	*Zn*	*Al*	$SO_4^{=}$	Cl^-	NO_3	*EC*
							Bluebird ore								
1.8	0.1	8.4	0.07	52	3.0	19.9	31	1.8	2.3	112	1.5	125	ND	1.5	1,450
2.1	22.5	31.6	9.0	16,250	0.2	1,500	2,625	0.5	0.8	14,560	347	63,000	38	77.5	48,000
2.2	20	18	22.6	10,000	<0.1	1,690	2,175	0.6	0.5	10,880	422	39,500	21	63	45,000
2.52	1.3	9.6	1.3	262	0.7	75	87.5	0.6	0.9	415	31	4,125	7.0	7.3	5,000
							Bunker Hill ore								
6.2	<0.02	13	<0.05	<0.1	0.9	34.5	12.9	0.9	<0.1	1.2	<0.5	60	ND	0.1	380
7.2	<0.02	87	<0.05	<0.1	3.3	55	2.7	2.1	<0.1	2.7	<0.5	375	ND	0.5	1,010
6.3	0.05	46.8	<0.05	<0.1	9.4	62.5	31.8	5.0	<0.1	28.9	<0.5	NA	NA	NA	NA
							Jersey ore								
6.7	<0.02	85.2	<0.05	<0.1	5.6	55.4	0.1	3.2	<0.1	<0.05	<0.5	NA	NA	NA	NA
6.0	0.1	97.1	<0.05	<0.1	54.0	250	18.0	7.4	0.6	31.8	<0.5	NA	NA	NA	NA
5.4	<0.02	74.3	<0.05	17.6	21.1	162	8.6	6.7	1.1	12.6	2.2	NA	NA	NA	NA[1]
5.5	0.6	97.0	0.1	132	17.0	157	115	8.5	1.9	402	2.8	NA	NA	NA	NA[1]
							Total upper level drainage (Cherry Ditch)								
2.7	0.3	12.7	0.2	98	0.6	42.5	37	0.8	0.5	160	4.6	1,500	1.0	4.0	2,350
							Composite upper level (Bluebird ore) drainage								
2.5	3.3	12.8	3.3	1,150	1.0	255	345	1.0	1.4	1,580	56	4,170	7.0	7.3	7,000
							Kellogg Tunnel drainage (total mine drainage)								
3.3	0.3	61	0.3	59	5.2	117	69	4.2	2.3	120	3.1	885	4.6	1.3	1,550

Note: NA = Not analyzed. ND = None detected. < = Below detection limit of equipment.

1. Drifts from Mac vein, which has considerable pyrite.

The major mineralogical difference between the three ore types is the presence of large quantities of pyrite in the Bluebird ore. Close examination along the high waterline of the drainage paths has revealed reddish-yellow ferric hydroxide deposits. The sulfuric acid produced is also able to leach heavy metals from any appropriate minerals subsequently contacted.

Water-quality data for water in the three ore types is given in Table 4.9. The availability of the pyrite and its ability to oxidize and form sulfuric acid is reflected by the water in the Bluebird-ore-type stopes. The quality of the drainage from Bluebird-type ore is a principal cause of the mine's total poor-quality drainage (see Table 4.9) and the concomitant high treatment costs.

Source of Ground-Water Recharge: During the initial collection of data, it was noted that certain areas discharge water only during periods that correspond to high surface runoff. This implies that the production of low-pH water is influenced by the path of water movement and by the quantity and pattern of ground-water recharge.

Recharge to the Bunker Hill Mine occurs through old workings in the upper levels of the mine (levels 3, 4, and 5). Hydrographs from the Kellogg Tunnel (the main drainage path from the mine), the Cherry Ditch (drainage into the Kellogg Tunnel from the upper levels of the mine above level 9), and Placer Creek, a nearby stream in the area, are presented in Figure 4.3. The difference in water volume between the Kellogg Tunnel and Cherry Ditch hydrographs represents the water pumped from levels 10 and below. Only a small variation in this pumpage volume is seen throughout the year. A similarity exists between the hydrographs from the Kellogg Tunnel, Cherry Ditch, and Placer Creek. The high flows in the mine correspond to the high flows in the stream resulting from snowmelt; little lag time is noted. The similarity of hydrographs implies a rather direct connection of the mine to the surface and associated ground-water recharge.

At least three areas and mechanisms for recharge to the mine have been isolated. They are:

1. A drift and stope area on level 4 (Small Hopes Tunnel) is hydraulically connected to the streambed of Milo Creek, a small perennial stream in the area. Stream wash materials are present in the drift, indicating that mining broke through to the streambed at some time in the past. During periods of high flow in Milo Creek, water infiltrates down from the stope into the drift. This recharge water is much less significant during periods of low flow.
2. The close proximity of the upper stopes to the land surface has induced surface fractures because of land subsidence, and the stopes and fractures have intersected previous faulted and fractured zones. This mechanism is prevalent on the East 5-level where the drift and stopes parallel Milo Creek. Mining breakthrough to the surface has not been reported, yet a continuous discharge occurs from these stopes.
3. Surface subsidence and related fractures resulting from mining by the caving method extend to the surface in the drainage area of Milo Creek. This is believed to be a primary mechanism allowing recharge to the upper levels of

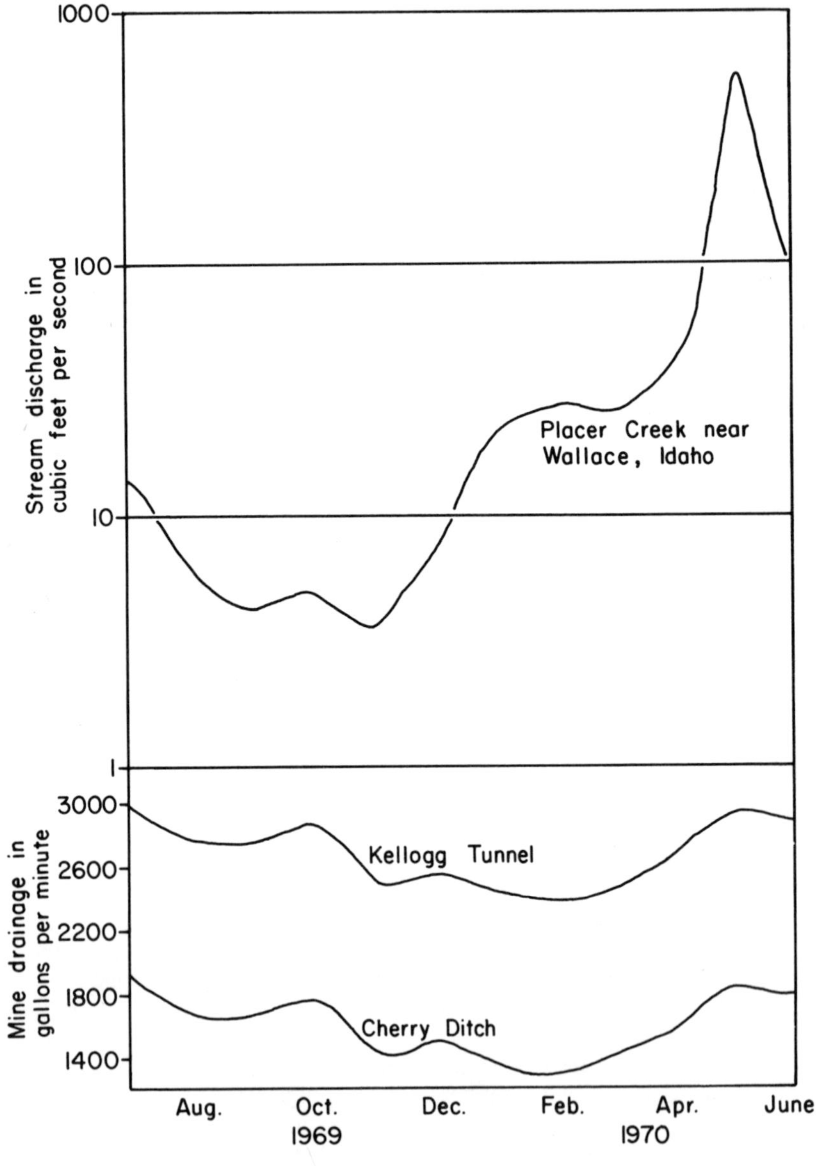

Figure 4.3. *Hydrographs of a surface stream and mine drainage.*

the mine. The caving area is located beneath three small valleys. Much of the runoff from this area is funneled by the nature of the topography into the depressions formed from the caving. No signs of surface outflow from the depressions are evident. The average water content of the snowpack at the head of the three small valleys is 30 inches as measured on April 1. Measured from U.S. Forest Service photographs, acreage covered by the snowpack is approximately 230 acres. This gives 575 acre-feet of potential recharge or 1,300 gallons per minute over the three high runoff months of April, May, and June. However, this runoff is never an evenly regulated flow during the three months. It tends to maximize in May.

Figures 4.4 and 4.5 show the principal underground workings of the upper levels of the Bunker Hill Mine with some updated features. The closeness of the workings to Milo Creek and the Bunker Hill reservoir dam are illustrated in this figure.

With the location of the underground workings and a possible surface recharge in mind, a dye tracer test was initiated in Milo Creek, 200 yards upstream from a small reservoir in the creek used for mine water supply. If the Small Hope's and East 5-level's stopes and drifts were connected to the surface stream, then the dye would be expected in the water. A fluorescent dye (Rhodamine WT) was used.

The results of the dye test show that the surface water took no longer than 30 minutes to reach the farthest sampling point. These observations verify the first two areas and mechanisms of recharge.

Following the dye injection, dye dilution procedures were initiated on Milo Creek to determine losses of water over different reaches of the creek. At this writing, the tests are in progress, but they will ultimately define the exact reaches through which the fractures, faults, drill holes, and underground openings are being recharged. With the recharge areas delineated, the Bunker Hill Company will be able to divert water by grouting, piping, and other methods and relocate the Bunker Hill reservoir in areas of near zero loss.

Drying up the recharge from Milo Creek will reduce a large portion of the yearly flow to the upper levels of the mine. In turn, the amount of acid salts washed from the upper levels and heavy metals leached by the acid-rich waters should be reduced. This reduction in water will be reflected at the lime treatment facility discussed in Appendix E through decreased operating cost.

This type of detailed ground-water investigation is sufficiently new to mining that its utility on a large scale cannot be evaluated. However, it appears to hold great promise, at least in hard-rock mines, wherein natural saturated hydraulic conductivity is heterogeneous and localized.

Dewatering Ground-Water Flow Systems

Source beds (aquifers) supplying leakage to deep mines may be dewatered during and after mining under favorable hydrogeologic conditions (Parizek 1971; Parizek and Tarr 1972; Ahmad 1970 and 1973). Data needed to determine the feasibility of dewatering by wells should be collected during

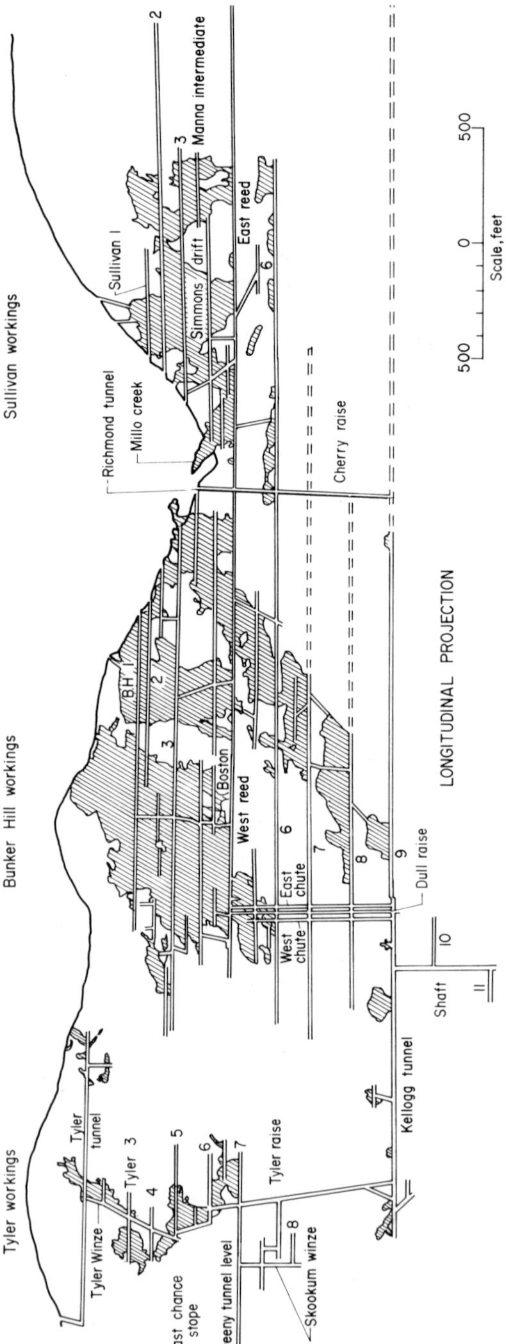

Figure 4.4. *Cross section showing principal underground workings of the Bunker Hill Mine.*

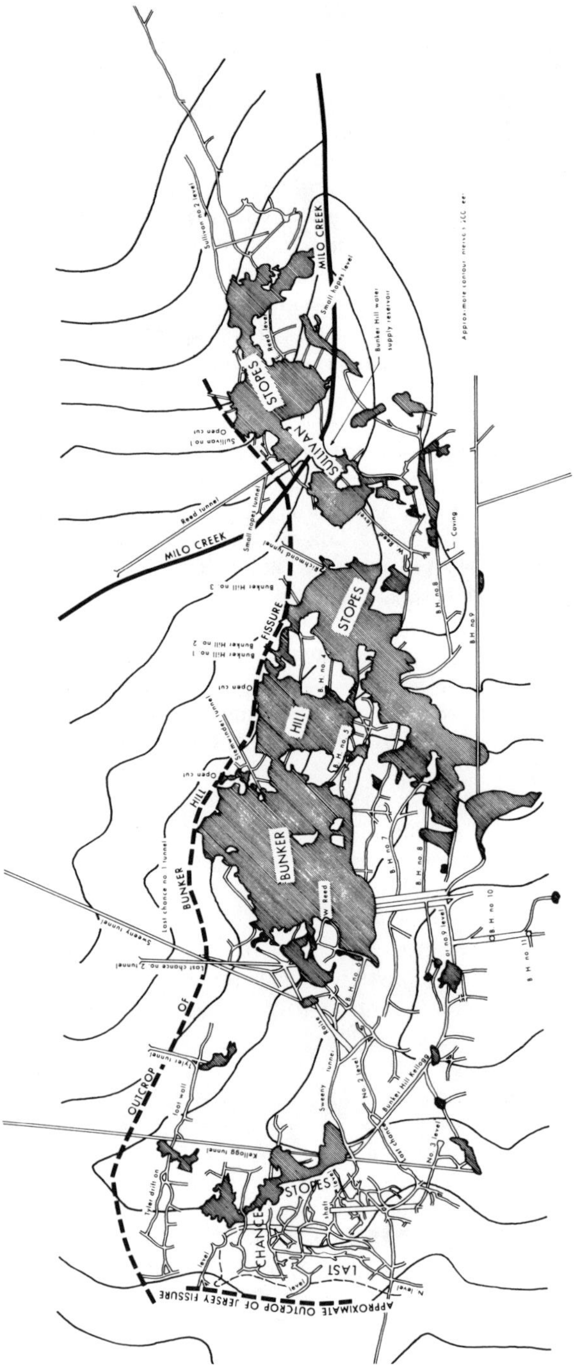

Figure 4.5. *Plan view of the Bunker Hill Mine.*

exploration and mining. Hydrogeologic data needed include hydraulic properties and the spatial distribution, thickness, and geometry of aquifers, aquicludes, and aquitards. Hydraulic boundaries, either permeable or impermeable, including faults, fracture patterns, and stratigraphic pinchouts, must be delineated. Hydraulic head distribution must also be mapped. These data should be collected so as to form the framework for model studies using numerical or electrical methods such as those described by Dominico (1972) and Walton (1970). The method is not appreciably different from the scientifically proven water-well field design procedures common to water supply projects.

The objective in the case of active mines is to get the ground water to the surface before it encounters and oxidizes sulfides in the atmosphere of the mine so that impending, expensive treatment costs can be averted. In the case of abandonment, the same reasoning prevails but economics change considerably. It behooves both types of operations to tie the dewatering project to a municipal or industrial water supply if at all possible.

Aquifer conditions favorable for dewatering by wells include thin permeable rocks directly above (or below) the coal bed (or ore body) with a high head difference between the aquifer and the layer to be mined, and limited leakage and recharge rates (Parizek 1971). Low-permeability rocks probably could not be dewatered adequately because too many wells would be necessary to achieve a large cone of depression.

The principle guiding the feasibility of dewatering by wells is that total pumpage must be adequate to deplete water in storage; then pumping must either dewater the aquifer or be equal to the rate of leakage from adjacent aquitards plus any recharge directly to the aquifer at its outcrop. Multiaquifer wells may be necessary if the zone to be mined is both overlain and underlain by aquifers.

So-called gravity or weeping wells may also be used to dewater aquifers with the added advantage of no pumping costs. Such wells depend on the existence of aquifers, above and below the sulfide source bed, which are separated by a confining stratum such as a clay or shale bed. However, the lower aquifer may be replaced by artificial drains in the mine opening. The hydraulic gradient must be directed downward in the portion of the ground-water flow system in the vicinity of the mining operation. The procedure involves drilling wells through the upper and lower aquifers with perforations opposite the aquifers. Water flows from the upper aquifer into the lower aquifer or into the drains (beneath the coal or other zone to be mined) without contacting the sulfides associated with the deposit. Ahmad and Parizek and Tarr proposed and examined the feasibility of weeping wells on several occasions. The concept has been portrayed by Parizek and Tarr as in Figure 4.6.

The data which must be generated to evaluate the feasibility of mine dewatering by wells at any particular site can be collectively referred to as ground-water flow system information. Brown and Parizek (1971) present an excellent example of a study that can be used as a guide for obtaining the kind of information needed to evaluate this method.

Lower cost may be the main reason for considering well dewatering as opposed to classical treatment of acid mine drainage. If the geometry of the aquifers, aquitards, and the rock body to be mined is sufficiently simple, Parizek concludes that annual cost in dollars per million gallons of water per day for wells with yields of 75 gallons per minute might approach $19,000 per year (costs in 1965 dollars). Similar costs for wells yielding 230 gpm and 760 gpm are $10,000 per year and $6,900 per year, respectively. Clearly, if a hydrogeologic environment permitted this approach, its consideration would be necessitated in an economical sense.

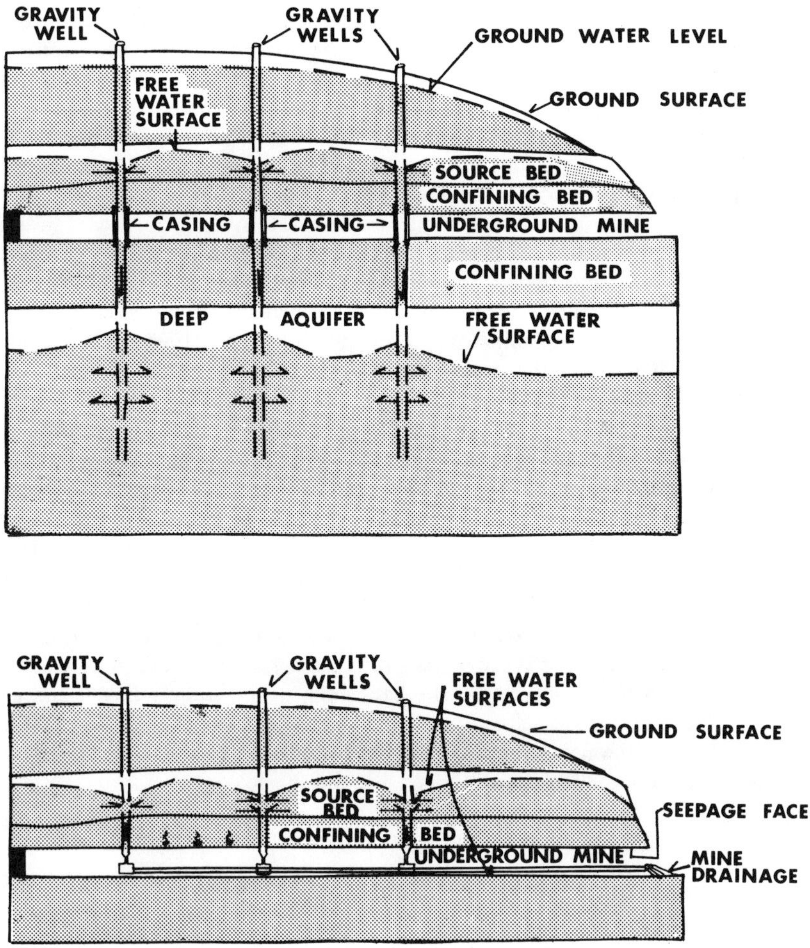

Figure 4.6 *Interception of water in aquifers. (This figure is modified after Parizek and Tarr 1972.)*

Containment of Acid Mine Drainage

Deep-Well Injection

The petroleum and chemical industries have used deep-well injection as a disposal technique for many years. Oilfield brines, in particular, have been discarded in this manner. More recently, the method has been utilized for pickling and other acid wastes by industries along the southern end of Lake Michigan and in Appalachia.

The method consists of injecting waste liquids into deep aquifers which are more or less confined by aquitards or aquicludes. Clearly, whether or not the method is safe is a function of the characteristics of the ground-water flow system. Therefore, the feasibility of the safe use of the method must be determined by thorough ground-water studies such as those referred to previously. In addition to hydrogeologic considerations, many states have approved legislation stating that waste water cannot be injected into strata containing water better in quality than the waste water. These laws mean that in general waste water can only be injected deep in the ground-water flow system because poor-quality water has a tendency to occur along very long ground-water flow paths where particle movement is slowest (for evidence, explanations, and additional references, see Toth 1963; Toth 1966; and Williams 1970). This deep occurrence of poor-quality ground water accounts for the high cost of installing wells for the injection of waste water into aquifers and in part for the deep-well injection method's relatively high total cost (see Table 4.5).

To make matters more pessimistic for deep-well injection, the author has observed that it is becoming more and more difficult to persuade the Environmental Protection Agency that point source waste discharge permits should be issued on the basis of deep-well injection as the best available control technology. Problems with plugging of the wells lend credibility to this claim. Nonetheless, the method has received thorough discussion in excellent volumes published under the support of the U.S. Atomic Energy Commission. One of them (AAPG 1968), for example, tabulates strata that have potential for deep-well injection in the Appalachian Basin, the major structural syncline in Pennsylvania, the Michigan Basin, the Salina Basin, the Denver Basin, and the San Juan Basin.

The major components required for the installation of a deep-well disposal system are the completed injection well and the surface pre-injection treatment system.

The completed injection well consists of the following major components:

1. The drilled, logged, and tested hole.
2. Casing (cemented in place).
3. Injection tubing.
4. Wellhead equipment.
5. Flow lines and valving.
6. Control system.

The pre-injection treatment system consists of:

1. Tankage for necessary treatment or processes.
2. Filters.
3. Pumps (low pressure for moving fluid through the treatment system and to the injection pumps).
4. Piping and valving.
5. Control system.
6. Chemical feed equipment.
7. Sludge-handling facilities.
8. Injection pumps (usually capable of high-pressure performance).

Impoundment

As shown in studies by Reilly (1960 and 1965) and Hall (1965), properly constructed lakes in abandoned coal strip mines can become sites for propagation of aquatic life instead of polluted pools. The objective of the design is to assure that all acid-forming minerals are covered by water in the lake occupying the mined-out area. This is most feasible in horizontal areas where the water table is shallow. In more rugged terrain, only dams are likely to provide continual impoundment.

This method is directed primarily at controlling runoff from stripped areas as opposed to underground acid mine drainage. The ultimate objective is analogous to sealing and flooding of abandoned underground mines.

Dilution of Acid Mine Drainage

Diluting acid mine drainage for the purpose of quality improvement has long been practiced in one of several forms. The most frequently used technique consists of allowing the drainage to enter stream channels, sometimes coupled with streamflow regulation. However, the passage by Congress of Public Law 92-500 restricts quality control to point sources of waste waters. Consequently, unless dilution is accomplished prior to release, it would not be possible to rely legally upon the method. Since this is not possible in most cases, dilution is no longer considered a reliable pollution control method.

Waste Water Treatment Techniques

The waste water treatment techniques available to the mineral resources industry for improving the quality of acid mine drainage are listed in Tables 4.5 and 4.6. Of the treatment techniques available at this writing, the Appalachian Regional Commission (1969) considers only neutralization and reverse osmosis (combined with deep-well injection) to be sufficiently well developed technologically and sufficiently promising economically to merit serious consideration. Lombardo (1973) and Temmel (1973) are inclined to exclude reverse osmosis and the author is inclined to agree with them.

Neutralization

Adjustment of pH with lime or limestone constitutes the most common method of treating acid mine drainage. Neutralization has been proven to be effective, and it is the least costly of the available chemical or biochemical treatment techniques. Neutralization is generally combined with an aeration or oxidation process to convert ferrous iron to ferric iron. The treatment operation described in Appendix E constitutes an excellent example of the treatment of a waste water composed primarily of acid mine drainage.

The objective of neutralization is to remove acidity and dissolved constituents from the acid mine drainage. Traditionally, iron and pH have been considered as the major objectionable parameters but more recent work has shown that other dissolved ions such as cadmium (see Table 4.3) constitute a threat to human health.

A similar statement holds for zinc with respect to aquatic life, particularly trout (see EPA 1973, for discussion of ramifications of the various toxic elements proposed by EPA). Consequently, since the publication of the classic study by the Appalachian Regional Commission (1969), the philosophy on treatment of acid mine drainage containing substances other than iron has changed and it is now considered essential to remove much more than iron.

As indicated in Chapter 3, lime or limestone can be used to adjust pH and remove sulfate when present above the saturation solubility of calcium sulfate. Aeration can be incorporated to oxidize soluble +2 ions to the less soluble +3 state. Soluble aluminum salts are precipitated also. However, in order to remove other constituents such as cadium and zinc, it may be necessary to raise the pH to the level of 9 or 10, followed by the addition of clean acid to the sediment-free effluent to neutralize back to the pH levels permissible under the waste discharge permit guidelines prior to discharge (see Chapter 1). This technique is utilized by the Bunker Hill waste treatment facility discussed in Appendix E. This method will result in the release of hard water containing higher than normal sulfates, but these constituents apparently will not be restricted by the EPA guidelines because they are not considered unreasonably harmful to the environment.

As pointed out by the Appalachian Regional Commission, neutralization treatments historically have been applied to mine drainage classes 1 and 2, since classes 3 and 4 generally have no acidity and lower iron concentrations. Unslaked lime, hydrated lime, and more recently limestone have been most widely used, primarily because of low cost.

The use of sodium-based alkalis such as soda ash or caustic also will remove acidity and iron. These alkalis are substantially more expensive, but there is one other major technical difference. Sodium sulfate is a highly soluble salt; thus there is no reduction in sulfate content. The insoluble materials are iron and aluminum oxides, and the quantity of insolubles separated would be less, since no sulfates would be removed.

The question of whether to neutralize with lime or limestone (or other alkalis) has traditionally been answered with slaked lime. Lombardo (1973) designates lime treatment as the "method of choice" throughout the industry.

However, Mihok et al. (1968) and Van Everdingen and Banner (1971) conclude that more widespread neutralization with limestone is feasible. They point out also that the sludge from limestone neutralization occupies lower volume and has a lower water content, a problem of no small concern when slaked lime is used (see Appendix E). The method of Mihok utilizes an autogenous mill to grind the limestone into balls. Some of the acid mine drainage to be treated is passed through the mill, and the remainder is combined with the mixture emanating from the mill. Additional details on the use of limestone as a neutralizing agent are noted by Bituminous Coal Research (1970 and 1971).

The use of slaked lime to achieve neutralization is well demonstrated and well documented (see Temmel 1973; Lombardo 1973; and papers in Appendix C, Appalachian Regional Commission 1969). An example of its use in the field is presented in this book in Appendix E; additional details, including costs and details of equipment and design parameters, are presented by Charmbury et al. (1967). Temmel concludes that sludge recycle and underground disposal can solve many of the associated sludge disposal problems. Several aspects of these studies are discussed in Chapter 3.

Other methods that can be used to improve the quality of acid mine drainage include neutralization with soda ash, caustic soda, and anhydrous ammonia. However, the costs associated with these methods drive them out of the economical picture as far as widespread use is concerned. They function in much the same manner as does hydrated lime and limestone. Some of these techniques are discussed in Chapter 3 also.

Table 4.10 presents the relative treatment cost data (in 1967 dollars) for each of these methods, as determined by Stephan and Lorenz (1968). These data illustrate why limestone and hydrated lime constitute the most widely used neutralizing agents. The matter has been summed up well by Wilmoth and Hill (1970). After extensive investigation, they concluded that studies utilizing lime, limestone, and soda ash to treat acid mine drainage having a high ferric-ferrous ratio (which can be achieved by aeration) have shown that:

1. Lime, limestone, and soda ash can neutralize mine drainage to pH 6.5 to 8.0, and iron to less than 7.0 ppm.
2. The major differences between soda ash neutralization and lime or limestone neutralization are that soda ash produces a water lower in hardness and calcium and high in sodium and sulfite. Other chemical parameters are comparable between the three neutralizing agents.
3. Chemical costs for treating by the three methods were: Soda ash, \$0.049, limestone, \$0.010, and lime \$0.005 per mg/l acidity per 1,000 gallons. (Wilmoth and Hill did not employ the method of Mihok et al. when evaluating limestone. This may explain the discrepancy between these treatment costs and those shown in Table 4.10.)
4. Limestone sludge is more desirable in that it occupies approximately two-thirds of the volume of lime sludge and has a higher solids content. It contains a large residual alkalinity which would be beneficial when disposed into an acid mine environment (although this residual alkalinity is expensive and of questionable value to the treatment process).

5. Lime is a very reactive material, and the neutralization reaction goes to completion in less than half an hour. The limestone reaction requires from 24 to 48 hours to go to completion and, therefore, requires a long detention time before discharge; however, aeration will reduce the detention time to one somewhat comparable to lime.
6. The limestone reaction is not very sensitive quantitatively, i.e., small changes in limestone feed rate or water quality do not cause large changes in product water quality; so the accuracy with which constituents are fed into the reactor need not be controlled with the precision required by lime. Accidental overtreatment is not the pollution problem with limestone that it might be with lime.
7. Lime is capable of producing higher pH values, which may be necessary in some cases to comply with permit guidelines. With limestone, pH's above 7.0 are essentially impossible to attain. In order to make the cost of limestone treatment competitive with lime, methods of improving the utilization efficiency must be found.
8. For all agents, the majority of the sludge will settle in one hour. But the suspended solids in the effluent may not satisfy permit guidelines.

Table 4.10 *Raw Material Costs for Neutralization of Sulfuric Acid in Acid Mine Water*

Neutralizing agents	*Use factor (100% alkali basis)*[1]	*Equivalent acid (lb of H_2SO_4/lb alkali)*	*Cost per ton*[2]	*Cost per ton (H_2SO_4 neutralized)*
CaO (100% basis)	1.00	1.75	–	–
Limestone (high calcium)				
Lump	0.56	.98	$ 1.60	$ 1.64
Pulverized	0.56	.98	3.50	3.57
Lime (hydrated)	0.76	1.33	19.25	14.47
Calcium carbonate				
(Air floated, 100% basis)	0.56	.98	13.50	13.77
Soda ash, 58% (Na_2O)				
Bags (C1)	0.52	.915	40.00	43.70
Bulk (C1)	0.52	.915	32.00	34.97
Caustic soda, 76% flake				
400 lb drums (works)	0.69	1.20	111.00	92.50
Ammonia (anhydrous)	1.65	2.88	92.00	31.94

Source: After Stephan and Lorenz 1968.

1. Use factor equals pounds CaO to which 1.0 lb alkali is equivalent.
2. Does not include freight costs.

9. No difference in the efficiency between dry feed and slurry feed methods of introducing the neutralizing agent was found for lime, limestone, and soda ash.
10. Dry feeding offers fewer operational difficulties than does slurry feeding and costs less, both initially and in cost of maintenance.

These points and others have been summarized as follows:

1. The advantages of lime treatment include:
   a. Removal of iron and aluminum salts.
   b. Removal of acidity.
   c. Reduction in sulfate ion concentration.
   d. Relative simplicity and ease of control.
   e. More readily available at all locations.
2. Disadvantages include:
   a. Addition of water hardness.
   b. Gypsum scale formation in plant and possibly in effluent.
   c. More difficult sludge-handling and dewatering problems.
   d. Possible to overtreat if careful control is not exercised.
3. The advantages of limestone treatment include:
   a. Chemical costs are lowest of all commercial alkalis (if transport distance is not too great).
   b. Sludge settling is more rapid than with lime.
   c. Sludge volume is lower than lime (2% vs. 12%).
   d. Removes acidity and iron.
4. Disadvantages include:
   a. Difficult to raise pH to high levels; therefore effluent may not be acceptable.
   b. Hardness is added to effluent.
   c. Sulfate removal no better than lime treatment.
   d. Rate of ferrous ion oxidation is slower because of absence of hydroxide formation (bicarbonate is formed instead).
   e. Bulk to be handled is greater than with lime.
   f. Grinding is necessary if the method of Mihok (1968) is to be used to prevent coating and concomitant waste of resource.

Wilmoth et al. (1972) report results of using combinations of lime and limestone to neutralize acid mine drainage. The tests were conducted at the Environmental Protection Agency's mine drainage field site near Norton, Virginia. The experiments showed that limestone is effective for neutralization at low values of pH, but efficiency decreases at values above 6.0 because of problems with carbonic acid and iron oxidation. Optimum results were attained by neutralization with limestone to pH 4.0 followed by any desired additional neutralization with lime. From 20 to 30 minutes were required for the limestone reaction and from 10 to 15 minutes for the lime reaction. Limestone process sludge was one-half that of lime and the percent solids were five times greater for the limestone.

Other Methods

Other methods of improving the quality of, or minimizing the production of, acid mine drainage include reverse osmosis, ion exchange, electrodialysis, evaporation, ozone oxidation, desulfating, sulfide iron removal, microbiological control, permanganate iron removal, and coating source areas with gels. Of these alternatives, the Appalachian Regional Commisssion (1969) considers only reverse osmosis to be sufficiently sound technologically and sufficiently promising economically to merit serious consideration. Consequently, reverse osmosis is the only treatment process considered in any detail here. The commission recommends that reverse osmosis be considered only in combination with deep-well injection to provide for the disposal of the brine. The product water can be collected and reused.

Considerable work has been done on chemical stabilization, and one such study is discussed subsequent to the discussion on reverse osmosis.

Reverse Osmosis

A reverse osmosis plant for acid mine drainage consists of pumps and filters for removal of suspended solids. Effluent from the filter is fed to the reverse osmosis unit where it is exposed in a pressure vessel to the membrane cells. The concentrated brine that discharges from the unit passes to a holding pond and is injected into a deep well or treated. As pointed out by the Appalachian Regional Commission (1969), tests under the cooperative support of the U.S. Bureau of Mines, the Office of Saline Water (both in the U.S. Department of the Interior), and the Environmental Protection Agency (see Rex Chainbelt 1972) have shown that the utilization of reverse osmosis with acid mine drainage feed is feasible and that product water can be brought to potable quality by a small increase in pH (see Table 4.11). On the basis of these tests, it is reasonable to conclude that the process would be most applicable for class 1

Table 4.11 *Reverse Osmosis Quality Data*

Content[1]	*Feed*	*Brine*	*Product*[2]	*Product*[2]	*Product*[2]
pH	2.30	2.10	4.0	4.75	3.35
Total acidity (as $CaCO_3$)	530	2,880	5.0	2.5	18.0
Dissolved solids	1,638	9,466	7.0	3.2	22.0
Total sulfates (as SO_4)	1,167	6,254	2.6	0.7	15.0
Calcium (as $CaCO_3$)	389	1,696	1	1	1
Magnesium (as $CaCO_3$)	245	1,394	1	1	1
Total iron (as Fe_2O_3)	156	1,188	trace	trace	0.46
Conductivity (mmhos at 25°C)	2,037	6,351	40	40	121

Source: After Appalachian Regional Commission 1969.

Note: Water recovery, 94.6%. Pressure, 400 psi.

1. In ppm (except pH and conductivity).
2. Variations in product water quality stem from three membrane types evaluated.

acid mine drainage waters, which are highly acidic and have low pH, though class 3 waters, which contain no iron and are alkaline, would be suitable (see Table 4.4).

Tests on low-concentration class 1 acid discharges have shown that gypsum deposition is not a problem and that, if oxygen is excluded, ferric iron deposition on the reverse osmosis membranes is avoided. It appears less feasible to employ reverse osmosis for other classifications of acid mine drainage. Class 2 water (partially oxidized and neutralized), which has a higher pH and relatively high concentrations of ferric iron (calcium carbonate and iron deposition), and class 4, which is neutral but high in iron and not oxidized, are both felt to offer the danger of membrane clogging not necessarily inherent in the class 1 water. Rex Chainbelt (1970) in fact observed clogging of a class 2 water in a pilot plant study. Obviously, capacity reductions induced by clogging significantly increase the operating costs of the reverse osmosis process, which are already high. Tests at the mine drainage treatment laboratory at Norton, West Virginia, have shown that biological fouling can also be a problem. It has been possible to keep the membranes free of biological growth by feeding copper sulphate to the feed stock at a concentration of 5 parts per million or by disinfection with ultraviolet light (Rex Chainbelt 1972).

The Appalachian Regional Commission concludes that reverse osmosis is not applicable to treatment of acid mine waters, other than for supplying potable water to a municipality or other uses. Under these conditions, the credit for fresh water production would make the cost of acid mine drainage treatment more economically attractive (or vice versa).

Silicate Treatment

Tyco (1971) has conducted extensive laboratory investigations of the effect of treating sulfide-containing refuse piles with sodium silicate and silica-alumina gel. The techniques tested were: (1) neutralization of the water with a dilute solution of sodium silicate, (2) development of a continuous gel on the refuse surface to seal off the entire pile from runoff water, and (3) development within the pile of a continuous silica-alumina gel to preclude percolation of water through the refuse and to minimize the effect of natural erosion of the gel surface. The investigators concluded that:

1. Neutralization of coal mine refuse by treatment with waterglass (sodium silicate) is an effective technique for preventing or minimizing acid mine drainage (AMD) generation for a period in excess of 120 inches of equivalent rainfall.
2. Sealing a coal mine refuse pile surface to water percolation with silicate-based gels is effective in preventing or minimizing AMD generation over a long period, perhaps in excess of 400 inches of equivalent rainfall.
3. Alumina-silica gels are more effective in prolonged sealing of the refuse than are silica gels because of the lower water solubility of the alumina-silica gel structures.

4. Surface gels are not stable to extremes in temperature. Both heating and freezing cause coagulation of the gel with the resultant loss in continuity. Such broken gels no longer seal the refuse, but the residual effects of the treatment minimize the generation of AMD over prolonged periods of time, similar to the effect of neutralizing the material with silicate.
5. In-depth gelation of refuse piles using alumina-silica gels should be effective in prolonged prevention or retardation of AMD generation, regardless of the environmental temperature variations.
6. The treatment of AMD water with sodium silicate and sodium aluminate in a holding pond is no more effective than the conventional application of lime. There is no advantage in settling rates or precipitate volume.

Tyco did not compute costs for this method but recommended an additional study to determine costs, depth of gelation control, quantity of chemicals needed, and demonstration of means of field application. The author is not aware of any subsequent reports, but it would appear that the method shows sufficient promise to merit pursuit by researchers.

SUMMARY OF COSTS AND INFLUENCING FACTORS

Stephan and Lorenz (1968) present a comprehensive survey of the costs of methods for controlling acid mine drainage. Their approach assumes that acid mine drainage is treated alone whereas, in reality, it frequently must be treated along with other wastes. Nevertheless, in the opinion of the author, it is among the best sources of cost data available. The estimates are in 1967 dollars, but these can be converted approximately to dollars current for any year with available conversion indices. Additional cost data are presented by Skelly and Loy (1973).

Stephan and Lorenz discuss costs for neutralization, ozone oxidation, flash distillation, reverse osmosis, ion exchange, deep-well injection, and drainage control methods (mine sealing, diversion, and impoundment). However, as shown in Tables 4.5 and 4.6, only neutralization, drainage control, mine sealing, and possibly reverse osmosis combined with deep-well injection have been considered sufficiently demonstrated and sufficiently cheap to merit consideration. Therefore, this section will be limited to these methods plus ground-water flow system dewatering. The costs for ground-water flow system dewatering were derived primarily from Parizek (1971).

Neutralization

Lime and Limestone

Costs on neutralization of acid mine drainage selected for presentation herein are derived from House Document 91-180 (see Appalachian Regional Commission 1969) and from Skelly and Loy (1973). The following factors have been determined to influence costs:

1. The use of unslaked lime with a slaker facility minimizes chemical costs but necessitates the cost of a lime slaker and safety precautions.
2. Hydrated lime reacts quickly and completely, thus reducing residence time, but it produces a high-volume sludge. Limestone offers an advantage in that a denser, lower volume sludge is formed. Dissolution of CO_2 causes reaction rates of limestone to be slower than those of hydrated lime.
3. Mixing and aeration are important operations in the use of lime, since the chemical reactions involve the formation of slightly soluble gypsum and the oxidation of ferrous iron. Suitable mixing and agitating and/or aeration facilities must be provided to preclude incomplete reactions and an unstable sludge and effluent. Oxidation of ferrous ions accelerates as the pH is raised.
4. Settling and/or thickening are as important as mixing and aeration, since the sludges formed are typically slow in settling. Charmbury et al. (1967) report

Use of lime for neutralization of acid mine drainage requires careful mixing and aeration. This West Virginia plant chemically treats 1.5 million gallons of acid mine water per day. Lime is added to the water automatically in a slurry mix tank; a rotary beater is used for agitation and a thorough mix.

Consolidation Coal Company photo

the successful use of a thickener to separate the iron oxide-gypsum sludge mixture, and subsequent centrifugation as a sludge dewatering process. They note that settling and thickening may represent a high percentage of the capital plant expenditures.

5. Sludge disposal cost may range from 13 to 15 percent of the total costs of a plant treating highly acidic mine drainage, to 11 to 13 percent for moderately acid discharges, to 7.5 to 10 percent for weakly acid discharges. However, these costs would be affected by land availability, soil type, underlying rock, ground-water conditions, and disposal method. Consequently, these estimates may be low.
6. Chemically, the reactions of limestone for neutralizing mine drainage are similar to the reactions of lime. Each removes acidity, each removes iron if combined with aeration, and each produces a sludge containing calcium sulfate and iron oxide.

However, the reaction of limestone or calcium carbonate with sulfuric acid produces CO_2, an acidic gas. In order to complete the reaction, this acidic gas must be expelled from solution, or the pH will not maximize and may decrease with time. In addition, since calcium carbonate is a solid, it tends to become coated with the insoluble products of the neutralization reaction, and thus requires over-neutralization to affect a sufficiently high effluent pH and satisfactory iron removal. If pH values above 8.0 are needed to reduce dissolved ions to guideline levels, use of lime may be required, thus increasing cost.

Tables 4.12 through 4.16 present House Document 91-180 costs for various segments of the treatment process, including sludge thickening and disposal (1967 dollars). The data in these tables are in part the basis for the following discussion. The remaining discussion is based on cost curves prepared by Charmbury et al. (1967), Holland et al. (1968), Skelly and Loy (1973), and on information in Appendix E.

Appalachian Regional Commission deduces (primarily from U.S. Bureau of Mines reports) that a one-million-gallon-per-day treatment plant using water of 2,000-parts-per-million acidity would require 16.6 tons per day of limestone. The limestone costs for such a plant would be \$50.00 per day. Using a power rate of \$0.015 per kilowatt-hour, the cost of generating a ton of fine limestone in the mill would be about \$0.67. For the above 1-mgd plant, the power costs would be \$11.12 per day. Reducing these numbers to an operating cost per 1,000 gal/day would suggest a cost of \$0.61 per 1,000 gal treated. If one assumes that labor costs, maintenance costs, and miscellaneous costs for a 900,000-gal-per-day lime treatment plant as presented by Holland et al. (1968) can be transposed to the above limestone plant, then \$0.05 (labor) + \$0.03 (maintenance) + \$0.03 (miscellaneous) are to be added, bringing the operating cost to \$0.171 per 1,000 gal for a l-mgd plant. Sludge disposal costs must be added to this. Holland et al. suggest \$0.035 per 1,000 gal of mine water treated for sludge costs. Here again, the use of limestone may reduce this cost because less volume is handled, requiring smaller lagoons. Thus, the investigators judged that sludge disposal costs might add \$0.025 per 1,000 gal to the above, for a total cost of \$0.196 per 1,000 gal, or about 75 percent of the costs estimated

Table 4.12 *Sludge Thickening as a Percent of Plant Capital*

Acidity + Fe (mg/l)	*mgd treated*	*~%*	*Total plant equipment costs*
4,000 + 800	0.240	13	$ 347,200
770 + 225	0.180	12	229,900
190 + 5.8	4.00	38	657,400
560 + 360	5.76	52	1,094,000
86 + 1.3	4.15	0	9,850
1,250 + 712	0.60	30	282,700

Source: After Appalachian Regional Commission 1969.

Table 4.13 *Capital Requirements and Equipment Costs for Hydrated Lime Treatment Plant at Six Treatment Sites*

Location	*Cost*	*mgd*	*Total net acidity*	*Cost/gpd*
a	$ 347,200	0.240	4,040	$0.440
b	229,900	0.180	770	1.275
c	657,400	4.00	190	0.164
d	1,094,000	5.76	560	0.190
e	9,850	4.15	86	0.00237
f	282,700	0.60	1,250	0.472

Source: After Appalachian Regional Commission 1969.

Note: Factors include water storage and transportation, control/filter building, aeration, thickening, sludge holding, and piping and site preparation.

Table 4.14 *Operating Costs for Hydrated Lime Treatment Plant*

Location	*Cost/1,000 gal*	*mgd*	*Total net acidity*
a[1]	$ 1.09	0.240	4,040
b[1]	0.72	0.180	770
c[1]	0.087	4.00	190
d[1]	0.226	5.76	560
e[2]	0.0068	4.15	86
f[1]	0.537	0.60	1,250

Source: After Appalachian Regional Commission 1969.

1. 20-year amortization at 4% interest.
2. 5-year amortization at 4% interest.

Table 4.15 *Operating Costs – Sludge Disposal for Hydrated Lime Treatment*

Cost/1,000 gal without sludge disposal	*mgd*	*Total net acidity*	*Sludge disposal cost/1,000 gal*
$0.530	0.3	2,800-4,000 acid, 900-1,200 Fe	$0.095
0.440	0.9	2,800-4,000 acid, 900-1,200 Fe	0.085
0.4125	2.7	2,800-4,000 acid, 900-1,200 Fe	0.0725
0.4050	8.1	2,800-4,000 acid, 900-1,200 Fe	0.0725
0.253	0.3	1,400 acid, 600-700 Fe	0.095
0.245	0.9	1,400 acid, 600-700 Fe	0.085
0.2175	2.7	1,400 acid, 600-700 Fe	0.0775
0.2135	8.1	1,400 acid, 600-700 Fe	0.0725
0.1910	0.3	600-700 acid, 322 Fe	0.085
0.1500	0.9	600-700 acid, 322 Fe	0.075
0.1320	2.7	600-700 acid, 322 Fe	0.0675
0.1240	8.1	600-700 acid, 322 Fe	0.0650

Source: After Appalachian Regional Commission 1969.

Note: Operating cost data from Holland et al. 1968.

by Holland for a similarly sized plant using hydrated lime. This discussion is summarized in Table 4.16.

This reasoning is the basis for the cost data presented in Table 4.5. Interestingly enough, the cost per thousand gallons of waste water treated by the plant described in Appendix E is $0.23 for an approximate 7-mgd plant. This is precisely in the middle of the range set by the commission (Table 4.5), without correcting for inflation.

Skelly and Loy have summarized more recent cost data in a more usable form. These are presented in Tables 4.17 and 4.18. They vary considerably from the data presented by the Appalachian Regional Commission. For that

Table 4.16 *Total Operating Cost per 1,000 Gallons*

Item	*Limestone*[1]	*Hydrated lime*[2]
Chemicals	$0.050	$0.115
Power for mill	0.011	–
Labor	0.050	0.050
Maintenance	0.030	0.025
Miscellaneous	0.030	0.030
Sludge disposal[2]	0.035	0.035
Total	$0.196	$0.255

Source: After Appalachian Regional Commission 1969.

Note: At 1 mgd of 2,000 ppm acidity.

1. After U.S. Bureau of Mines data.
2. After Holland et al. 1968.

matter, there is an eightfold variation in the operating costs presented by Skelly and Loy. The reasons for these great differences are not obvious, but the author would judge a reasonable treatment cost to approximate \$0.25 per 1,000 gal for a 10-mgd plant treating what is normally called acid mine drainage with lime. Noteworthy, also, is the fact that we cannot be certain that any of the above costs provide for effluent of a quality that will satisfy guidelines such as those in Table 1.1.

Table 4.17 *Summary of Capital Costs for Conventional Lime Neutralization Process*

Company and location	*Design flow rate (m^3/day)*	*Total acidity (mg/l)*	*Capital costs* Total cost	*Capital costs* Cost/m^3
Bethlehem Mines Co. No. 58-A	908	4,080	\$ 347,200	\$ 382.38
Bethlehem Mines Co. No. 58-B	1,136	8,150	423,200	372.53
Young & Son	681	770	229,900	337.59
Morea Strip	15,140	190	657,400	43.42
Blue Coal Corp. Loomis no. 4	21,802	560	1,094,000	50.18
Duquesne Light Co. Warwick no. 3	2,271	1,250	229,700	101.14
West Virginia University	1,136	3,500	–	–
School of Mines	1,136	1,400	–	–
Mine no. 1	1,136	650	–	–
	3,407	3,500	–	–
	3,407	1,400	–	–
	3,407	650	–	–
	10,220	3,500	–	–
	10,220	1,400	–	–
	10,220	650	–	–
Duquesne Light Co. Warwick no. 2	11,446	1,560	582,000	50.85
Commonwealth of Pa. Slippery Rock Creek treatment plant	11,446	240	750,000	65.53
Rausch Creek Mine drainage plant	37,850	–	1,747,380	46.17
Mountaineer Coal Co.	2,725	250	120,000	44.04

Source: After Skelly and Loy 1973.

Note: m = meters.

Table 4.19 presents reagent costs for the combination lime-limestone treatment process reported by Wilmoth et al. (1972), discussed previously. These data illustrate the potential economies of reagent cost for the combination process, at least at the pilot plant stage. However, additional data will be necessary before total costs can be compared.

Table 4.18 *Summary of Operating Costs for Conventional Lime Neutralization Process*

Company and location	*Design flow rate (m^3/day)*	*Total acidity (mg/l)*	*Operating cost*	
			Annual cost	*Cost per 1,000m^3 (per mg/l acidity)*
Bethlehem Mines Co No. 58-A	908	4,080	$ 95,250	$0.070
Bethlehem Mines Co. No. 58-B	1,136	8,150	140,000	0.041
Young & Son	681	770	47,400	0.248
Morea Strip	15,140	190	126,571	0.121
Blue Coal Corp. Loomis no. 4	21,802	560	475,000	0.107
Duquesne Light Co. Warwick no. 3	2,271	1,250	117,500	0.113
West Virginia University	1,136	3,500	68,448	0.047
School of Mines	1,136	1,400	44,236	0.076
Mine no. 1	1,136	650	30,223	0.112
	3,407	3,500	172,463	0.040
	3,407	1,400	108,405	0.062
	3,407	650	73,913	0.091
	10,220	3,500	477,968	0.037
	10,220	1,400	290,723	0.056
	10,220	650	196,607	0.081
Duquesne Light Co. Warwick no. 2	11,446	1,560	209,715	0.032
Commonwealth of Pa. Slippery Rock Creek treatment plant	11,446	240	51,000	0.051
Rausch Creek Mine drainage plant	37,850	–	–	–
Mountaineer Coal Co.	2,725	250	–	–

Source: After Skelly and Loy 1973.
Note: m = meters.

Table 4.19 *Estimated Chemical Operating Costs for Combination Limestone-Lime Treatment Process*

	Typical chemical costs cents/3.785 m^3 (cents/1,000 gal) [1]			
			Limestone - Lime	
Final pH	*Lime only*	*Limestone only*	*cost*	*% savings*
6.5	2.61	–	1.94	25.7
6.5	–	3.44	2.37	31.1
9.0 (ferric)	3.13	–	2.46	21.3
9.0 (ferrous)	4.43	–	3.77	14.9

Source: After Wilmoth et al. 1972.

1. The acid mine drainage used had an average quality of: pH, 2.8; acidity, as $CaCO_3$, 430 mg/l; iron, total, 92 mg/l.

Other Alkalis

Other alkaline chemicals could be used for neutralization of acid mine drainage, but few have had actual field use. The principal reason is cost (see Table 4.10), but this is not the exclusive factor.

The following considerations are also to be taken into account:

1. Soda ash, caustic, magnesium oxide:
   a. These alkalis are water soluble, and thus fast reactors.
   b. The sludge produced is composed of iron and aluminum oxides.
   c. The effluents could be higher in dissolved solids and, in the case of highly or moderately acidic waters, this increase could exceed guideline limits, thereby requiring additional treatment.
2. Ammonia:
   a. The comments already made on soda ash and caustic would apply, and ammonia as a neutralizer would add nutrients to the water.

Therefore, none of these methods appears to be competitive with lime or limestone for neutralizing acid mine drainage.

Reverse Osmosis

The primary cost-determining factors in the reverse osmosis process are plant treatment capacity; concentration of dissolved solids in the influent, because it controls the membrane flux and operating pressure; acidity, which affects construction and labor costs during plant construction; brine disposal method; and membrane life.

Several pilot plants with acid mine drainage feed have shown the process to be technically feasible. Sufficient experience with lime neutralization and deep-well disposal is available to determine their reliability. But only limited cost data for various-size plants and with varying influent concentrations are available for acid mine drainage.

The cost data presented here are derived primarily from the Appalachian Regional Commission (1969) and Goff and Gloyna (1970).

The Appalachian Regional Commission cost estimates for reverse osmosis-acid mine drainage processing appear highly speculative and not comprehensive in scope or justification. But the various data from these references were pieced together to create a more comprehensive composite than had been published previously.

The commission presents *estimated* costs for reverse osmosis processing based on seawater desalination, reverse osmosis processing with deep-well disposal, and reverse osmosis processing with lime neutralization utilizing both hydrated lime and limestone. The elements of capital cost are shown first, then capital costs for the above referenced processes, followed by the elements of the operating costs and specific operating costs of these processes.

As already noted, the capital costs of a reverse osmosis plant vary according to the capacity of the plant, the quality of the acid mine drainage, and the method of brine disposal. The capital costs for reverse osmosis plants increase four times as rapidly as plant capacity for larger plant sizes. A plant to treat highly acidic water costs about 60 percent more than a plant to treat a less acidic water. Costs for brine neutralization facilities are also variable. They are reported to increase twice as fast as the degree of acidity in the input water. Total plant costs for the smaller scale plants are dominated by the brine disposal facility cost (from 43 to 55 percent); this cost decreases to between 7 and 9 percent for the large-scale plants.

These variations in capital requirements are reflected in the unit amortization and operating costs for such a facility (Table 4.20). The deep-well disposal technique for brines favors the large-scale plants. The total unit costs are less for this combination at the 10-mgd scale than in the neutralization technique combinations. The principal reason for these high costs is that brine disposal adds an economic burden. Standard desalinization plants usually discharge brines back into the ocean and do not have to bear this burden.

The commission observes that costs of usable or product water from this process are high because the volume of product water for weak, moderate, and strong acidity solutions is 80, 60, and 50 percent of the respective feedwater volume. For the reverse osmosis plant of 10 mgd with deep-well disposal of brine, the \$0.48 per 1,000 gal of feedwater is approximately equivalent to \$0.96 per 1,000 gal of product water.

A comparison of the costs of the neutralization and reverse osmosis treatment techniques (Table 4.20) clearly shows that neutralization costs are significantly lower than those of reverse osmosis. However, even the neutralization costs are still appreciable in terms of using the technique for the purpose of improving stream water quality. A limestone neutralization facility, having a capacity of 10 mgd, and treating moderately acid water, would cost \$1,900 per day, or \$694,000 per year. A lime neutralization plant of the same capacity and treating the same water would cost \$2,900 per day or \$1,059,000 per year.

Table 4.20 *Comparative Costs for Mine Drainage Treatment by Neutralization and Reverse Osmosis*

Treatment process	*Plant size*								
	0.3 mgd			*1.0 mgd*			*10 mgd*		
	Feedwater acidity[1]			*Feedwater acidity*			*Feedwater acidity*		
	Weak	*Moderate*	*Strong*	*Weak*	*Moderate*	*Strong*	*Weak*	*Moderate*	*Strong*
Neutralization:									
Lime with sludge disposal	0.28	0.41	0.65	0.22	0.33	0.52	0.19	0.29	0.48
Limestone with sludge disposal	0.22	0.24	0.38	0.18	0.21	0.31	0.15	0.19	0.26
Reverse osmosis:									
With lime	0.78	1.44	1.93	0.50	0.93	1.45	0.38	0.71	1.08
With limestone	0.71	1.02	1.53	0.46	0.69	1.04	0.31	0.46	0.69
With deep-well disposal[2]	1.04	1.13	1.35	0.58	0.63	0.71	0.29	0.40	0.48

Source: After Appalachian Regional Commission 1969.

Note: Costs in dollars per 1,000 gal of acid mine drainage treated.

1. Range of feedwater acidity (in ppm):

	Neutralization	*Reverse osmosis*
Weak	600- 700	500
Moderate	1,400-1,700	1,500
Strong	2,800-4,800	3,500

2. Based on 7,000-ft-deep well.

Goff and Gloyna indicate a cost of $0.59 per 1,000 gal for a 1-mgd plant treating brackish water and $0.38 per 1,000 gal for a 10-mgd plant treating brackish water. However, these figures are in terms of fresh water output. It is essentially impossible to compare satisfactorily these figures with those of the Appalachian Regional Commission for reverse osmosis because of unknown differences such as film disposal technique and suspended solid input. However, it is possible to compare them with the costs of neutralization as presented by the commission. Comparison reveals that reverse osmosis is simply not economically feasible on a competitive basis.

Deep-Well Injection

Estimated cost data for deep-well injection combined with reverse osmosis are presented in summarized form in Table 4.20. The following discussion constitutes a breakdown of the costs in that table, plus some of the rationale used by the Appalachian Regional Commission (1969) in arriving at the costs.

The factors involved in cost determinations for deep wells are:

1. Fixed amortization of capital equipment costs.
2. Energy required for pumping.
3. Labor and maintenance.
4. Equipment repairs or replacement.
5. Well workovers required for maintaining reasonable injection pressures, or to correct plugging, or to replace down-hole equipment.
6. Chemicals required for pretreatment.
7. Sludge hauling or disposal costs.
8. Injection pressure variations affecting pumping costs.
9. Cost of standby well and injection equipment to be used in event the active well has to be shut down for repair or equipment replacement.
10. Exploration costs to find acceptable receiving aquifer (not mentioned by the commission).

The commission used the summary of oilfield brine disposal costs from 81 wells as presented in Table 4.21 in estimating orders of magnitude for various depths and injection rates. These costs include 20-year amortizaton at 4 percent interest. It was assumed that for acid wastes these operating costs would be 10 to 20 percent too low due to the greater corrosivity.

The factors taken into account in these costs include repairs, labor, chemicals (for lime and alum coagulation, sedimentation, chlorine), pumping energy, administrative expense, and equipment and replacement costs.

Capital cost factors for existing acid waste disposal wells in Appalachia are shown in Table 4.22.

If the proper geologic conditions are available for installation of a deep-well disposal system, no effluent will result. However, a residue would result from pre-injection treatment and would have to be discarded in some manner. Sludge from the surface treatment system would have to be discarded terrestrially.

Direct extrapolation from petroleum brine disposal capital costs to capital costs for acid mine drainage disposal was not considered feasible by the Appalachian Regional Commission. For this reason, only disposal wells that were in the Appalachia area and that were handling acid wastes were considered for the cost data shown in the commission's report. Using this approach,

Table 4.21 *Oil Field Brine Deep-Well Disposal System Operating Costs*

Injection rate (gpd)	*Injection pressure (psi)*	*Depth (ft)*	*Cost ($/1,000 gal)*
1,000	900	3,000	$16.47
		7,000	27.71
		12,000	58.41
	260	3,000	16.33
		7,000	27.39
		12,000	57.41
	0	3,000	15.92
		7,000	27.14
		12,000	58.16
100,000	900	3,000	0.424
		7,000	0.563
		12,000	0.898
	260	3,000	0.357
		7,000	0.500
		12,000	0.758
	0	3,000	0.335
		7,000	0.477
		12,000	0.815
2,000,000	900	3,000	0.180
		7,000	0.217
		12,000	0.283
	260	3,000	0.119
		7,000	0.158
		12,000	0.223
	0	3,000	0.094
		7,000	0.131
		12,000	0.204

Source: After Appalachian Regional Commission 1969.

Note: Includes 20-year amortization 4% annual interest as presented by the commission. These data were not converted to current interest rates by the author because of the recent trend toward wide fluctuations in interest rates.

Table 4.22 *Capital Costs, Acid Waste Disposal Wells, Appalachia Region*

Well location	*Eastern Ohio*	*Western Pennsylvania*		*Southwestern New York*	
Total depth (ft)	3,300	4,800	6,000	3,060	4,300
Design injection rate (gpm)	100	100	100	300	100
Design injection pressure (psi)	700	300	4,000	(*)	(*)
Total system cost	$564,000	$960,000	$770,000	$502,000	$447,000
Surface equipment	400,000	400,000	220,000	125,000	147,000
Completed well	164,000	560,000	550,000	377,000	300,000
Cost averages,					
Per ft	$ 170.00	$ 200.00	$ 128.00	$ 164.00	$ 104.00
Per psi	806.00	3,200.00	192.00	(*)	(*)
Per 1,000 gal	3.92	6.66	5.34	1.16	3.10

Source: After Appalachian Regional Commission 1969.

Note: Data are from five plants (one located in eastern Ohio, two in western Pennsylvania, and two in southwestern New York).

* Undetermined

the commission deduced that all factors relating to constructing and maintaining the well and preparing the water for injection add up to a cost of $150 per foot of well depth in the depth range of 3,000 to 5,000 feet. Presumably this figure is in terms of 1967 dollars. Unlike oilfield brine disposal, a high percentage of the cost is involved in the surface treatment facilities and pumping equipment.

In order to relate operating costs to injection rate, it was necessary to assume a life of the well based on ultimate acceptance capacity. Twenty years was the period used, and injection was assumed to be continuous at a constant rate over that period. Twice the service, power, and maintenance costs observed for brine disposal were assumed for acid injection. These assumptions permitted estimates of the total operating cost for given well depths and pressures (based on Table 4.21). These costs are shown in Table 4.23.

The actual pressures to be encountered for any given injection rate cannot be estimated but must be determined for each injection well. The value of 900 pounds per square inch shown in Table 4.23 was chosen to establish an injection-rate operating cost and has no relationship to the injection capacities shown. Nonetheless, the effect of capacity is clearly illustrated since the fixed capital charge per day is a major charge. Increasing the amount of fluid injected per day directly reduces the cost per thousand gallons.

Well Dewatering

Seaber and Hollyday (1965) present costs for constructing and operating water wells (1964 dollars). Parizek (1971) used these data to estimate the costs of dewatering a ground-water flow system as a means of preventing acid mine drainage. Seaber and Hollyday used an unrealistically low interest rate (4 percent) by present standards; however, standard conversion indices take into account interest rate changes, and so these costs can be converted to any year of interest to the reader by use of a conversion index.

Table 4.23 *Acid Deep-Well Disposal System, Estimated Operating Cost*

Injection rate (gpd)	*Depth (ft)*	*Injection pressure (psi)*	*Cost ($/day)*	*Cost ($/1,000 gal)*
1,000	3,000	900	$105.30	$105.30
10,000	3,000	900	108.00	10.80
100,000	3,000	900	135.00	1.35
1,000,000	3,000	900	405.00	0.41
1,000	7,000	900	244.30	244.30
10,000	7,000	900	247.00	24.70
100,000	7,000	900	274.00	2.74
1,000,000	7,000	900	544.00	0.54

Source: After Appalachian Regional Commission 1969.

Note: 20-year amortization and 4% interest is included as presented by the commission. These data were not converted to current interest rates by the author because of the recent trend toward wide fluctuations in interest rates.

Factors pertinent to the cost of dewatering by wells include pumping equipment costs, drilling costs, hydrogeologic environment, pumping lift, number of wells required, well diameter, and engineering and administrative costs. Seaber and Hollyday consider annual costs to include payments to retire costs incurred because of the above factors and, in addition, annual power usage and maintenance costs.

In the analysis, wells capable of 500-gpm yields were assumed to average 400 ft in depth, have a 20-ft static water level, contain 40 ft of 8-in.-diameter casing, be 8 in. in diameter, and have a pumping lift (drawdown) of 200 ft to achieve the desired yield. One exploratory well was assumed to be necessary for every production well in order to locate wells capable of a 500-gpm yield. Seaber and Hollyday's 1964 average price for an 8-in.-diameter well in soft rock, such as shale, was $1,800 and $200 for the casing. A 24-hour pumping test for aquifer analysis and designing the deep-well turbine pumping unit was estimated at $500. Motor, column, shaft, pump, and strainer costs for a unit capable of delivering 500 gpm from a 200-ft pumping level were estimated at $4,500. Estimated cost of land and rights-of-way was $1,000 per well and $1,500 per pumphouse. Wiring, metering, and piping added up to $1,500 per well. Contingencies were estimated as 10 percent of the estimated equipment, and construction costs and engineering and administration were estimated as 15 percent of total construction costs and contingencies.

Annual costs were computed by Seaber and Hollyday as a single end-of-year payment covering interest on the initial investment and payments to a depreciation fund. Annual power costs were based on a 24-hour-per-day use and 75 percent wire-to-motor efficiency. Annual maintenance costs were estimated as 4 percent of the cost of equipment, and the well's life was assumed to be 25 years without pump replacement.

On this basis, Parizek notes that the total annual cost in dollars per million gallons of water per day delivered to the surface for wells with poor yields (75 gpm each) might approach $19,000 per year. The cost would approach $10,000 per year for wells with 239-gpm yields and $6,900 per year for wells with up to 760-gpm yields. Parizek sums up his estimates by concluding that a dewatering system pumping 2.17 mgd would cost from $15,000 to $25,000 per year, depending on the number of wells required (to be determined by the hydrogeologic environment). Parizek obtained these figures by projecting the 1964 costs to 1971 by a 4 percent annual increase in prices. Using the midrange figure of $20,000 and correcting for inflation at 5 percent per year for two years brings the 1973 cost to less than $0.03 per 1,000 gal of water pumped.

Personal experience of the author with municipal pumping systems suggests that the actual cost of such an operation may approach $0.05 to $0.06 per 1,000 gal. Nevertheless, the figure is far below that of any available treatment scheme. Thus, it is clear that dewatering should receive prime consideration as a means of preventing acid mine drainage, at the very least while the mine is operating. The selection or rejection of the method ultimately will depend on the characteristics of the hydrogeologic environment, which must be determined in advance of production by a thorough hydrogeologic analysis.

Mine Sealing

Stephan and Lorenz (1968) report costs for mine sealing as obtained from three sources. These data are summarized in Table 4.24. According to Stephan and Lorenz, the Department of Mines, West Virginia, estimates for 1966 range from \$850 to \$2,650 per seal. The Federal Water Pollution Control Administration reported estimates of \$600 to \$2,000 per seal, the estimated average being about \$1,300 per seal. The U.S. Bureau of Mines estimates for sealing certain abandoned mine openings were about \$5,000 per seal. However, these estimated costs included extensive preconstruction work required before the seals could be installed. Skelly and Loy (1973) report that double bulkhead seals of the type discussed herein under "Mine Sealing" cost from \$7,500 to \$15,000 per seal without grout curtains. But costs may go as high as \$50,000.

These costs clearly do not include the cost of the hydrogeologic investigation so essential for the determination of the feasibility of mine sealing at any specific site. To implement mine sealing without such investigations would constitute a considerable risk to nearby wells, springs, basements, excavations, highway cuts, streams, and downstream water uses. Therefore, the cost should be added to the cost of mine sealing. Unfortunately, at the present time the author is not aware of the availability of this type of economic data. Thus the cost of mine sealing must be viewed as quite variable and poorly defined.

Cost of Control of Ground-Water Recharge

Control of ground-water recharge consists of: (1) controlling the flow path of ground water in a mine so that contact with high-content sulfide zones is precluded, (2) prevention of ground-water recharge at the surface by diversion of surface runoff around high-rate recharge zones, and (3) in some cases grouting. These objectives are sought through the use of pipes, open ditches, impoundments, and grout curtains. Stephan and Lorenz (1968) provide limited data on the cost of these techniques for the purpose of acid mine drainage control, even though costs are quite variable due to variable topography and hydrogeologic conditions. These data are presented in Table 4.21 (presumably in 1966 dollars).

Open pipe flumes have been estimated to cost about \$6.00 per linear foot. The cost of diverting streams has been estimated at \$1.51 per linear foot. Costs on construction of impoundment dams at the end of the last cut in strip mining operations have been estimated at \$125 to \$500 per acre-foot. Impoundment basins,* which may be required for diversion, are usually earthen basins that hold water for one of several possible purposes. Such basins can impound 2.5 to 5 mgd and cover a surface area of 5 to 50 acres. Cost estimates on construction of the basin (in 1966 dollars) range from \$500 to \$1,000 per acre (Stephan and Lorenz 1968). The following data, which perhaps are more usable, were obtained from Skelly and Loy (1973).

* These data should also be applicable to the cost of constructing and maintaining the structures discussed previously under the section entitled "Impoundment."

Table 4.24 *Cost Data for Acid Mine Water Drainage Control Measures*

Method	*Unit*	*Cost*	*Comments*
Mine sealing			
Wet and dry mine seals	Per seal	$850-2,650	Department of Mines, West Virginia
Mine bulkheads	Per opening	1,100	Active mine, Pennsylvania
Average of 8,092 seals	Per opening	1,070	Weighted average, using ENR Index
Mine sealing	Per seal	600-2,000	FWPCA, average estimate: $1,300
Mine sealing	Per seal	5,000	USBM figures; abandoned mine; extensive work required for preparing and making seal
Drainage diversion			
Pipe flume	Lin ft	6	Drainage from bench to stream
Ditch	Lin ft	0.04-0.40	Open ditch above high wall
Stream diversion	Lin ft	1.51	Soil Conservation Service estimate
Impoundments:			
Ponds (5-30 acres)	Surface, acre	1,000-500	Data from Pennsylvania and West Virginia. The larger the impoundment, the smaller the cost
Dams	Acre foot	125-500	Impoundment dams constructed at the end of last strip pit

Source: After Stephan and Lorenz 1968.

Note: Derived from Handbook of Pollution Control Costs in Mine Drainage Management, U.S. Department of Interior, Federal Water Pollution Control Administration, December 1966.

Diversion ditches cost from $1.30 to $3.90 per cubic meter ($1.00 to $3.00 per cubic yard). Dikes range from $0.45 to $0.85 per cubic meter ($0.35 to $0.65 per cubic yard). A 91.5 centimeter (36 inch) 1/2 section of bituminized fibre pipe averages $32.80 per linear meter ($10.00 per linear foot) in place. A 45.7 centimeter (18 inch) corrugated metal pipe is $26.00 per linear meter ($8.00 per linear foot) in place. Concrete costs approximately $39 per cubic meter ($30 per cubic yard) in place. Asphalt paving ranges from $2.40 to $6.00 per square meter ($2.00 to $5.00 per square yard) in place. Dumped rock costs from $2.60 to $7.80 per cubic meter ($2.00 to $6.00 per cubic yard) and riprap ranges from $13.00 to $34.00 per cubic meter ($10.00 to $25.00 per cubic yard). Jute matting costs $0.70 to $2.40 per square meter ($0.60 to $2.00 per square yard).

Skelly and Loy estimate the cost of sealing a drill hole to range between $100 and $1,200 with an average somewhere in the order of $600 per hole. They conclude that grout curtains cost from $25 to $80 per linear foot, depending on hole depth, drilling conditions, and geology.

REFERENCES

AAPG 1968. *Subsurface Disposal in Geologic Basins–A Study of Reservoir Strata*, ed. John Galley, Memoir 10, American Association of Petroleum Geologists, Tulsa, Okla.

Ahmad, M. U. 1970. A Hydrological Approach to Control of Acid Mine Pollution from Lake Hope. *Ground Water*, Vol. 8, No. 6.

Ahmad, M. U. 1973. Coal Mining and its Effect on Water Quality. In *Ground Water Pollution*, pp. 17-51. St. Louis: Underwater Research Institute.

Appalachian Regional Commission 1969. *Acid Mine Drainage in Appalachia.* House Document 91-180, Publication No. 35-4320, U.S. Govt. Printing Office, Washington, D.C.

Baker, R. A., and Wilshire, A. G. 1970. *Microbial Factor in Acid Mine Drainage Formation.* Document 14010 DKN 11/70, U.S. Govt. Printing Office, Washington, D.C.

Bituminous Coal Research 1970. *Limestone Treatment of Acid Mine Drainage.* Document PB 195 282, National Technical Information Service, Dept. of Commerce, Washington, D.C. (Also published as FWQA Research Series 14010 EIZ.)

Bituminous Coal Research 1971. *Studies of Limestone Treatment of Acid Mine Drainage–Part 2.* Document PB 208 326, National Technical Information Service, Dept. of Commerce, Washington, D.C. (Also published as EPA Research Series 14010 FOA.)

Brown, R. L., and Parizek, R. R. 1971. *Shallow Ground Water Flow Systems Beneath Strip and Deep Coal Mines at Two Sites, Clearfield County, Pennsylvania.* Special Report 84, Dept. of Environmental Resources, Commonwealth of Pennsylvania.

Charmbury, H. B., Maneval, D. R., and Girard, L. 1967. *Operation Yellowboy–Design and Economics of a Lime Neutralization Mine Drainage Treatment Plant.* SME, AIME Preprint 67F35.

Colmer, A. R., and Hinkle, M. E. 1947. The Role of Micro-organisms in Acid Mine Water. *Science*, Vol. 106, pp. 253-256.

Dominico, P. A. 1972. *Concepts and Models in Ground Water Hydrology.* New York: McGraw-Hill Book Co.

EPA 1973. *Toxic Pollutant Effluent Standards.* Part 129, Subchapter D, Water Programs; Chapter 1–Environmental Protection Agency; Title 40–Protection of the Environment, August 31. U.S. Environmental Protection Agency, Washington, D.C.

Goff, D. L., and Gloyna, E. F. 1970. Reverse Osmosis Developments. In *Water Quality Improvement by Physical and Chemical Processes*, ed. E. F. Gloyna and W. W. Eckenfelder, Jr., pp. 312-326. Austin: Univ. of Texas Press.

Goldberg, E. F., and Power, G. 1972. *Legal Problems of Coal Mine Reclamation.* Document 14010 DMO 03/70, U.S. Govt. Printing Office, Washington, D.C.

Hall, E. P. 1965. *The Sheban Project.* Proceedings, First Symposium on Acid Mine Drainage Research, Mellon Institute, Pittsburgh. Monroeville, Pa.: Bureau of Coal Research.

Halliburton Company 1970. *New Mine Sealing Techniques for Water Pollution Abatement.* Document 14010 DMO 03/70, U.S. Govt. Printing Office, Washington, D.C.

Hawley, J. R., and Shikaze, K. H. 1971. *The Problem of Acid Mine Drainage in Ontario.* Proceedings, Third Annual Meeting of Canadian Mineral Processors. Mines Branch, Dept. of Energy, Mines and Resources, Ottawa, Ontario, Canada.

Hill, R. D. 1968. *Mine Drainage Treatment–State of the Art and Research Needs.* Report to FWPCA, U.S. Dept. of the Interior, Cincinnati, Ohio.

Holland, C. T., Corsaro, J. L., and Ladish, D. J. 1968. *Factors in the Design of an Acid Mine Drainage Treatment Plant.* Proceedings, Second Symposium on Coal Mine Drainage Research, Mellon Institute, Pittsburgh. Monroeville, Pa.: Bureau of Coal Research.

Lombardo, J. L. 1973. State of the Art–Acid Mine Drainage Control. Paper presented at American Mining Congress 1973 Convention, Denver; available from American Mining Congress, Washington, D.C.

Lorenz, W. C., and Stephan, R. W. 1969. *Factors that Affect the Formation of Coal Mine Drainage in Appalachia.* In Appalachian Regional Commission Report, Part 2, p. C-C-1 (see reference, Chapter 3, to Appalachian Regional Commission, 1969).

Lusk, B. E. 1973. New Surface Mining Technology to Minimize Environmental Disturbance. Paper presented at American Mining Congress 1973 Convention; available from American Mining Congress, Washington, D.C.

Mihok, E. A., Duel, M., Chamberlain, D. E., and Selmeczi, J. G. 1968. *Mine Water Research–The Limestone Neutralization Process.* U.S. Bureau of Mines Report of Investigations 7191.

Ohio State University Foundation 1971. *Acid Mine Drainage Formation and Abatement.* Document PB 199 835, National Technical Information Service, Dept. of Commerce, Washington, D.C.

Parizek, R. R. 1971. *Prevention of Coal Mine Drainage Formation by Well Dewatering.* Special Research Report SR-82, Dept. of Environmental Resources, Commonwealth of Pennsylvania.

Parizek, R. R., and Tarr, E. G. 1972. *Mine Drainage Pollution Prevention and Abatement Using Hydrogeological and Geochemical Systems.* Proceedings, Fourth Symposium on Coal Mine Drainage Research, Mellon Institute, Pittsburgh. Monroeville, Pa.: Bureau of Coal Research.

Ralston, D. R., Trexler, B. D., Wai, C. M., Renison, W. T., and Rickman, Jr., F. E. 1973. *Solutions to Problems of Pollution Associated with Mining in Northern Idaho.* Contract Completion Report No. HO 122070, U.S. Bureau of Mines, Dept. of the Interior, Washington, D.C.

Reilly, J. D. 1960. The Ecology of Water Areas Associated with Coal Strip Mined Lands in Ohio. *Ohio Journal of Science*, Vol. 60, No. 2, pp. 108-121.

Reilly, J. D. 1965. *Limnology of Acid Mine Impoundments.* Proceedings, First Symposium on Acid Mine Drainage Research, Mellon Institute, Pittsburgh. Monroeville, Pa.: Bureau of Coal Research.

Rex Chainbelt 1970. *Treatment of Acid Mine Drainage by Reverse Osmosis.* Document 14010 DYK 03/70, U.S. Govt. Printing Office, Washington, D.C.

Rex Chainbelt 1972. *Reverse Osmosis Demineralization of Acid Mine Drainage.* Document 14010 FQR 03/72, U.S. Govt. Printing Office, Washington, D.C.

Schmidt, J. W., and Conn, K. 1969. *Abatement of Pollution from Mine Waste Waters.* Proceedings, First Annual Meeting of the Canadian Mineral Processors, pp. 52-74. Mines Branch, Dept. of Energy, Mines and Resources, Ottawa, Ontario, Canada.

Seaber, P. R., and Hollyday, E. F. 1965. *An Appraisal of the Ground Water Resources of the Lower Susquehanna River Basin.* U.S. Geological Survey Open File Interim Report.

Skelly and Loy Inc. 1973. *Processes, Procedures and Methods to Control Pollution from Mining Activities.* EPA Document 430/9-73-011, U.S. Environmental Protection Agency, Washington, D.C.

Smith, E. E., and Shumate, K. S. 1970. *Sulfide to Sulfate Reaction Mechanism.* Document 14010 FPS 02/70, U.S. Govt. Printing Office, Washington, D.C.

Stephan, R. W., and Lorenz, W. G. 1968. *Survey of Costs on Methods for Control of Acid Mine Drainage Pollution.* Report by U.S. Bureau of Mines, Pittsburgh. (Included in Appalachian Regional Commission, Appendix C, 1969.)

Syracuse University 1971. *Inorganic Sulfur Oxidation by Iron Oxidizing Bacteria.* Document 14010 DAY 06/71, U.S. Govt. Printing Office, Washington, D.C.

Temmel, F. M. 1973. Case History on Acid Mine Drainage Control. Paper presented at American Mining Congress 1973 Convention, Denver; available from American Mining Congress, Washington, D.C.

Toth, J. 1963. A Theoretical Analysis of Ground-water Flow in Small Drainage Basins. *Journal of Geophysical Research,* Vol. 68, No. 10, pp. 4795-4812.

Toth, J. 1966. *Ground-water Geology, Movement, Chemistry and Resources Near Olds, Alberta.* Bulletin, 17, Research Council of Alberta, Edmonton.

Tyco Laboratories Inc. 1971. *Silicate Treatment for Acid Mine Drainage, Prevention.* Document 14010 OLI 02/71, U.S. Govt. Printing Office, Washington, D.C.

Van Everdingen, R. O., and Banner, J. A. 1971. *Precipitation of Heavy Metals from Natural and Synthetic Acidic Aqueous Solutions during Neutralization with Limestone.* Technical Bulletin 35, Inland Water Branch, Dept. of Energy, Mines and Resources, Ottawa, Ontario, Canada.

Walton, C. W. 1970. *Ground Water Resource Evaluation.* New York: McGraw-Hill Book Co.

Williams, R. E. 1970. Ground-water Flow Systems and the Accumulation of Evaporite Minerals. *Bulletin, American Association of Petroleum Geologists*, Vol. 54, No. 7, pp. 1290-1295.

Wilmoth, R. C., and Hill, R. D. 1970. *Neutralization of High Ferric Iron Acid Mine Drainage.* Document 14010 ETV 08/70, U.S. Govt. Printing Office, Washington, D.C.

Wilmoth, R. C., Scott, R. B., and Hill, R. D. 1972. *Combination Limestone-Lime Treatment of Acid Mine Drainage.* Proceedings, Fourth Symposium on Coal Mine Drainage Research, Mellon Institute, Pittsburgh. Monroeville, Pa.: Bureau of Coal Research.

5. *Bauxite refining and aluminum smelting*

SOURCES OF INFORMATION

Much of the information presented in this chapter is condensed from a United States government document prepared by Battelle Columbus Laboratories (Battelle 1973). Battelle researched the origin and disposal of wastes in the aluminum industry as authorized by EPA under the requirements of Public Law 92-500. The document is unique because it is the only available source of the kind and of the amount of data it contains. Other sources of information are referenced in the traditional fashion.

REFINING OF BAUXITE
(Standard Industrial Classification [SIC] 2819)

Bauxite is the only ore of aluminum used in the United States. Nine plants, of which one is in the Virgin Islands, exist currently (see Table 5.1). Aluminum oxide (Al_2O_3), alumina, is the compound extracted from the ore. Bauxite is hydrated aluminum oxide combined with impurities including iron oxide, various silicates, and quartz. Bauxite is the residuum of the leaching of strata high in aluminum oxide.

The Bayer process or the combination process dissolves the impure alumina in the ore in caustic soda (NaOH) to form soluble sodium aluminate. The solution is cooled, diluted, and hydrolyzed to precipitate aluminum hydroxide which is filtered and calcined to alumina (see Figure 5.1).

The reactions are:

(monohydrate) $$Al_2O_3 \cdot H_2O + 2NaOH \longrightarrow 2NaAlO_2 + 2H_2O$$

(trihydrate) $$Al_2O_3 \cdot 3H_2O + 2NaOH \longrightarrow 2NaAlO_2 + 4H_2O.$$

Table 5.1 *Operating Companies, Locations, Capacities, and Date of Operation of U.S. Bauxite Refining Plants*

Company	Plant location	Date of first commercial operation	Approximate plant capacity (tons of alumina)		
			(Metric tons/day)	(Short tons/day)	(Metric tons/year)[1]
Aluminum Company of America	Mobile, Alabama	1938	2,270	2,500	820,000
	Pt. Comfort, Texas	1959	3,200	3,500	1,150,000
	Bauxite, Arkansas	1952	900	1,000	325,000
Kaiser Aluminum & Chemical Corporation	Baton Rouge, Louisiana	1942	2,800	3,100	1,100,000
	Gramercy, Louisiana	1960	2,400	2,650	865,000
Reynolds Metals Co.	Corpus Christi, Texas	1953	3,600	4,000	1,300,000
	Hurricane Creek, Arkansas	1942	2,100	2,300	755,000
Ormet Corporation	Burnside, Louisiana	1958	1,550	1,700	560,000
Martin-Marietta Aluminum Co.	St. Croix, Virgin Islands	1967	900	1,000	325,000
		Total	19,720	21,750	7,200,000

Source: After Battelle 1973.

1. Basis 360 days/year operation (98.6% of capacity).

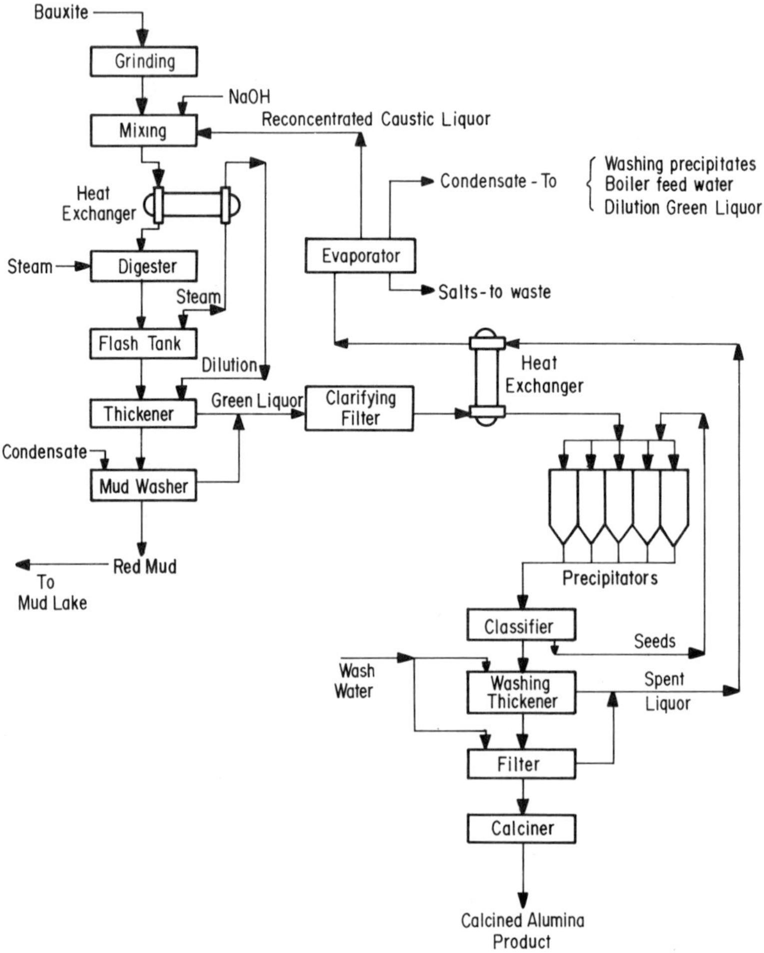

Figure 5.1. *Generalized diagram of the Bayer process (after Battelle 1973).*

As shown in Figure 5.1, the ground ore is mixed with a caustic solution, and heat and pressure are applied by steam heating of jacketed autoclaves or by direct injection of steam. Conditions are varied with ore composition but they are generally as follows:

- trihydrate forms: solution contains 100 to 150 g/l of Na_2O at temperatures of 120 to 170°C at 3.40 to 4.75 atmospheres (50 to 70 psi) pressure.
- monohydrate forms: solution contains 200 to 300 g/l of Na_2O at temperatures of 200 to 250°C at pressures as high as 35 atmospheres (500 psi).

Bauxite ores contain different proportions of the monohydrate and trihydrate forms and operating conditions are adjusted to obtain optimum processing. Most bauxites processed in the United States are predominately trihydrate which permits the use of the lower concentrations, temperatures, and pressures. According to Battelle (1973), minor differences in processing among refineries do not significantly influence effluents. The high-energy-consuming Bayer process maximizes heat recovery; heat exchangers are used extensively. All possible NaOH is recovered from the process residues for reuse which causes buildup of soluble contaminants. Excessive concentrations of contaminants in the sodium aluminate liquor can interfere with alumina precipitation. Some contaminants can be eliminated by contacting the spent liquor with red mud residues in a holding tank (discussed subsequently); another approach is to pass a portion of the mother liquor through a salting-out evaporation stage.

The Bayer process produces a slurry containing $NaAlO_2$ in aqueous solution and undissolved solids known as red muds. Tables 5.2 and 5.3 present the constituents of red mud.

Red muds from different bauxites have slightly different characteristics. The yield of red mud from Surinam bauxite is low (approximately one-third ton per ton of alumina product) and the mud is reportedly amenable to filtration and effective washing on a filter. The residue is relatively easy to handle and disposal area requirements are moderate. More red mud is produced by Jamaican bauxite (about one ton per ton of alumina). Its physical characteristics make filtration expensive.

The muds are washed by countercurrent decantation through numerous—sometimes as many as seven—stages. The NaOH recovered is concentrated by evaporation and recycled.

The red mud usually is moved as a waterborne slurry to a red mud storage lake. Jamaican red mud has poor settling properties so that its disposal has posed greater problems.

Table 5.2 *Red Mud Insoluble Solids*

Constituent	*Percent*
LiO	11.0
SiO_2	5.5
Al_2O_3	12.0
Fe_2O_3	49.5
P_2O_5	2.0
CaO	8.0
Na_2O	3.5
TiO_2	5.0
MnO_2	1.0
Miscellaneous	1.5

Source: After Rushing 1973.

Note: Specific gravity = 3.6.

Table 5.3 *Red Mud Slurry Soluble Solids*

Constituent	*Amount*
Al_2O_3	2.5 g/kg liq.
NaOH	3.7 g/kg
Na_2CO_3	1.6 g/kg
Na_2SO_4	0.4 g/kg
NaCl	0.7 g/kg
$Na_2C_2O_4$	0.1 g/kg
Specific gravity	1.008
pH	10.5
BOD	6 ppm
COD	148 ppm

Source: After Rushing 1973.

In the combination process applied to high-silica Arkansas bauxites, the red mud is further treated to extract alumina and to recover sodium. Red mud is combined with limestone (effectively $CaCO_3$) and sodium carbonate, and sintered at 1,100 to 1,200°C. Silica is converted to calcium silicate and the residual alumina to sodium aluminate. The sintered products are leached for additional sodium aluminate which is filtered and either returned to the main stream for precipitation or precipitated separately. The residual solids (brown muds) are slurried to a waste lake.

In the combination process, lime and soda ash substitute for NaOH according to the following reaction:

$$Ca(OH)_2 + Na_2CO_3 \longrightarrow 2NaOH + CaCO_3.$$

From either process, purified sodium aluminate solution is cooled by heat exchangers to 50 to 60°C prior to entry into precipitation vessels. Seed material is added and composition and agitation are controlled so as to precipitate alumina trihydrate in a form amenable to separation and washing. Precipitation requires one to three days. The precipitated trihydrate (aluminum hydroxide) is dewatered and calcined at 1,200°C to drive off water, break down the trihydrate alumina to anhydrous alumina, and transform the alumina to the crystalline form (alpha) which is most suitable for electrolytic reduction.

In the United States, all bauxite refining is conducted by primary aluminum producers. The nine U.S. bauxite refineries are owned by five primary aluminum producers, and the majority of the refiners are integrated with primary production. Location of plants and their capacities are presented in Table 5.1.

Water Use

The subsequent summary of water use, water balance, and air pollution features has been prepared by the author on the basis of information presented by Battelle (1973). The reader can obtain additional detail from Battelle.

The previous discussion shows that the refining of bauxite can be characterized by leaching of Al_2O_3 from bauxite ores by a hot, pressurized, caustic solvent with subsequent precipitation, drying, and calcination of the pure Al_2O_3 and the discard of the residue (red or brown mud). Water use includes water for the solvent, washing of precipitates and other steps of the chemical extraction process, considerable use for heat exchange and for temperature control. The major waste feature is the red or brown mud lakes which are somewhat analogous to tailings ponds in a flotation-concentrator operation. A mud lake is a receiver of solid wastes, a receiver and reservoir of process water, a sink for evaporation and seepage, and a collector of precipitation. Water transports the waste portion of the ore to the lake.

Even though all United States plants use the Bayer process or the combination process, they differ with respect to waste treatment and management schemes. Location, climate, type of ore, and waste management philosophy all contribute to differences. At this writing seven of the nine United States bauxite refineries report practicing total impoundment of mud wastes, and the remaining two will do so shortly.

A generalized water-flow diagram for a bauxite refinery practicing total impoundment is shown in Figure 5.2. The diagram portrays intake water

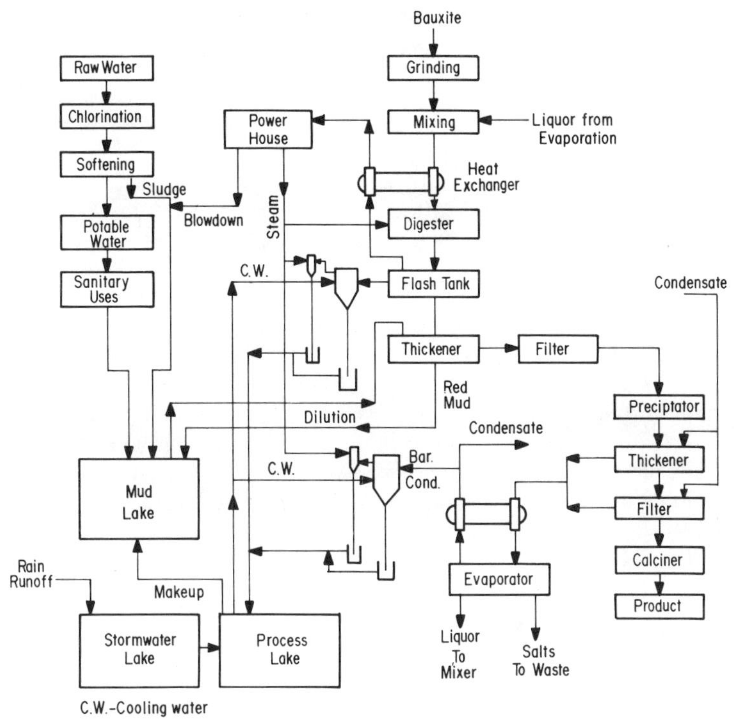

Figure 5.2. *Generalized diagram of water circuit for Bayer plant employing total impoundment (after Battelle 1973).*

treatment, sanitary uses, and boiler plant water streams. The major waste stream is to the red mud lake. The return flow is part of a closed water circuit in which alkalinity is critical.

The red mud residue ranges between 0.33 ton/ton of Al_2O_3 produced (Surinam bauxite) to 2.0 tons (Arkansas bauxite). At 17 to 20 percent solids, the water going to the red mud lake with the mud approximates 1.9 million to 18 million liters/day (500,000 to 4.8 million gallons/day or 350 to 3,300 gallons/minute). Not all water returns to the plant, in part because the terminal density of the settled solids may range from 35 to 75 percent. This entrapment is one of the means by which water is removed from the recycle circuit.

The other major waste waters in aluminum refining are:

- Spent-liquor
- Condensates
- Barometric condenser cooling water
- Miscellaneous cooling water streams
- Miscellaneous waste streams

Spent Liquor

The pregnant liquor is cooled, diluted, and sent to the precipitators after separation from the mud residue by filtration or countercurrent decantation and polishing by filtration. Fine seeds of hydrated alumina are then added and, upon further cooling for one to three days, hydrolysis occurs and the alumina precipitates as a hydroxide. Only about half of the alumina is allowed to precipitate because, if precipitation exceeds 50 percent, coprecipitation of contaminants increases to unacceptable levels (see Table 5.4).

After filtration of the alumina trihydrate, the spent liquor is heated, concentrated, and returned to the digester. Thus, the spent liquor is not a waste stream as such, although some wastes may be withdrawn from it. In order to minimize contaminants in the process liquor which might retard precipitation or precipitate with the product, a portion of the spent liquor may be passed through a "salting-out" evaporation where it is reduced to low volume in order to eliminate such contaminants, particularly sulfates. In order to prevent redissolution of sulfates, the sulfate slurry is normally discarded in an abandoned mud lake or in a landfill as a solid waste.

Condensates

Numerous evaporation operations are conducted in bauxite refineries. The high-temperature, high-pressure slurry from the digesters is flashed to atmospheric pressure for filtration or thickening. The hot clarified pregnant liquor may be vacuum-flashed to cool it for precipitation. The spent liquor is evaporated to concentrate it. All of these operations produce steam, requiring condensation in a heat exchanger against a process stream which needs heat.

Table 5.4 *Relationship Between Precipitation of $Al(OH)_3$ and SiO_2*

Percent $Al(OH)_3$ precipitated	*Corresponding percent SiO_2 precipitated*
50	4
75	8
80	12
87	25
94	80
100	100

Source: After Hayward 1952.

These condensates are utilized for boiler-feed water, product washing, and washing of red muds.

Barometric Condenser Cooling Water

Barometric condensers are large consumers of water, and the water management scheme used depends on water availability, the water balance, and effluent discharge regulations.

In a system where stringent water conservation measures are practiced, the barometric condenser water loop is closed with the red mud lake stream. Where water conservation is not necessary a once-through scheme may be employed. However, the latter scheme may have to be abandoned under the constraints of Public Law 92-500.

Water use by barometric condensers is characteristically several thousand liters per minute for each condenser. Water use may reach 20,000 to 40,000 liters/ton (5,000 to 10,000 gallons/ton of product (Battelle 1973). Theoretically, the overhead vapors to be condensed should approximate distilled water. But Battelle reports that sometimes entrainment occurs and alkaline and aluminum constituents appear in the effluent from barometric condensers. If this water is discharged to a receiving stream already containing magnesium or calcium compounds, these may precipitate as white hydroxides. But entrainment should be minimal from a well-designed and operated evaporator, with unsatisfactory levels being reached only during breakdown. De-misters in the evaporator vapor space further minimize entrainment.

Miscellaneous Cooling Water Streams

The cooling water streams in a bauxite refinery include air compressor aftercooler discharges and cooling water from rotary calcining kilns. These are characteristically noncontact services, and the only pollution is thermal. They are frequently once-through systems and the effluent is discharged to the environment, but disposal with the red mud is possible.

Cleaning Acid Wastes

The thermal efficiency of a bauxite refinery depends on the efficiency of the heat exchangers which transfer heat from hot to cold process streams. Many streams contain substantial quantities of dissolved solids, and scaling of exchanger surfaces is a recurring problem. Acid cleaning is universally employed, generally with sulfuric acid, although small quantities of inhibited hydrochloric acid or acetic acid are used. Scaling also occurs on filtration equipment and filter cloths; similar cleaning procedures are used. Normally, the spent acid contains sulfates with low to moderate free acid concentrations. These sulfate solutions commonly are directed to active or to abandoned mud lakes where neutralization occurs. The quantity of sulfuric acid used by one plant averaged about 500 kg (1,000 lb) of acid per day (Battelle 1973).

Water Balance in a Bauxite Refinery

The feasibility of total impoundment is a function of the water gains and losses in a bauxite refinery.

Water Gains

Almost all plants have a fresh water intake. Fresh water may be needed for drinking, for boiler feed water, for barometric condensers, for washing precipitated aluminum hydroxide, or for makeup water.

Precipitation is a more or less uncontrollable water gain. Bauxite refineries cover large areas, and runoff can affect the water balance significantly. Some rainfall can be diverted, but rain on mud lakes enters the water circuit.

In most cases, imported ores do not contribute water since these ores are dried before shipment. Domestic ore (Arkansas) is not dried before processing, so water is derived from it.

Water Losses

Water losses in a bauxite refinery may arise from:

- Drying and calcining
- Red mud
- Evaporative cooling of green liquor
- Evaporation of spent liquor
- Evaporation from lakes
- Seepage
- Red mud calcination (combination process)

Drying and Calcining: Precipitated aluminum hydroxide, $Al(OH)_3$, is filtered from the spent liquor, washed to displace the entrapped NaOH, dried, and calcined. Assuming a 50 percent solids cake from the filter, a ton of water is evaporated for each ton of alumina. Thus from 520 to 3,600 metric tons (575 to 4,000 tons) of water are lost daily from each U.S. bauxite plant.

Red Mud: The red or brown muds entrap water because at equilibrium the mud in a lake may contain 50 percent or more moisture. Approximately one ton of water is lost for each ton of mud. This figure ranges from 0.33 to 2.0 tons mud/ton Al_2O_3 produced for various ores.

Evaporative Cooling of Green Liquor: Evaporation is ubiquitous in bauxite refineries. Initially it occurs in cooling the bauxite slurry from the digesters. The slurry from the digesters is flashed to atmospheric pressure in several stages; the vapors released are condensed and reused for boiler feed water, for cake washing, or for dilution of the pregnant liquor before precipitation. It normally is not discarded (see Hudson 1963). The slurry must cool before going to the precipitators; the heat is recovered. The steam from the flash tanks is used to heat process liquor recycled to the digesters. Some of the liquor is diverted through the mixer, where the liquor and bauxite are combined. The final cooling of the green liquor before filtration or countercurrent decantation may occur in an evaporator operating under vacuum. The vapors are condensed and the mixture of cooling water and condensate from the barometric condenser may enter a closed circuit, with the effluent going to the lake or cooling tower from which the condenser cooling water is drawn. But the barometric condenser may use once-through cooling water which is discharged from the plant to a waterway.

The pregnant liquor filtrate from the red mud separation stage also needs further cooling prior to precipitation. Some cooling may be done in heat exchangers against the cold spent liquor en route to the evaporators, but evaporative cooling also is used. The effluents from barometric condensers generally go to a reservoir for recycle.

Concentration of Spent Liquor: Heat is expended to concentrate the spent liquor filtrates derived from separation of the aluminum hydroxide from the mother liquor. Typically, evaporators concentrate the liquor from about 150 g/l to about 170 g/l Na_2CO_3 (113 g/l to 128 g/l NaOH). The vapors from the first stages are condensed against a cold process stream, but the last stage utilizes high vacuum and a barometric condenser. Depending on the water balance and the availability of water for barometric condensers, the effluent is discharged to a waterway or recycled to an impoundment lake in the plant water circuit. No appreciable carryover of caustic material from an evaporator to a barometric condenser should occur (Battelle 1973).

Evaporation from Lakes: Red mud lake areas may vary from 40 to 800 hectares (100 to 2,000 acres). A plant may have additional process lakes, clear-water lakes, or storm-water lakes. These lakes may be net evaporators or net collectors. In more arid locations along the Texas Gulf Coast, they lose water; farther east, in southern Louisiana or Alabama, they may gain it.

Seepage: Seepage from lakes can be a source of water loss if preparation of the bottom and injection of wastes do not provide an impervious layer. Most red mud lakes have a collection ditch surrounding the lake dike in which some or all of the seepage is collected and pumped back to the lake. But the ditch should also be monitored and, under some circumstances, wells should be located around the lake, to insure that seepage does not reach and enter the local ground-water flow system through the bottom of the lake.

Red Mud Calcination: A somewhat unusual water loss occurs during the red mud calcination peculiar to combination-process bauxite refineries. The red mud from the first stage is mixed with soda ash and lime, then calcined. The red mud underflow from the countercurrent decantation unit (at approximately 25 percent solids) generally is filtered to about 40 to 50 percent solids to reduce the water content and the evaporative load. About a ton of water per ton of red mud processed is lost via this route.

Air Pollution Control

The dust from the calcination of alumina is the principal air pollution problem in a bauxite refinery. Electrostatic precipitators have been used for control, but not always adequately. New designs of precipitators may help. Baghouses are also used for cleanup downstream from electrostatic precipitators. No United States plants report using wet scrubbers on this operation. However, if air quality standards require their use, they will become a source of waste water. Where wet scrubbers are used on other dust-producing operations (lime kilns or conveyor transfer points), the effluent normally either recycles to the process or is included with the main red mud flow. Relative to the large volumes of red mud, Battelle (1973) considers these waste streams to be insignificant, but observations by the writer lead to the conclusion that careless handling and maintenance will have to be avoided if waste discharge permits are not to be violated.

Characteristics of Red Mud Wastes

As shown in Table 5.5, from 0.33 to approximately 1.0 ton of red mud per ton of alumina produced is generated by bauxite refining. In the case of brown mud from Arkansas bauxite, wastes are 2.0 to 2.5 tons per ton of alumina. The mud stream generally issues from washing thickeners and is pumped to a disposal lake. The mud is red because of its iron oxide content, which may be as high as 50 percent for Arkansas or Jamaican bauxite.

Table 5.5 *Mud Production Rates for Various Bauxites*

	Type of bauxite ore		
	Surinam	*Jamaica*	*Arkansas*
Tons mud/ton Al_2O_3 produced	0.33	1.0	2.0
Tons mud/1,000 tons/day Al_2O_3	330	1,000	2,000
Volume of mud (m^3/d)[1]	205	624	1,248
Volume of mud (m^3/yr)	73,800	224,640	497,280
Volume of mud (acre ft/yr)	60	182	364

Source: After Battelle 1973.

Note: Production rates are for 1,000 metric tons of alumina production/day.

1. Density of settled mud taken at 1,600 kg/m^3 (100 lb/ft^3).

Table 5.6 *Screen Analysis of Red Mud*

Screen mesh		*Percent dry solids*
	+10	0.0
–10	+20	0.2
–20	+50	0.8
–50	+100	0.8
–100	+200	0.8
–200	+325	1.9
–325		95.5

Source: After Rushing 1973.

Table 5.2 presents a chemical analysis of the insoluble solids in Jamaican red mud. Rushing (1973) reports that some 23 elements have been identified by analysis, but 98.5 percent of the material is accounted for by the oxides of eight elements plus water and carbon dioxide.

Red mud displays a complex chemical makeup which varies with the ore and with plant practices. Principal soluble constituents in a typical Jamaican red mud liquor at 17 percent solids are shown in Table 5.3. One of the characteristics of Jamaican red mud is its fine grain size. Rushing reports a particle size distribution as shown in Table 5.6. This table explains in part the tendency of the mud to retain so much water and it is the basis for some of the problems involved in containing it. The other explanation for the high water content is the fact that some of the clay-size particles are clay minerals which tend toward hydration.

Rushing reports that, of the 95 percent minus-325-mesh (44 microns), about 60 percent lies between 5 and 44 microns, and some 35 percent is less than 5 microns in size. Rushing states that:

> The small particle size of the red mud is similar to a material of about 3 percent fine sand, 62% silt, and 35% clay, but since few true clay minerals are present, the red mud will present physical properties of silty fines. Red mud slurry is moderately thixotropic in that the apparent viscosity decreases or the fluidity increases as the cumulative shear rate increases; it acts as a Bingham plastic in that a yield stress must be exceeded before flow commences.
>
> Jamaican red mud will reach a maximum compaction of about 35% solids if allowed to settle and compact below a layer of water. If the mud slurry is allowed to dry in air, surface cracking will start at about 28% solids. Desiccation fissuring will continue on air drying with a volume shrinkage. The volume of one ton of red mud solids at 80% solids will be 1/4 the volume required for one ton of mud at 35% solids. The air-dried mud will reslurry at solids content less than 60%; however, if the drying is continued to solids content greater than 60%, the desiccated mud agglomerates and will not reslurry although there may be some parting of the lumps at fissure or crack planes.

Considerable effort has been devoted to developing a use for red mud. One possibility is to utilize its high iron content (about 50 percent as Fe_2O_3). Fursman et al. (1970) describe a process based on sintering it with carbon and limestone and melting the sinter in an electric furnace to produce poor-quality iron. The process has been further developed by Guccione (1971), but it has not yet achieved commercial application.

Investigators have researched the applicability of red mud to the manufacture of portland cement, bricks, and roads, but in the United States no markets have been developed. However, its availability upon total impoundment offers the advantage of storage for future uses.

Pollution Control

Water

Battelle (1973) concludes that the most critical waste water constituents of bauxite refinery effluents include:

- Alkalinity
- pH
- Total dissolved solids
- Total suspended solids
- Temperature
- Sulfate

The waste streams are essentially inorganic, so oxygen demand is generally insignificant.

As shown by Table 5.7 these constituents are derived primarily from:

1. Red mud: produced at rates of 500 to 3,600 metric tons/day, and associated with 15 to 100 tons/day of caustic material, plus about half as much sodium aluminate.
2. Barometric condenser water: up to 12,000 gpm and dissipating as much as 38 MM kg-cal/hr (150 MM BTU/hr) of process heat.
3. Spent sulfuric acid: 500 to several thousand kg/day of sulfate, containing some free acid.
4. Salts, mostly sulfates, from salting-out evaporators: variable, but possibly up to several thousand kg/day.
5. Miscellaneous plant wastes: primarily alkaline, and highly variable in quantity, from spills, cleanups, and washings.

Battelle concludes that current technology for the treatment and control of bauxite refining processing wastes does not, except for total impoundment, provide a single scheme that is applicable to all bauxite ores and all geographical areas. Most plants have historically impounded their main process wastes, i.e., the gangue material in the ore (red or brown mud). Compared to this waste, which may represent from 500 to 3,600 metric tons of solids per day, plus an equal or greater quantity of water, all other processing wastes are inconsequential. In arid climates, the tendency is toward total impoundment; in high-precipitation regions, the tendency is the reverse because the water

Table 5.7 *Characterization of Principal Waste Streams from U.S. Bauxite Refineries*

Waste	*Quantity*	*Characterization*
Red mud	500-3,600 tons/day (dry basis)	15-20% solids
	1,000-7,200 tons/day (wet, settled)	5-12 g/l soda
	3,000-20,000 tons/day (slurry at 18% solids)	2-5 g/l aluminum pH = 12.5
Spent cleaning acid	Variable, 5-10 tons/week intermittently discharged	Na_2SO_4, plus some free H_2SO_4 HCl or $HA\bar{c}$ may also be used pH = 0
Salts from salting-out evaporator	Variable, estimated up to several thousand kg/day	Na_2SO_4 – alkaline pH = 12.5
Barometric condenser, cooling water	Millions of l/hr	Temp. rise of up to 15°C (25°F) may contain traces entrained alkali
Boiler and cooling tower blowdown	Variable, thousands of l/day	Dilute alkaline solutions pH = 12.5
Water softener sludge	One to few tons/day	Lime and suspended solids from intake water
Sanitary waste	375 l/day/capita	BOD 70 g (0.15 lb)/day/capita

Source: After Battelle 1973.

balance is positive. However, total impoundment has been proven feasible in some areas and it has generally been recognized by the bauxite refining industry that the refinery water circuit can operate as closed or nearly closed.

Some United States plants currently recycle all process waters and impound all solid wastes. Several others completely impound the red mud and parts or all of various other significant streams (Battelle 1973).

The constituents in the water circuit that inhibit total recycle (zero discharge) are dissolved solids and, in some instances, heat. The effects of buildup of constituents in process water do not appear to be completely understood. As in other industries, there is a tendency, and reasonably so, not to recycle water if constituent behavior is not understood from long experience.

In the case of heat, the cooler the water the less the treatment needed and it is easier to achieve a vacuum on a barometric condenser. Heat dissipation in a closed system may not be sufficient to provide adequate cooling of barometric condenser effluents for reuse. Thus some operations now use once-through cooling water. Alternatives, both costly, are cooling towers or larger lakes.

Construction of Mud Lakes

There are two approaches to constructing mud lakes. The dikes may be constructed of imported material in a manner similar to an earth dam so that the complete lake is available from the beginning, and additional dike construction is not required during the life of the lake. The construction of this kind of dike has been described by Rushing (1973) and summarized by Battelle (1973). Figure 5.3, part (a) summarizes this type of dike construction.

An alternative approach is to build initially a low dike and to allow the red mud to form an embankment as disposal continues. The construction of this type of dike is outlined schematically in Figure 5.3, part (b). The red mud is cycloned into the pond in a manner similar to that described more thoroughly in Chapter 11. Miners (1973) describes the use of such a method in greater detail. On the basis of research conducted by the author and others, as

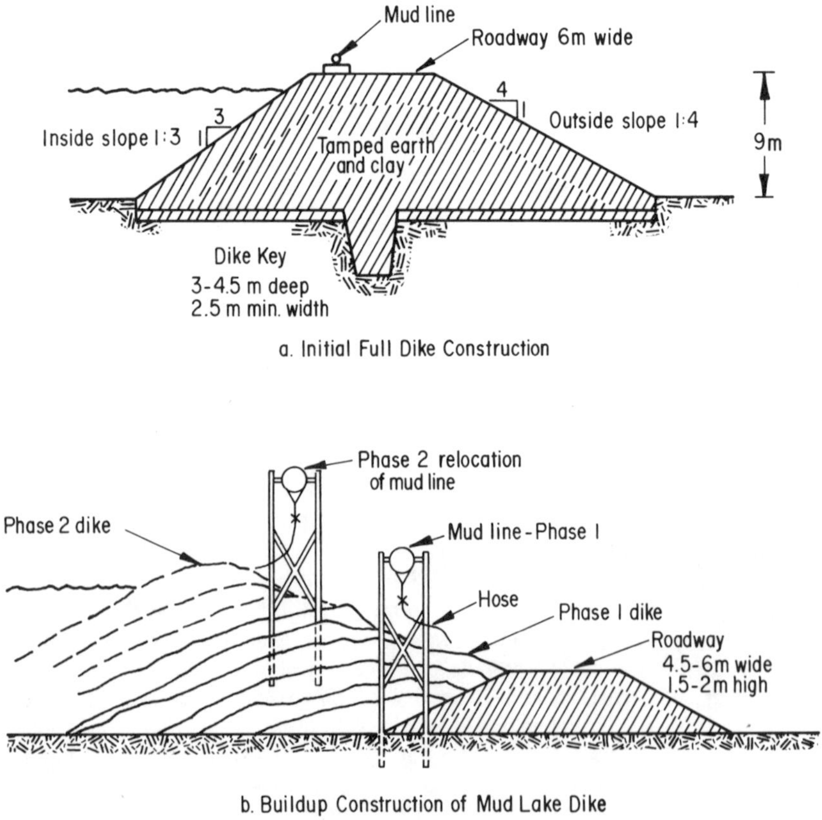

Figure 5.3. *Mud Lake dike construction (after Battelle 1973).*

discussed in Chapter 11, the success of this type of dike construction depends on the presence of a coarse sand fraction in the red mud. Thus, it should be more practicable for Surinam bauxite than for Jamaican bauxite. Great care should be taken in the use of this method because such embankments can fail if they are constructed of material that is too fine. The previously quoted discussion of Rushing suggests that this is frequently the case. Jamaican red mud can pose special problems because of its small particle size and somewhat thixotropic character. Its settling properties are poor. As mentioned previously, it is reported to reach a compaction of only 35 percent solids (or less) if allowed to settle and compact in water (Rushing 1973). If Jamaican mud can be spread in thin layers, inches rather than feet deep, Rushing reports that it will dry out satisfactorily, given a location in a net evaporation climate. Once past a critical moisture content, somewhere in the 60 percent solids range, the mud does not resuspend when rewetted. However, adoption of this approach to the disposal of muds like Jamaican red mud hinges on the existence of an arid or semiarid climate and the availability of large tracts of land for the disposal area.

One alternative, investigated by Good and Fursman (1968) utilizes centrifugal dewatering of Jamaican red mud. Their study suggests that the solids content can be increased to 40 percent, and that approximately 40 percent of the slurry liquid can be recovered as a clear effluent for recycle to the plant. Good and Fursman conclude that recovery of the alumina and NaOH is approximately a break-even financial operation. With this procedure, greater consolidation and dewatering of Jamaican red mud may be possible.

The Kaiser Aluminum & Chemical Corporation has implemented a so-called DREW system for land disposal of bauxite residue at its Gramercy, Louisiana, operation (Vogt 1974). The acronym stands for "drainage and evaporation of water." This system differs from the unusual tailings disposal methods discussed in chapter 11 in that a sand drainage blanket is installed beneath the red mud disposal area. Vogt reports that 50 percent consolidation of red mud can be achieved by the method and that impoundment area reclamation appears feasible.

In all mud lake construction, ground-water contamination must be precluded by a bottom that is as impervious as possible or by seepage collection. Tests should be made to evaluate the bottom, and impermeable materials should be inserted if appropriate. The permeability distribution in all existing mud lakes also should be determined so that leakage can be estimated and monitored if necessary.

Mud lakes are multi-purpose facilities. They may serve as cooling ponds and water reservoirs and as reservoirs for other minor waste streams from the plant; these may include boiler and cooling tower blowdowns and treated sanitary waste effluents. If NaOH concentrations are not excessively high, they can serve to some extent as an additional mud washing stage.

Depending on storage area and economics of slope stability achievement, it may also be necessary to thicken the mud as described in Chapter 3 prior to disposal. This practice is common to the treatment of tailings by the copper industry. Appendix E gives an example of this in the lead-zinc industry.

A system to permit use of red mud tailings as land fill instead of discharging them into the Mississippi River is under construction at a Louisiana alumina plant. Here drainage pipe is placed in a trench at bottom of storage pond; later it will be covered with coarse permeable gravel bed.

Kaiser Aluminum & Chemical Corporation photo

Proposed Treatment Technology

Table 5.8 summarizes the processes that Battelle (1973) considers to be the best practicable control technology currently available for the waste streams discussed in the preceding sections. It would be advisable to include techniques for prevention of pond seepage to this list.

COST OF IMPOUNDING AND RECYCLING BAUXITE REFINERY WASTES

Capital costs of impoundment and recycle include the following major cost items (Battelle 1973):

1. Cost of land.
2. Cost of construction of reservoir(s).
3. Cost of equipment and facilities.

The cost of land constitutes a large portion of an impoundment scheme, where large tracts are needed. Estimated land costs for mud lake sites range from $200/hectare ($500/acre) to $1,200/hectare ($3,000/acre). However, according to Battelle, most refineries already have ample land acquired at lower cost.

Reported costs for plants processing all three types of bauxite ore are presented in Table 5.9. Battelle based the unit capital costs on the estimated total alumina production represented by the capacity of the impoundment lakes. Unit operating costs were determined by apportioning reported operating cost expenditures over the annual alumina capacity of the plant. Based on these data, Battelle concluded that impoundment of the processing wastes from Surinam bauxite requires a capital investment of $0.25 to $0.50/ton of mud impounded and an operating cost of about $0.90 to $1.30/ton of mud impounded.

For Jamaican ore, which has higher mud yield, capital costs decrease to about $0.20/ton of mud and operating costs to about $0.50/ton. Lowest costs, on a mud basis, are associated with mud-prolific Arkansas bauxite. Capital costs are in the order of $0.10 to $0.20/ton of mud. Operating costs are about $0.10 to $0.25/ton of mud.

It is the opinion of the author that costs additional to these will be incurred by some plants because of seepage control and because the water balance will be positive in some areas, thereby precluding total recycle of mud lake water. This water is likely to have a high pH and contain dissolved solids, some of which will be sufficiently objectionable to require removal.

Regulations Promulgated for Bauxite Refining

The U.S. Environmental Protection Agency used the information in Battelle (1973) and the reviews thereof by industry and other public agencies as a basis for promulgation of the following regulations applicable to the refining of

Table 5.8 *Summary of Effluent Reductions Achieved for Bauxite Refinery Process Wastes Using Best Practicable Technology Currently Available*

Waste stream	*Parameters*	*Best practicable (Level I) technology currently available*	*Effluent reduction achieved*
Red mud	TSS, TDS, alkalinity	Impound and recycle aqueous phase; concentrate if necessary	No discharge
Spent cleaning acid	TDS, sulfates, pH	Impound in red mud lake	No discharge
Salt slurry from salting-out evaporator	TDS, sulfates, alkalinity	Impound in red mud lake or landfill	No discharge
Barometric condenser cooling water	TDS, heat, alkalinity	Cool and recycle	No discharge
Barometric condenser cooling tower blowdown	TDS	Impound in red mud lake	No discharge
Hose-down and clean-up streams	TSS, TDS, alkalinity	Recycle to process (insufficient data)	No discharge

Source: After Battelle 1973.

Table 5.9 *Summary of Waste Disposal Cost Data*

Item	*Plant*							
	A	*C*	*D*		*E*[b]	*F*		*B*
			Old pond[a]	*New pond*[a]		*Old pond*[c]	*New pond*	
Plant capacity (metric tons/yr)	820,000	325,000	755,000	755,000	1,300,000	560,000	560,000	1,150,000
Bauxite type	Surinam	Arkansas	Arkansas	Arkansas	Jamaican	Surinam	Surinam	Jamaican
Mud ratio[d]	0.33	2.0	2.0	2.0	1.0	0.33	0.33	1.0
Lake capacity (metric tons dry mud)	$4.2x10^6$	$11.1x10^6$	$7.5x10^6$	$11.7x10^6$	$53.5x10^6$ [e]	$3.0x10^6$	$1.7x10^6$	–
Capital cost ($)	$1.13x10^6	$0.9x10^6$	$0.69x10^6$	$2.06x10^6$	$9.52x10^6$	$1.26x10^6$	$0.82x10^6$	–
Unit capital cost ($/metric ton mud)	0.26	0.08	0.09	0.18	0.18	0.52	0.48	0.30
Annual cost ($/yr)	$264,000	75,600	373,000	–	571,000	166,000	–	–
Mud (metric tons/yr)	205,000	660,000	1,520,000	–	1,160,000	187,000	–	–
Unit annual costs ($/metric ton mud)	$1.29	0.11	0.24	–	0.49	0.89	–	0.35
($/metric ton alumina)	0.43	0.22	0.48	–	0.49	0.30	–	–

Source: After Battelle 1973.

Note: Mud basis, in 1971 dollars.

a. Construction costs of old pond were expensed as incurred; new pond on capitalized basis.
b. Exemplary plant, zero discharge of pollutants.
c. Annual costs represent average costs for two old ponds.
d. Pounds mud/lb of alumina produced.
e. Very large lake; estimated capacity includes 20-year life still remaining.

bauxite by the Bayer or the combination process. The regulations are subject to alteration in individual cases by a regional administrator of EPA, should evidence deem it necessary.

Effluent Limitations Guidelines Representing the Degree of Effluent Reduction Attainable by the Application of the Best Practicable Control Technology Currently Available

(a) Subject to the provisions of paragraph (b) . . . below . . ., the following limitations establish the quantity or quality of pollutants or pollutant properties, . . . which may be discharged by a point source . . . after application of the best practicable control technology currently available: there shall be no discharge of process waste water pollutants to navigable waters.

(b) During any calendar month there may be discharged from the overflow of a process waste water impoundment either a volume of process waste water equal to the difference between the precipitation for that month that falls within the impoundment and the evaporation within the impoundment for that month, or, if greater, a volume of process waste water equal to the difference between the mean precipitation for that month that falls within the impoundment and the mean evaporation for that month as established by the National Climatic Center, National Oceanic and Atmospheric Administration, for the area in which such impoundment is located (or as otherwise determined if no monthly data have been established by the National Climatic Center).

Effluent Limitations Guidelines Representing the Degree of Effluent Reduction Attainable by the Application of the Best Available Technology Economically Achievable

(a) Subject to the provisions of paragraph (b) . . . below . . ., the following limitations establish the quantity or quality of pollutants or pollutant properties, . . . which may be discharged by a point source . . . after application of the best available technology economically achievable: there shall be no discharge of process waste water pollutants to navigable waters.

(b) [Same as section (b) above.]

Standards of Performance for New Sources

(a) Subject to the provisions of paragraph (b) . . . below . . ., the following standards of performance establish the quantity or quality of pollutants or pollutant properties, . . . which may be discharged by a new source subject to the provisions of this subpart: there shall be no discharge of process waste water pollutants to navigable waters.

(b) [Same as section (b) above.]

PRIMARY ALUMINUM SMELTING (SIC 3334)

This section pertains to the production of primary aluminum by the electrolysis of alumina obtained from bauxite refineries. Effluents from wet scrubbers used for air pollution control constitute the major waste water from this industry. This water contains fluorides, suspended solids, and organic constituents. Treatment of scrubber effluent or conversion to dry scrubbers are possible corrective devices. Combination with bauxite refinery wastes in red mud lakes should be considered, depending on water balance and location.

Sources of Information and data

Battelle (1973) has compiled an inventory of the primary aluminum industry collected as a part of Contract 68-01-1518 under the authorization of Public Law 92-500. Items covered in the investigation include process details, air pollution control systems, waste water treatment methods, and ancillary operations. Aluminum reduction plants with exemplary waste water control or treatment were identified from U.S. Army Corps of Engineers discharge permit applications, through consultation with a committee of the United States Aluminum Association, and with primary aluminum company representatives. Battelle personnel then visited 10 of the 31 primary aluminum plants, including the exemplary plants identified and others selected as representative of various specific industry practices, to obtain information covering waste control and treatment technologies and associated costs, identification and concentration of waste water constituents, and discharge rates. Three plants were selected for an additional on-site analysis because their waste water handling practices represent the best technology generally available to the aluminum industry. The field work included sampling of internal streams and outfalls.

The data obtained were assessed to identify the sources and volumes of waste water produced, and the concentration of constituents contained in the discharge. Other sources of information are referenced in the standard fashion.

Reduction of Alumina to Primary Aluminum

Battelle (1973) reports that the United States has 31 alumina reduction plants with a total annual capacity of about 4.5 million metric tons (5 million short tons) of which about 60 percent is provided by the three largest companies. The availability of inexpensive electrical power is a major consideration in site selection and accounts for the concentration of plants in the Pacific Northwest and in the Tennessee Valley. The energy consumed annually at full production is reported to be in the range of 80 to 100 billion kilowatt-hours.

The basic electrolytic process for reducing alumina to aluminum is shown in Figure 5.4. The principal materials used in the primary aluminum industry include alumina, cryolite, pitch, petroleum, coke, and aluminum fluoride.

The electric cell, or pot, is the heart of the aluminum plant. It consists of a steel container lined with refractory brick with an inner liner of carbon. Cells are aligned in rows called a potline. The energy supply consists of several hundred volts with 60,000 to 100,000 amperes of direct electrical current. The carbon liner on the bottom of the furnace constitutes the cathode of the cell when covered with molten aluminum. The anode is baked carbon. The electrolyte is cryolite (80 to 85 percent by weight), calcium fluoride (5 to 7 percent), aluminum fluoride (5 to 7 percent), and alumina (2 to 8 percent). The composition of the bath varies as electrolysis proceeds, so alumina is added intermittently to maintain the proper concentration of dissolved alumina. The fused salt bath usually is operated at a temperature of about 950°C. A cell reduces the aluminum from the apparent trivalent state to the liquid metal

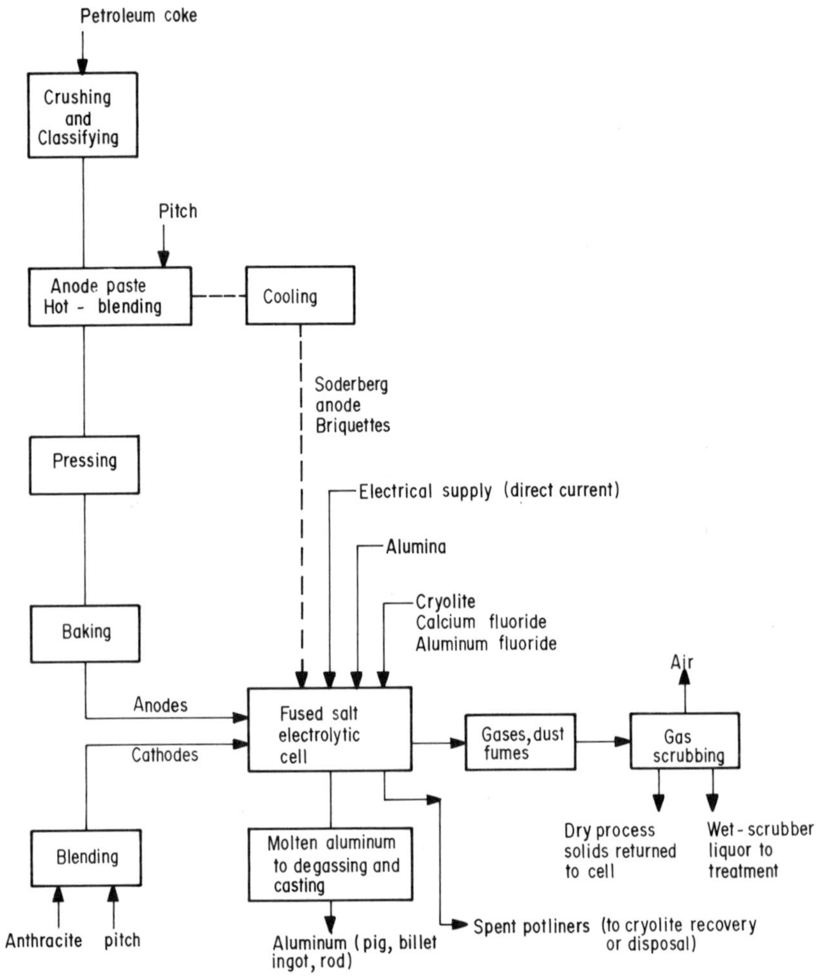

Figure 5.4. *Process diagram for the electrolytic production of aluminum (after Battelle 1973).*

state at the cathode. Oxygen, present in the bath in the divalent state, reacts with the carbon anode and surrounding constituents to form a mixture of 75 percent carbon dioxide and 25 percent carbon monoxide, which gradually consumes the carbon anode. The electrolytic aluminum reduction cell continually consumes alumina and the carbon anode, and evolves gaseous reaction products.

The airborne emissions from aluminum reduction include carbon dioxide and carbon monoxide along with some volatile fluoride compounds, sulfur oxides, and fine dust from the cryolite, aluminum fluoride, alumina, and carbonaceous materials used in the cell. The removal of this fume from the working area, along with cell cooling, involves air and water quality control which concerns design of the building, hoods, ducts, dust collectors, cyclones, and gas scrubbers. The air pollution control measures produce a major effluent if wet gas scrubbers are used.

The consumed anode material, about 0.5 kg of anode per kg of aluminum produced, is replaced continuously (Soderberg anodes) or intermittently (prebaked anodes). The raw materials for anodes (coke and pitch) are prepared in anode paste plants, the product of which is anode paste consisting of a mixture of high-grade coke (petroleum and pitch coke) and pitch (sometimes tar).

The anode paste preparation involves roughing, crushing, screening, calcining, grinding, and mixing. Therefore, dust control equipment is needed during this phase of the process.

Prebaked Anode System

In the prebaked anode method, the anodes are baked and graphitized by heating for up to 30 days at a maximum temperature of 1,100°C (2,000°F). The flue gases from this process contain particulate carbon, tar vapors, sulfur compounds, and fuel-combustion products. Fluorine compounds also may be present depending on anode stub recycle practice. The tars are formed from cracking, distilling, and oxidizing the pitch used as a binder. When the fuel gases are treated by wet scrubbing or electrostatic precipitation, the effluent contains tars and oils, sulfates, particulate matter, and in some cases, fluorides.

Soderberg Anode System

To construct the Soderberg continuous anode, the anode paste is packed into a rectangular metal shell suspended above the electrolysis cell. In this paste the concentration of aggregate coke is of the order of 55 percent, and consists primarily of coarse and medium-size fractions. As the anode paste descends through the anode shell, it is gradually baked. At some 50 centimeters (20 inches) above the molten bath, the anode mass becomes a fully baked, monolithic anode. The tars and oils characteristic of anode baking, together with other fumes, are evolved by the cell.

Electrolyte

The electrolyte in aluminum reduction cells dissolves alumina and provides a molten bath with a melting temperature lower than that required to melt alumina, and low enough to preclude aluminum carbides. Natural or artificial cryolite, a double fluoride of sodium and aluminum, is the major constituent in aluminum reduction cells.

Cell liners fail every two to three years and must be discharged as a solid waste. Battelle (1973) reports that the disposal amounts to some 1,200 cubic meters (about one acre foot) of liner waste per United States plant per year.

Waste Water Production

Primary aluminum smelters need water for sanitary purposes, boiler feed, cooling circuits applied to DC power equipment, metal casting operations, furnace cooling in the anode plant, and miscellaneous cooling equipment. Cooling waters may be cooled in towers, passed through several units in series, or used on a once-through basis, with various discharge practices. The major contaminating use of water is in fume scrubbing (if wet systems are used) associated with the anode plant, casting operations and the aluminum reduction cells.

After surveying the United States industry, Battelle (1973) concluded that its approaches to fume control include:

1. No control.
2. Wet scrubbing using once-through water (discharged with or without treatment).
3. Wet scrubbing with recirculation of water and reclamation of contained fluorides and alumina, often through precipitation of cryolite.
4. The use of dry scrubbing systems using no water, with reclamation of fume components.

Current economic and environmental pressures have brought much of the industry to practices listed in paragraphs 3 and 4 above. The dry fume scrubbing method reportedly is being installed in some of the plants recently under construction, and has been or is being installed to replace wet scrubbers in some of the older plants.

Dry fume scrubbing reduces plant water requirements to sanitary, boiler feed, and cooling water. Dry scrubbers minimize air pollution, decrease discharge of process-contaminated waste water, and have the potential for reclamation and recycle of formerly wasted materials. However, clogging problems may be caused by the hydrocarbons evolved during anode baking.

Figure 5.5 shows water use, treatment, and disposal for primary aluminum plants using wet scrubbers for air pollution control. Treatment of intake water (1) ranges from chlorination to full clarification and treatment. Stream (2) is made potable and the effluent is discharged to sewage. Stream (3) is makeup water for a closed-loop cryolite recovery stream (8), which may or may not include a pot room secondary air scrubber. The materials added during cryolite recovery vary. Some cryolite recovery systems are sophisticated (and proprietary) while others recover or treat water with disposal or recycling of the solids. The bleed stream (9) controls sulfate in the recycle circuit. Some plants recover cryolite in which case stream (9) represents once-through discharge. Stream (4) through a cast house furnace air scrubber is common but, according to Battelle, several plants plan to eliminate the stream by changes in degassing techniques which minimize noxious fumes, or by the installation of a dry collection system for alumina and hydrogen chloride. Streams (5) and (6) are not ubiquitous because dry processes are often employed. However, where there is a liquid effluent, the carbon particulates usually are settled in ponds.

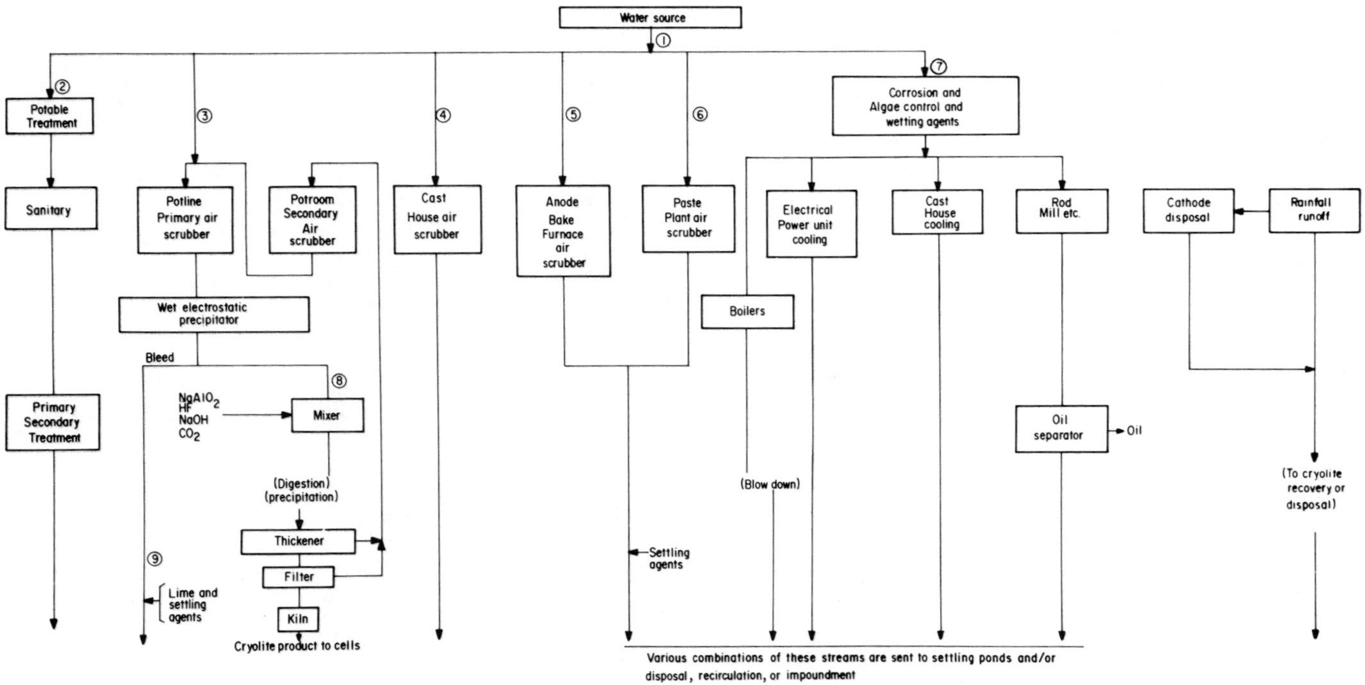

Figure 5.5. *Schematic composite flow diagram for plant using wet scrubbing. In a specific plant, any particular stream or unit may not exist as alternative technology is applied (after Battelle 1973).*

Segments of stream (7) are treated to promote wetting and to inhibit corrosion and algae growth. The chief effluent constituent is oil from rod mills or other fabrication units. The use of commercial oil separators is common.

On the basis of the above discussion Battelle identifies the potential sources of waste water as:

1. Wet scrubbers.
   a. Primary potline
   b. Secondary pot room.
   c. Anode bake plant.
   d. Cast house.
2. Cooling water.
   a. Casting.
   b. Rectifiers.
   c. Fabrication.
3. Boiler blowdown.

Primary Potline Air Wet Scrubbers

The wet scrubbers which collect fumes and dust from the electrolytic cells produce most of the aquatic pollutants from primary aluminum plants. Carbon dioxide, carbon monoxide, and hydrogen fluoride are generated in each cell. In addition, cryolite vaporized from the molten bath, sulfur oxides produced from sulfur impurities in the anode, and dust from all materials handled at the cell (alumina, cryolite, and fluorides of calcium and aluminum) contribute to the scrubber load. In plants using a Soderberg anode system in which the anode paste mix is baked at the cells, volatile hydrocarbons and additional oxides of sulfur also are collected. The efficiency of scrubbers in collecting water-soluble gases and dust from the primary air (that collected from the cells) is ordinarily above 96 percent. Thus Battelle concludes that the quantity of materials collected in the liquor should be directly proportional to the production of aluminum metal.

Secondary Pot Room Air Wet Scrubbers

Cells are hooded but some fumes and dust escape. Thus, some plants exhaust pot room air through wet scrubbers. The constituents of this scrubber liquor are similar to those from the primary scrubbers, but they reportedly constitute less than 10 percent of the total volume. The large volume of air handled in secondary scrubbers is detrimental to capture efficiency; thus improved hooding at the cells is often preferred over scrubbing of pot room air.

Anode Bake Plant Air Wet Scrubbers

The flue gas from anode bake furnaces is commonly treated in wet scrubbers. The effluent contains particulate carbon, acid, tars, oils, and sulfur oxides. Fluorides may be present depending on anode stub recycle practice. Such a stream cannot be processed through a recovery system because it would return solids to the electrolyte cells. Therefore, it usually is added to other effluent streams, treated to promote settling, and diverted to settling ponds. The overflow from these settling ponds is ordinarily mixed with other plant effluent streams.

Cast House Air Scrubbers

Molten aluminum from the cells is degassed by bubbling chlorine plus nitrogen gas through the melt. As the off-gas is scrubbed, an acidic liquor containing dissolved chlorides and suspended alumina is produced.

Cooling Water

Cooling water is used for aluminum casting, for electrical and mechanical equipment, and in anode preparation. Chlorine is added to minimize algae growth. The major fraction of cooling water flows in closed systems or is discharged without treatment.

Table 5.10 presents illustrative effluent quality data as supplied to Battelle (1973) by one of 11 companies. The effluent loadings as calculated from similar effluent concentration and flow data supplied by 10 companies are presented in Table 5.11. The control and treatment technology currently practiced by each of 11 plants is presented in Table 5.12.

Of the eight plants reporting fluoride values in Table 5.11, five (D, H, I, J, and K) are in the range of 0.5 to 1.0 kg/metric ton Al (1 to 2 lb/ton Al) and

Table 5.10 *Concentrations of Selected Constituents in Illustrative Influent and Effluent Water, Primary Aluminum*

Constituent	*Influent concentration*[1] *(mg/l)*	*Effluent concentration*[1] *(mg/l)*	*Net concentration (mg/l)*
Alkalinity	59	174	115
Chemical oxygen demand	2	30	28
Total solids	335	1,065	730
Dissolved solids	322	797	475
Suspended solids	9	268	259
Sulfate	24	80	56
Chloride	104	270	166
Cyanide	0	0.05	0.05
Fluoride	0	15	15
Aluminum	0	0.2	0.2
Calcium	32	83	51
Copper	0	0.05	0.05
Magnesium	11	30	19
Nickel	0.01	0.02	0.01
Sodium	52	140	88
Zinc	0.04	0.02	Neg
Oil and grease	4	10	6
Phenol	–	–	–

Source: After Battelle 1973.

Note: Volume = 1.72 mgd.

1. From a company report.

Table 5.11 *Quantities of Selected Constituents in Water Effluent from Selected Primary Aluminum Plants in the United States*

	A		*B*		*C*		*D*		*F*	
Constituent	N		G^1		N		G		N^2	
Alkalinity	0.01	(0.02)	–	–	–	–	1.1	(2.1)	–	–
Chemical oxygen demand	0.6	(1.2)	3.5	(7.1)	0.3	(0.6)	0.8	(1.7)	12.7	(25.5)
Total solids	0.5	(1.0)	–	–	8.9	(17.8)	10.8	(21.7)	–	–
Dissolved solids	0.45	(0.9)	–	–	4.4	(8.7)	10.5	(21.0)	23.8	(47.6)
Suspended solids	0.55	(1.1)	1.1	(2.2)	3.8	(7.5)	0.38	(0.8)	16.5	(33.1)
Sulfate	–	–	4.2	(8.4)	–	–	4.6	(9.2)	–	–
Chloride	–	–	50.3	(100.6)	0.078	(0.2)	1.5	(2.9)	–	–
Cyanide	–	–			–	–	–	–	–	–
Fluoride	–	–	0.6	(1.2)	5.0	(10.0)	0.34	(0.7)	15.3	(30.6)
Aluminum	–	–	–	–	0.17	(0.34)	0.03	(0.07)	2.7	(5.5)
Calcium	–	–	–	–	–	–	–	–	–	–
Copper	–	–	–	–	0.001	(0.002)	–	–	0.005	(0.01)
Magnesium	–	–	–	–	0.31	(0.6)	–	–	–	–
Nickel	–	–	–	–	–	–	–	–	0.024	(0.04)
Sodium	–	–	–	–	0.47	(0.9)	2.1	(4.3)	–	–
Zinc	–	–	–	–	0.015	(0.03)	0.0017	(0.003)	–	–
Oil & grease	0.1	(0.2)	0.5	(1.0)	0.45	(0.9)	0.04	(0.08)	1.2	(2.4)
Phenol	–	–	–	–	–	–	–	–	–	–

	G		H		I		J		K	
Constituent	N		G		N[3]		N		N	
Alkalinity	–	–	–	–	2.6	(5.3)	11.5	(23.0)	–	–
Chemical oxygen demand	–	–	–	–	0.65	(1.3)	0.6	(1.2)	–	–
Total solids	–	–	–	–	16.8	(33.7)	34.8	(69.6)	1.8	(3.5)
Dissolved solids	–	–	9.2	(18.5)	10.9	(21.9)	32.5	(65.1)	1.6	(3.1)
Suspended solids	9.3	(18.7)	2.2	(4.5)	5.9	(11.9)	1.9	(3.8)	0.22	(0.4)
Sulfate	14.0	(28.0)	0.045	(0.09)	1.3	(2.6)	1.27	(2.5)	5.0	(9.8)
Chloride	–	–	–	–	3.8	(7.6)	1.7	(3.4)	0.7	(1.4)
Cyanide	–	–	–	–	0.001	(0.002)	0.004	(0.01)	0.001	(0.003)
Fluoride	9.8	(19.6)	0.55	(1.1)	0.35	(0.7)	1.1	(2.2)	0.96	(1.9)
Aluminum	1.3	(2.7)	0.18	(0.3)	0.005	(0.01)	0.46	(0.9)	0.076	(0.15)
Calcium	3.7	(7.3)	–	–	1.2	(2.3)	2.0	(4.0)	–	–
Copper	0.002	(0.004)	–	–	0.001	(0.002)	0.008	(0.016)	0.0018	(0.004)
Magnesium	0.1	(0.2)	0.17	(0.3)	0.45	(0.9)	0.21	(0.4)	–	–
Nickel	0.017	(0.03)	–	–	0.00025	(0.0005)	–	–	0.0016	(0.003)
Sodium	1.5	(3.1)	0.57	(1.1)	2.0	(4.0)	17.2	(34.4)	3.3	(6.7)
Zinc	0.003	(0.007)	–	–	–	–	0.047	(0.09)	0.0013	(0.003)
Oil & grease	0.23	(0.47)	–	–	0.15	(0.3)	0.20	(0.4)	0.19	(0.4)
Phenol	0.067	(0.13)	–	–	–	–	–	–	–	–

Source: After Battelle 1973.

Note: In kg/metric ton (lb/ton) of AP produced.

N = Net values. Concentration of each constituent in intake water subtracted from concentration in effluent and the difference used to calculate values given.

G = Gross values. Data for intake water not available.

1. Data reported as "not to exceed" so quantities are limits, not actually present.
2. Does not include effluent from separate cryolite manufacturing facility operated on primary plant site.
3. Zero discharge to navigable waters. Plant effluent is diverted to adjacent alumina refining plant for use as makeup water.

Table 5.12 *Treatment Employed by U.S. Aluminum Smelters*

Plant	*Anode type*	*Control or treatment applied*
A	PB	Dry scrubbing
B	VSS	Lime/recycle
C	PB	Lime/once-through
D	PB	Cryolite/recycle
E	PB	Cryolite/recycle
F	HSS	None
G	VSS	Lime/once-through
H	PB	Cryolite/recycle
I	HSS	Cryolite/recycle
J	HSS	Cryolite/recycle
K	PB	Cryolite/recycle

each of these five plants practices cryolite precipitation and recycle. Plants C and G practice once-through lime precipitation and report effluent loadings of 5 to 10 kg/metric ton Al (10 to 20 lb/ton Al). Plant F practices no water treatment and no recycle and the effluent loading is 15 kg/metric ton Al (30 lb/ton Al).

The data in Tables 5.10 and 5.11 were verified by Battelle by collection and analysis of additional samples. These combined data were used to conclude that fluoride, suspended solids, oil and grease (tars), pH, and free cyanide are the major pollutants in primary aluminum effluents. Converted to milligrams per liter, fluoride content ranged from 10 to 30, suspended solids from 5 to 30, oil and grease from 1 to 10, and cyanide levels from 0.002 to 0.036 mg/l in United States plant discharges at the time of the Battelle survey.

Proposed Treatment Technology

Battelle (1973) concluded that the technologies for controlling waste water in the primary smelting of aluminum include dry fume scrubbing, recycle of water to wet scrubbers after precipitation of fluorides and neutralization and alkaline chlorination, if pH and cyanide levels necessitate it. Treatment methods for reducing pollutant concentrations include cryolite precipitation, precipitation by lime or alum, absorption on activated alumina or hydroxylapatite, and reverse osmosis.

Table 5.13 presents current and potential treatment and control technologies. Battelle concluded that the most significant reduction in current discharge volume is obtainable by converting wet fume scrubbers to dry fume scrubbers or by treating and recycling the water from wet scrubbers. Dry scrubbing was observed to be feasible by contacting pot gases with dry alumina to absorb pollutants and subsequently collecting particulates by fabric filtration. The system was deemed applicable to gases collected immediately above the pot but not to dilute pot room ventilation air. The alumina subsequently is fed back to the cells. The process uses no water.

Table 5.13 *Summary of Present and Potential Control and Treatment Technologies*

Wastewater source	*Present practice*	*Possible added control*	*Possible added treatment*
Pot (primary) wet scrubber	Discharge without treatment	Convert to dry scrubbing	–
Pot (primary) wet scrubber	Discharge without treatment	Install cryolite precipitation plus recycle	Install lime treatment of bleed stream
Pot (primary) wet scrubber	Lime and settle once-through	Install recycle	Install alumina adsorption
Pot (primary) wet scrubber	Cryolite precipitation with recycle	–	Install lime treatment of bleed stream
Pot room (secondary) wet scrubber	Discharge without treatment	Install cryolite precipitation plus recycle	–
Pot room (secondary) wet scrubber	Lime and settle once-through	Install recycle	Install alumina adsorption
Cast house wet scrubber	Settle	Convert to alternate degassing	–
Anode bake plant wet scrubber	Settle	Recycle	Flocculate and aerate
Paste plant wet scrubber	Settle	Recycle	–
Cast house cooling	Discharge without treatment	Close loop	Cooling tower
Rectifier cooling	Discharge without treatment	Convert to air-cooled rectifiers	Cooling tower
Rainfall runoff	Discharge without treatment	Route to cryolite recovery and recycle	–

Source: After Battelle 1973.

The dry scrubbing process (Figure 5.6) utilizes hoods and ducts to collect and deliver the gases from the pots to an operating unit, possibly a cyclone-type device to separate coarse particulates, a reactor section in which the gases are exposed to alumina, and a fabric filtration stage, from which the gases are released to the atmosphere, usually through a stack. Associated equipment includes fans, alumina delivery, storage and removal devices, and baghouse auxiliary equipment. Three commercial variations of the process are reported to exist, with differing mechanisms, principally in the contactor stage. In one type of dry scrubber, the mixing of gas with alumina occurs in a fluidized bed,

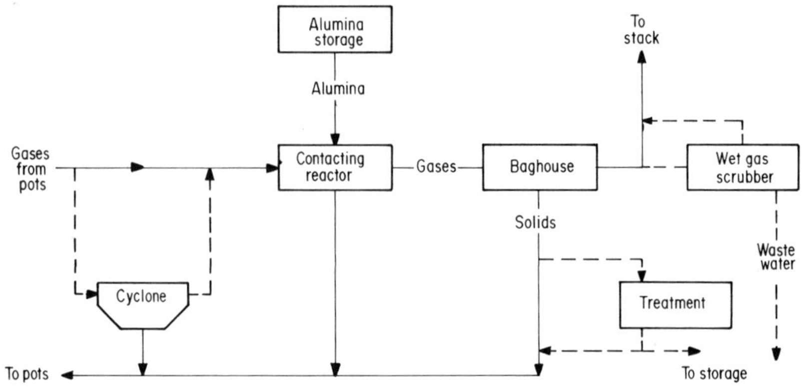

Figure 5.6. *Diagram of dry gas-scrubbing process elements. Dashed lines indicate alternative arrangements (after Battelle 1973).*

with the fabric filters, or a baghouse at the top of the same chamber. In another design, the air is blown upward through a venturi throat, into which alumina is injected downward. The solid and gas mix in the throat and in the column above the throat. The gases and eluted solids are sent to the baghouse. In the third design, the collected gases are drawn rapidly through a horizontal duct as alumina is injected down into the moving gases. Turbulent mixing of gases and solids occurs; the gases subsequently are drawn through a baghouse.

All three designs are reported to be in operation on a commercial scale with proven applications on prebake anode and vertical stud Soderberg anode cells. No production-scale application to horizontal stud Soderberg-type operations exists in the United States, although one producer reported to Battelle that a test unit serving several pots of this type is being evaluated.

The dry scrubbing system depends on the adsorption of fluoride compounds on the alumina. Operation is strongly dependent on the surface area of the alumina and the exposure or contact time. Adsorption decreases rapidly after the formation of the first layer.

Companies surveyed by Battelle indicated that dry scrubbing is being installed in the United States on both new and existing plants. When a dry system is properly operated with efficient hooding, relatively stringent atmospheric emissions limits were reported to be satisfied, without the use of water. Thus, the dry scrubbing process is of major significance to water pollution control at primary aluminum smelters.

Battelle emphasizes that water from wet scrubbers can be renovated for continuous return to the wet scrubber and that dissolved fluoride in primary potline and secondary pot room wet scrubber water can be precipitated and settled. Treatment reduces the suspended solids and oil and grease content at the same time. Fluoride is removed from the scrubber effluent by precipitation as cryolite or as calcium fluoride. For cryolite precipitation, sodium aluminate (or NaOH and hydrated alumina) is added; lime slurry is used for calcium fluoride precipitation. After precipitation, the slurry is sent to clarifiers or thickeners.

The treatment of wet scrubber liquors to recover cryolite removes sufficient fluoride to permit recycle of the treated effluent. The fluoride is in a form that usually can be returned to the aluminum electrolysis bath. The value of the recovered cryolite constitutes a cost credit to treatment. However, sulfate buildup has been demonstrated to prevent complete recycle of the liquor. Sulfur impurities derive from coke and pitch used in anode preparation; they are converted to sulfur oxides during electrolysis and are collected in the scrubber water as sulfates. If all the liquor were recycled, the solubility of sodium sulfate would be exceeded. Therefore, a bleed must be maintained to keep the sulfate concentration below saturation. This bleed stream is relatively low in volume, but contains the major portion of the fluoride effluent from the entire plant. The recycle system consists of utilizing the clarified overflow from the thickener tanks as the scrubbing medium (shown in Figure 5.7). The liquor from the scrubber, containing about 0.5 g/l fluoride, is reacted with sodium

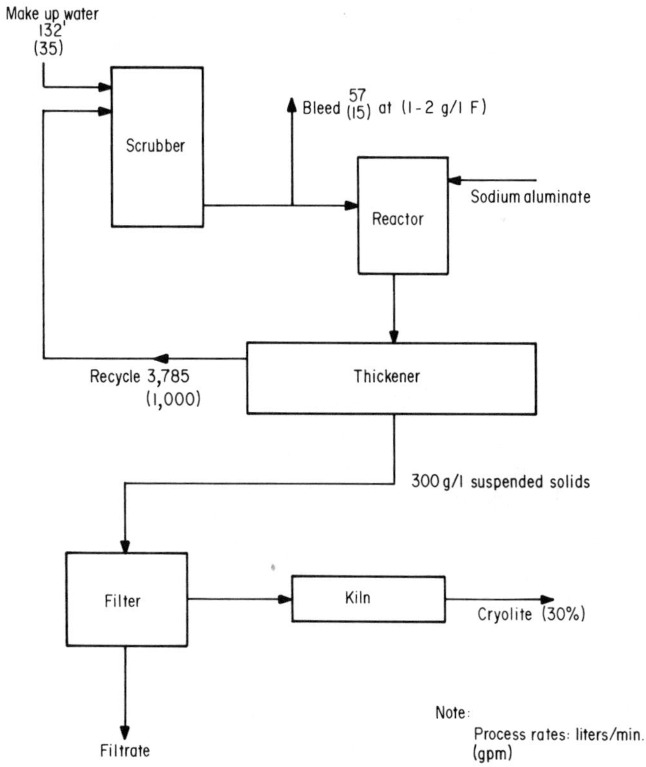

Figure 5.7. *Process schematic recycle system for fluoride removal–250 ton/day aluminum (after Battelle 1973).*

aluminate to form cryolite. This stream is sent to the thickener for suspended solids removal. These suspended solids contain cryolite (about 30 percent by weight), carbon (5 to 15 percent), and other insoluble solids which have been collected by the wet scrubbers. The supernatant clarified liquor from the thickener is recycled to the scrubbers. The slurry from the bottom of the thickener tank at a solids concentration of from 200 to 500 g/l is filtered to remove some of the liquid, and the cake (about 60 percent solids) is dried in a kiln or multiple hearth furnace. If the cryolite is pure enough, it can be returned to the reduction cells; if not, it is landfilled as a solid waste. In some cases, the filtrate is treated to precipitate fluoride as CaF_2. A recycle stream gives better control than a once-through system (0.5 to 1.5 kg/metric ton [1 to 3 lb/ton] of Al versus about 10 kg/metric ton [20 lb/ton]). Solids settle out faster as the concentration of the suspension increases (Battelle 1973).

Fluorides can also be removed from streams that cannot be recycled by the use of aluminum sulfate or by adsorption onto activated alumina. The reader should refer to Culp and Stoltenburg (1958) and to Zabban and Jewett (1967) for details of each method. Battelle provides a limited summary of these methods. Both produce solid wastes.

Oil and grease are present in the waste water as a result of baking of the anode. About one-half to two-thirds of the oil reportedly is adsorbed onto various precipitated solids—thus the thickening operations can be considered as a partial means of control.

Proposed Best Practicable Control Technology Currently Available

Battelle (1973) used the foregoing information to conclude that the best practicable control technology currently available for wet effluents from the primary aluminum industry is the treatment of wet scrubber water and other fluoride-containing effluents to precipitate the fluoride, followed by settling of the precipitate and recycling of the clarified liquor to the wet scrubbers. The two precipitation methods currently available are cryolite precipitation and precipitation with lime. Both are considered technologically sound.

The technology for cryolite precipitation as applicable to the primary aluminum industry is described by Battelle as:

1. Segregation of fluoride-containing waste waters for treatment. These include potline scrubber water, pot room scrubber water, anode bake plant scrubber water, used-cathode disposal liquor or runoff from used-cathode storage area, and storm-water runoff if contaminated with fluoride.
2. Recycling clarified liquor after precipitation of cryolite. Total recycle is not possible. A bleed from the system is required to prevent sulfate buildup in the recycled liquor.
3. Minimizing the volume of the bleed stream. This assures that the quantity of pollutants discharged does not exceed the effluent limitations.
4. Providing a holding pond or lagoon, if necessary, to further settle solids in the bleed stream. Aeration of the lagoon may be necessary to oxidize oil and grease.

Lime precipitation is considered to require:

1. Segregation of fluoride-containing waste waters as listed for precipitation of cryolite.
2. Recycling clarified liquor after precipitation of calcium fluoride with a bleed as necessary to maintain the quality of the recycle stream.
3. Minimizing the volume of the bleed stream.
4. Providing a holding pond or lagoon, if necessary, to minimize the discharge of suspended solids, and oil and grease.

On the basis of its investigation, Battelle concluded that alternative control technologies which can be employed are dry fume scrubbing, total impoundment, and reuse of effluent by a companion operation.

Dry fume scrubbing of primary potline gases eliminates the major sources of water pollutants from the primary aluminum plant. However, it was concluded that dry fume scrubbing is still developmental for certain Soderberg anode-type installations and anode bake plant gases; therefore, it is not considered feasible for all primary aluminum smelting plants at present. But where applicable, dry fume scrubbing constitutes exemplary control practice.

One plant reported to Battelle that it achieves zero discharge of waste water pollutants through total impoundment of aqueous wastes and one primary aluminum plant reported that it currently achieves zero discharge of pollutants by sending all effluent water to a nearby plant for use as makeup water.

Table 5.14 presents the water quality guidelines which Battelle concluded are possible to achieve with implementation of the best practicable control technology currently available.

Table 5.14 *Recommended Loading Rates Achievable under Best Practicable Control Technology Currently Available*

	Effluent limitations			
	Single-day maximum[1]		*30-day average*[2]	
Effluent characteristic	*(kg/kkg Al)*	*(lb/ton Al)*	*(kg/kkg Al)*	*(lb/ton Al)*
Fluoride	2	4	1	2
Suspended solids	3	6	1.5	3
Oil & grease	0.5	1	0.25	0.5
Cyanide	0.01	0.02	0.005	0.01
pH range 6-9				

Source: After Battelle 1973.

Note: Effluent limitations are defined as kilograms of pollutant per metric ton of aluminum produced or pounds of pollutant per short ton of aluminum produced.

1. The single-day maximum is the maximum value for any one day.
2. The 30-day average is the maximum average of daily values for any 30 consecutive days.

Proposed Best Available Technology Economically Achievable

Battelle (1973) concluded that application of the best practicable control technology currently available results in a relatively low-volume, high-concentration bleed stream and that the best available technology economically achievable is the lime or cryolite precipitation treatment of such a bleed stream to reduce the discharge of fluoride. In order to implement this technology, the following procedures were recommended:

1. Restriction of the volume of fluoride-containing effluent to be treated to approximately 5,000 liters/metric ton of aluminum (1,200 gallons/short ton) and treating the stream with lime or calcium chloride to reduce the fluoride concentration to a final value of approximately 10 mg/l.
2. Alternatively, volumes as high as 50,000 liters/metric ton of aluminum (12,000 gallons/short ton) can be treated to a final fluoride concentration of 1 mg/l. Treatment to 1 mg/l final concentration was considered to require adsorption methods that are not state-of-the-art methods in the primary aluminum industry but could be adapted from related water conditioning applications.

Table 5.15 presents the concomitant recommended loadings subject to best available technology economically achievable.

New Source Technology

Battelle (1973) concluded that new source technology should consist of dry scrubbing of potline air, control and treatment of fluoride-containing waste streams by recycle and treatment of any necessary bleed stream by lime precipitation, and the treatment of cast house cooling water and other streams, as required, for oil and grease removal with a gravity separator or aerated lagoon. The standards of performance attainable through the application of this technology are presented in Table 5.16.

Cost

Present Practices

Battelle (1973) presents data on capital investment and on operating costs for present control practices as obtained from selected aluminum companies. Battelle modified the data in the following way to put all on a common basis:

1. The capital investment reported was changed to 1971 dollars by the use of the Marshall and Steven's Index.*
2. The operating cost was recalculated to reflect common capitalized charges. To do this, the annual operating cost was calculated as follows:
   a. Operating and maintenance: as reported by the aluminum companies.

* Quarterly values of this index appear in *Chemical Engineering*.

b. Depreciation: 5 percent of the 1971 capital.
c. Administrative overhead: 4 percent of operating and maintenance.
d. Property tax and insurance: 0.8 percent of the 1971 capital.
e. Interest: 8 percent of the 1971 capital.
f. Other: as reported by aluminum companies.

The following procedures were used to estimate capital and operating costs of other processes which could be applied to water treatment. Equipment costs were estimated from data in Peters and Timmerhaus (1968) and Eckenfelder (1970). The total capital investment was calculated as this cost plus:

Component	Percent of equipment cost
Installation	50
Piping	31
Engineering	32
Electrical services	15
Contractor's fee	5
Contingency	10

The operating cost was calculated by estimating labor and raw material requirements, and adding the following items:

Component	Percent of investment cost
Maintenance	5.0
Depreciation	5.0
Tax and overhead	0.8
Interest	8.0

Table 5.15 *Effluent Limitations Attainable by Implementation of Best Available Technology Economically Achievable*

Effluent characteristic	*Single-day maximum*[1]		*30-day average*[2]	
	(kg/kkg Al)	*(lb/ton Al)*	*(kg/kkg Al)*	*(lb/ton Al)*
Fluoride	0.1	0.2	0.05	0.1
Suspended solids	0.2	0.4	0.1	0.2
Oil & grease	0.03	0.06	0.015	0.03
Cyanide	0.01	0.02	0.005	0.01
pH range 6-9				

Source: After Battelle 1973.

Note: Effluent limitations are defined as kilograms of pollutant per metric ton of aluminum produced or pounds of pollutant per short ton of aluminum produced.

1. The single-day maximum is the maximum value for any one day.
2. The 30-day average is the maximum average of daily values for any 30 consecutive days.

Table 5.16 *Effluent Limitations for New Sources*

Effluent characteristic	*Single-day maximum*[1]		*30-day average*[2]	
	(kg/kkg Al)	*(lb/ton Al)*	*(kg/kkg Al)*	*(lb/ton Al)*
Fluoride	0.05	0.1	0.025	0.05
Suspended solids	0.1	0.2	0.05	0.1
Oil & grease	0.03	0.06	0.015	0.03
Cyanide	0.01	0.02	0.005	0.01
pH range 6-9				

Source: After Battelle 1973.

Note: Effluent limitations are defined as kilograms of pollutant per metric ton of aluminum produced or pounds of pollutant per short ton of aluminum produced.

1. The single-day maximum is the maximum value for any one day.
2. The 30-day average is the maximum average of daily values for any consecutive 30 days.

The economic data obtained from the procedures are summarized in Table 5.17.

Potline (Primary) Gas Scrubber Water: It was concluded that the effluent from gas scrubbers on the potline can be controlled by dry scrubbers which eliminate water use, or by wet scrubbers whereby the scrubber water is recycled with precipitation of the fluoride picked up by the scrubber. This latter method requires a small bleed stream of 10 to 20 gpm of effluent water.

In the Battelle survey, three companies quoted a total investment for conversion to dry scrubbing in the range of \$36 to \$112 per annual ton of aluminum produced. Nielsen and Kielback (1972) report a capital cost of \$30 to \$40/annual ton, while Cook and Swany (1973) report a 1970 cost of \$60/annual ton for primary control of prebake plants, and \$33/annual ton for vertical stud Soderberg. These costs include both the collection system and the primary removal equipment. Battelle (1973) settled on \$40/annual ton as an average for its cost calculations. Operating cost data are relatively sparse for dry scrubbing because of the small percentage of plants (about 4 out of 33) utilizing dry scrubbers. Rush et al. (1973) use an operating cost of \$10.20/ton for control of prebake potline gases and a profit of \$0.55/ton for vertical stud Soderberg-type plants. The profit reflects the credit calculated for recovered fluoride. Battelle used the value of \$10.20/ton as representative of dry scrubbing operating costs because more prebake plants are in use than vertical stud Soderberg types.

Battelle obtained cost information on the use of a recycle wet scrubber system from the aluminum companies it surveyed. As shown in Table 5.17 companies H, I, and K use wet scrubbers for primary potline pollution control. Recycled solution is used for the scrubbing operation. Reported capital costs vary from \$3.03 to \$12.40/annual ton and operating costs vary from \$2.10 to \$4.53/ton. These costs include only the water control circuit which consists of the chemical addition tank, thickener, cryolite recovery equipment, and associated pumps and piping. The operating costs, in general, do not include any credit for recovered fluoride as, in general, the cryolite is not of high enough

Table 5.17 *Cost Data for Control and Treatment of Waste Water from Primary Aluminum Production*

Plant	Level in discharge			Aluminum production (ton/day)	Potline (primary)		Pot room (secondary)		Anode bake		Cast house cooling		Rectifier cooling water
	Fluoride (kg/ton)	Susp. solid (kg/ton)	Oil & grease (kg/ton)		Capital ($/annual ton)	Operating ($/ton)	Capital ($/annual ton)	Operating ($/ton)	Capital ($/annual ton)	Operating ($/ton)	Capital ($/annual ton)	Operating ($/ton)	
A	0	0.04	0.015	249	Dry scrubbing		Not used		Not used		(1.6)	(0.4)	Air
B	(d)	(d)	(d)	436	(d)	(d)	Not used		Soderberg		(d)	(d)	Generators
C	9	9	0.5	245	12.5	1.24 (2.5)[c]	Not used		0.53	0.10	Untreated		Untreated
D	0.35	0.4	0.04	220	Not used		23.2	7.06 (8.78)[c]	1.04	0.20	Untreated		–
E	2.08	3.13	0.35	655	Dry scrubbers		5.4	3.15 (3.22)[c]	0.38	0.07	Untreated		Air
F	18.6	82[b]	1.2	663	Untreated		Untreated		Soderberg		Untreated		(d)
G	8.8	10	1.3	218	2.32	(d)	Untreated		Soderberg		(d)	(d)	Untreated
H	0.53	2.65	–	598	6.06	1.45 (2.10)[c]	Not used		Not used		(d)	(d)	Untreated
I	0.35[a]	6.2[a]	0.25[a]	284	12.4	1.0 (4.34)[c]	Not used		Soderberg		Untreated		Generators
J	1.06	1.62	0.22	455	–	–	Not used		Soderberg		Untreated		Untreated
K	1.20	1.90	0.24	259	3.03	4.33 (4.53)[c]	Included with potline		Not used		(0.43)[c]	(0.06)[c]	With cooling water

Source: After Battelle 1973.

Note: Ton is metric. Not used: No wet-type pollution control device in use. Untreated: A wet scrubber is used for air pollution control, but the water is discharged untreated.

a. Recycled to adjacent alumina plant.
b. 65 from cryolite plant.
c. Cost estimated, either to put costs on a common basis or on equipment descriptions obtained.
d. Insufficient information obtained to permit a cost estimate.

quality to be recycled to the potline; it is landfilled as a solid waste. One company did calculate a credit equal to about $2/ton of aluminum produced.

Pot Room (Secondary) Gas Scrubber Water: Five of the 11 companies surveyed by Battelle (1973) currently practice air pollution control of pot room air. Of these five, three (D, E, and K) reported practicing water control. Cost information was contributed to the survey by two companies, D and E. Plant D reported a capital cost of $23.20/annual ton with an operating cost of $8.80/ton. These costs include the recycle water circuit only and consist of chemical addition, thickening, filtration, and kiln operation on the recovered cryolite, and a treatment operation using $CaCl_2$ to precipitate fluoride from the cryolite filtrate stream. No credit for recovered cryolite was taken in operating costs. In the second case (plant E), capital costs were reported to be $5.40/annual ton capital, and operating costs were reported to be $3.15/ton. These costs also include only the recycle water control circuit. However, in this case the cryolite filtrate stream was not heated.

Anode Bake Plant: Water control by recycling water through the anode bake plant scrubber is not practiced, according to reports. However, one plant reported that it runs its bake plant at an acceptable air emission level without wet scrubbers on the exhaust gases by a sophisticated control of anode firing and new flues in the exhaust. Costs are indirectly incurred in labor, improved operating procedures, or plant improvements.

Cast House Cooling: Cooling towers are used to control effluent water from the cast house ingot cooling operation. A portion of the cooling water is bled off in order to prevent the buildup of dissolved and suspended solids, as well as the oil and grease. One company (plant A) reported treating the bleedstream (150 gpm) in an aerated lagoon with a 15-day retention time, reducing the hydrocarbon content by 95 percent. The cost of performing this operation, which includes the cooling tower, lagoon and associated piping, engineering, and services, was calculated to be $1.60/annual ton capital and $0.40/ton operating (Battelle, 1973).

Costs of Additional Treatment Processes

As mentioned previously, Battelle (1973) concluded that dry scrubbing of primary air system can result in no discharge of pollutants and that additional control measures can be added to the wet scrubbing processes to effect a reduction in the amount of pollutants discharged to the waste streams.

Potline Scrubber Water Treatment: It was concluded that the choice of additional treatment schemes to treat scrubber water effluent depends primarily on whether a recycle system or once-through system is in use (see Table 5.18 for a summary of additional treatment costs). In a recycle system, additional control of fluorides, suspended solids, and oil and grease can be realized by treatment of the filtrate stream from the cryolite and the bleed stream from the scrubber with time on each. Costs for this treatment were calculated to be $1.50/annual ton capital and $0.64/ton operating. It was observed that the low volume of effluent to be treated (about 120 liters/minute [30 gpm]), and high concentration of fluoride (about 1,000 mg/l) account for the low cost.

Table 5.18 *Costs of Various Alternatives for Fluoride Removal*

Process alternative	*Discharge fluoride, (kg/ton)*	*Capital cost, ($/annual ton)*	*Operating cost ($/ton)*
Dry scrubbing	0	$40	$10.2
Wet scrubbing–once-through	5	7.4	2.5
Wet scrubbing–recycle	1	10	4.6
Recycle with bleed and filtrate treatment	0.05	11.5	5.2
Once-through and alum treatment	1	18.3	11
Once-through and activated alumina treatment	0.25	9.7	6.3
Once-through and hydroxylapatite treatment	0.25	21.9	16.5
Once-through and reverse osmosis treatment	0.8	–	24.5

Source: After Battelle 1973.

Note: Ton = metric ton; values are 10 percent lower for short ton.

The addition of treatment for the effluent from a once-through potline and pot room scrubber after lime treatment (if practiced) was reported to be more costly than the previous treatment of recycle effluents because of large volumes of water with low concentrations of fluorides and other pollutants.

For the purposes of cost calculations, Battelle (1973) assumed a plant size of 227 metric tons/day (250 short tons/day). The flow rate of water from the once-through scrubbing system was assumed to be 83,300 liters/ton (20,000 gal/short ton) with a concentration of 35 mg/l fluoride.

Alum Treatment: Battelle (1973) calculated that the addition of alum treatment would add about $11.00/annual ton capital and $8.40/ton operating, including a mixing tank, flocculant tank, clarifier, and pumps. A 37-meter (121-foot) diameter clarifier accounted for 84 percent of the $372,000 equipment cost. Alum at a rate of $69/ton accounted for about 69 percent ($700,000/year) of the operating cost.

Activated Alumina: Battelle (1973) calculated the costs for the activated alumina adsorption process to be $2.70/annual ton capital and $3.80/ton operating. Capital costs included two alumina adsorption towers (one for adsorption, one for regeneration), a mixer for the treatment of the spent reagent solution, a thickener, the initial charge of alumina, and associated pumps and piping. The cost of the sulfuric acid used for regeneration was 50 percent of the $313,000/year operating cost.

Regulations Promulgated for Primary Aluminum Smelting

On the basis of the information presented by Battelle (1973) and reviews thereof by industry and public agencies, the U.S. Environmental Protection Agency promulgated the following regulations for primary aluminum smelting by the Hall-Heroult process. Exceptions for special cases can be granted.*

* See *Federal Register*, Vol. 39, No. 68, Part II, April 8, 1974, pp. 12821-12826.

Effluent Limitations Guidelines Representing the Degree of Effluent Reduction Attainable by the Application of the Best Practicable Control Technology Currently Available

The limitations establishing the quantity or quality of pollutants or pollutant properties which may be discharged by a point source after applying the best practicable control technology currently available are shown in Table 5.19.

Effluent Limitations Guidelines Representing the Degree of Effluent Reduction Attainable by the Application of the Best Available Technology Economically Achievable

The limitations establishing the quantity or quality of pollutants or pollutant properties which may be discharged by a point source after application of the best available technology economically achievable are shown in Table 5.20.

Standards of Performance for New Sources

The standards of performance establishing the quantity or quality of pollutants or pollutant properties which may be discharged by a new source are shown in Table 5.21.

SECONDARY ALUMINUM SMELTING (SIC 3341)

Sources of Data and Information

Battelle (1973) has conducted, as part of Contract 68-01-1518, a survey of waste disposal and production by the secondary aluminum smelting industry. Sources of information for the survey included pollution abatement offices, representatives of 61 of the 85 United States companies, the Aluminum Recycling Association, and the U.S. Clean Water Committee. This report is the only one of its kind in existence; hence it is heavily utilized as a source of information for this section. Other sources of information include the U.S. Bureau of Mines and the published literature.

Description of Processes

Secondary aluminum smelting recovers, remelts, and processes aluminum scrap (as opposed to alumina) to produce aluminum metal or alloy. The four operations are:

1. Collecting, sorting, and transporting.
2. Presmelting preparation.
3. Charging, smelting, and refining.
4. Pouring of the product line.

Figure 5.8 shows a generalized flowsheet of a secondary aluminum operation displaying these four categories. Battelle (1973) observed that collecting, sorting, and transporting use no water; therefore no effluent is produced. The

Table 5.19 *Effluent Limitations Promulgated for Primary Aluminum Smelting*

Effluent characteristic	*Effluent limitations*			
	Maximum for any one day	*Average of daily values for 30 consecutive days shall not exceed*	*Maximum for any one day*	*Average of daily values for 30 consecutive days shall not exceed*
	Metric units (kg/kkg of product)		*English units (lb/1,000 lb of product)*	
F	2.0	1.0	2.0	1.0
TSS	3.0	1.5	3.0	1.5
pH	Within the range of 6.0 to 9.0			

Note: Effluent limitations guidelines representing the degree of effluent reduction attainable by the application of the best available technology economically achievable.

Table 5.20 *Effluent Limitations Promulgated for Primary Aluminum Smelting*

Effluent characteristic	*Effluent limitations*			
	Maximum for any one day	*Average of daily values for 30 consecutive days shall not exceed*	*Maximum for any one day*	*Average of daily values for 30 consecutive days shall not exceed*
	Metric units (kg/kkg of product)		*English units (lb/1,000 lb of product)*	
F	0.1	0.05	0.1	0.05
TSS	0.2	0.1	0.2	0.1
pH	Within the range of 6.0 to 9.0			

Note: Effluent limitations guidelines representing the degree of effluent reduction attainable by the application of the best available technology economically achievable.

Table 5.21 *Effluent Limitations Promulgated for Primary Aluminum Smelting*

Effluent characteristic	*Effluent limitations*			
	Maximum for any one day	*Average of daily values for 30 consecutive days shall not exceed*	*Maximum for any one day*	*Average of daily values for 30 consecutive days shall not exceed*
	Metric units (kg/kkg of product)		*English units (lb/1,000 lb of product)*	
F	0.05	0.025	0.05	0.025
TSS	0.1	0.05	0.1	0.05
pH	Within the range of 6.0 to 9.0			

Note: Standards of performance for new sources.

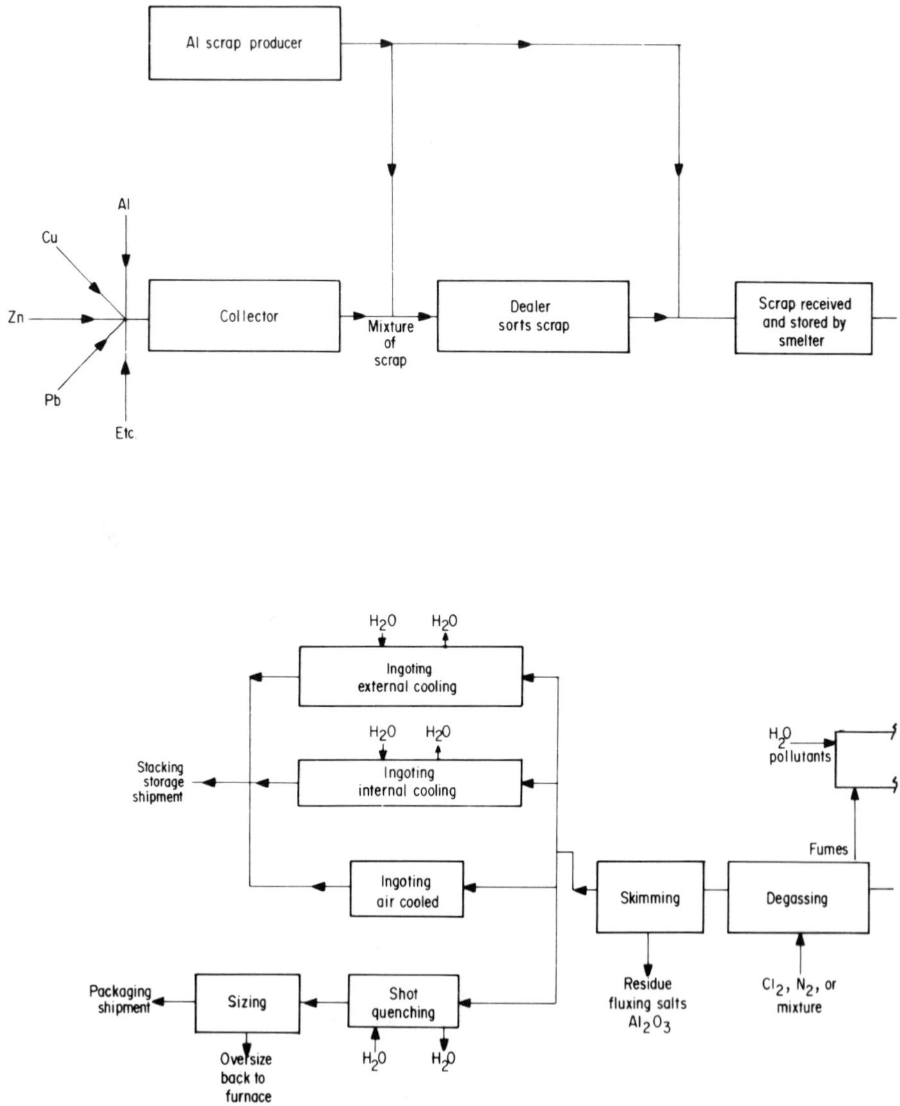

Figure 5.8. *The total secondary aluminum process (after Battelle 1973).*

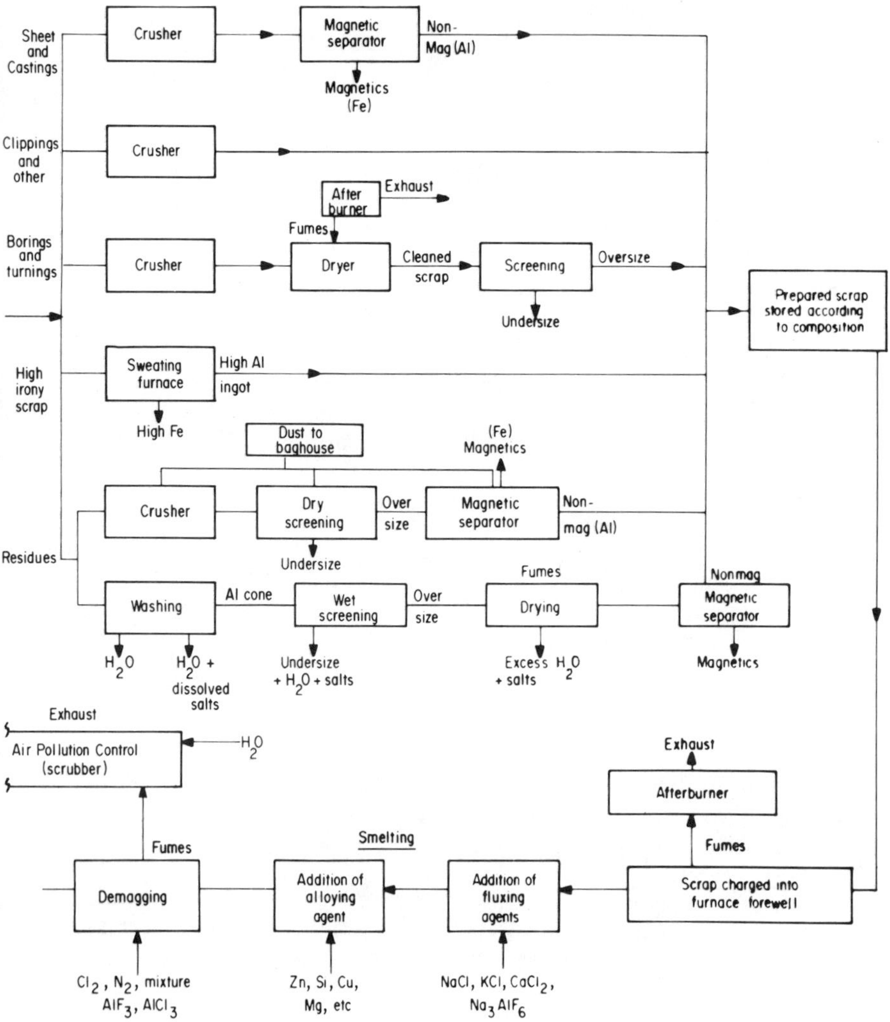
Sheet and Castings
Crusher
Magnetic separator
Non-Mag (Al)
Magnetics (Fe)
Clippings and other
Crusher
After burner
Exhaust
Fumes
Borings and turnings
Crusher
Dryer
Cleaned scrap
Screening
Oversize
Undersize
Prepared scrap stored according to composition
High irony scrap
Sweating furnace
High Al ingot
High Fe
Dust to baghouse
(Fe) Magnetics
Crusher
Dry screening
Over size
Magnetic separator
Non-mag (Al)
Undersize
Residues
Washing
Al cone
Wet screening
Over size
Fumes
Drying
Nonmag
Magnetic separator
H_2O
H_2O + dissolved salts
Undersize + H_2O + salts
Excess H_2O + salts
Magnetics
Exhaust
Air Pollution Control (scrubber)
H_2O
Exhaust
Afterburner
Fumes
Smelting
Fumes
Demagging
Addition of alloying agent
Addition of fluxing agents
Scrap charged into furnace forewell
Cl_2, N_2, mixture AlF_3, $AlCl_3$
Zn, Si, Cu, Mg, etc
NaCl, KCl, $CaCl_2$, Na_3AlF_6

scrap raw materials are either solids or residues. Solids are considered to include borings and turnings, new clippings and forgings, old castings and sheet, and aluminum containing iron. Residues are considered to include dross and skimmings from foundry melting operations, from fabricators and from the primary aluminum industry, and slag from secondary smelters. Secondary smelters recycle a metal that otherwise would have to be discarded in a solid waste. This also conserves energy because only 5 percent of the energy needed to produce virgin aluminum is required to produce an equal amount of secondary aluminum.

Of the 69 United States secondary aluminum smelters surveyed by Battelle, 23 reported processing residues. In addition to 10 to 30 percent metallic aluminum, residues contain oxides, carbides, fluxing salts, and other contaminants. Liberation of aluminum is accomplished by either wet or dry processes.

In dry processing, the scrap is crushed, then screened to remove fines; iron and steel are removed magnetically. Dust evolves and constitutes a source of air pollution which is usually controlled by baghouses, but some plants report practicing wet dust collection. The effluent from the wet dust collectors used for air pollution control may become a source of water pollution.

Of the 23 United States plants processing residues, six reported to Battelle that they use wet techniques whereby the raw material is first fed into a long rotating drum. Water is passed through the drum to wash and flush the feed and liberate the aluminum. The washed material is screened, dried, and separated magnetically and the nonmagnetics are sent to the smelter. Fine particulates and dissolved salts are sources of water pollution. The quality of four residue processing waste waters is presented in Table 5.22. Battelle elected to specify suspended solids, pH, and ammonia as critical pollutant parameters from wet residue processes.

Smelting

The smelting of segregated aluminum scrap with reverberatory furnaces consists of charging scrap into the furnace, addition of fluxing agents and alloying agents, mixing, removal of magnesium (demagging), degassing, and skimming. A smelter may not incorporate all seven steps and may not follow the above order.

Charging, fluxing, alloying, and mixing are dry processes with only minimal air pollution problems which do not employ wet scrubbers. Battelle (1973) concluded that alloying and mixing produce no solid wastes; charging and fluxing do but these processes are not sources of water pollution. Magnesium removal, skimming, and pouring and cooling, on the other hand, were observed to produce contaminated effluents.

Magnesium Removal (Demagging)

Scrap aluminum is reported to average about 0.3 to 0.5 percent magnesium, while alloys produced average about 0.1 percent magnesium. Therefore, after the furnace is charged and the melt brought to the desired chemical composition, it is necessary to remove magnesium. Chlorine or chlorinating agents, such as anhydrous aluminum chloride or chlorinated organics, or aluminum fluoride are used for demagging. Demagging constitutes a major source of air pollution, which in turn is converted to a water pollution problem if wet scrubbers are utilized.

Chlorination, the preferred demagging method, is conducted at temperatures between 760°C and 816°C (1,400°F and 1,500°F). The reaction requires about 3.5 kg per kg of magnesium removed. Elemental chlorine gas is fed to the bottom of the melt. It rises through the melt and reacts with magnesium and aluminum to form chlorides which float to the surface and combine with fluxing agents to be skimmed off. Magnesium reduces aluminum chloride in the melt. Early in the demagging cycle, the principal reaction product is magnesium chloride. As magnesium is removed and is less available for reaction with chlorine, the reaction of chlorine with aluminum becomes more significant; the reduction of the aluminum chloride by magnesium diminishes; and the production of aluminum chloride, a volatile compound, occurs. The aluminum chloride evolves as fumes, making ventilation and air pollution control equipment necessary. Control of the air polluting fumes is frequently done by wet scrubbing which creates a source of water pollution.

The character of the effluent from wet scrubbing of fumes from the chlorination process is presented in Table 5.23. Cadmium, copper, and pH would be considered problem parameters.

When aluminum fluoride is employed as the demagging agent, it reacts with the magnesium to form magnesium fluoride which in turn combines with the flux on top of the melt where it is skimmed off. About 4.3 kg of aluminum fluoride are required per kg of magnesium removed. Air pollutants consist of gaseous fluorides or fluoride dusts which are controlled by either dry or wet scrubbing.

Battelle (1973) elected to specify pH, and total suspended solids as the critical pollutant parameters for fume scrubbing (wet scrubbers) for air pollution control of the demagging process.

Skimming

The impure semisolid fluxing agent known as slag or dross is removed from the surface of the melt, usually with a perforated ladle that permits molten metal to pass. The slag is placed in pans to cool or it is cooled in an internally water-cooled "dross cooler."

Once cooled, the slag is either stored until shipped to a residue processor, reprocessed by the company, or is dumped as a solid waste. Battelle (1973) concluded that, if stored in the open, the slag is a source of ground and surface water contamination because of abundant soluble salts (NaCl, KCl, $MgCl_2$). Dross cooling also generates fumes which are a source of air pollution. The cooling, as well as reactions in the smelting, produce nitrides and carbides of aluminum which, after reacting with water or water vapor in the air, release hydrocarbons and ammonia to the atmosphere. The ammonia also may become a component of water pollution.

Table 5.22 *Character of Selected Waste Water from Residue Processing*

	Plants					
	D-3[a]	*D-6*[b]	*D-4*[c]		*D-8*[d]	
Parameter	*Loading (kg/mton Al)*	*Conc. (mg/l)*	*Conc. (mg/l)*	*Loading (kg/mton)*[e]	*Conc. (mg/l)*	*Loading (kg/mton)*[e]
Alkalinity	6.47	314	586	102	500	–7.5[f]
COD	0.97	2,045			29	0.17
Total solids	–	–	24,264	5,144	17,800	326
Total dissolved solids	13.51	12,920	–	–	17,400	324
Total suspended solids	0.121	4,961	15	1.5	159	–5.6
Sulfate		1,100	47	1.5	151	1.8
Chloride	0.319	6,492	15,465	3,264	8,903	150
Cyanide		0.04			0.05	0
Fluoride	0.129	2.9	8.7	1.81	16.5	0.38
Ammonia	0.33	0.75	350	73	0.30	–0.03
Aluminum	0.002	0.3	16.4	3.5	28	–1.49
Calcium		58.8	23	–7.4	48	0.17
Copper	<0.001	0.174	0.070	0.008	0.137	0.003
Magnesium	–	32.5	6	3.9	76	1.39

Nickel	–	1.2	0.240	0.009	0.20	0
Sodium	–	2,560	11,600	2,528	3,103	46.2
Potassium	–	1,087	6,470	1,407	4,302	102
Zinc	0	0.015	0.10	0	0.198	–9.1
Cadmium	–	0.05	0.002	0	0.005	–0.001
Lead	–	0.20	0.020	0.004	0.028	–0.001
Manganese	–	0.16	0.045	0.002	0.060	0
Chlorine residue	–		–	–	–	–
Oils & grease	0.053	55.4	0	0	0.5	0
Phenols[g]		–	–	–	0.03	0
pH	8.68	8.3	9.09		9.2	
Nitrates	0.032					

Source: After Battelle 1973.

a. Calculated from U.S. Corps of Engineers; concentration data not given.

b. From residue milling solid waste washing; tonnage values of residue waste processed not available; loading cannot be calculated. Water flow is 151 l/minute.

c. Data from 7-month and 9-month average and verification data from state: metals verified composite of 18 samples collected over a period of 6 days.

d. Represents composite of 9 samples collected over 3 days. Milling waste stream is blended with scrubber waste stream.

e. Loading calculated as:

$$[\text{conc. effluent (mg/l)} - \text{conc. intake (mg/l)}] \times \frac{\text{quantity of water used (l)}}{\text{quantity of Al recovered from residue (mton)}}$$

f. Negative values indicate that the process reduced the concentration of this parameter and are derived from reported analytical values.

g. In parts per billion.

Table 5.23 *Character of Waste Water from Chlorination Fume Scrubbing – No Treatment*

	C-7[a]		*D-6*[b]	
Parameter	*Conc. (mg/l)*	*Loading (grams/kg of Mg)*[c]	*Conc. (mg/l)*	*Loading (grams/kg of Mg)*[c]
COD	123	12.1	536	95.8
Total solids	2,910	301	–	–
Total dissolved solids	1,885	194	10,500	1,856
Total suspended solids	225	22.3	480	83.0
Sulfate	11	0.51	481	84.4
Chloride	4,420	446	8,671	1,560
Cyanide	<0.02	0	–	–
Fluoride	0.24	–0.08[d]	0.7	–0.324
Aluminum	472	50.9	6.12	0.615
Calcium	0.12	–0.215	990	176
Copper	0.25	0.02	1.31	0.236
Magnesium	41.2	3.86	55.8	9.81
Nickel	0.050	0.003	0.74	0.106
Sodium	3.11	–0.007	770	32.7
Potassium	–	–	206	37.1
Zinc	0.952	0.091	3.58	0.64
Cadmium	0.066	0.006	0.30	0.054
Lead	0.061	0.004	0.24	0.025
Manganese	0.449	0.049	2.34	0.349
Chlorine residue	0.257	0.027	–	–
Oils & grease	13.9	0.590	6.24	0.403
Phenols[f]	20.7	–0.002	–	
pH	2.1	–	1.0	–

Source: After Battelle 1973.

a. Average of three composite samples.
b. Average of five composite samples.
c. Loading calculated as:

$$[\text{conc. effluent (mg/l)} - \text{conc. intake (mg/l)}] \times \frac{\text{quantity of water used (l)}}{\text{quantity of Mg removed (kg)}}$$

d. Negative numbers indicate that the process apparently reduced the concentration of this parameter, and are derived from the reports of analytical results.
e. Analytical methods from *Standard Methods for the Examination of Water and Waste Water*, Water Pollution Control Federation 13th Ed. 1971.
f. In parts per billion.

Pouring and Cooling

After a furnace has been charged, the specified composition attained by blending and demagging, and the melt degassed and skimmed, the liquid metal is cooled to about 732°C (1,350°F) for pouring to produce ingots or other products. Ingots frequently are cooled with water sprayed on both molds and hot metal as they travel along a conveyor track above a casting pit. Ingots may also be cooled by forcing water through passages in the mold, in which case water does not contact the hot aluminum metal. In some cases, the molds are cooled during passage through a cooling tunnel where cooling occurs via a water mist-air mixture which generates no waste water. Eleven of 69 plants canvassed by Battelle (1973) reported that they currently air-cool their ingots.

Another product known as shot, a deoxidant, is produced by the industry. Shot is manufactured by pouring molten metal onto a vibrating feeder with openings in the bottom which allow the molten metal to drop through into a water bath. The droplets solidify in the water and are dried, sized, and packed for shipment. The oversize shot is recharged into the furnace.

Battelle noted that cooling water may be totally evaporated, sent to a cooling tower and recirculated, or treated or discharged after one or two uses. Recirculation requires removal of sludge from the cooling tower and pit, which produces a solid waste. Sludges are usually landfilled.

The characteristics of the cooling water reported to Battelle by one plant are presented in Table 5.24. Suspended solids, phenols, oils and grease, lead, manganese, and chemical oxygen demand (COD) are considered to be the most detrimental constituents present. Battelle elected to specify lead, manganese, oil, and grease and total dissolved solids as critical pollutant parameters. Alkalinity, fluoride, calcium, magnesium, nickel, ammonia, and sulfate were rejected as critical pollutant paramenters because they were not added to the stream by the plants inspected.

The sources of sludge in cooling water were observed to include airborne solids from ambient air during spray cooling, buildup of hydrated alumina from chemical reaction with the molten aluminum, and debris and dust from the plant floor. Flux salt buildup (NaCl) occurs in recirculated water used for shot cooling.

Proposed Control and Treatment of Waste Water

On the basis of the previous discussion, Battelle (1973) concluded that the three main processes producing polluted effluents are metal cooling, fume scrubbing, and residue processing. It was concluded also that current practice for the treatment of residue milling waste water can control only the amounts of suspended solids, pH, fluoride, heavy metals, COD, and ammonia and that treatment for only aluminum, magnesium, copper, nickel, zinc, cadmium, lead, and manganese is practicable for fume scrubber water. No mention is made of pollutants subject to effluent limitations for cooling water, but presumably at least pH control would apply also.

Waste Water from Metal Cooling

As noted in Table 5.24, the major pollutants in the waste water generated during the cooling of ingot molds and shot containing molten alloy are oils and greases and suspended and dissolved solids. However, lead, manganese, and COD are also present. Battelle (1973) believes that the amount of waste water generated from metal ingot cooling can be minimized by recirculation and

Table 5.24 *Character of Cooling Waste Water (Plant C-7)*

Parameter	Intake water		Effluent concentrations in samples, (mg/l)				Net loadings in waste water[b] (gram/mton)		
	Conc. (mg/l)	Loading[a] (gram/mton)	6	15	16	Average	Average	Min.	Max.
Alkalinity	8	5.43	6	16	–	7.3	–	–	5.37
COD	4	2.71	1,000	365	2,252	1,206	815	245	1,525
Total solids	86	58.4	234	188	3,222	1,215	766	69.2	2,128
Total dissolved solids	73	49.5	78	118	548	248	119	3.4	322
Total suspended solids	6	4.07	102	64	2,620	929	626	39.4	1,774
Sulfate	6	4.07	10	15	21	15.3	6.3	2.7	10.2
Chloride	6	4.07	16	12	401	143	93.0	4.07	268
Cyanide	<0.02	<0.013	0.05	<0.02	<0.02	0.03	0.006	0	0.020
Flouride	1.04	0.706	0.84	1.36	0.16	0.78	–	–	0.217
Aluminum	0.01	0.007	0.01	0.08	32.0	10.7	7.26	0	21.7
Calcium	2.38	1.62	2.60	3.10	1.07	2.26	–	–	0.489
Copper	0.037	0.025	0.037	0.037	0.325	0.133	0.065	0	0.195
Magnesium	1.95	1.32	1.07	1.07	1.37	1.17	–	–	–
Nickel	<0.02	<0.014	0.043	<0.02	<0.02	0.03	0.007	0	0.016
Sodium	3.19	2.16	2.81	2.91	5.06	3.59	0.271	–	1.27
Zinc	0.031	0.021	0.231	0.038	1.555	0.61	0.393	0.005	1.03
Cadmium	<0.009	<0.006	0.009	<0.009	0.027	0.015	0.004	0	0.012
Lead	<0.026	<0.018	0.026	0.052	1.147	0.46	0.261	0	0.76
Manganese	<0.010	<0.006	0.042	0.042	0.229	0.10	0.061	0.022	0.149
Chlorine residue	<0.02	<0.014	0.02	0.02	<0.02	0.02	0	0	0
Oils & grease	6.3	4.28	255.6	64.3	5,180	1,833	1,240	39.4	3,511
Phenols[c]	30	0.020	35	3	260	99.3	0.043	–	0.156
pH	6.7		3.4	7	6.1	5.5	5.5	3.4	7

Source: After Battelle 1973.

a. $\frac{\text{(conc., mg/l) (water used, l/day)}}{\text{(metric tons poured, kkg/day)}} \times 10^{-3}$ gram/mg = loading in grams/metric ton of alloy, where water used is 39,700 l/day (average) and metric tons poured is 58.5 kkg/day (average).

b. Minus intake load.

c. In parts per billion.

cooling and that waste water discharge could be eliminated by either total consumption through regulated flow, air cooling or recirculation, except for smelters producing deoxidizer shot.

Of 58 secondary smelters canvassed which generate cooling waste waters, 15 reported that they recirculate the water continuously with no discharge. Seven others reported that they recycle the cooling water, but discharge the holding tanks periodically, usually at six-month intervals; discharge is to permit sludge removal (solid waste) from cooling towers and pits. The solid waste is disposable by landfilling or other appropriate methods as discussed in Chapter 11.

Installation of a recirculation system was observed to involve the construction of a cooling tower, possible enlargement of the cooling pit, an auxiliary holding tank so that sludge can be removed without discharge, associated plumbing, and necessary pumps. Maintenance on the recirculation system is largely due to sludge buildup.

Battelle observed that 3 out of 58 smelters generating cooling waste water have adjusted the flow rates so that the water is essentially totally evaporated by the hot ingots, thereby generating no waste water. Specially designed nozzles produce a water-mist spray that reduces the steam-to-metal interface. It was observed that this approach requires longer conveyors than does the routine water cooling process.

Of 69 secondary smelters canvassed by Battelle, 13 reported air cooling ingots. The hot ingots are conveyed through an air tunnel which is approximately twice the length of ordinary water cooling conveyors. Maintenance was noted to be higher on air-cooled systems because of the longer conveyor, the added heat load on the lubricants, and additional blower motors. In some cases a water mist is added to the air to improve the cooling rate, but the water is completely evaporated, so no effluent occurs.

Alternative treatment schemes are considered to include skimming off oil and grease, grease traps, and a combination of recirculation of cooling water and periodic treatment of it to remove solids. Solids (hydrated alumina) could go to a landfill and be handled like any other sludge.

Waste Water from Fume Scrubbing

Fumes formed during chemical magnesium removal are collected and scrubbed to reduce air emissions. The discharge from these wet fume scrubbing devices contains most of the volatile metal salts entrained in the gas flow. Aluminum chloride and magnesium chloride are the principal constituents. Battelle (1973) observed that when aluminum fluoride is used for magnesium removal, the principal volatile products are silicon tetrafluoride and hydrogen fluoride. In both cases, the air pollutants are transformed to water pollutants. Control alternatives to neutralization and sedimentation (see Table 5.25) are considered to include dry fume control via the Derham process, the Alcoa process, or the Teller process; or treatment of scrubber effluent. The following process descriptions, all oriented toward waste production, are condensed from the Battelle report. Generally the processes produce either a solid waste or a recycleable solid.

Table 5.25 *Effect of Neutralization and Settling on Scrubbing Waste Water Loading*

	Waste loadings, gram of pollutant/kg of Mg removed							
	Plant C-7 (Case I)			Plant C-7 (Case II)			Plant D-6	Plant D-8
Parameter	Before treatment	After treatment	Net effect	Before treatment	After treatment	Net effect	No treatment	Alkali treatment, no settling
Alkalinity	–	1.8	–	–	5.47	–	–	2,754
COD	20.1	6.09	–14.0	2.58	6.84	4.26	95.8	1.52
Total solids	684	999	315	77.4	450	372	–	4,864
Total dis. solids	453	710	257	42.0	382	340	1,856	3,772
Total sus. solids	45.1	284	239	16.9	66	49.1	83	1,193
Sulfate	1.03	.115	–.92	.402	.402	0	89.4	41
Chloride	775	443	–332	200	234	34	1,560	851
Cyanide	0	0	0	0	.047	.047	–	0
Fluoride	–.108	.053	.161	–.064	–.053	.011	–0.324	–0.3
Aluminum	124	66.7	–57.3	13.5	15.1	1.6	0.615	184
Calcium	–.260	–.260	0	–.182	–.182	0	176	–0.21
Copper	.024	.010	–.014	.017	.007	–.01	0.236	0.01
Magnesium	6.03	2.01	–4.02	2.86	1.30	–1.56	9.81	6.02
Nickel	.008	.008	0	.002	.005	.003	0.106	0
Sodium	.002	261	261	0	143	143	32.7	1,919
Zinc	.140	.058	–.082	.053	.036	–.017	37.1	1.36
Cadmium	.009	.005	–.004	.004	.002	–.002	0.64	0.12
Lead	.009	.009	0	0	.006	.006	0.054	–0.02
Manganese	.132	.011	–.121	.011	.006	–.005	0.025	0.01
Chlorine residue	.078	0	–.078	0	4.2	4.2	0.349	–
Oil & grease	–.242	.426	.668	1.85	3.5	1.65	–	0
Phenols[1]	–.003	–.002	.001	–.001	–.002	–.001	0.403	0.02
pH	1.7	9.2		1.7	9		1.0	9.5

Source: After Battelle 1973.

1. In parts per billion.

The Derham Process: It was concluded that the Derham process facilitates magnesium removal, with chlorine, from secondary aluminum melts with a minimum of fume generation and water use. Magnesium chloride, the reaction product of magnesium removal, is entrapped in a liquid flux. The flux subsequently is used in the melting operations. A separate bath with a special flux cover, and means to circulate the molten metal to and from that separate bath, are the principal components of the system.

The molten metal entering the treatment unit is treated with gaseous chlorine to achieve magnesium removal. By maintaining a relatively thick cover of molten salt on the bath in the treatment unit, the emissions of aluminum chloride to the atmosphere usually produced by demagging are reported to be almost completely arrested. As the flux cover becomes saturated with magnesium chloride, it is removed and may be used as a flux in the main melting furnace or it may become a solid waste.

Gaseous effluents from the treatment unit are blended with the combustion gas effluent and released to the stack. In some cases emission control requirements may be satisfied by blending the gases. In situations requiring particulate control with baghouses, the chloride emissions, although hygroscopic, are sometimes dilute enough not to impede baghouse operation, i.e., blinding.

Battelle (1973) concludes that overall performances of the Derham process generally is satisfactory in terms of meeting most air pollution restrictions, but that a backup scrubber may be necessary under stringent regulations and/or transient process conditions; however, the loading should be very low.

The Alcoa Process: The Aluminum Company of America reported to Battelle (1973) that it is evaluating a fumeless demagging process. It claims 100 percent efficiency in chlorine utilization for magnesium removal, and it recovers molten magnesium chloride as a product. Battelle notes that at present it is being used in England and for captive scrap processing and that the unit is installed between the holding furnace and a casting machine and removes magnesium continuously as the metal flows through it.

The Teller Process (In addition to Battelle 1973, see Teller 1972a): Baghouses sometimes are ineffective or inefficient in the removal of fumes from demagging operations because plugging occurs during collection of submicron particulates. These particles enter the weave and cause blinding.

The Teller process reportedly overcomes the problem of blinding by having the bags precoated with a solid to absorb effluent gases as well as particulates, supposedly without plugging. Upon saturation, the coating is removed along with the collected dust by vibration and a fresh coating is applied. If not recycleable, the collected particulates and spent coating constitute a solid waste and are discarded in a landfill. Thus the liquid state in the usual sequence gaseous waste ⟶ liquid waste ⟶ solid waste is omitted. Battelle reports that a prototype has been installed in one demagging facility where its performance is being evaluated.

Battelle noted also that the application of this system to the case of fluoride emissions from glass furnaces requires simultaneous filtration and chromatographic absorption and baghouse recovery. A chromatographic solid is injected into the gas duct which is subsequently separated from the gas in a baghouse. The solid serves as an absorbent for acid gases and as a precoat to prevent bag blinding. It repeatedly breaches rather than blocks the interstices and becomes the filter; the bag surface serves only as a support. The chromatographic material forms a monomolecular layer on a reactive carrier. In one case, the carrier cost was estimated to be \$30 per metric ton. In the absorption of hydrogen fluoride, it reportedly can provide one transfer unit in 0.0254 cm (0.01-in.) depth of the chromatographic material. With an injection rate of 0.454 to 0.908 kg per 280 cu m (1 to 2 lb per 10,000 cu ft) of gas, 80 to 90 percent removal of hydrogen fluoride was observed in the duct injected and 99 percent in the baghouse collector.

Battelle stresses that, in order to apply the Teller process to specific secondary aluminum operation, the nature and the variability of the emission with the types of scrap and/or the ratio of scrap types being charged, as well as the rate of magnesium removal must be established.

Treatment of Wet Fume Scrubber Effluent: The waters from wet fume scrubbing of emissions produced by the chlorine process are highly acidic due to the hydrolysis of aluminum chloride and magnesium chloride. When aluminum fluoride is used, scrubbing produces dissolved fluorides which are removable by treatment. Battelle (1973) believes that neutralization to a pH of 6.0 to 8.5 along with settling will precipitate most of the aluminum and magnesium as hydroxide (see Table 5.25). Coprecipitation of heavy metal hydroxides also occurs. The effectiveness of neutralization is diminished if too much alkali is added since resolubilization of aluminum hydroxide occurs at about pH 9. Sludges that cannot be recycled are discarded as solid wastes.

Waste Water from Residue Milling

Waste water generated during wet milling of slag residues from secondary smelters is high in dissolved salts. These concentrations are lower in drosses or skimmings from foundries. Waste water currently is treated in settling ponds wherein the suspended materials are removed. Few dissolved solids are removed. Battelle (1973) believes that this settling pond technology can be expanded to include three to four stages with partial recycle of the sludge and the recycle of the clear supernatant from the fourth stage to the mill; pH control must account for reduction of ammonia. Polyelectrolyte addition enhances settling.

The alternative to wet residue milling and resulting waste water treatment is dry milling of the residues. Of the 23 United States residue processors, 17 reported to Battelle that they practice dry milling. Impact mills, grinders, and screens are used to remove the metallic aluminum from the nonmetallics. The dust formed in these operations is collected by baghouses. The baghouse dust and the nonmetallic fines (mostly impure aluminum oxides) from the screening constitute the solid waste from the operation. Good solid waste disposal practices should be employed to prevent leaching of this material by runoff (see Chapter 11).

A countercurrent milling and washing approach now in use may reduce the solid waste disposal problem. The final recovered metal is washed with clean water which provides low-salt feed to the reverberatory furnaces. The waste water without suspended solids is suitable for salt recovery by evaporation and crystallization using furnace waste heat. Ultimate disposal would contend with dirt, trace metals, and insolubles (solid wastes) from the brine which should contain very low levels of soluble salts. Such salt recovery operations reportedly exist in England and Switzerland where the salts recovered help pay for the operation because they are reusable as fluxing salts in the secondary aluminum industry. According to Battelle's survey, such a system has not been employed in the United States.

Levels of Treatment and Associated Costs

Capital Investment Calculations

Battelle (1973) has estimated costs and reduced them to a common year for various appropriate waste water treatment technologies which would reduce air and water pollution. Equipment and total capital investment costs were obtained from secondary aluminum processors. These costs were changed to 1971 dollars by the use of the Marshall and Steven's Index.* In addition, where cost data were not available, equipment costs were estimated from published data (Peters and Timmerhaus 1968). The total capital investment was then calculated as this cost plus:

Component	Percent of equipment cost
Installation	50
Piping	31
Engineering	32
Electrical services	15
Contractor's fee	5

Operating Cost Calculations

Operating costs were obtained from secondary aluminum processors during the same survey. These included costs of raw materials and maintenance costs. This information was used as follows:

Raw material cost:	as reported
Maintenance cost:	as reported or taken as 5% of total plant cost
Depreciation:	10% of total capital
Interest:	8% of total capital
Tax and insurance:	1% of plant cost

* Quarterly values of this index appear in *Chemical Engineering.*

Proposed Best Practicable Control Technology Currently Available for Waste Water from Metal Cooling, and Costs

Battelle (1973) concluded that the best practicable control technology currently available for metal cooling waste water should result in no discharge of waste water pollutants. The achievement of this limitation depends on complete recycle, reuse, or consumption of all water within the process. The investigators recommend that this be achieved through:

1. Air cooling of ingots.
2. Total consumption of cooling water for ingot cooling.
3. Recycle or reuse of cooling water for deoxidizer-shot or ingot cooling.

With recirculation, the degree of sludge and oil removal would be influenced by plant procedures and the care used in controlling housekeeping. It was concluded that implementation of the air-cooling method or the total evaporation-cooling method (the air-cooling method with water mist added to assist the air cooling) requires:

1. The addition of ingot molds to the lengthened conveyor line.
2. The installation of blowers.
3. In the case of total evaporation cooling, the addition of special nozzles, flow meters, and controls to existing water lines.

And it was concluded that implementation of a recycle system for ingot cooling requires:

1. The addition of a cooling tower, holding tanks, and pumps to existing water-cooling facilities.
2. Provisions for oil and grease removal.
3. Provisions for sludge removal, dewatering, and disposal.

Of the 58 plants canvassed, 31 (54 percent) reported that they cool ingots by one of the above methods.

On the basis of these observations, capital costs were estimated to be about $0.43 per annual metric ton of aluminum alloy to convert an existing once-through cooling system to a recirculation system. An operating cost of $0.15 per ton was estimated, but this does not include savings resulting from the lowered fresh water use. Conversion to an air-cooled ingot line from a water-cooled line was estimated to require an investment of $9.20 per ton. Operating costs would be $1.09 per ton with no credit being claimed for water savings.

Battelle noted that there are four possible associated effects of these procedures. These effects are on atmospheric heating, energy, air quality, and solid waste production:

1. An incremental addition to the thermal load of the plant by thermal radiation from air cooling of ingots will occur.
2. Added electrical energy requirements of about 11 kilowatt-hours per ton

would be needed for air-cooling operations.

3. Negligible impact on air quality is anticipated from water evaporation either from consumptive water-mist cooling or from sludge drying.
4. Solid waste disposal of dried sludge would have a minor impact because of very small amounts accumulated, and its nontoxic character (Al_2O_3). Oil and grease collected during recycled water-cooling operations would be discarded through responsible waste oil disposal contractors.

Proposed Best Practicable Control Technology Currently Available for Waste Water from Fume Scrubbing, and Costs

Battelle (1973) recommended the effluent limitations in Table 5.26 as being achievable by the application of the best practicable control technology currently available to waste water generated by fume scrubbing during magnesium removal with chlorine. The recommended effluent limitation based on the application of the best practicable control technology currently available to waste water generated during magnesium removal with aluminum fluoride is to produce no discharge of process waste water pollutants.

The values in Table 5.23 were derived by the investigators as follows:

1. The 30-day average value for total suspended solids is the average of the values given in Table 5.22, namely 284 gm/kg and 66 gm/kg. Battelle considers these two values to be the most representative available. The "after treatment" values are higher than the suspended solids values in the untreated waste because neutralization produces fine particles of reaction products which add to the suspended solids values.
2. Similarly, the 30-day-average value of oil and grease is the average of the two values (0.4 and 3.5 grams/kg) from the same effluent values.
3. The 30-day-average value for chemical oxygen demand is the average of the two values (6.1 and 6.8 grams/kg) for the same effluents.
4. The 30-day-average ranges of pH are those estimated to optimize coprecipitation of heavy metals such as copper and amphoteric elements such as zinc and aluminum.

Battelle judged the best practicable control technology currently available for control for the discharge of pollutants contained in fume scrubber waste water to be the following:

1. When chlorination is used for magnesium removal, technology consists of adjustment of the scrubber effluent pH to between 7.5 and 8.5 followed by settling for solids removal. Prior adjustment of the pH of the scrubber liquor so that the resultant effluent from the scrubber is at a pH of 7.5 to 8.5 followed by settling for solids removal was considered equally practicable.
2. When aluminum fluoride is used for magnesium removal, technology consists of adjustment of the scrubber effluent pH to between 7.5 and 8.5 followed by settling for solids removal. The treated effluent is recycled continuously. Solid fluorides are removed continuously.

Table 5.26 *Recommended Effluent Limitations for Treated Fume Scrubber Waste Water Generated During Chlorine Demagging*

Effluent characteristic	*Effluent limitations*[1] *30-day average*[2] *(gm/kg of Mg removed)*	*(lb/lb of Mg removed)*
Total suspended solids	175	0.175
Oil & grease	2	0.002
Chemical oxygen demand	6.5	0.0065
pH	Within the range of 7.5 to 9.0	

Note: Recommended by Battelle 1973, based on its interpretation of the best practicable control technology currently available.

1. Effluent limitations are defined as the weight of the indicated constituent in grams (pounds) discharged per kilogram (pound) of magnesium removed from the metal treated.
2. 30-day average is the maximum average of daily values for any consecutive 30 days.

Battelle observed that the effluent from treatment of fume scrubber water from the chlorine-magnesium removal process pH adjustment) cannot be recycled continuously due to buildup of sodium chloride. But partial recycle of the clarified effluent reduces water consumption and treatment costs.

Neutralization and sedimentation of chloride scrubber waste water requires tanks for pH adjustment, polyelectrolyte addition, and sedimentation, as well as associated pumps, controls, and plumbing.

Implementation of continuous recycle of fluoride scrubber waste water would require liquid storage and pumping capabilities. Provisions for removing solid wastes would be required also.

The adjustment of pH to 7.5 to 8.5 combined with sedimentation was judged to remove aluminum and magnesium as hydroxides from chloride fume scrubber waste water. Some removal of heavy metals as hydroxides would be expected also. At a pH of 9.0, or greater, aluminum hydroxide and other amphoteric metal pollutants would dissolve. For this reason, Battelle recommended that pH not exceed 8.5.

An adjustment of pH to 7.5 would also reduce fluoride solubility. The limited solubility of the neutral fluoride salts in water would permit recycling of the supernatant solution.

Of the 29 plants Battelle surveyed which use wet scrubbers to control air emissions, 20 (69 percent) practice some form of pH adjustment. Of these 20 plants, 15 (51 percent) remove solids by settling.

Battelle estimated that a capital cost of about \$2.75 per annual metric ton of aluminum alloy produced would be required to install a pH adjustment-sedimentation facility to treat chloride scrubber effluent. An operating cost of

$1.50 per metric ton was estimated for such an installation. Capital expenditures would be lower for plants already neutralizing the scrubber effluent.

For plants using aluminum fluoride for magnesium removal, treatment of scrubber waste water would require, in addition to neutralization and settling, a means of recirculating the scrubber water continuously and continuous solids removal. An estimated capital investment of $9.90 per annual metric ton and an operating cost of $2.45 per metric ton were projected for these facilities.

Battelle expects this treatment technology to have only one significant impact on nonwater elements of the environment. This impact, if it occurred, would be due to the reliance on the land for additional disposal of solid waste from the treatment. But the solid wastes are primarily inorganic and non-leachable except for the solid waste from fluoride recovery. It should be discarded in a carefully designed, well-monitored landfill to prevent the contamination of surface or subsurface waters by lixiviation.

The foregoing method of expressing and classifying effluent limitations was justified by Battelle in the following way:

> Effluent limitations specify the quantity of pollutants which may be discharged from a point source after the application of the best practicable control technology currently available. This quantity must be relatable to a unit of production so that the effluent limitations can be broadly applied to plants in the same subcategory.
>
> The amount of pollutant generated during the removal of magnesium from a given heat is dependent upon the amount of magnesium present in the scrap and the final magnesium content in the metal produced. Judicious selection of scrap entering the melt will reduce this difference, the length of time required for chemical treatment, and the amount of chemical required for reducing the magnesium content to the desired level. These variables in turn establish the amount of pollutant entering the scrubber water. There are variabilities in the amount of magnesium removed for a unit weight of chemical agent. Frequently these are dependent on the furnace operator's techniques and/or on the plant practice and therefore are not suited for a production unit. An invariant production unit suitable for determinations of pollutant loadings is the amount of magnesium removed relative to the amount of metal produced. This can be determined from the percent magnesium contained in the charge before magnesium removal and the resultant magnesium content.
>
> The application of this guideline requires the reporting of the number of pounds of magnesium removed based on the magnesium content of the melt before magnesium removal, the magnesium content of the product metal, and the net weight of the metal treated for magnesium removal. These data are currently a part of company records. The flow rate of the discharge water stream from the scrubber system, and the analyses of the pollutants in that stream, are also required.

Proposed Practicable Control Technology Currently Available for Waste Water from Residue Milling

Battelle (1973) recommended the effluent limitations in Table 5.27 as achievable with the application of the best practicable control technology currently available to residue milling waste waters. These recommendations were derived by Battelle as follows:

1. The 30-day-average value for total suspended solids is that reported for plant D-4 in Table 5.22. This value was selected because it was based on

verified, seven-to-nine-month averages of sampling, and is otherwise considered a valid value on the basis of plant operations and raw material variation.

2. The value of fluoride is derived from data for plant D-8 in Table 5.22. It is based on nine composite samples taken over a three-day period.
3. The value for ammonia is based on the concentration of ammonia in the effluent from a plant using exemplary milling practice (0.3 mg/l, plant D-8, Table 5.22) and on the calculated loading on the associated flow (200 gpm, or 1,090,080 liters/day) and production (37.8 metric tons/day). This use of concentration reflects the chemistry of the reaction during alkaline wet milling.
4. The limitation value for aluminum was derived in the same manner as the ammonia value, i.e., using the concentration of 28 mg/l for plant D-8 in Table 5.22. The flow and production were the same as in paragraph 3, giving a value of 1.0 kg/metric ton of metal recovered.
5. The values of ammonia, aluminum, copper, and pH are interrelated. The pH specified is to be achieved with reagents other than ammonia. However, if an ammonia loading were not specified, the specified pH value could be present due to a high ammonia content. Further, ammonia and copper interact to form chemical complexes whose presence would not necessarily be reflected in the measurement of pH. Aluminum is specified to prevent under- or over-alkalization.
6. The value of chemical oxygen demand specified is that listed for plant D-3 in Table 5.22 (0.97 rounded to 1 kg/metric ton). The source of COD in the effluent was not fully documented.

The investigators consider the best practicable control technology currently available for control of the discharge of pollutants contained in waste water from residue milling to be:

> A settling treatment of three to four stages with partial recycle of the sludge and the clear supernatant from the fourth stage to the mill. Adjustment of the intake water pH is necessary to reduce ammonia levels in the waste water during milling.
>
> When milling is done without pH adjustment of the intake water, ammonia remains in solution as a pollutant. To aid the settling of the milling wastes, a polyelectrolyte is frequently added to reduce the level of suspended solids. Recirculation of the sludge in the last settling pond to the mill will reduce the overall sludge content of the final pond.

The utilization of this technology was reported by three of six wet residue processing plants surveyed in the United States.

Settlement alone may remove suspended solids but it will not remove most dissolved solids. Evaporation and crystallization is considered a practical alternative for dissolved solids removal. But the cost of salt recovery (for flux cover use) was observed to exceed the price of the salt recovered by this method. The alternative to discharge was considered to be total impoundment.

Battelle expects no added impact on nonwater elements of the environment by the adaptation of settling for removal of suspended solids. Currently solid

Table 5.27 *Recommended Effluent Limitations for Treated Waste Water from Residue Milling*

Effluent characteristic	Effluent limitations[1] 30-day average[2] (kg/kkg of metal recovered)	(lb/ton of metal recovered)
Total suspended solids	1.5	3.0
Fluoride	0.4	0.8
Ammonia	0.01	0.02
Aluminum	1.0	2.0
Copper	0.003	0.006
Chemical oxygen demand	1.0	2.0
pH	Within the range of 7.5 to 9.0	

Note: Recommended by Battelle 1973, based on its interpretation of the best practicable control technology currently available.

1. Effluent limitations are expressed in kilograms (pounds) of the indicated constituent discharged per metric ton (short ton) of metal value recovered from the processing of the residues.
2. 30-day average is the maximum average of daily values for any consecutive 30 days.

wastes from wet residue milling operations are discarded terrestrially.

It was estimated that this technology would require a capital cost of about \$8.70 to \$15.30 per annual metric ton of alloy recovered as molten metal and an operating cost of \$3.30 to \$10.90 per annual metric ton. Variations in the cost are dependent upon the rate of water use for milling and the solids content of the residue.

Battelle justifies the foregoing method of expressing and classifying effluent limitations as follows:

> Effluent limitations specify the quantity of pollutant which may be discharged from a point source after the application of the best practicable control technology currently available. This quantity must be related to a unit of production so that the effluent limitations can be broadly applied to various plants in the same category, regardless of their production capacity.
>
> The amount of pollutants in the waste waters from residue milling largely depends on the source of the residue. Residues from primary smelters and foundries (dross, skimmings) contain little, if any, soluble salts and up to 40 percent recoverable metal. Residues from secondary smelters (slags) contain high levels of soluble salts (KCl, NaCl) and as little as 5 to 10 percent metal.
>
> The production unit used for effluent limitations is the amount of molten metal recovered from the residue. The information required for the application of this guideline includes the weight of metal produced (currently a matter of routine record), the rate of flow of the effluent from the residue milling operation, and the concentrations of the pollutants in that flow.

Proposed Best Available Treatment Control Technology Economically Achievable for Waste Water from Metal Cooling, and Costs

Battelle (1973) concluded that, on the basis of the preceding information, the application of the best available technology economically achievable for cooling waste waters should produce no discharge of process waste water pollutants to navigable waters. The best available technology economically achievable is identical to the best practicable control technology currently available as described previously. No additional costs are expected.

Proposed Best Available Treatment Control Technology Economically Achievable for Waste Water from Fume Scrubbing, and Costs

Battelle (1973) concluded that the best available technology economically achievable is the use of in-process and end-of-process controls and treatment to prevent discharge of waste water pollutants. One of the following approaches was recommended:

1. The use of currently available processes for fumeless chlorine magnesium removal (Derham process).
2. Using a combination of AlF_3 for demagging and continuous recycling of scrubbing water from emission and effluent control systems.
3. Using a combination of AlF_3 for demagging and a coated baghouse system for air pollution control (Teller process).

The subsequent summaries are discussed in greater detail in preceding sections and in the Battelle report.

The Derham and Alcoa processes remove magnesium from molten aluminum scrap with a minimum of fume generation by the efficient use of chlorine. No water is used for fume control but a backup scrubber may be required.

The removal of magnesium with AlF_3 is advantageous in that it permits fume scrubber effluent to be continuously recycled because the fluoride salts are relatively insoluble and can be settled out. This approach is not possible if chlorine is used because of the dissolved solids buildup in the scrubber water.

The removal of magnesium with AlF_3, along with fume control with the coated baghouse (Teller process), employs chemically treated filters (bags) which remove the pollutants from the exhaust, thereby eliminating scrubbers and their waste water.

The cost of achieving the described pollution control status for fine emission control will depend on which of the three techniques is used. Battelle estimates that the Derham process for magnesium removal would involve a capital expenditure of $3.40/annual metric ton of capacity and an operating cost of $2.50/metric ton. The Alcoa process was estimated to require a capital cost of $5.90/annual metric ton and an operating cost of $2.90/metric ton (with no credit being taken for selling the magnesium chloride). The capital cost of AlF_3 for magnesium removal combined with continuous recirculation of scrubber water for emission control was estimated to be $14.00/annual metric ton and \$5.40/metric ton operating costs. The chemically treated baghouse system (Teller process) for elimination of air emissions during magnesium removal with AlF_3 was estimated to require a capital expenditure of about \$27.70/annual metric ton of capacity and an operating cost of \$7.30/metric ton.

Practicability of the Derham process, it was noted, is demonstrated by its present use in the industry. The process was reported to be under license or operating at four plants in the United States and in four plants outside the United States. However, Battelle observes that the Derham and Alcoa processes will require extensive research and development to overcome limited capacity (Alcoa) and to reduce reliance on backup scrubbers (Derham).

Two large U.S. plants reported using AlF_3 for demagging along with continuous recirculation of scrubber water to attain air emission and effluent control.

The chemically treated baghouse (Teller system) for dry air pollution control during AlF_3 demagging has not yet been demonstrated from an air quality standpoint. One major plant in the secondary industry reported that it has installed the system and is presently evaluating its effectiveness.

It was concluded that the Derham process causes no known nonwater quality environmental problems. However, it was noted that the residues resulting from its application may be too high in soluble salts for economic processing by residue milling techniques for metal recovery and could present a solid waste disposal problem.

Battelle expects application of AlF_3 with continuous scrubber water recirculation to result in a solid waste disposal problem. Fluoride salts precipitated and settled from the scrubbing water are somewhat soluble and could be leachable in a landfill. Therefore, disposal in a landfill would require the appropriate precautionary site selection and precautionary engineering and management measures discussed in Chapter 11. Application of chemically treated baghouse systems for dry air pollution control also produces a solid waste. The bag coating and the collected dust and fumes may contain fluoride salts that are slightly soluble and leachable to ground water.

Proposed Best Available Treatment Control Technology Economically Achievable for Waste Water from Residue Milling, and Costs

Battelle (1973) concluded that the best available technology economically achievable for waste water from residue milling is the replacement of present wet milling operations by totally dry milling methods. In dry milling, the residue is crushed and the contained salts are screened out as undersized waste material. But the dry operation is extremely dusty and the dust must be captured to avoid air pollution. Baghouses are appropriate.

It was estimated that the capital cost of preventing discharge of process waste water pollutants from the milling of residues would be about \$130 per annual ton of aluminum production capacity. This is the cost of building a new plant. The change from wet to dry milling would involve a complete process change. Operating costs were not obtained, but such costs would be higher than those for wet processing.

Battelle concluded that the cost of recovery of salts from waste water from residue milling is dependent on the type of residue being processed. Estimated capital cost to evaporate the water from low-salt content residues is \$16/annual metric ton of aluminum. Operating costs are \$24/metric ton. When high salt-content residues are processed, the estimated capital costs are \$200/annual metric ton and the operating costs are \$124/metric ton.

The practicability of dry processing of residues for aluminum recovery is demonstrated by the fact that 15 out of 12 United States plants utilize totally dry mill procedures and generate no waste water.

It has been shown that dry milling and wet milling of residues generate large quantities of solid wastes, ranging from 2.3 to 9.0 tons per ton of aluminum recovered, depending on the grade of the residue. Generally this solid waste from dry milling contains the soluble salts that were washed out during wet milling. Therefore, solids should be discarded in an acceptable landfill to prevent contamination of ground water and surface water.

Recovery of salts by evaporation from wet milling waste water was estimated by Battelle to require additional thermal energy of 8.6×10^6 kg cal/ton for the low-salt residue waste water and 176×10^6 kg cal/metric ton for the high-salt residue waste water (based on metric tons of aluminum recovered).

Regulations Promulgated for Secondary Aluminum Smelting

On the basis of information presented by Battelle (1973) and reviews thereof by industry and public agencies, the U.S. Environmental Protection Agency promulgated the following regulations covering discharges from the recovery, processing, and remelting of aluminum scrap. Exceptions may be granted by an EPA regional administrator in special cases.

Effluent Limitations Guidelines Representing the Degree of Effluent Reduction Attainable by the Application of the Best Practicable Control Technology Currently Available

(a) The following limitations establish the quantity or quality of pollutants or pollutant properties . . . which may be discharged by a point source . . . after application of the best practicable control technology currently available and which uses water for metal cooling: there shall be no discharge of process waste water pollutants to navigable waters.

(b) The following limitations establish the quantity or quality of pollutants or pollutant properties . . . which may be discharged by a point source . . . after application of the best practicable control technology currently available and which uses aluminum fluoride in its magnesium removal process ("demagging" process): there shall be no discharge of process waste water pollutants to navigable waters.

(c) The . . . limitations [shown in Table 5.28] establish the quantity or quality of pollutants or pollutant properties . . . which may be discharged by a point source . . . after application of the best practicable control technology currently available and which uses chlorine in its magnesium removal process. . . .

(d) The . . . limitations [shown in Table 5.29] establish the quantity or quality of pollutants or pollutant properties . . . which may be discharged by a point source . . . after application of the best practicable control technology currently available and which processes residues by wet methods. . . .

Table 5.28 *Effluent Limitations Promulgated for Secondary Aluminum Smelting*

Effluent characteristic	*Effluent limitations*	
	Average of daily values for 30 consecutive days shall not exceed	*Average of daily values for 30 consecutive days shall not exceed*
	Metric units (kg/kkg magnesium removed)	*English units (lb/1,000 lb magnesium removed)*
TSS	175.0	175.0
COD	6.5	6.5
pH	Within the range of 7.5 to 9.0	

Note: Effluent limitations guidelines representing the degree of effluent reduction attainable by the application of the best practicable control technology currently available. Chlorine is used in the magnesium removal process.

Table 5.29 *Effluent Limitations Promulgated for Secondary Aluminum Smelting*

Effluent characteristic	*Effluent limitations*	
	Average of daily values for 30 consecutive days shall not exceed	*Average of daily values for 30 consecutive days shall not exceed*
	Metric units (kg/kkg of product[1])	*English units (lb/1,000 lb of product)*
TSS	1.5	1.5
F	0.4	0.4
Ammonia (as N)	0.01	0.01
Al	1.0	1.0
Cu	0.003	0.003
COD	1.0	1.0
pH	Within the range of 7.5 to 9.0	

Note: Effluent limitations guidelines representing the degree of effluent reduction attainable by the application of the best technology currently available. Residues are processed by wet methods.

1. Product is hot aluminum.

Effluent Limitations Guidelines Representing the Degree of Effluent Reduction Attainable by the Application of the Best Available Technology Economically Achievable

The following limitations establish the quantity or quality of pollutants or pollutant properties . . . which may be discharged by a point source . . . after application of the best available technology economically achievable: there shall be no discharge of process waste water pollutants to navigable waters.

Standards of Performance for New Sources

(a) The following standards of performance establish the quantity or quality of pollutants or pollutant properties which may be discharged by a new source: there shall be no discharge of process waste water pollutants to navigable waters.

(b) Application of the factors listed in section 306(b) of PL 92-500 may require variation from the standard of performance set forth in this section for any point source subject to such standard of performance and which uses chlorine in the magnesium removal process ("demagging" process). If variation is determined to be necessary for any such source, the discharge of process waste water pollutants shall be allowed from the magnesium removal process only, and such source shall be subject to effluent limitations no less stringent than those required by paragraph (c) above.

REFERENCES

Battelle 1973. Development Document for *Proposed Effluent Limitations Guidelines and New Source Standards of Performance for the Bauxite Refining, Primary Aluminum Smelting and Secondary Aluminum Smelting Industry.* EPA Contract Completion Report No. 68-01-1518, EPA Document 440/1-73-019 a, b, and c, Supt. of Documents, U.S. Govt. Printing Office, Washington, D.C.

Cook, G. C., and Swany, C. R. 1973. Evaluation of Fluoride Recovery Processes, Alcoa Smelters. Paper No. A71-37, Metallurgical Society of AIME.

Culp, R. L., and Stoltenburg, H. A. 1958. Fluoride Reduction at LaCrosse, Kansas. *Journal of American Water Works Association*, Vol. 50, p. 423.

Eckenfelder, W. W., Jr. 1970. *Water Quality Engineering for Practicing Engineers.* New York: Barnes and Noble, Inc.

Fursman, O. C., Mauser, J. E., Butler, N. O., and Stickney, W. A. 1970. *Utilization of Red Mud Residues from Alumina Production.* U.S. Bureau of Mines Report of Investigations 7454.

Good, R. C., and Fursman, O. C. 1968. *Centrifugal Dewatering of Jamaican Red Mud.* U.S. Bureau of Mines Report of Investigations 7140.

Guccione, E. 1971. Red Mud, a Solid Waste can now be Converted to a High Quality Steel. *Engineering & Mining Journal*, Vol. 172, No. 9, pp. 136-137.

Hayward, C. R. 1952. *An Outline of Metallurgical Practice.* 3rd ed. New York: Van Nostrand.

Hudson, L. K., 1963. Recent Changes in the Bayer Process. In *Extractive Metallurgy of Aluminum, Vol. 1, Alumina*, ed. by D. Gerard and P. F. Stroup. New York: Interscience Publishers.

Miners, K. A. 1973. Alcan's Experience with Disposal of Red Mud Tailings from the Bayer Ore Process. In *Tailing Disposal Today*, Proceedings, International Tailing Symposium, Tucson, Ariz., ed. C. L. Aplin and George O. Argall, Jr., pp. 53-62. San Francisco: Miller Freeman Publications.

Nielsen, K., and Kielback, A. W. 1972. Recent Developments in Dry Scrubbing Techniques. Proceedings, Symposium, 101st AIME meeting on Environmental Control, San Francisco.

Patterson, J. W., and Minear, R. A. 1971. *Waste Water Treatment Technology.* Report No. 11 E Q 71-4, from Illinois Institute of Technology to State of Illinois, Institute for Environmental Quality.

Peters, M. S., and Timmerhaus, K. D. 1968. *Plant Design and Economics for Chemical Engineers.* 2nd ed. New York: McGraw-Hill Book Co.

Rush, D., Russell, F. C., and Iverson, R. E. 1973. Air Pollution Abatement of Primary Aluminum Potlines; Effectiveness and Cost. *Journal of the Air Pollution Control Association*, Vol. 23, No. 2, p. 98.

Rushing, J. C. 1973. Alumina Plant Tailings Storage. Paper No. A73-58, Metallurgical Society of AIME, Chicago, Illinois.

Teller, A. J. 1972a. Air Pollution Control. *Chemical Engineering.* Deskbook issue.

Teller, A. J., 1972b. Control of Emissions from Glass Manufacture. *Ceramic Bulletin*, Vol. 51, No. 8.

Vogt, M. F. 1974. Development Studies on Dewatering Red Mud. Paper presented at 103rd annual meeting of AIME, Dallas, Texas; also available from M. F. Vogt, Kaiser Aluminum and Chemical Corp., Gramercy, La.

Zabban, W., and Jewett, H. W. 1967. *The Treatment of Fluoride Wastes.* Proceedings, 22nd Purdue Industrial Waste Conference, Purdue Univ., pp. 706-716.

6. *Phosphate fertilizer industry*

SOURCES OF INFORMATION

A considerable portion of the information presented in this chapter is condensed from or based on Environmental Protection Agency Contract Completion Report No. 68-01-1508. This report is referred to in this chapter as Davy Powergas (1973).

Other sources of information include the U.S. Bureau of Mines, the published literature, and the author's observations.

STATUS OF INDUSTRY

The phosphate fertilizer industry employs a series of processes beginning with the extraction of a mineral from an ore body and ending with the production of one or more forms of phosphate fertilizer. Its inclusion in this book is justified by the fact that the source of the phosphate ultimately is a mineral resource, as opposed to the atmosphere in the case of nitrogen fertilizers. Phosphate plants are sometimes associated with other metallurgical operations such as smelters because the latter are a source of by-product sulfuric acid from air pollution control devices.

Prior to 1955, phosphate was the nation's most popular plant nutrient. Superphosphate with a P_2O_5 content of some 20 percent was the common form, but subsequently more concentrated forms of phosphate have become readily available. These are merchant grade phosphoric acid, triple superphosphate, and ammonium phosphate. These forms require greater production of phosphoric acid, which in turn requires larger sulfuric acid plants. Approximately 2.8 tons of concentrated sulfuric acid are required for each ton of P_2O_5 as phosphoric acid. Most United States phosphate operations are located near Florida phosphate rock sources and, less importantly, near sources in North Carolina, Idaho, and Utah. Davy Powergas (1973) reports 39 wet process phosphoric acid plants operating in 15 states with capacities ranging from 41,000 to 360,000 metric tons per year P_2O_5, with five new plants currently in design and construction stages.

The two types of triple superphosphate (TSP at 46 to 48 percent P_2O_5) produced are nonuniform, pulverized run-of-pile (ROP) TSP and a hard uniform pelletized material known as granular TSP or GTSP. The latter represents the more advanced technology.

Monoammonium and diammonium phosphate production is on the increase, at an annual rate of about 20 percent for the past decade. Davy Powergas reports 53 operating ammonium phosphate plants located in the United States ranging from 9,000 to 550,000 metric tons per year in capacity.

Interestingly enough, the increased demand for fertilizer stimulated by increased agricultural productivity has improved the economic picture for sulfur dioxide removal for air pollution control purposes. Sulfuric acid is an integral part of the fertilizer production process, and at this writing it is not in abundant supply. Two years previous, the picture looked less optimistic.

PHOSPHATE FERTILIZER PROCESSES

Davy Powergas (1973) has subdivided the phosphate fertilizer industry into eight separate processes for purposes of origin and disposal of wastes. These are: sulfuric acid production, phosphate rock grinding, wet process phosphoric acid production, phosphoric acid concentration, phosphoric acid clarification, normal superphosphate production, triple superphosphate production, and ammonium phosphate production. A ninth process might legitimately be added and labeled phosphate rock extraction and concentration. The resultant waste problems of this ninth category are discussed at the end of this chapter and in the subsequent chapter on tailings disposal.

Additional information provided on the above categories consists of the fact that all the phosphate fertilizer processes have sulfuric acid, phosphate rock, or phosphoric acid in common; so the effluents from all the processes contain fluoride (F) and phosphorus (P). Ammonia is the only source of nitrogen in ammonium phosphate process effluents. Most operations combine process effluents into a contaminated waste water recycle system which has no discharge most of the time. However, a positive water balance occurs when precipitation becomes too great, whereupon the discharge must be treated.

Sulfuric Acid

Prior to 1971, most of the sulfuric acid used for phosphate fertilizer production was produced in "single absorption"-type contact sulfuric acid plants. Exit gas from a single absorption plant generally contains SO_2 in excess of the air pollution control standard established by EPA (1.81 kilograms [4.0 pounds] per ton of 100 percent acid produced). Since 1971, EPA has been recommending a second absorption step; hence the name "double absorption" process. A second absorption tower is installed between the first and final sulfur dioxide to sulfur trioxide conversion steps.

The details of the single absorption and double absorption processes are presented in a variety of chemical engineering texts, and they are summarized in limited detail by Davy Powergas (1973).

Phosphate Rock Grinding

A portion of the phosphate rock as mined and separated is too coarse for direct use. This portion is ground to the particle size required for optimum phosphoric acid plant process efficiency. Pneumatic devices force the material through the grinding and sizing circuits. Air is exhausted from the system to prevent moisture precipitation, and bag-type air cleaners remove dust before discharge to the atmosphere. This size reduction process does not use water. The product of the grinding goes to the acidulation process, and ultimately waste gypsum is produced from it. The waste gypsum is pumped to gypsum ponds for disposal.

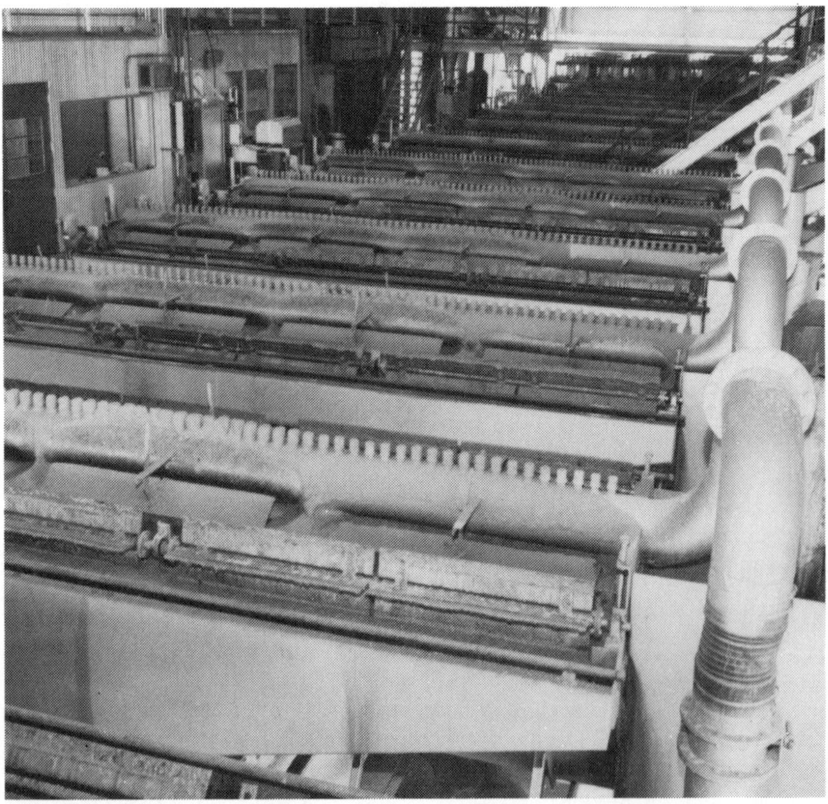

Flotation cells are used to process the coarser portion of phosphate rock after it has been cycloned to remove the slime. Tailings from the flotation circuit are discarded into a tailings pond. *American Cyanamid Company photo*

A second type of waste is produced by the portion of the phosphate rock that does not enter the grinding circuit described above. This portion is too fine to be retained on the screen which segregates the phosphate pebble product that enters the dry grinding circuit. The material that passes this screen is cycloned to remove the slime which is discarded in a slime pond. The coarser portion emanating from the cyclone is processed by flotation to produce a concentrate high in the mineral apatite (calcium phosphate with fluorine). The waste from the flotation circuit is discarded in a tailings pond. Thus, the waste products from phosphate rock are gypsum, flotation tailings, and slimes. Disposal of the latter two components is discussed at the end of this chapter and in the chapter on tailings disposal. The Davy Powergas report deals only with the gypsum pond.

Phosphate Rock Digestion and Filtration (Acidulation)

Acidulation consists of reacting the apatite in phosphate rock, usually with sulfuric acid, although other acids can be utilized. The solid waste product is gypsum because the calcium phosphate is converted to calcium sulfate. Gaseous and liquid effluents are also produced.

The materials needed to produce phosphoric acid are ground phosphate rock, sulfuric acid, and water. Phosphate rock is mixed with the sulfuric acid after dilution to 55 to 70 percent. Undesirable constituents in the wastes are contributed by the reaction of the acid with fluorine, iron, aluminum, silica, and uranium compounds also present in phosphate rock. Fluorine emanates from the reaction vessel and other plant equipment as hydrofluoric acid or gaseous silicon tetrafluoride. The latter compound hydrolizes in moist air to form fluosilicic acid and silica. Wet scrubbers collect these gases and their effluents are contaminated. The solid waste gypsum also contains fluoride, and the two are contaminants in a phosphoric acid plant effluent (see Williams 1973 and Mink, Williams, and Wallace 1971 for discharge data).

After the sulfuric acid converts the apatite to gypsum, the mixture of phosphoric acid and gypsum is filtered to separate the particulate gypsum from the phosphoric acid (approximately 30 percent P_2O_5). Davy Powergas (1973) observed that production of each ton of P_2O_5 as phosphoric acid produces approximately five tons of gypsum, which is discarded in a gypsum pond via a closed-circuit circulation system. Disposal of the solid and liquid portions of this slurry constitutes a major waste disposal problem.

Phosphoric Acid Concentration

The P_2O_5 concentration of the phosphoric acid from the sulfuric acid acidulation process (26 to 30 percent P_2O_5) is increased by evaporation of water until a 40 to 54 percent P_2O_5 solution is reached. Low-pressure steam is the energy medium. Dissolved fluorine and other impurities in the evaporated water are carried to the barometric condensers where they contaminate the condenser water.

Phosphoric Acid Clarification

Iron, aluminum silicates, gypsum, and fluorosilicates must be removed from the 54 percent phosphoric acid before it is converted to another form of phosphate. In the United States, these constituents are precipitated primarily by temperature control and precipitation. Separation of precipitated solids is accomplished by gravitational methods. In Europe and Mexico, solvent extraction techniques are used. Indirect cooling water is used in this process as is water in minor quantities for equipment washing.

Normal Superphosphate

The necessary raw materials for normal superphosphate production are 65 to 75 percent sulfuric acid and ground phosphate rock. The exothermic and rapid reaction between these two materials is:

$$\underset{\text{Phosphate}}{Ca_3(PO_4)_2} + \underset{\text{Sulfuric acid}}{2H_2SO_4} + \underset{\text{Water}}{H_2O} \longrightarrow \underset{\text{Gypsum}}{2CaSO_4 \cdot 2H_2O} + \underset{\text{Normal superphosphate}}{Ca(H_2PO_4) \cdot H_2O}$$

While fluid, the mixture is pumped to an enclosed conveyor or a storage space called a den. In the den the material congeals relatively quickly, during which time crystallization progresses and gases are emitted. Retention time in the den ranges from one to four hours, after which the material is stored as a porous mass for a period of three to eight weeks to complete the chemical reaction between the rock and acid.

Wet scrubbers collect the gases emitted during the reaction of the sulfuric acid and the calcium phosphate. The scrubber effluent is contaminated with the same elements present in the effluent from the phosphoric acid production process described previously. Davy Powergas (1973) recommends handling both these effluents with the contaminated gypsum pond water. As will be shown subsequently, treatment of this water is in order if the water balance is positive and it is necessary to have a discharge. However, Koziorowski and Kucharski (1972) report that in Europe sodium silicofluoride is recovered from these wastes and sold as a by-product of the production of superphosphate.

Triple Superphosphate

Triple superphosphate (TSP) contains about 46 to 48 percent P_2O_5. As pointed out previously, the two manufacturing processes produce a material designated as run-of-pile (ROP) and a granular triple superphosphate (GTSP). ROP TSP is a nonuniform pulverized material and its manufacture creates air pollution problems. GTSP is a hard, uniform pellet produced by equipment that facilitates collection and treatment of dust and fumes.

Both processes consume ground phosphate rock and phosphoric acid. The idealized chemical reaction is:

$$Ca_3(PO_4)_2 + 4H_3PO_4 + 3H_2O \longrightarrow 3Ca(H_2PO_4)_2 \cdot H_2O$$

Phosphate rock	Phosphoric acid	Triple superphosphate (monocalcium phosphate)

The ROP process is similar to the normal superphosphate process with the exception that in the TSP process phosphoric acid, rather than sulfuric acid, is used as the acidulating acid. The 46 to 54 percent P_2O_5 phosphoric acid and phosphate rock are reacted in a cone mixer. The slurry is discharged and begins to solidify in less than 30 seconds. Solidification and gas evolution occur on a conveyor enroute to the curing area. The material is broken up as it discharges from the conveyor, after which it is stored on a curing pile to cure for two to four weeks before shipping.

In the GTSP process, the phosphoric acid is 40 percent instead of 46 to 54 percent. The phosphoric acid and ground phosphate rock are agitated together, but the lower strength acid facilitates slower solidification and a more complete reaction. After one to two hours of agitation, the slurry is contacted with dry GTSP granules in a pug mill or rotating drum. The slurry-coated GTSP granules are hot rotary dried, which accelerates the chemical reaction essentially to completion and evaporates excess water. Under- and over-sized granules from the drier are reused as seed for another slurry. Product-size granules are cooled and stored or shipped.

Effluent from scrubbers used for control of air pollution is the major waste product. Dust from granulation in the GTSP process is collected in baghouses without water use (Davy Powergas 1973).

Ammonium Phosphate

The production of diammonium phosphate is estimated to have increased at a rate of 19.8 percent annually for the past ten years. Grade is determined by its content of up to 55 percent P_2O_5 and 18 percent nitrogen.

Ammonium phosphate is produced from ammonia and wet process phosphoric acid. Sulfuric acid is used to produce monoammonium phosphate grade 16-20-0. Nitrogen content is added by ammoniation during neutralization of the phosphoric acid. The chemical reactions are:

$$H_3PO_4 + NH_3 \longrightarrow NH_4H_2PO_4 + \text{Heat}$$

Phosphoric acid	Ammonia	Diammonium phosphate

and

$$H_2SO_4 + 2NH_3 \longrightarrow (NH_4)\,2SO_4 + \text{Heat}$$

Sulfuric acid	Ammonia	Ammonium sulfate

The second reaction occurs only in the production of grade 16-20-0. It is simultaneous with the monoammonium phosphate reaction (Davy Powergas 1973).

Methodology is similar to the GTSP process. Gaseous or liquid ammonia is reacted with 30 to 40 percent P_2O_5 phosphoric acid. A slurry forms which is contacted with dry recycled granules in a pug mill or rotating drum. Slurry-coated granules are hot rotary dried. Dried granules are sized, and off-sizes are used as recycle granules for slurry. Product-size granules are cooled and sold.

SOURCES OF WASTE WATER AND EFFLUENT CONTROL TECHNOLOGIES

Davy Powergas (1973) defines seven categories of waste water sources characteristic of the phosphate fertilizer industry; not all sources constitute effluents. The seven sources are:

1. Water treatment plant effluent.
2. Closed-loop cooling tower blowdown.
3. Boiler blowdown.
4. Recycled gypsum pond water.
5. Process makeup water.
6. Non-point source discharges (include surface waters from rain or snow that become contaminated and leaks and spills).
7. Contaminated water from gypsum pond (treated and released when water balance is positive).

Sulfuric Acid Water Treatment Plant Effluents

Plant steam pressure (9.2 to 52.0 atmospheres) determines, in part, the quality required for raw water for a sulfuric acid plant. If good-quality water is not available, feed water is treated by removing suspended solids prior to conventional hot lime-zeolite water treatment. Such treatment produces a waste effluent stream separate from other parts of an acid plant.

Davy Powergas (1973) recommends that accidental breaks and leaks be handled by the installation of emergency catch basins and automatic and detection shutoff devices on blowdown streams. Lime treatment for the emergency storage facilities was also recommended for maximum safety.

Closed-Loop Cooling Tower Blowdown

The cooling water requirements and normal blowdown quantities as estimated by Davy Powergas (1973) are listed in Table 6.1.

Closed-loop cooling systems cool water by evaporation, which concentrates impurities and the chemicals required to inhibit scale growth, corrosion, and bacteria growth. Blowdown maintains impurities at an acceptable operating level. The quality of the blowdown, as shown in Table 6.2, varies with the makeup water impurities and inhibitor chemicals used. Chromate (CrO_4 or

Table 6.1 *Cooling Water Use and Discharge*

Process	*Circulation requirement*		*Discharge requirement*	
	(l/kkg)	*(gal/ton)*	*(l/kkg)*	*(gal/ton)*
Sulfuric acid (per ton 100%) H_2SO_4	75,000-83,000	18,000-20,000	1,670-2,500	400-600
Rock grinding (per ton rock)	33-625	8-150	33-625[a]	8-150[a]
Phosphoric acid (per ton P_2O_5)	0-19,000	0-4,500	0-19,000[a]	0-4,500[a]
Phos. A. conc (per ton P_2O_5)	0	0	0	0
Phos. A clarification (per ton P_2O_5)	690-3,200	165-770	690-3,200[a]	165-770[a]
Normal super (per ton product)	0	0	0	0
Triple super (per ton product)	0	0	0	0
Ammon. phosphate (per ton product)	0	0	0	0

Source: Modified after Davy Powergas 1973.

a. Noncontaminated; only temperature increase in discharge water.

Cr_2O_7) has been observed in the effluent because of its use as a corrosion inhibitor. Davy Powergas notes that it is toxic to plant and animal life and can be removed by treatment with an oxidizing agent such as SO_2, followed by elevation of pH to 8.5. The oxidizing agent converts the hexavalent chromium to trivalent chromium. Elevation of pH to 8.5 causes the trivalent chromium to precipitate as chromic hydroxide.

Cooling is a no-contact process, so the only cooling water contamination from process liquids is through mechanical leaks in heat exchangers. Such contamination occurs periodically, and monitoring equipment is used for detection of leaks.

United States plants treat cooling tower blowdown separately or combined with other plant effluents for treatment. Plants utilizing chromate or zinc treatment compounds generally treat the blowdown stream separately to minimize treatment costs, according to Davy Powergas.

Boiler Blowdown Water

Sulfuric acid plants contain the only significant steam generation equipment in a phosphate complex.

Table 6.2 *Contaminants in Cooling Water*

Contaminant	*Concentration (mg/l)*
Chromate	0-250
Sulfate	500-3,000
Chloride	35-160
Phosphate	10-50
Zinc	0-30
TDS	500-10,000
SS	0-50
Biocides	0-100

Source: After Davy Powergas 1973.

Boiler blowdown flow rates reported by Davy Powergas (1973) are 1,300 to 1,670 l/kkg (310 to 400 gal/ton). Table 6.3 portrays typical contaminants and their concentration ranges. Oxidation plus neutralization and sedimentation is an available technology for removal of these constituents. It may, in fact, be feasible to combine boiler blowdown water with cooling tower blowdown water prior to treatment, or both may be combined with the contaminated gypsum pond water prior to treatment.

Davy Powergas does not define separate effluent treatments or limitations for boiler blowdown. No reason is stated, but presumably it is because boiler blowdown falls under a separate category of industrial activity.

Recycled Gypsum Pond Water

Contaminated water from the gypsum pond is used to supply most contact water needs of a phosphate fertilizer complex. The majority of United States plants reportedly impound and recirculate all water that has direct contact with any of the process gas or liquid streams.

These contact processes include wet process phosphoric, phosphoric acid concentration, normal superphosphate, triple superphosphate, and ammonium

Table 6.3 *Contaminants in Sulfuric Acid Plant Boiler Blowdown Water*

Contaminant	*Concentration (mg/l)*
Phosphate	5-50
Sulfite	0-100
TDS	500-3,500
Zinc	0-10
Alkalinity	50-700
Hardness	50-500
Silica (SiO_2)	25-80

Source: After Davy Powergas 1973.

phosphate production. Davy Powergas (1973) concluded that the water requirements of barometric condensers, gypsum sluicing, gas scrubbing equipment, and heat exchangers can all be supplied by contaminated water. Each time the water is reused the contaminant level is increased. Concentrations of 8,500 mg/l fluoride and over 5,000 mg/l phosphorus are common; pH levels dip as low as 1.0. Table 6.4 shows contaminated water use by process. Use of such poor-quality water requires noncorrosive equipment. This recycle stream produces an effluent only when the pond water balance becomes strongly positive. Treatment is discussed subsequently.

Process Makeup Water

Process water is defined by Davy Powergas (1973) as fresh water, untreated except for suspended solids removal. Fresh water use is held to a minimum because all makeup water eventually ends up in the contaminated water system which in turn means additional costly treatment if the water balance goes positive.

Process water requirements obtained during the Davy Powergas study are presented in Table 6.5. There is customarily no discharge except into the contaminated water recycle system. As shown in Table 6.5, most fresh water is utilized in the production of sulfuric acid.

Davy Powergas concludes that it is possible to use contaminated water for dilution of sulfuric acid, thereby making the total water balance more negative, which minimizes the need for treatment. In addition, the P_2O_5 recovery of the phosphoric acid complex is increased by that amount of P_2O_5 in the contaminated water. However, it is also pointed out that the two available methods of accomplishing sulfuric acid dilution with pond water are proprietary and that one method is considered a trade secret and the other is protected by patent. It was reported that either process can be added to existing plants or included in the design of a new facility.

Table 6.4 *Rate of Use of Recycled, Contaminated Water by Process*

Process	*Rate of water use*	
	(l/kkg)	*(gal/ton)*
Sulfuric acid	0	0
Rock grinding	0	0
Wet process phosphoric acid	16,400-20,800	3,800-5,000
NPK process-nitric acid acidulation	1,000-2,300	240-540
Phosphoric acid concentration	2,500-2,600	550-570
Phosphoric acid clarification	690-1,040	225-250
Normal superphosphate	940-1,040	225-250
Triple superphosphate	660-1,040	158-250
Ammonium phosphate	5,000-6,500	1,200-1,500

Source: After Davy Powergas 1973.

Table 6.5 *Rate of Use of Process (Fresh) Water*

Process	*Rate of water use*	
	(l/kkg)	*(gal/ton)*
Sulfuric acid	63-83	15-20
Rock grinding	0	0
Wet process phosphoric acid	0	0
Phosphoric acid concentration	0.8-1.6	0.2-0.4
Phosphoric acid clarification	0	0
Normal superphosphate	0	0
Triple superphosphate	0	0
Ammonium phosphates	0	0

Source: After Davy Powergas 1973.

The trade secret procedure reportedly involves the mechanical means by which the dilution is made so as not to create a plugging problem. The patented method is described as involving redesign of the phosphoric acid reactor cooling system to remove the heat load formerly removed by the sulfuric acid dilution cooler.

The patented process was reported to have been developed by United States Steel Agricultural Chemicals Corporation. It involves sulfuric acid dilution by a two-step procedure in a manner radically different from current practice. The details of process control, vessel design, and materials construction were all reported to be proprietary information.

Non-Point Source Discharges

Dry fertilizer material scattered over a phosphate plant is dissolved by precipitation. The magnitude of this contaminant source is a function of dust containment, housekeeping, precipitation rate, and the design of the general plant drainage facilities. Davy Powergas (1973) obtained no data on this source of water during its study.

Pond Seepage

The gypsum waste product is delivered to a gypsum pond along with the contaminated water in the form of a slurry. The ponds are surrounded by embankments. The Davy Powergas (1973) report contains a brief discussion of the prevention of seepage from these ponds. Prevention of seepage should be pursued vigorously to prevent pollution of ground water.

Under present practice, the bases of the dikes called starter dams usually are formed of soil from the immediate surroundings. Raising the dikes normally requires that gypsum be moved from inside the pond to the top of the earthen base. The design of such a system is described by Toland (1971).

In some cases, these dikes permit seepage of contaminated water through them. It is necessary to collect and reimpound seepage, primarily because of the fluoride present in it, for reasons discussed previously. Seepage collection and reimpoundment are facilitated by construction of a seepage collection ditch around the perimeter of the gypsum pond, as depicted in Figure 6.1, along with a pump station at the collection point of the seepage ditch to move the collected seepage water back into the gypsum pond. The permeability of the pond and embankments should be minimized also by the use of segregation of fines as described by Williams, Kealy, and Mink (1973). Leakage should be routinely monitored by wells installed in and beneath the embankments.

Contaminated Water (Gypsum Pond Water) Treatment System

The contaminated water treatment system effluent is the major discharge from a phosphoric acid complex. Discharge from this system is kept minimal in order to minimize treatment costs. Several United States plants reported to Davy Powergas (1973) that they had had no treatment and no discharge for several years.

The need to treat and discharge water depends on the contaminated water inventory. If an excess occurs, discharge occurs for the period of time required to achieve balance. This period is reported to be two to four months per year in some cases, depending on the rainfall and evaporation ratio. The water balance calculations undoubtedly ignore losses by recharge of the ground-water flow system by pond water because no estimates of such losses have been reported.

Contaminated water treatment systems generally have capacities of 2,085 to 4,170 l/min (500 to 1,000 gpm). The most common treatment system is the two-stage liming process. This is a variation of the process as discussed in Chapters 3 and 4.

According to Davy Powergas, at least two stages of neutralization are needed to remove fluoride and phosphate. Fluoride occurs principally as fluosilicic acid, along with small amounts of sodium and potassium fluosilicates. Phosphate is present principally as phosphoric acid along with minor calcium phosphate.

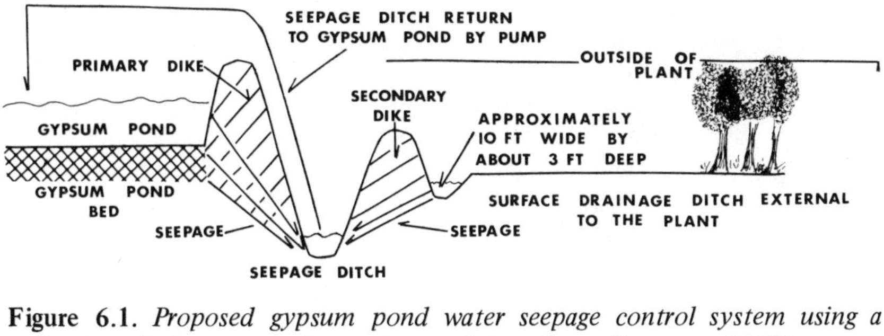

Figure 6.1. *Proposed gypsum pond water seepage control system using a collection ditch around the perimeter (after Davy Powergas 1973).*

The first of the two treatment stages provides neutralization to elevate pH from 1.0 to 4.0. Prior to stage 1, the contaminated water contains up to 9,000 mg/l fluoride and up to 6,500 mg/l phosphorus. Success is reported to be heavily dependent on mixing efficiency at the point of lime addition and the consistency of pH control. At pH of about 4, the fluoride precipitates in a settling basin, principally as calcium fluoride (CaF_2) according to:

$$\underset{\text{Fluosilicic acid}}{H_2SiF_6} + \underset{\text{Lime}}{3CaO} + \underset{\text{Water}}{H_2O} \longrightarrow \underset{\text{Calcium fluoride}}{\underline{3CaF_2}} + \underset{\text{Water}}{4H_2O} + \underset{\text{Silica}}{\underline{SiO_2}}$$

The overflow has a pH level of 4.0 and contains 30 to 60 mg/l fluoride and up to 5,500 mg/l phosphorus. The pH is again increased to 6.0 or above, whereupon calcium compounds, primarily dicalcium phosphate, plus additional quantities of CaF_2 precipitate. The primary reactions are:

$$\underset{\text{Phosphoric acid}}{2H_3PO_4} + \underset{\text{Lime}}{CaO} + \underset{\text{Water}}{H_2O} \longrightarrow \underset{\text{Monocalcium phosphate}}{\underline{Ca(H_2PO_4)_2}} + \underset{\text{Water}}{2H_2O}$$

and

$$\underset{\text{Monocalcium phosphate}}{Ca(H_2PO_4)_2} + \underset{\text{Lime}}{CaO} + \underset{\text{Water}}{H_2O} \longrightarrow \underset{\text{Dicalcium phosphate}}{2CaHPO_4} + \underset{\text{Water}}{2H_2O}$$

This mixture is allowed to settle for removal of $CaHPO_4$ and minor amounts of CaF_2. The overflow water is reported to contain 15 to 30 mg/l fluoride and 30 to 60 mg/l phosphorus at a pH of 6.0 to 8.0. The removal of phosphorus depends on the final pH and quality of the neutralization facilities. Neutralization to pH 9.0 to 11.0 reduces phosphorus to 15 to 30 mg/l, but this degree of neutralization exceeds the permissible pH discharge guideline; so pH would have to be reduced subsequently by acid addition as described in Appendix E.

If an ammonium phosphate plant is present, some form of nitrogen is likely to be present in the gypsum pond water. NH_3 is the most probable form. Davy Powergas notes that double lime treatment will not remove it, and that there is no good method for doing so. It was concluded that the best method to keep the NH_3 level low is to prevent its entry into the water system by reinjecting ammonium phosphate plant gas scrubber water and wash water into process streams.

Koziorowski and Kucharski (1972) report the neutralization of these wastes with limestone and lime in Czechoslovakia and the Soviet Union, respectively, but no details or effluent treatment levels are given.

POLLUTANT PARAMETERS

On the basis of the previous information, Davy Powergas (1973) defined the pollution parameters for the phosphate fertilizer industry to be pH, phosphorus, fluorides, ammonia, and suspended solids. pH was selected for obvious reasons. Phosphorus and ammonia were selected because they can be plant nutrients. Fluoride is harmful to plant and animal life.

DEEP-WELL INJECTION

The author has encountered one United States phosphate plant that discards its pond overflow by the deep-well injection method.* This procedure may also be promising for other plants located in the appropriate hydrogeologic environment (see discussion on deep-well injection in Chapter 4).

The U.S. Industrial Chemicals Company (USIC), Douglas County, Illinois, is engaged in the manufacture of various petrochemical and agricultural chemicals. Some of its operations, in particular the operation of a phosphoric acid unit, produce waste water that contains appreciable concentrations of phosphoric and sulfuric acids, as well as fluorides, carbon, and other salts. These waters, along with the by-product gypsum, are contained in surface pits with clay keyed dikes and moats to prevent escape.

Efforts have been made to minimize the production of waste waters by reuse within the processing units. But the accumulation of rainwater in the pit system exceeds both the natural evaporation and the plant's ability to use water.

After review of the problem, the deep-well disposal method was selected as the safest, surest, and fastest solution to the disposal problem. It is reported that the Cabot Corporation (adjoining the USIC property to the east) has successfully discarded acid waste water in a deep well for several years.

The property and well are located on the west flank of the Tuscola anticline, a part of the LaSalle anticlinal belt of the Illinois basin. The disposal formation is a creviced, vugular dolomite containing saline water. Several hundreds of feet of impervious strata are present at the location, which would be expected to preclude any migration of wastes vertically. The disposal formation is capped with approximately 1,000 feet of Prairie DuChien dolomite.

The well was drilled and 9-5/8-in., 32.3-lb, H-40 casing was run to 850 ft prior to cementing back to surface. The casing contains a float and guide shoe and centralizers to provide a complete encirclement and permanent seal of cement around the casing. The hole was deepened to 5,000 ft and 7-in., 20-lb and 23-lb, J-55, STC casing was inserted and cemented to the surface. The last 50 sacks of cement were Howco Latex cement to provide greater protection around the casing shoe. The float collar, float shoe, scratchers, and centralizers

* The source of this information is Dr. Robert Bergstrom, head of the Section of Ground Water Geology and Geophysical Exploration, Illinois State Geological Survey, Urbana, Illinois. The information is contained in a disposal well permit issued by the Illinois EPA, August 4, 1970.

were used to provide a seal of the casing from bottom to top. The hole was then deepened approximately 300 additional feet until the lost circulation (cavernous) zone, known to exist in the lower Knox or Eminence-Potosi formation, was penetrated.

Plastic-coated 4-in. tubing was inserted and a plastic-coated packer was set near the bottom of the casing. The casing-tubing annulus was filled with fresh water, kept under pressure and monitored with a permanent recording pressure gauge to signal any leaks that might occur.

USIC uses a plastic-lined centrifugal pump capable of discharging 350 gpm through 4-in. plastic line to transfer injected fluid from settling pond to well site.

COSTS

Costs of effluent control technology as estimated by Davy Powergas (1973) are summarized in Table 6.6. Table 6.6 shows costs and effluent quality levels for the treatment technologies discussed in the preceding sections, including costs for implementing the process alterations necessary to eliminate the treatment of gypsum pond water by driving the water balance negative. As noted earlier, this is accomplished by proprietary processes which facilitate the use of contaminated water for sulfuric acid dilution. The treatment method (Table 6.6) listed as "pond water used for sulfuric acid dilution to reactor system" covers the cost of adding the system to an existing plant. The lower row with the same title covers the cost of installing the system in a new plant. The costs in the tables are self-explanatory in light of the preceding discussion. As pointed out by Davy Powergas (1973), costs necessary to bring the effluent levels to a pH of 6.0 to 9.0, a fluoride content of 15 to 30 mg/l, and a phosphorus content of 30 to 60 mg/l will be in the order of \$1.90 per thousand gallons. Increasing the pH level to 9.0 to 11.0 in order to further reduce dissolved constituents will bring the costs to an estimated \$3.00 per thousand gallons. Even this high cost does not include the subsequent cost of reducing pH back to 7.0 or 8.0.

In the opinion of the author, these cost estimates may be somewhat exaggerated. Evidence for this opinion is contained in Appendix E. The Bunker Hill Company costs for treating a waste water with similar pH are an order of magnitude lower, although admittedly the flow rate is greater and only one treatment facility and only one treatment method are required for the Bunker Hill operation. Important also is the fact that the costs of the treatment operation do not include costs of settling pond seepage control.

PHOSPHATE SLIMES

A major semisolid waste produced by the phosphate rock mining and milling industry consists of the mixture of water and solids known as phosphate slimes. The slimes are analogous to mill tailings as discussed in Chapter 11 in that they are in part the fine-grained waste product of a flotation milling circuit. The handling costs for these wastes and other types of mill wastes would be expected to be similar, and the problems are without doubt similar.

Table 6.6 *Water Effluent Treatment Costs, Phosphate Subcategory*

Treatment alternative	Annual						Effluent quality				
	Investment[1]	Interest on money[2]	Depreciation[1]	Operating & maintenance costs (excluding energy and power)[1]	Energy and power costs[1]	Total annual costs[1]	Energy (MMkwh/year)	Raw waste load (liters/sec)	Raw waste load (gpm)	Resulting effluent level (mg/l)	Supplemental data
Sulfuric acid	$232,760	$ 17,470	$ 23,280	$ 9,310	$ 5,000	$ 55,060	.31	63.0	1,000	pH 6.0-9.0	
Pond water treating	349,600	26,220	34,960	13,990	$0.05/1,000 gal treated	–	–	63.0	1,000	pH 6.0-9.0, F 30 p 40, N 40	See note
Gypsum pond water seepage control	163,680	12,375	16,370	6,550	5,000	40,295	.31	252	4,000	No discharge	
Additional liming	–	–	–	–	–	–	–	–	–	pH 6.0-9.0, F 15, P 30	
DAP self-contained process (existing)	312,800	23,460	31,280	58,510	344,650	457,900	21.5	–	–	No discharge	This is an evaporator to concentrate feed stream phosphoric acid (A)
Pond water used for sulfuric acid dilution to reactor system ("added to" case)	310,960	23,320	31,000	12,440	168,500	235,260	10.5	–	–	No discharge	Cost of add on to existing plant (B)
Pond water used for sulfuric acid dilution to reactor system (new plant)	110,400	8,280	11,040	4,420	168,500	192,240	10.5	–	–	No discharge	Cost of adding this system to a new plant (B)
Sulfuric acid dilution with pond water	368,000	27,600	36,800	14,720	22,400	101,520	1.4	–	–	No discharge	(B)

Source: After Davy Powergas 1973.

Note: Raw material $1.40 per 1,000 gallons treated; annual costs $0.50 per 1,000 gallons treated; total overall cost (raw material plus annual costs) $1.90 per 1,000 gallons treated.

Size basis: (A) 900 kkg/day (1,000 ton/day) DAP plant
(B) 900 kkg/day (1,000 ton/day) P_2O_5 plant

1. All cost figures are for August 1971.
2. Ten-year period, straight line method.

Table 6.7 *Approximate Mineralogical Weight Composition of Phosphate Slimes*

Mineral	*Percent*
Carbonate fluorapatite	20-25
Quartz	30-35
Montmorillonite	20-25
Attapulgite	5-10
Wavellite	4- 6
Feldspar	2- 3
Heavy minerals	2- 3
Dolomite	1- 2
Miscellaneous	0- 1

Source: After IMCO 1971.

About 30 percent of the rock mined becomes slimes. Most mill wastes are characterized by a fine-grain-size distribution, but they differ in that phosphate slimes contain more clay minerals (Table 6.7) and thus retain more water at equilibrium with gravity. Deposits with only 15 percent solids are not uncommon.

Failures of phosphate slime disposal areas in Florida have received much notoriety. They reflect the poor slope stability characteristics of a material containing considerable fines and a concomitant high water content.

Considerable effort is being expended currently to develop an economical technique for lowering the water content, thereby increasing slope stability. One such effort is under way at the U.S. Bureau of Mines' Tuscaloosa, Alabama, Metallurgy Research Laboratory under a cooperative project with the Florida phosphate industry. Feld (1974) states in the project's first quarterly report for 1974 that the objective of the project is "to effect an economic curtailment of water and land pollution by developing techniques for dewatering phosphate slime tailings." The following points were extracted from the Feld report:

1. Of 15 samples examined, terminal solids varied from 6 to 19 percent after six months of settling.
2. Attapulgite, a hydrated magnesium silicate clay mineral, has been determined to be an extremely important factor in controlling the settlement of the slimes. The relative attapulgite content of the slimes has been observed to be directly related to the settling rate. High-terminal solids were obtained with samples having a low-attapulgite content, while low-terminal solids were obtained with samples having a high-attapulgite content.
3. Electrophoretic mobilities and cation exchange capacities were measured for 15 samples of Florida phosphate slimes from throughout the Florida land-pebble phosphate district. In all samples, most of the particles did not move in an applied electric field, indicating an electrophoretic mobility of zero, and in addition that most of the slime particles were already in a flocculated condition. Of the particles that did move in an applied electric field, all were

negatively charged. Mobilities ranged from 0.24 to −1.93 micrometers/second per volt/centimeter. The cation exchange capacities ranged only from 17 to 45 milliequivalents per 100 grams of slime solids, indicating a low-montmorillonite content.

4. Flocculation tests have been conducted on a limited basis to identify the most effective flocculant and to establish optimum conditions for rapid settling of phosphatic slimes. Results of these tests have shown that the polyacrylamide group is extremely effective in promoting primary agglomeration of the slime solids; however, a secondary treatment following or concurrent with flocculation is considered necessary to further increase the percent solids of the slimes. Consequently, research is being expanded not only to include other groups of flocculants and reagents, but also to develop a technically feasible secondary treatment process to improve thickening.

The Tuscaloosa Metallurgy Research Center has prepared a U.S. Bureau of Mines Information Circular entitled, "The Phosphate Slimes Problem—A Review and Bibliography."*

Slope stability is not the only problem associated with the phosphate slimes. Revegetation is difficult for several reasons, not the least of which is dust.

Some investigators have attempted to find uses for phosphate slimes. All uses must be compatible with the composition of the slimes as shown in Tables 6.7 and 6.8. IMCO (1971) has demonstrated, for example, that the slimes can be used for light-weight aggregate but that it is not economically competitive at present. In the opinion of the author, prospects for economical utilization of the slimes do not look promising.

PROPOSED BEST PRACTICABLE CONTROL TECHNOLOGY CURRENTLY AVAILABLE AND PROPOSED GUIDELINES

Davy Powergas (1973) recommended that discharge guidelines be based on rainfall conditions rather than on a production basis. This recommendation stemmed from the fact that a gypsum pond does not need to have an effluent unless precipitation is sufficient to drive the recycle water balance positive.

Gypsum Pond Contaminated Water System

Double lime treatment of gypsum pond water was recommended; it reportedly has been in industrial use for some 15 years. First-stage treatment takes the pH to 3.5 to 4.0. Second-stage treatment takes the pH to 6.0 to 9.0, which reduces the phosphate (as P) concentration to 30 to 60 mg/l and the fluoride (as F) concentration to 15 mg/l, or fewer. Radium 226 is precipitated to a low concentration by lime treatment to a pH of 8.0. Pond design and operation to leave enough freeboard to contain a 10-year storm was recommended. Operation to maintain the necessary freeboard can include proper treatment and release of water.

* The publication was undergoing review and unavailable at this writing. It contains 445 references and lists 174 patents relevant to the phosphate slimes problem.

Table 6.8 *Chemical Analysis of Phosphate Slimes, Dry Basis*

	Amount (%)				
Constituent	*A*	*B*	*C*	*D*	*E*
SiO_2	27.9	32.6	34.1	41.8	29.0
Al_2O_3	16.5	16.2	16.2	7.2	16.8
CaO	16.5	18.4	16.6	11.7	11.9
P_2O_5	14.2	11.8	12.5	9.8	13.1
MgO	1.85	2.3	2.0	1.8	1.9
Fe_2O_3	6.60	4.8	4.0	3.3	7.2
MnO	0.08	–	–	–	–
TiO_2	0.64	–	–	–	–
C	0.55	1.3	0.9	–	–
CO_2	1.10	1.4	2.0	–	–
Na_2O	0.36	–	–	–	–
K_2O	0.61	–	–	–	–
H_2O^+	6.48	12.8[F]	14.1[F]	12.6[F]	16.3[F]
H_2O^-	4.37	–	–	–	–
F	1.38	2.7	1.8	2.2	2.2

Source: After IMCO 1971.

A. IITRI analysis of slime stored at Noralyn, Florida.
B. 1-year-old pond, ~14% solids, M.H. Stanczyk and I.L. Feld, *Electro-Dewatering Tests of Florida Phosphate Rock Slime,* U.S. Bureau of Mines RI 6451 (1964).
C. Undetermined age, ~17% solids.
D. Undetermined age, ~30% solids.
E. 2-year-old pond, 27% solids.
F. Loss on ignition.

Ammonium Phosphate Self-Contained Process

This technology limits ammonia N as a contaminant in phosphate fertilizer process effluent. The recommended treatment is a process change which adjusts the process water balance to permit absorption of all process effluent back into the process by a combination of minimizing process effluent quantity, followed by an increase in the phosphoric feed acid concentration to a level that will permit reuse of the effluent. A limited number of production plants currently practice this technology.

Sulfuric Acid Plant Effluent Control

The recommended technology provides for a monitoring system to signal an emergency and for facilities for contaminated water isolation and subsequent treatment as identified in a preceding section, entitled "sulfuric acid water treatment plant effluents."

Recommended Effluent Limitations Guidelines

The following limitations describe the quantity or quality of pollutants or pollutant properties which were recommended for discharge after application of the best practicable control technology currently available as defined by Davy Powergas (1973).

1. Subject to the provisions of paragraphs 2, 3, and 4 below there shall be no discharge of process waste water pollutants into navigable waters.
2. An impoundment, which is designed, constructed and operated so as to contain the precipitation from the 10-year, 24-hour rainfall event as established by the National Climatic Center, National Oceanic and Atmospheric Administration for the area in which such impoundment is located, may discharge that volume of process waste water which is equivalent to the volume of precipitation that falls within the impoundment in excess of that attributable to the 10-year, 24-hour rainfall event, when such event occurs.
3. During any calendar month, there may be discharged from an impoundment either a volume of process waste water equal to the difference between the precipitation for that month that falls within the impoundment and the evaporation for that month or if greater, a volume of process waste water equal to the difference between the mean precipitation for that month that falls within the impoundment and the mean evaporation for that month as established by the National Climatic Center, National Oceanic and Atmospheric Administration for the area in which such impoundment is located (or as otherwise determined if no monthly data have been established by the National Climatic Center).
4. Any process waste water discharged pursuant to paragraph 3 of this section shall comply with each of the requirements in Table 6.9.

PROPOSED BEST AVAILABLE TECHNOLOGY ECONOMICALLY ACHIEVABLE

On the basis of the information summarized in previous sections, Davy Powergas (1973) concluded that the best available technology economically achievable consists of wet process phosphoric acid–gypsum pond water dilution of sulfuric acid. It was concluded that this technology would drive the pond water balance negative except for very unusual rainfall events and that ordinarily no pond discharge should occur. A discharge would be allowed under the following condition: A process waste water impoundment, which is designed, constructed and operated so as to contain the precipitation from the 25-year, 24-hour rainfall event as established by the U.S. National Weather Service for the area in which such impoundment is located, may discharge that volume of precipitation that falls within the impoundment in excess of that attributed to the 25-year, 24-hour rainfall event, when such event occurs.

PROPOSED NEW SOURCE PERFORMANCE STANDARDS

Davy Powergas (1973) recommended that new source performance standards be identical to the standards for the best available technology economically achievable as discussed in the preceding section.

Table 6.9 *Proposed Limitations for Discharges from Gypsum Ponds for Phosphate Fertilizer Industry Practicing Best Practicable Control Technology*

Effluent characteristic	*Effluent limitations*			
	Maximum for any one day	*Average of daily values for 30 consecutive days shall not exceed*	*Maximum for any one day*	*Average of daily values for 30 consecutive days shall not exceed*
	Metric units (mg/l)		*English units (ppm)*	
Total phosphorus (as P)	70	35	70	35
Fluoride	30	15	30	15
TSS	50	25	50	25
pH	Within the range 6.0 to 9.0		Within the range 8.0 to 9.5	

DISCHARGE REGULATIONS PROMULGATED BY EPA

Effluent Limitations Guidelines Representing the Degree of Effluent Reduction Attainable by the Application of the Best Practicable Control Technology Currently Available: The limitations promulgated by EPA for this category are those proposed by Davy Powergas (1973) as discussed above.

Effluent Limitations Guidelines Representing the Degree of Effluent Reduction Attainable by the Application of the Best Available Technology Economically Achievable: The limitations promulgated by EPA for this category are those proposed by Davy Powergas (1973) as discussed above.

New Source Performance Standards: The new source performance standards promulgated by EPA are those proposed by Davy Powergas (1973) as discussed above.

REFERENCES

Davy Powergas 1973. Development Document for *Proposed Effluent Limitations Guidelines and New Source Performance Standards, Basic Fertilizer Chemicals Segment of the Fertilizer Manufacturing Industry.* EPA Document 440/1-73-011, Supt. of Documents, U.S. Govt. Printing Office, Washington, D.C.

Feld, I. L. 1974. *Phosphate Slime Dewatering.* First Quarterly Report, Tuscaloosa, Metallurgy Research Center, U.S. Bureau of Mines, Tuscaloosa, Alabama, pp. 9-20.

IMCO 1971. *Utilization of Phosphate Slimes.* International Minerals and Chemical Corp. and U.S. Govt. Printing Office Document 14050 EPU 08/71, Stock No. 5501-0097, Washington, D.C.

Koziorowski, B., Kucharski, J. 1972. *Industrial Waste Disposal.* Translated from the Polish by J. Bandrowski. New York: Pergamon Press.

Miners, K. A. 1973. Alcan's Experience with Disposal of Red Mud Tailings from the Bayer Ore Process. In *Tailing Disposal Today*, Proceedings, International Tailing Symposium, Tucson, Ariz., ed. C. L. Aplin and George O. Argall, Jr., pp. 553-562. San Francisco: Miller Freeman Publications.

Mink, L. L., Williams, R. E., and Wallace, A. T. 1971. *Effect of Industrial and Domestic Wastes on the Water Quality of the Coeur d'Alene River.* Idaho Bureau of Mines and Geology Pamphlet 149.

Toland, G. C. 1971. Gypsum Tailings Pond Design, Kimberly, British Columbia. Preprint of Paper 71-I-300 presented at Fall Meeting, AIME, SME, Seattle; also available from Dames and Moore Inc., Salt Lake City, Utah.

Williams, R. E. 1973. Modernization of Tailings Disposal Practices in the Coeur d'Alene District, North Idaho. In *Tailing Disposal Today*, Proceedings, International Tailing Symposium, Tucson, Ariz., ed. C. L. Aplin and George O. Argall, Jr., pp. 634-673. San Francisco: Miller Freeman Publications.

Williams, R. E., Kealy, C. D., and Mink, L. L. 1973. Effects and Prevention of Leakage from Mine Tailings Ponds. *Transactions, SME, AIME*. Vol. 254, pp. 212-216.

7. Iron and steel industry and coke manufacturing

SEGMENTS CONSIDERED AND SOURCES OF INFORMATION

The iron and steel industry is among the most comprehensive industries with respect to number and types of processes employed. Operations extend from the mining of coal and the mining and milling of iron ore at one extreme to the processing and shaping of steel products at the other. Clearly, it is not possible to cover all aspects of the generation and disposal of wastes from these operations in a single chapter; this chapter is restricted to coke manufacturing, burden preparation, iron and steel making, degassing, and casting.

The waste production and disposal aspects of these segments of the industry were the subject of a recent study by the Cyrus W. Rice Corporation for the U.S. Environmental Protection Agency under authorization by Congress through Public Law 92-500 (see Rice 1974). This document is unique in that no other study of its kind is available. Consequently, much of the information presented in this chapter is derived from the Rice report. The chapter on iron and steel in Koziorowski and Kucharski (1972) is a source of supporting information, and other references also are cited.

PROCESSES

The steel industry as covered here consists of: (1) the conversion of coal to coke, (2) combining coke, iron ore, and limestone in a blast furnace to produce pig iron, and (3) the conversion of pig iron into steel in an open hearth, basic oxygen, or electric furnace. Subsequently, the steel is subjected to a high vacuum for degassing. Casting follows. The flow sheet for a steel mill is shown in Figure 7.1.

Coke plants are ordinarily operated along with steel mills to produce coke and by-products. These include coke oven gas, coal tar, crude and refined light oils, ammonium sulfate, anhydrous ammonia, ammonia liquor, naphthalene.

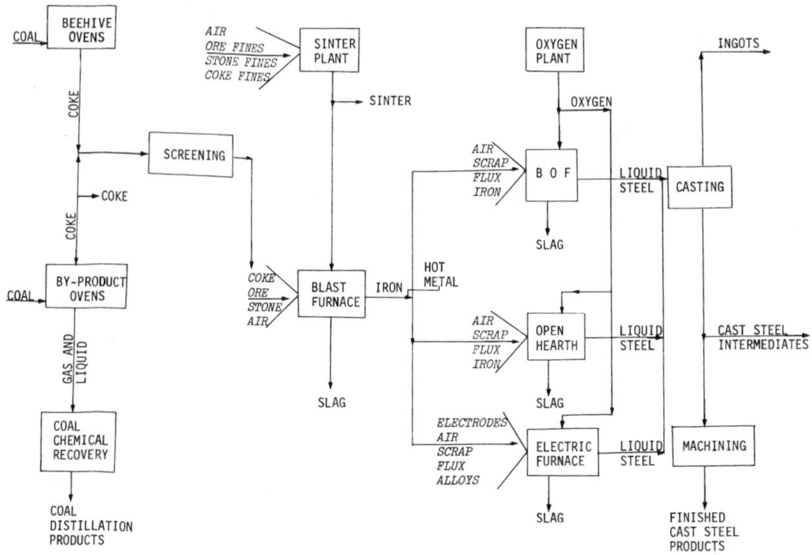

Figure 7.1. *Generalized flow sheet for a steel mill (after Rice 1974).*

A by-product coke plant heats bituminous coal in reducing ovens; the residue in the oven is the coke; the volatile components are processed to produce the products listed above.

Molten iron normally is produced by a blast furnace. The furnace is charged with iron ore, limestone, and coke via its top, and heated air is blown into the bottom. Coke combustion provides the heat to activate the reducing reactions that take place. The molten limestone in the furnace combines with the ore impurities. According to Rice (1974), 2 tons of ore, 0.5 ton of coke, 0.5 ton of limestone, and 3.5 tons of air can be considered to produce approximately 1 ton of iron, 0.5 ton of slag, and 5 tons of blast furnace dust and gas which contains the fines from the burden carried out by the blast; these fines constitute the flue dust. Molten iron is tapped from the bottom of the furnace; the molten slag floating on the iron also is withdrawn periodically. Blast furnace flue gas is burned to preheat the air blast to the furnace.

Auxiliary equipment includes stoves to preheat the air, dry dust catchers for recovery of flue dust, wet scrubbers to wash remaining flue dust with water, and secondary scrubbers such as electrostatic precipitators.

Steel is defined as any alloy of iron containing less than 1.8 percent carbon. Steel is produced from pig iron and/or scrap by the basic oxygen process (BOP), the open hearth furnace, or the electric arc furnace. All three of the steel-making processes refine the pig iron from the blast furnace blended with scrap, or scrap alone. Alloy elements are added as ordered, whereas unwanted elements are oxidized down to specified levels.

Blast furnaces and certain other processes produce fume, smoke, and gases as airborne wastes. Other impurities combine with the slag which floats on the surface of the molten metal. Waste water is produced by the wet scrubbers when used on the furnaces, and also by slag operations.

Some 30 percent of the steel produced in the United States is made in open hearth furnaces although the basic oxygen process is becoming predominant. Open hearth furnaces in the United States range in capacity from 10 to 600 tons per heat.

In the open hearth process, the steel-making ingredients are charged as needed into the front of the furnace through movable doors. Heat is supplied by liquid or gaseous fuel and hot air. The molten steel is tapped from the furnace when appropriate constituent levels have been achieved. Tapping occurs some 8 to 12 hours after the first charge, depending on the addition of oxygen.

The major drawback of the open hearth process is the long time required to produce a heat. A basic oxygen furnace is up to eight times as fast.

The basic oxygen process occurs in a vertical cylinder surmounted by a truncated cone. Water-cooled oxygen is supplied through a pressurized tube (lance) mounted above the center of the vessel. Scrap molten iron and a flux are charged to the vessel and oxygen is introduced through an oxygen lance. A turbulent reaction occurs which mixes molten metal and hot gases causing impurities to burn off. The metal is refined after 18 to 22 minutes of oxygen application. Alloys are added as needed, after which the steel is withdrawn.

Basic oxygen process steel production in the United States increased from about 600,000 tons in 1957 to 64 million tons in 1971. Basic oxygen production will continue to increase at the expense of open hearth furnaces.

The electric arc furnace produces high-quality steel, particularly high-carbon stainless steel. Electric furnaces range up to 30 feet in diameter and produce from 2 to 400 tons per cycle in 1.5 to 5 hours (Rice 1974).

An electric furnace cycle consists of the meltdown, the molten-metal period, the boil, the refining period, and the pour. Heat is generated by an electric arc passing from electrodes to the furnace charge.

The economic position of the nation's steel industry depends largely on imports. In the first two months of 1972, foreign steel accounted for one-seventh of the nation's apparent steel consumption. Imports are to a large extent from Japan where the non-open hearth processes were initiated earlier than in the United States. The import-export picture improved greatly in 1973 and, barring energy shortages, similar results were expected in 1974.

SOURCES AND CHARACTERISTICS OF WASTES AND WASTE WATERS

For purposes of production of waste water (mainly from air pollution control devices but also from coke by-product recovery) Rice (1974) categorizes iron- and steel-making processes according to Table 7.1. Elaborate descriptions of these processes and waste sources are contained in the Rice report. Subsequent sections of this chapter constitute summaries only; the reader should consult the original report for greater detail.

Table 7.1 *Subcategorization of the Steel-Making Operations of the Iron and Steel Industry*

By-product coke
Beehive coke[1]
Sintering
Blast furnace (iron)
Blast furnace (ferromanganese)
Basic oxygen furnace (semi-wet air pollution control methods)
Basic oxygen furnace (wet air pollution control methods)
Open hearth furnace operation
Electric arc furnace (semi-wet air pollution control methods)
Electric arc furnace (wet air pollution control methods)
Vacuum degassing
Continuous casting

Source: After Rice 1974.

1. Waste production by the beehive coke process is not discussed here because use of the process is declining rapidly.

By-Product Coke Process

As noted previously, by-product coke plants consist of ovens for heating bituminous coal to drive off volatile components. The volatile components are recovered and processed and the residue is coke. The ovens are narrow, silica brick rectangular prisms arranged side by side in batteries. Heat is supplied by burning about 40 percent of the gas produced by the coking process in flues between the oven walls. The remaining gas is used as fuel elsewhere. The gases from the ovens escape through pipes on top of the ovens and pass into a collection main. At the end of a heat, the doors are removed from each end of an oven and the red-hot residue is pushed into a quenching car. The residue is quenched or cooled by water sprays. Rice (1974) reported that about 2,100 liters of water per ton of coke are required, some 35 percent of which is evaporated in the quench tower. The remainder flows to a settling basin for sedimentation of fine coke particles. They are removed periodically for other uses. The clarified water is recycled to the quenching tower but the settling basin may overflow if the water balance becomes positive. The overflow becomes a source of waste water.

The coal tars must be extracted from the gases that leave the oven and enter the collection main. The hot gases are withdrawn from the collection main and sprayed with water (called flushing liquor) which saturates the gas. This water condenses much of the tar in the raw gas and the mixture flows down the suction main to a decant tank. Gas is then sucked through primary coolers where the temperature is further reduced by indirect application of cooling water.

The condensate from the primary coolers is pumped to the decanter and mixed with the tar and flushing liquor from the collecting main. The tar and liquor are separated by gravity and the lighter tar is stored. A portion of the liquor is reused for flushing liquor. Some of the water from this process originates from the moisture in the coal. This excess liquid, called ammonia liquor, is removed from the decant tank and pumped to storage for later distillation.

The crude coal tar collected by the tar extraction process is often sold or the tar may be refined in a continuous distillation unit. The four fractions, ordinarily produced at progressively higher temperatures, are light oils, middle or creosote oils, heavy oils, and anthrocene oil. The light oils are mixed with sulfuric acid and neutralized with caustic soda, after the first crude fractionization, and then redistilled. In the case of the middle oil fraction, after naphthalene removal, phenols and other acids are extracted with a caustic solution, neutralized, and then fractionally distilled. The waste water produced contains a variety of compounds from process water. Cooling and condenser water from distillation processes are also produced.

The ammonia extraction process follows the tar extraction process. In the United States ammonia is recovered by the semi-direct and indirect methods. Approximately 85 percent of coke plant ammonia produced is recovered as ammonium sulfate by the semi-direct process. The balance is recovered as ammonia liquor using the indirect process (Rice 1974).

In the semi-direct ammonia recovery process, the ammonia absorber, or saturator, succeeds the tar extractor. The gas passes through a dilute sulfuric acid solution in a closed system from which ammonium sulfate is crystallized and dried for sale.

As noted previously, some ammonia is dissolved in the flushing liquor and some is scrubbed from the gas with water. An ammonia still is used to concentrate the excess flushing liquor which is also called the ammonia liquor.

The ammonia liquor from these two sources is steam distilled for removal of ammonia and other volatiles. From the free leg ammonia, hydrogen sulfide, carbon dioxide, and hydrogen cyanide are distilled and returned to the gas stream. The fixed leg receives lime to decompose ammonium salts. Ammonia is liberated and returned to the gas stream. Recovery of ammonia from the gas is accomplished as ammonium sulfate in the saturators.

Dephenolizers recover phenol from the ammonia liquor as sodium phenolate. Liquid extraction and vapor recirculation are usually employed.

After passing through the ammonia absorption process, the coke oven gas enters final spray coolers in which the water dissolves soluble constituents and flushes out the insoluble naphthalene. The naphthalene is skimmed off and the water flows to a cooling tower to prepare it for reuse in the final spray cooler. Rice concluded that a properly designed closed recirculation system discharges little or no waste water because the cooling tower evaporation rate should equal the rate of moisture condensation from the gas and that, in an open system, final cooler water can be the largest source of contaminated waste water in a by-product plant.

From the final spray coolers, the gas moves through gas scrubbers where the crude light oils are absorbed out. They contain additional by-products which are separated and recovered. The gas is then treated for hydrogen sulfide removal or stored for use in energy production.

After collection in the gas scrubbers, the light oils are separated from the absorbent by steam distillation; the absorbent is cooled and reused in the gas scrubbers. The steam distillation vapors are condensed in the light oil condenser. They then flow to the light oil decanter for segregation of light oil and condensed water. Rice reported that indirect cooling generally is used in the absorbent cooler and light oil condenser and no waste water is produced. The water separated from the light oil is reported to be a major source of waste water.

After passage through the light oil scrubbers, H_2S is sometimes removed from the coke oven gas by the carbonate process which produces sulfur.

In plants that have no ammonia stills or ammonia saturates, the Keystone process recovers anhydrous ammonia. Rice reports that waste water produced is similar to that from the conventional light oil decanter.

On the basis of this information, Rice concluded that the most significant liquid wastes discharged from a by-product coke plant are excess ammonia liquor (varying from straight flushing liquor to still waste), final cooling water overflow, light oil recovery wastes, and indirect cooling water. Small volumes of water may also result from coke wharf drainage, quench water overflow, and coal pile runoff. Also, Rice noted that the following points are pertinent:

1. Indirect (noncontact) cooling water is not normally considered waste, but leaks in coils or tubes may contribute a significant source of pollution.
2. Direct contact of the gas in the final cooler with sprays of water dissolves soluble gas components and flushes out condensed naphthalene. This final cooler water becomes so highly contaminated that most plants cool and recirculate it. When a closed recycle system is not used, this waste water may exceed the raw ammonia liquor as a source of high-contaminant loads.
3. Condensed steam from the stripping operations and cooling water constitute the bulk of liquid wastes discharged. Flow of light oil recovery wastes varies from 2,100 to 6,300 l/kkg (500 to 1,500 gal/ton) of coke at plants which discharge cooling once-through water to 125 to 625 l/kkg (30 to 150 gal/ton) where cooling water is recycled. Effluent from the light oil recovery plant contains phenol, cyanide, ammonia, and oil.
4. The quenching of coke requires about 2,100 liters of water per metric ton of coke (500 gallons per ton). Approximately 35 percent of this water is evaporated by the hot coke; the steam and the constituents it carries discharge from the quench tower. During quenching most of the fire is extinguished, but enough heat must remain in the coke mass to evaporate the water trapped within the coke lumps. Quench station runoffs are collected and coke fines are settled and recovered for other mill uses. The clarified water is recirculated to the quench tower. Evaporative losses are continuously made up. Historically, contaminated waste waters have been used as makeup to quenching operations, but the practice is objectionable from an air pollution standpoint and various studies indicate that metal corrosion in the vicinity of quench stations using contaminated makeup water is accelerated. Further disadvantages accrue in subsequent blast furnace operations if coke quenched with contaminated waste water is charged

to the furnace. The practice increases the pollution potential of the blast furnace gas wash waters. Rice recommended that future quenching operations should utilize total recycle of quench wastes, with only fresh water makeups.

5. Coke wharf drainage and stock pile runoff constitute a minor but nuisance pollutant. These areas reportedly are generally trenched and the waste waters do not enter a receiving stream. Characteristics of coke by-product plant wastes reported by Rice are presented in Table 7.2. The samples are composites of the aforementioned sources.

Burden Preparation

Sintering

A sintering plant agglomerates and recycles fines to the blast furnace. Iron-bearing fines such as mill scale and dust from basic oxygen, open hearth, and blast furnaces are blended with fine iron ore and limestone. Sintering blends the iron-bearing components and limestone with coke fines which act as fuel. The mixture is ignited and, after ignition, a downdraft of air burns the

Table 7.2 *Characteristics of Two Sintering Plant Wastes and Four By-Product Coke Plant Wastes*

	Sintering plants	
Characteristics	*H*	*J*
Flow (l/kkg)	434	1,423
Suspended solids (mg/l)	4,340	19,500
Oil & grease (mg/l)	504	457
Fluoride (mg/l)	0.644	–
Sulfide (mg/l)	188	64.4

	By-product coke plants			
Characteristics	*A*	*B*	*C*	*D*[1]
Flow (l/kkg)	580	530	154	19,200
Ammonia (mg/l)	1,900	1,380	7,330	39
BOD_5 (mg/l)	1,550	1,280	1,120	12
Cyanide (mg/l)	102	110	91	7.7
Oil & grease (mg/l)	–	240	101	2.1
Phenol (mg/l)	450	350	910	6.1
Sulfide (mg/l)	–	629	197	4.2
Suspended solids (mg/l)	–	36	421	23

Source: After Rice 1974.

1. Concentrations are low due to the addition of the final once-through cooler stream which contained significant cyanide.

coke to fuse the bed. Carbon dioxide is released by the limestone, and a large part of the sulfur, chloride, and fluoride are driven off as gases. Oils are vaporized and removed with the gases.

The hot sinter is crushed as it emanates from the sinter machine, whereupon it is screened, then air cooled. After cooling, the sinter is sized so that the blast furnace charge is size-controlled at all times. The minus-0.6-centimeter fraction is recycled to the initial sinter step.

Dust control requires that each transfer point be hooded and dedusted. The submicron-sized dust particles are recycled to the beginning of the process.

Rice (1974) concluded that the areas of pollution in the sintering plant are the dust control equipment and the dust, the volatilized gases and the oil in the process gases. The sulfur in the process gas derives primarily from the coke fines which contain more than twice the sulfur of the larger coke. Chloride derives from the blast furnace dust and fluoride derives from the fluorspar. The limestone also may supply some fluoride. Some older United States sinter plants reportedly utilized wet scrubbers; others were equipped with cyclone-type dust collectors. New plants employ fabric-type dust filters because of the lower power cost and because of additional water pollution problems inherent in disposing of wet scrubber effluent, it was reported to Rice. Table 7.2 presents characteristics of sinter plant wastes.

Pelletizing and briquetting are also part of burden preparation but Rice (1974) observed that no plants in the United States utilizing these processes were scheduled to go on line until 1974; therefore their waste production characteristics were not investigated. However, Rice noted that it has been proven that all fine steel plant wastes can be recycled to the blast furnace after application of a cement binder and curing to insure a calcium silicate bond which will withstand the blast furnace forces. Koziorowski and Kucharski (1972) indicate agreement. Rice also indicates that no such plant is being planned at this time in the United States.

A blast furnace feed producing plant consists of grinding the coarser waste material fine enough to pelletize (at least 50 percent minus-325 mesh). The ground and already fine waste is bound with a cement binder, and the mixture is pelletized on a pelletizing disc. The pellets are then distributed on a curing belt to a depth of about 12 centimeters. The humidity over the curing belt is controlled at near saturation and the temperature is increased from 20°C to 90°C in approximately three hours. The partially cured pellets are then transferred to a curing bin where they cure for another 24 hours. The pellets are screened at 0.6 centimeter and the fines are recycled through the process. This process precludes pollution by emitting only filtered air (Rice 1974).

Iron Manufacturing (Blast Furnaces)

The majority of today's iron is produced in blast furnaces which reduce iron ore (iron oxide) to metallic iron. Iron ore, limestone, and coke are charged into the furnace where the coke burns and produces carbon monoxide gas, which combines with the ore to form carbon dioxide gas and metallic iron. The coke supplies the heat for the reaction and maintains the iron in a molten state.

The solid raw materials enter the top of the furnace and hot air is blown into the bottom. Liquid iron and slag are tapped from the bottom of the furnace.

Silica (SiO_2) is the major impurity of iron ore and coke. Limestone removes this silica in the furnace. At the high furnace temperature, the lime combines with the silica to form molten slag. The slag floats on the iron and is skimmed off periodically.

According to Rice (1974), approximately 3.5 kkg (3.8 tons) of air are blown through the furnace for each kkg (1.1 tons) of iron. Prior to injection this air is compressed to 3 or 4 atmospheres and heated to between 800°C and 1,000°C. Steam-turbine-driven compressors furnish compressed air. These turbines may be the back-pressure type, the extracting type, or the condensing type. If the steam is condensed, large volumes of cooling water are passed through the turbine condensers. Rice considers the liquid wastes associated with this area to be similar to those from utility power generating stations. Heat is considered to be the major contaminant.

Much cooling water is needed to operate a blast furnace. Most plants reportedly do not recycle cooling water completely, but in some water-short areas recirculating cooling systems are used. In order to conserve water, many plants use a portion of the cooling water from the furnace in the gas cleaning operations. Rice noted that, aside from noncontact cooling water, no waste water should emanate from a furnace.

The gases emanating from the top of a blast furnace are hot, dust laden, and rapidly moving. The gases consist primarily of nitrogen, carbon dioxide, carbon monoxide, and water vapor with trace amounts of hydrogen cyanide which is the product of a reaction between the nitrogen in the air and the hot coke in the furnace. Its concentration is a function of temperature; a very hot furnace produces more cyanide than a cooler one. In the furnace's reducing atmosphere, none of the normal oxides of nitrogen or sulphur occur; but traces of hydrogen sulfide do occur (Rice 1974).

Furnace gases are first cleaned for use as fuel by passage through a dust collecting settling chamber to settle out the larger particles. No liquid wastes are reported. Subsequently the gas is passed through wet scrubbers and coolers. All or part of the gas may also be passed through electrostatic precipitators for further cleaning. After cleaning, the gas is burned to make steam to drive compressors and to heat the air for the furnace. According to Rice, the effluent from the gas scrubbers and coolers constitutes the major portion of the direct-contact waste water from the blast furnace.

This effluent normally is thickened for removal of settleable solids. The thickener sludge is filtered and the filter cake and other fine iron oxide particles are sent to the sinter plant for agglomeration and subsequent recycle. The water collected by the filter is returned to the thickener.

Coke delivered to the blast furnace contains phenol and nitrogen compounds. Their concentrations are reported to be much higher in coke that has been quenched with a coke plant waste stream. These compounds evaporate from the coke in the top of the furnace and emanate from the furnace with the top gas. Rice noted that some are transferred to the water in the gas scrubbers, creating a source of water pollution which is transferred back to the coke during water recirculation.

Air cooling and slag granulation are the two common processes for handling the slag from a blast furnace. In the granulation process, slag flows into a pit of water. High-pressure streams of water disassociate the column of slag as it is poured into the pit. The rapid cooling causes the slag to expand and crack into small particles. This process generates steam which emanates into the atmosphere with an odor of hydrogen sulfide. Rice reports that the use of this process has declined in recent years because some of the low-density slag floats and is difficult to keep out of the receiving waterways and because of air pollution by the steam plume.

In the air-cooled slag process, molten slag is poured into dry pits for slower cooling. But a limited amount of water is sprayed on the hot slag to accelerate cooling. Rice reports observing that most of the water sprayed on the hot slag evaporates but, if excess water is used, it may drain from the pit as a contaminated liquid. The composition of the overflow from the slag-granulating operation and the drainage from the air-cooling pits are similar except for concentration. The effluents from slag operations contain reduced sulfides.

In summary, Rice observed that large volumes of water are required to operate a blast furnace. Most of the water is used for noncontact cooling of the blast furnace hearth and shell, the stove burners, and to condense steam that drives the compressors for the hot air blasts. This water is heated by 1°C to 5°C; otherwise, if it remains segregated, it should be discharged uncontaminated. However, air pollution control requires use of water for contact cooling and cleaning of blast furnace gases. Slag quenching also requires water. These waste waters contain suspended solids and chemicals from the blast furnace gas stream and the slag. The blast furnace gas scrubbing water constitutes the majority of the contact waste water from the blast furnace area. Minor, intermittent waste water discharges result from runoff from raw material storage and general yard drainage as well as various cleanup waters in the plant. Table 7.3 presents the characteristics of raw waste waters from four blast furnace plants studied by Rice.

Table 7.3 *Characteristics of Four Iron Blast Furnace Plant Wastes*

	Plants			
Characteristics	*L*	*M*	*N*	*O*
Flow (l/kkg)	22,500	8,050	14,000	13,000
Ammonia (mg/l)	1.41	3.91	9.75	12.3
Cyanide (mg/l)	1.44	0.858	–	–
Phenol (mg/l)	0.578	–	0.530	0.0853
Suspended solids (mg/l)	1,720	651	307	1,170
Fluoride (mg/l)	0.454	0.044	2.16	–
Sulfide (mg/l)	4.34	38.8	0.448	–

Source: After Rice 1974.

Steel mill operator directs charge of molten pig iron into basic oxygen furnace.
Inland Steel Company photo

Steel Manufacturing

The electric arc furnace, the open hearth furnace and the basic oxygen furnace all manufacture steel.

The basic oxygen process was introduced in the United States in the early 1950s and is replacing the outdated open hearth process. In 1972, the basic oxygen process accounted for 56 percent of steel production, the open hearth for 26.3 percent, and the electric arc furnace for 17.7 percent in the United States (Rice 1974).

All three methods use the same type of charge materials and produce the same waste products including slag (a solid waste), smoke, fume, and waste gases.

The charge comprises hot metal (iron), scrap steel, limestone, burnt lime (CaO), fluorspar (CaF_2), dolomite, iron ores (oxides of iron), and iron-bearing pellets and mill scale. Alloying materials, such as ferromanganese and ferro-silicon, added to the ladle or furnace bath bring the steel composition to required specifications.

The wastes from the raw material handling systems are dust, fumes, and smoke. They become waterborne wastes when wet dust collector systems are employed to control air pollution. They become solid wastes if the scrubber effluents are treated and produce sludge or if they are collected dry.

Basic Oxygen Process

The basic oxygen furnace* produces steel in a pear-shaped, refractory-lined, open-mouth vessel charged with a mixture of hot metal, scrap, and fluxes. Pure oxygen is injected into the furnace through a water-cooled, copper-tipped steel lance.

Rice (1974) pointed out that the basic oxygen furnace utilizes four main water systems with the characteristics listed below.

1. Oxygen lance cooling water system, which may be a once-through or a closed recirculating system. The effluent from the once-through system generally displays temperature increase of 11°C to 17°C at a discharge rate of 30 to 93 liters per second.
2. Furnace trunnion ring cooling water system, which is generally once-through with an aqueous discharge 22°C hotter than entry water. According to reports, these cooling systems are being added to existing shops to reduce thermal stresses and concomitant warping of the heavy fabricated steel plate trunnion rings. Water use rates range from 13 to 26 l/sec.
3. Hood cooling water system. The furnace hood affords the burning of CO gas and conveys fumes to the gas cleaning system. The CO gases are produced by oxygen from the oxygen lance combining with the carbon in the melt. A cooling water system cools the hood. The three types of hoods used are

* The reader should consult the Rice (1974) report for a much more detailed treatment of this section, particularly with respect to waste production.

water-cooled plate panel, water tube hood, or steam generating hood. The most significant contaminant in any effluent produced is heat.

4. Fume collection cooling water system. The fume collection system and type of hood cooling system utilized depend on capital cost, operating costs, plant location, availability of resources (power and water), and available pollution abatement equipment. The fume collection systems in United States plants range from completely dry precipitators to semi-wet and wet high-energy venturi scrubbers. Any aqueous discharge from the semi-wet scrubber system is water ranging in temperature from 82°C to 88°C and containing suspended iron oxides (Fe_2O_3, FeO), fluxing materials and lime.

Rice notes further that wet high-energy venturi scrubber fume collection systems generally employ steam generating hoods close-coupled with quenchers. As the hot gases exit from the hood, they are quenched from 150°C to 83°C. Sometimes large self-contained cooling towers are added to further reduce gas temperatures to 43°C. The cooling water can be recycled after chemical treatment. Makeup water is added to compensate for evaporation loss, blowdown, and cooling tower drift.

Table 7.4 summarizes waste concentrations for five plants studied by Rice. Only critical parameters are shown.

Open Hearth Process

Fuel as oil, coke oven gas, natural gas, pitch, creosote and tar is burned at one end of an open hearth furnace to melt scrap and for other process requirements. The hot gases produced travel the length of hearth above the charge, where they are conducted into a flue and downward to a regenerator brick chamber. These bricks absorb heat and cool the gases to between 650°C and 750°C. A system of valves that reverse gas flow and air flow facilitates preheating of incoming furnace combustion air by the heat in these bricks. In some plants, the gases leaving the brick chambers pass through waste heat boilers which reduce the waste gas temperature to between 260°C and 315°C.

Rice (1974) considers the major wastes from the open hearth process to be

Table 7.4 *Characteristics of Five Basic Oxygen Process Steel-Making Plant Wastes*

	Plants				
Characteristics	*K*	*S*	*T*	*U*	*V*
Flow (l/kkg)	542	4,270	2,570	3,040	1,080
Fluoride (mg/l)	–	–	10.9	2.36	2.76
Suspended solids (mg/l)	321	180	3,730	396	5,330

Source: After Rice 1974.

Note: Net plant raw waste loads.

Table 7.5 *Characteristics of Two Open Hearth Plant Wastes*

Characteristics	*Plants*	
	W	*X*
Flow (l/kkg)	2,530	2,290
Suspended solids (mg/l)	388	3,885
Fluoride (mg/l)	21.4	16.2
Nitrate (mg/l)	20.2	33.2
Zinc (mg/l)	2.06	880

Source: After Rich 1974.

Note: Net plant raw waste loads.

slag, iron oxides ejected as dust, and waste gases. The gases consist of air, carbon dioxide, and water vapor. Also included are oxides of sulfur and nitrogen produced by certain fuels and oxides of zinc if galvanized steel scrap is used. Fluorides may be emitted as gaseous and particulate matter. Fluorspar (CaF_2) is usually the source. Iron oxide fume (dust) is reportedly generated at the rate of some 12.5 kilograms per metric ton of steel. The gas and dust generation rate is considered to be fairly constant over the heat cycle except during oxygen lancing, whereupon the rate increases.

Older United States plants originally did not have gas cleaning equipment and the fume and gases emanated from the waste heat stacks and produced air pollution.

Some newer operations utilize dust collection units and some older plants have been equipped with them, it was reported. Precipitators or venturi scrubbers are used. The precipitators generally are dry systems producing no waterborne effluents. The venturi scrubbers discharge an effluent and, because of the presence of sulphur oxides, it is acid.

Table 7.5 summarizes the characteristics of open hearth waste waters from two plants examined by Rice. Only critical parameters are shown.

Electric Arc Process

The heat for melting the scrap, charge, and fluxes in an electric arc furnace is supplied by an electric current which passes through the scrap or steel bath via three carbon electrodes inserted through the furnace roof. Pure oxygen may be blown across the bath to hasten oxidation which, in turn, reduces electrical current consumption.

Rice (1974) classified the waste products produced as smoke, slag, carbon monoxide and dioxide gases, and oxides of iron emitted as fume. But other contaminants such as zinc from galvanized scrap are reported to be released if appropriate scrap is utilized. Oxides of nitrogen and ozone are released during arcing. Generally, 5 kilograms of dust per metric ton of steel are produced but it is reported that 15 kilograms per metric ton may emanate if inferior scrap is used. The wastes are airborne but become waterborne if wet fume scrubbers are employed. The three types of dust collectors used are baghouses, wet scrubbers, and dry precipitators. Rice reported the utilization of four methods for exhausting fume from electric furnaces:

1. Plant rooftop or furnace building extraction. The plant rooftop or building extraction system utilizes sealing of the mill buildings and installation of exhaust hoods in rooftop trusses for exhausting the fume generated by furnaces. A baghouse collector cleans the exhaust gases. This system requires huge volumes of exhaust air—36,500 cu m or 1.3 x 106 cu ft per minute for five 45-kkg (50-ton) furnaces—and large baghouses; but the system is reportedly adaptable to electric furnaces with the double slagging practice. It also reportedly captures most fugitive emissions from furnace tapping and slagging.
2. Local fume hoods. The second type of furnace exhaust control consists of hoods fitted adjacent to furnace door openings, electrode openings, and around junctures between the roof and furnace shell. A baghouse collector captures fume, smoke, and gases as they bleed through the furnace openings. These systems are reported not to be effective when hinged-type furnace roofs are in an open position during scrap charging.
3. Water-cooled roof elbow. Water-cooled elbow systems are generally tightly fitted to the furnace roofs, and the hot gases are exhausted from the furnace interior through the cooled elbow. A combustion air space between the water-cooled elbow and the gas cleaning ductwork provides combustion air for any carbon monoxide gases emitted from the furnaces.
4. Fourth-hole extraction. The fourth-hole system utilizes a fourth refractory-lined hole in the furnace roof. A space between the fourth-hole exhaust port and the gas cleaning ductwork provides combustion air for carbon monoxide gases emitted from the furnace.

Both water-cooled elbow and fourth-hole exhaust systems are compatible with all three types of dust collectors. If baghouses are used, a spray chamber is added to the gas cleaning system to cool the gas temperature to 135°C. If precipitators are used, a spark box is added to cool gases to about 260°C but, if high-energy venturi scrubbers are employed, the gases are quenched to their saturation temperature. The spray chamber, spark box, and quenchers discharge a liquid effluent.

In summary, Rice observed that the electric arc furnace has two main sources of waste water production. These are (1) the cooling-water system for the roof ring, electrode ring, and door cooling which generally is a once-through system but can be a closed recirculating system; and (2) fume cooling and cleaning effluents. The effluent from type (1) is heated cooling water, generally with a temperature increase of 17° to 22°C. The fume collection systems range from a completely dry system to semi-wet high-energy venturi scrubbers. The dry fume collection system consists of baghouses with local exhaust or rooftop exhaust hoods. There is no effluent from these systems. Cooling air is drawn into the hoods to cool the hot gases to approximately 135°C for baghouse cleaning.

Table 7.6 *Characteristics of Four Electric Furnace Plant Wastes*

	Plants			
Characteristics	*Y*	*Z*	*AA*	*AB*
Flow (l/kkg)	406	1.01	1,250	751
Fluoride (mg/l)	–28.7	–	14.8	11.3
Suspended solids (mg/l)	863	77.4	2,160	42,800
Zinc (mg/l)	13	–	405	5,637

Source: After Rice 1974.

Note: Net plant raw waste loads.

In the semi-wet system, a spark box or spray chamber conditions the hot gases for either a precipitator or a baghouse. Generally a spark box is used with a precipitator system and a spray chamber is used with a baghouse. The spark box cools the gases to 200°C whereas spray chambers cool gases to 135°C. A water-cooled elbow connected to the furnace roof is used for the exhaust ductwork. The effluent from the water-cooled elbow is heated cooling water. The systems are reported to be generally once-through with temperature increases of 17°C to 22°C.

Wet, high-energy, venturi scrubber, fume collection systems use the water-cooled elbow to extract gases from the electric arc furnace. Table 7.6 summarizes the characteristics of the effluents from four plants studied by Rice. Only critical constituents are shown.

Vacuum Degassing

In the vacuum degassing process, a high vacuum reduces the hydrogen, carbon, and oxygen content, and improves steel cleanliness. Rice (1974) subdivides vacuum degassing facilities into three major categories:

1. Recirculating degassing. Metal is forced into a refractory-lined chamber, exposed to a vacuum, then discharged from the chamber.
2. Stream degassing. Falling streams of molten metal are exposed to a vacuum, then collected under vacuum in ingot molds or ladles.
3. Ladle degassing. The teeming ladle is positioned inside a chamber where the metal is placed under a vacuum and stirred by argon gas or electrical induction.

A four- or five-stage steam jet ejector with barometric condensers creates the vacuum. Heat is furnished by electric carbon heating rods. Rice identifies the wastes from vacuum degassing as cooling water and condensed steam with iron oxide fumes and CO gases entrained along with cooling water from steam condensing. Table 7.7 presents the characteristics of two degassing plant effluents that Rice analyzed. Only critical constituents are shown.

Table 7.7 *Characteristics of Two Degassing Plant Wastes*

	Plants	
Characteristics	*AC*	*AD*
Flow (1/kkg)	3,750	813
Suspended solids (mg/l)	23.2	70.7
Zinc (mg/l)	2.01	7.76
Manganese (mg/l)	5.72	13.3
Lead (mg/l)	0.471	1.39
Nitrate (mg/l)	25.3	3.03

Source: After Rice 1974.

Note: Net plant raw waste loads.

Continuous Casting

Steel not formed in ingot molds is normally handled by continuous casting wherein billets, blooms, slabs, and other shapes are cast directly from the teeming ladle. Ingot, mold preparation, soaking pit, and stripping stages are bypassed. The steel ladle is suspended above a refractory-lined container with several nozzles in the bottom. The container regulates the flow of molten steel from teeming ladles to the continuous casting molds by nozzle orifice size, ferrostatic head, or stoppered nozzles which can stop the flow. The casting molds are made of water-cooled, chrome-plated copper. As the hot steel flows through the container nozzles into the casting mold, a hard steel exterior shape is formed. The center remains liquid. After the cast product is discharged from the mold, it enters a spray chamber for additional cooling.

Rice (1974) concluded that the continuous casting process has three main water systems with the following characteristics:

1. Mold cooling water. The mold cooling water system generally is a closed, recirculating noncontact system utilizing heat exchangers or evaporative cooling. The cooling water temperature rise is approximately 6°C which provides for minimal differential thermal expansion of the mold. A surge tank facilitates addition of potable water for makeup and/or chemical treatment.
2. Machinery cooling water. The machinery cooling water system is generally an open, recirculating, noncontact system with induced-draft cooling towers. The cooling water temperature rise across the machinery is reported to be approximately 14°C. The aqueous discharge from the machinery cooling water system is cooling tower blowdown.
3. Spray cooling water. The spray cooling water system employs contact water spray cooling of the cast product. As the cast product (slabs, blooms, or billets) emerges from the molds, the sprays further cool, harden, and thicken the skin on the cast product. The waste products can be characterized as iron oxide scale, oils, heat, and some volatiles.

Table 7.8 summarizes the net plant raw waste loads for two plants studied by Rice. Only the critical parameters are presented.

Ingot Casting

Waste waters from ingot casting generally are produced by water use in the mold cleaning and cooling processes. This water may constitute a source of pollution but, according to Rice (1974), the water generally is controlled to prevent a poor working environment and the area has no effluent.

Pig Casting

Some water is used to wash the pig casting area and clean spray equipment, and water is required to cool the pigs. Some plants divert this runoff to a settling basin which is periodically cleaned out. Rice (1974) concluded that, because of the small volume of water and the intermittent nature of the pigging operation, there should be no effluent from this operation.

Slagging

Molten blast furnace slag is dumped into pits, open at one end to permit removal after quenching. It is quenched and cooled for transport to a disposal site or a slag processing plant. It is sold for some purposes, including road and driveway metal, but in some cases it constitutes a solid waste and must be discarded as such. It has been used to fill in lakes and swamps along the south end of Lake Michigan, for example. According to Rice (1974), quenching produces little runoff; most of the water is evaporated but, as the slag temperature is lowered, some excess quench water may remain. Normally it collects in the bottom of the pit and is evaporated by the next hot slag charge, according to reports. However, Rice also noted that some of the excess quench water may infiltrate into the ground as a subsurface discharge. The water is highly alkaline and high in dissolved solids; therefore Rice recommended concrete-lined slag quenching pits to prevent pollution of ground water. Some plants reportedly already use concrete-lined pits.

Table 7.8 *Characteristics of Two Continuous Casting Plant Wastes*

	Plants	
Characteristics	*AE*	*AF*
Flow (1/kkg)	17,100	6,172
Suspended solids (mg/l)	7.87	74.0
Oil & grease (mg/l)	20.5	22.0

Source: After Rich 1974.

Note: Net plant raw waste loads.

SIGNIFICANT POLLUTANTS IN STEEL INDUSTRY WASTES

Rice (1974) concluded that, on the basis of its study of the steel industry, the major waste water constituents that are of pollutional significance include ammonia, BOD_5, cyanide, phenol, oil and grease, suspended solids, and heat. Other parameters, such as chloride, are present in significant amounts but were not established as control parameters because their presence in effluents is not as significant and the cost of treatment and technology for removal is considered to be beyond the scope of best practicable or best available technology. Rice also concluded some parameters cannot be designated as control parameters until sufficient data are made available on which to base effluent limitations or until sufficient data on treatment capabilities are developed.

Rice observed that its studies indicate that consideration should be given to including additional control parameters in certain subcategories because of the quantities sometimes present and the pollutional significance of the material. These parameters include sulfide, fluoride, nitrate, zinc, lead, and manganese.

Rice's rationale for selection of the waste parameters follows.

Coke Making

The principal liquid wastes in coke making originate from the ammonia liquor, coke quenching effluents, benzol plant decant waters and final cooler waters. They contain phenols, cyanide, BOD_5, ammonia, sulfide, suspended solids, and oil.

Sintering

The dust produced from the sintering plant operation is frequently recovered by wet washers on the exhausts of hoods and building ventilators. This waste water is produced as a result of air pollution control and occupational health and safety measures. These waste waters may contain significant amounts of suspended matter, oil, sulfide, and fluoride, depending on the materials that are a part of the sinter mix.

Iron Making

The principal waste water produced by blast furnace operations is wash waters for the exit gases (for removal of suspended matter) and noncontact cooling water. The coke may also produce pollutants that were prevalent in the coke-making waste waters. Therefore, iron-making blast furnace waste waters may contain ammonia, cyanide, phenol, suspended solids, and sulfide. The ferromanganese furnace will contain manganese in addition to the normal parameters inherent in the typical iron-making furnace.

Steel Making

Waterborne wastes from steel making result from scrubbing of the gas stream to prevent air pollution and from noncontact cooling. Basic oxygen and electric furnace waste waters may contain suspended solids and fluorides from fluorspar, one of the basic raw materials in steel making. The open hearth, due to the nature of its scrap mix, will also contain zinc; nitrates may be present because of the huge volumes of air that are required with the open hearth method to provide better combustion.

Vacuum Degassing

In the vacuum degassing process, steel is subjected to a high vacuum in an enclosed refractory-lined chamber. Steam jet ejectors with barometric condensers produce the vacuum. In the refining process, certain alloys are added which may be drawn into the gas stream. In addition, the system is purged with nitrogen so as to have no residual CO. Therefore, the waste water products from this operation are condensed steam and waste water containing suspended solids, zinc, manganese, lead, and nitrates.

Continuous Casting

Waste waters from the continuous casting operations result from spraying scale from the surface of the steel. Therefore, continuous casting waste waters may contain suspended matter and oil. The mold cooling and machine cooling systems are usually closed systems and the water picks up only heat.

CONTROL AND TREATMENT TECHNOLOGIES

Rice (1974) conducted studies in each of the previously described subcategories at plants that were deemed to have achieved exemplary performance levels through their treatment technology. After an extensive literature search, Rice concluded that the published literature on this subject is sketchy and not comprehensive. Most of the publications encountered by the author deal with small segments of the total problem and, without on-site investigations, it would be impossible to piece them together in a comprehensive, meaningful fashion. For this reason, the Rice report is particularly important to the academic community, for it is here that future personnel must be trained to cope with the industry's waste production and disposal problems in an integrated fashion.

Table 7.9 presents a summary of waste water treatment practices available to or employed by steel plants. The table shows the status of treatment techniques currently employed in the steel industry as well as the effect of proposed technologies on gaseous and solid waste production. The waste waters are those resulting from direct process contact and from air pollution control as discussed in the preceding sections. Rice also presents a detailed breakdown of the costs of the treatment technologies described in Table 7.9. However, it is too lengthy to present here. Only summaries of discussions of the data pertinent to Public Law 92-500 are presented in the table. Summarized costs of recommended best practicable and best economically achievable technologies are presented in subsequent sections.

Table 7.9 *Iron and Steel-Making Control and Treatment Technology*

Treatment and/or control methods	*Resulting effluent levels for critical constituents (mg/l)*		*Status and reliability*	*Problems and limitations*	*Implementation time*	*Land requirements*	*Environmental impact other than water*	*Solid waste generation and primary constituents*
Subcategory: By-product coke [1]								
A. Ammonia liquor treatment via free still only; dephenolizer; settling pond for solids; light oil recovered for sale to outside contractors; quench water recycles with no blowdown; final cooler water recycles with blowdown to dephenolizer; crystalizer barometric condenser water once-through to settling pond.	NH_3 Phenol CN^- BOD_5 $S^=$ O&G SS pH	1,000 5 90 300 25 20 50 6-9	Widely practiced in industry. Subject to upsets from slug loads. Fair.	Requires constant attention to maintenance & housekeeping. Heated discharges.	6 months	1 acre (200x200 ft)	Quenching with contaminated water releases volatiles to air.	Coke fines are usable in plant. Solids to landfill.
Alternative 1 – Physical/chemical								
B. To (A) add lime and steam to fixed leg of ammonia still; neutralize prior to settling.	NH_3 Phenol CN^- BOD_5 $S^=$ O&G SS pH	125 2 30 150 10 15 50 6-9	Used by some plants in industry. Good.	Same as in (A). Lime addition requires care in handling.	6 months	1 acre (200x200 ft)	Volatile compounds released to air.	Same as in (A), with additional sludge from lime addition.

C. To (B), add aeration; aggressive chemical oxidation; neutralization; break point chlorination; clarification and/or filtration; carbon adsorption. Recycle crystallizer effluent through final cooler water recycle system.	NH_3 Phenol CN^- BOD_5 $S^=$ O&G SS pH	10 0.5 0.25 20 0.3 10 10 6-9	Chemical oxidation practiced at some blast furnace (iron) plants; other technology from chemical, refining & water treatment industries. Very good.	Part of technology untested on coke plant wastes. Very close control of intermediate steps must be practiced.	1-3 years	1½ acres (200x300 ft)	Volatile compounds released to air.	Same as (A), with additional sludge from neutralization steps.
Alternative 2 – Biological								
B. To (A), add lime and steam to fixed leg of ammonia still; abandon dephenolizer; neutralize; add single-stage bio-oxidation for phenol removal.	NH_3 Phenol CN^- BOD_5 $S^=$ O&G SS pH	125 1 20 100 1.0 10 50 6-9	Used by some plants in industry. Good.	Same as in (A). Lime addition requires care in handling.	6 months	1 acre (200x200 ft)	Volatile compounds released to air.	Same as in (A), with additional sludge from neutralization steps.
C. To (B), add aeration; multistage biological treatment; neutralization; and filtration; recycle crystallizer effluent through final cooler water recycle system.	NH_3 Phenol CN^- BOD_5 $S^=$ O&G SS pH	10 0.5 0.25 20 0.3 10 10 6-9	Single-stage biological oxidation practiced at some coke plants; other technology from chemical, refining & water treatment industries. Very good.	Part of technology untested on coke plant wastes. Very close control of intermediate steps must be practiced.	1-3 years	1½ acres (200x300 ft)	Volatile compounds released to air.	Same as (A), with formation of biological sludges added.
D. As an option to (A), (B), and (C), distillation of all partly detarred gases and liquids by controlled combustion. No liquid discharges.	NH_3 Phenol CN^- BOD_5 $S^=$ O&G SS pH	0 0 0 0 0 0 0 6-9	Used at some coke plants. Effective elimination of waste load from water, but transfers load to air.	Can be done only at plants where impact on air quality can be tolerated. Of limited application.	8-12 months	1/2 acre (100x200 ft)	High impact on air quality.	Formation of ashes.

Table 7.9 *(continued)*

Treatment and/or control methods	*Resulting effluent levels for critical constituents (mg/l)*		*Status and reliability*	*Problems and limitations*	*Implementation time*	*Land requirements*	*Environmental impact other than water*	*Solid waste generation and primary constituents*
Subcategory: Sintering[2]								
A. Aqueous discharge from scrubber through classifier to thickener "once-through." Overflow to sewer, underflow through vacuum filters to sinter plant or landfilled, filtrate recycled to thickener.	SS O&G $S^=$ F^- pH	40 45 65 30 8-10	Widely practiced, usually in conjunction with blast furnace operations. Dependable system.	No reduction of heat load or dissolved chemicals.	18 months	1 acre (200x200 ft)	Air: Particulate 0.1 lb/1,000 lb exhaust gases.	Solids consumed internally.
B. Same as (A) except with chemical polymer flocculation in thickener.	SS O&G $S^=$ F^- pH	20 45 65 30 8-10	Usually included with blast furnace treatment system. Improves solids removal.	No reduction of heat load or dissolved chemicals.	18 months	1 acre (200x200 ft)	Air: Particulate 0.1 lb/1,000 lb exhaust gases.	Solids consumed internally.
C. Same as (B) except thickener overflow recycled to scrubber system with blowdown. Oil skimmer added to thickener. Add neutralization of blowdown.	SS O&G $S^=$ F^- pH	50 10 20 50 6-9	Recycle increases certain constituent concentrations but reduces loads.	No reduction of heat load. Increase in dissolved chemical concentrations.	18 months	1 acre (200x200 ft)	Air: Particulate 0.1 lb/1,000 lb exhaust gases.	Solids consumed internally.
D. Same as (C) except blowdown treated through improved settling with aeration, lime treatment for F^-, neutralization, and sedimentation.	SS O&G $S^=$ F^- pH	25 10 0.3 20 6-9	$S^=$ & F^- removals practiced in other industries successfully. Process must be monitored.	Requires close attention to treatment systems.	18 months	1½ acres (200x300 ft)	Air: Particulate 0.1 lb/1,000 lb exhaust gases.	Solids consumed internally, and other solids to landfill.

E. Same as (D) except additional F^- removal via activated alumina treatment (see Chapter 5).	SS O&G $S^=$ F^- pH	10 3 0.3 5 6-9	F^- removal demonstrated on pilot scale; technology subject to scaling up to full size.	Requires close attention to treatment systems.	18 months	1½ acres (200x300 ft)	Air: Particulate 0.1 lb/1,000 lb exhaust gases.	Solids consumed internally, and other solids to landfill.
Subcategory: Blast furnace (iron)[3]								
A. Once-through solids removed via thickener and vacuum filter. Polymer added to improve settling.	SS CN^- Phenol NH_3 $S^=$ F^- pH	50 2.0 1.0 10 4 5 7-9	Widely used, SS removal efficiency depends upon sludge level & filter schedule.	Removes suspended solids, and a minor portion of volatiles.	12-18 months	1/2 acre (100x200 ft)	Volatiles lost through surface evaporation.	Iron oxide sludge to sinter plant or landfill.
B. To (A), add recycle over cooling tower, discharge blowdown only.	SS CN^- Phenol NH_3 $S^=$ F^- pH	50 15 4 125 6 40 6-9	Used in steel industry. Reliable if properly spared. Sludge level controls solids overflow.	Removes most suspended solids plus much of chemical load, although concentrations increase.	18-24 months	3/4 acre (150x200 ft)	Water spray & volatiles to atmosphere.	Iron oxide sludge to sinter plant.
C. To (B), add treatment of blowdown via alkaline chlorination; precipitation of fluorides with lime; neutralization, filtration and carbon adsorption.	SS CN^- Phenol NH_3 $S^=$ F^- pH	10 0.25 0.5 10 0.3 20 6-9	Alkaline chlorination used at some plants. Carbon adsorption used in other industries. Treatments subject to equipment failures.	May require batch treatment of blowdown to assure performance. High operating costs.	18-24 months	3/4 acre (150x200 ft)	Increased demand for chlorine, causing increase in pollution from chlorine production & power supply.	Iron oxide sludge to sinter plant. Sludge from neutralization step to landfill.

Table 7.9 *(continued)*

Treatment and/or control methods	*Resulting effluent levels for critical constituents (mg/l)*	*Status and reliability*	*Problems and limitations*	*Implementation time*	*Land requirements*	*Environmental impact other than water*	*Solid waste generation and primary constituents*
Subcategory: Blast furnace (ferromanganese)[4]							
A. Once-through gas cooler discharge; closed recycle of venturi scrubber discharge through thickener, and vacuum filter. Polymer added to aid settling.	SS 100 Phenol 1 CN^- 100 NH_3 200 $S^=$ 120 Mn 16 pH 8-10	Used in this industry. Requires attention to recycle system.	High dissolved solids in recycled water; pick-up of volatiles from scrubber recycled water in gas cooler water.	18-24 months	3/4 acre (150x200 ft)	Volatiles are lost to atmosphere.	Filter cake not reusable in process. Must go to landfill.
B. Closed recycle of venturi scrubber as in (A); *separate* recycle of gas cooler water over cooling tower, with pH control. Blowdown to sewer, and to makeup for venturi scrubber recycle system.	SS 100 Phenol 4 CN^- 30 NH_3 200 $S^=$ 30 Mn 16 pH 6-9	Used in the past in this industry. Requires constant attention to separate recycle systems.	High concentrations of dissolved material due to recycling; blowdown loads are reduced, but concentrations are high.	18-24 months	1 acre (200x200 ft)	Volatiles are lost to atmosphere over the cooling tower.	Filter cake not reusable in process. Must go to landfill.
C. Same as in (B), with treatment of gas cooler system blowdown via alkaline chlorination; neutralization; filtration; and carbon adsorption.	SS 10 Phenol 0.5 CN^- 0.25 NH_3 10 $S^=$ 0.3 Mn 5 pH 6-9	Part of technology used at some iron-making blast furnaces; other systems tested on pilot scale. Requires attention to details. Very good.	High operating costs. May require batch treatment of blowdown to assure performance.	18-24 months	1 acre (200x200 ft)	Increased demand for chlorine, causing increase in pollution from chlorine production and power supply.	Additional sludges formed during neutralization.

Subcategory: Basic oxygen furnace (semi-wet air pollution control methods)[5]								
A. Thickener with polymer and/or magnetic flocculation "once-through"; overflow to sewer, underflow through vacuum filters, filter cake recycled to sinter plant or landfill filtrate recycled to thickener.	SS F^- pH	50 20 10-12	Widely used in steel industry. Good system.	Must control surges to system; no reduction of heat load.	15 months	1/4 acre (100x100 ft)	Air: Particulate 0.1 lb/1,000 lb exhaust gases.	Solids consumed internally or used as landfill.
B. Same as (A) except overflow recycled to process spray system through recycle pump system. No aqueous discharge.	SS F^- pH	0 0 –	Practiced by many plants in steel industry. Very good.	Requires more attention than once-through systems.	15 months	1/4 acre (100x100 ft)	Air: Particulate 0.1 lb/1,000 lb exhaust gases.	Solids consumed internally or used as landfill.
Subcategory: Basic oxygen furnace (wet air pollution control methods)[6]								
Fume collection system with boiler hoods								
A. Aqueous discharge from primary scrubber to classifier to thickener. "Once-through," overflow to sewer, underflow through vacuum filters, filter cake recycled to sinter plant or landfilled, filtrate recycled to thickener.	SS F^- pH	80 30 6-9	Widely practiced in industry. Good.	No reduction of heat load must control surges.	18 months	1 acre (200x200 ft)	Air: Particulate 0.1 lb/1,000 lb exhaust gases.	Solid waste consumed internally.
B. To (A), add magnetic and/or chemical polymer flocculation.	SS F^- pH	40 30 6-9	Widely practiced in industry. Very good.	Same as (A).	18 months	1 acre (200x200 ft)	Air: Particulate 0.1 lb/1,000 lb exhaust gases.	Solid waste consumed internally.

Table 7.9 *(continued)*

Treatment and/or control methods	*Resulting effluent levels for critical constituents (mg/l)*		*Status and reliability*	*Problems and limitations*	*Implementation time*	*Land requirements*	*Environmental impact other than water*	*Solid waste generation and primary constituents*
Subcategory: Basic oxygen furnace (wet air pollution control methods) (continued)								
C. To (B), add thickener overflow recycle system with blowdown; neutralization of blowdown stream.	SS F^- pH	50 50 6-9	Widely practiced in industry. Very good.	Dissolved material is concentrated by recycle.	18 months	1 acre (200x200 ft)	Air: Particulate 0.1 lb/1,000 lb exhaust gases.	Solid waste consumed internally. Additional sludges from neutralization to landfill.
D. To (C), add blowdown treatment via settling with coagulation; lime treatment and neutralization.	SS F^- pH	25 20 6-9	Used in controlling steel and other industry wastes. Excellent.	Lime addition requires care in handling; adds to solids wastes generation problem.	18 months	1½ acres (200x300 ft)	Air: Particulate 0.1 lb/1,000 lb exhaust gases.	Solid waste consumed internally. Additional sludges to landfill.
E. To (D), add activated alumina treatment; filtration (see Chapter 5).	SS F^- pH	10 5 6-9	Used in water treatment. Excellent	Technology untested on steel plant wastes; requires attention to all preceding steps.	18 months	1½ acres (200x300 ft)	Air: Particulate 0.1 lb/1,000 lb exhaust gases.	Solid waste consumed internally. Additional sludges to landfill.

Subcategory: Open hearth[7]								
A. Aqueous discharge from primary quencher to classifier to thickener, *"once-through"* overflow to sewer underflow through vacuum filters, filter cake recycled to sinter plant or landfilled. Filtrate returned to thickener.	SS F^- NO_3 Zn pH	80 20 35 220 3-7	Currently used in steel industry. Fair.	No reduction of heat load; must control surges.	18 months	1 acre (200x200 ft)	Air: Particulate 0.1 lb/1,000 lb exhaust gases.	Solid waste consumed internally.
B. Same as (A) but with thickener magnetic and/or chemical flocculation.	SS F^- NO_3 Zn pH	50 20 35 200 3-7	Currently used in steel industry. Good.	No reduction of heat load; must control surges; polymer feed must be maintained.	18 months	1 acre (200x200 ft)	Air: Particulate 0.1 lb/1,000 lb exhaust gases.	Solid waste consumed internally.
C. Same as (B) except thickener overflow recycled to scrubber system with blowdown. A lime addition to recycled water is required from a process standpoint.	SS F^- NO_3 Zn pH	50 100 150 25 6-9	Widely used in industry. Good.	No reduction of heat load; must control surges. pH chemical feed must be maintained.	18 months	1 acre (200x200 ft)	Air: Particulate 0.1 lb/1,000 lb exhaust gases.	Solid waste consumed internally or disposed to landfill.
D. Same as (C) except blowdown treated through lime precipitation of fluoride and zinc; neutralization, sedimentation, and denitrification.	SS F^- NO_3 Zn pH	25 20 45 5 6-9	Used in water treatment & chemical industries. Needs testing on steel wastes. Very good.	No reduction of heat load; must control surges. pH chemical feed must be maintained. High volume of solids formed.	18 months	1½ acres (200x300 ft)	Air: Particulate 0.1 lb/1,000 lb exhaust gases.	Solid waste consumed internally or disposed to landfill.
E. Same as (D) with additional blowdown treatment using activated alumina and filtration.	SS F^- NO_3 Zn pH	10 5 45 3 6-9	Same as (D).	Same as (D).	18 months	2 acres (200x400 ft)	Air: Particulate 0.1 lb/1,000 lb exhaust gases.	Solid waste consumed internally or disposed to landfill.

Table 7.9 *(continued)*

Treatment and/or control methods	*Resulting effluent levels for critical constituents (mg/l)*		*Status and reliability*	*Problems and limitations*	*Implementation time*	*Land requirements*	*Environmental impact other than water*	*Solid waste generation and primary constituents*
Subcategory: Electric arc furnace (semi-wet air pollution control methods)[8]								
A. Discharge from spark box or flame trap to classifier to thickener; overflow recycled to spark box or flame trap; underflow through vacuum filters, filtrate returns to thickener; sludge to sinter or landfill.	SS F^- pH	0 0 –	Currently practiced by steel plants of this type. Excellent.	No reduction of heat load. Spray system requires much maintenance.	12 months	1/8 acre (50x100 ft)	Air: Particulate 0.1 lb/1,000 lb exhaust gases.	Solids consumed internally or used as landfill.
Subcategory: Electric arc furnace (wet air pollution control methods)[9]								
A. Aqueous discharge from scrubber & separator through classifier to a thickener. *"Once-through"* thickener overflow to sewer, underflow through vacuum filters, filter cake recycled to sinter plant or landfilled, filtrate recycled to thickener.	SS F^- Zn pH	100 20 16 6-9	Used in steel industry. Good. Minimum maintenance and downtime.	No reduction of heat load. Must control surges. Most EAF plants have no sinter plants nearby.	18 months	1 acre (200x200 ft)	Air: Particulate 0.1 lb/1,000 lb exhaust gases.	Solid wastes consumed internally or used as landfill.
B. Same as (A) but with thickener magnetic and/or chemical polymer flocculation.	SS F^- Zn pH	50 20 16 6-9	Used in steel industry. Good.	Same as (A).	18 months	1 acre (200x200 ft)	Air: Particulate 0.1 lb/1,000 lb exhaust gases.	Solid wastes consumed internally or used as landfill.

C. Same as (B) except thickener overflow recycled to scrubber system with blowdown.	SS F^- Zn pH	50 75 10 6-9	Widely used in steel industry. Very good.	Same as (A).	18 months	1 acre (200x200 ft)	Air: Particulate 0.1 lb/1,000 lb exhaust gases.	Solid wastes consumed internally or used as landfill.
D. Same as (C) except blowdown treated with lime addition, neutralization, and sedimentation.	SS F^- Zn pH	25 20 5 6-9	Currently in use by some plants in other industries; technically transferable. Excellent.	Same as (A).	18 months	1½ acres (200x300 ft)	Air: Particulate 0.1 lb/1,000 lb exhaust gases.	Solid wastes consumed internally or used as landfill.
E. Same as (D) except additional treatment of blowdown with activated alumina and pressure filtration.	SS F^- Zn pH	10 5 3 6-9	Used in water treatment industry; technically transferable. Excellent.	Same as (A).	18 months	1½ acres (200x300 ft)	Air: Particulate 0.1 lb/1,000 lb exhaust gases.	Solid wastes consumed internally or used as landfill.
Subcategory: Vacuum degassing[10]								
A. Scale sump or settling basin for solids removal. "Once-through" overflow to sewer. Solids recycled to sinter plant.	SS Pb Mn NO_3^- Zn pH	100 2.5 15 80 20 6-9	Used in steel industry.	Surges must be controlled. No reduction in heat load.	18 months	1 acre (200x200 ft)	Gases pass off to atmosphere.	Solids consumed internally.
B. Same as (A) except overflow recycled via cooling tower to degassing unit with blowdown to sewer.	SS Pb Mn NO_3^- Zn pH	50 2.0 10 175 15 6-9	Used in steel industry.	Surges must be controlled. No reduction in heat load.	18 months	1 acre (200x200 ft)	Gases pass off to atmosphere.	Solids consumed internally.

Table 7.9 *(continued)*

Treatment and/or control methods	*Resulting effluent levels for critical constituents (mg/l)*		*Status and reliability*	*Problems and limitations*	*Implementation time*	*Land requirements*	*Environmental impact other than water*	*Solid waste generation and primary constituents*
Subcategory: Vacuum degassing (continued)								
C. Same as (B) except blowdown is treated by lime addition; coagulation/flocculation; anaerobic denitrification; neutralization; and final clarification.	SS Pb Mn NO_3^- Zn pH	25 0.5 5 45 5 6-9	Some treatment methods used in this and related industries. Denitrification is not necessary where N_2 is not used in the process. Very good.	Surges must be controlled. No reduction in heat load. Denitrification untested on steel plant wastes.	18 months	1/2 acre (100x200 ft)	Gases pass off to atmosphere.	Solids consumed internally. Additional solids from lime treatment to landfill.
D. Same as (C) except for final treatment of blowdown via pressure filtration.	SS Pb Mn NO_3^- Zn pH	10 0.3 3 45 3 6-9	Used in steel industry. Very good.	Surges must be controlled. No reduction in heat load.	18 months	1/4 acre (100x100 ft)	Gases pass off to atmosphere.	Solids consumed internally. Additional solids to landfill.
Subcategory: Continuous casting[11]								
A. Recycle system with scale pit; overflow recycled via flat bed filter to cooling tower to caster spray system with blowdown to sewer. Oil skimming at scale pit surface.	SS O&G pH	50 15 6-9	Used in this industry. Good. Scale and oil removal facilities must be maintained.	No reduction in heat load. Pit must be kept clean to prevent solids buildup and washover.	12 months	1/8 acre (50x100 ft)	None	Solids consumed internally. Oil sold for reprocessing or incinerated.

B. Same as (A) except blowdown treatment by pressure filtration.	SS O&G pH	10 10 6-9	Widely used in this industry. Excellent. Scale and oil removal facilities must be maintained.	No reduction in heat load. Pit must be kept clean to prevent solids buildup and washover.	15 months	1/4 acre (100x100 ft)	None	Solids consumed internally. Additional solids to landfill.

Source: After Rice 1974.

Note: Treatment and control methods in each subcategory are listed in order of increasing effectiveness.

1. Well-maintained plants employing recycle of waste water were estimated to display the following water use rates:

	(l/kkg of coke)	*(gal/ton of coke)*
Ammonia liquor (treated in free fixed stills)	104	25
Steam condensate	75	18
Benzol plant wastes	125	30
Blowdown from final cooler recycle	83	20
Condensate from desulfurizers (where used)	104	25
Barometric condenser effluent:		
a. Once-through (BPCTCA)	342	82
b. Recycle with blowdown (BATEA)	20	7

2. Rate of waste water production was estimated at 209 l/kkg (25 gal/1,000 lb) of sinter.
3. Waste water production rates were estimated at 521 l/kkg (6.25 gal/1,000 lb) of iron produced.
4. Waste water production rates were estimated to be 1,043 l/kkg (125 gal/1,000 lb) of ferromanganese produced.
5. In the semi-wet process, a spark box or a spray chamber using slightly more spray water than can be evaporated is used to condition gases for further cleaning, producing a highly contaminated but small volume of waste water. Because recycle permits no discharge, no flow rates are given.
6. In the wet process, high-energy scrubbers or wet gas washers are used to cool and condition furnace gases, producing much larger volumes of moderately contaminated waste waters than are common to the semi-wet systems. Waste water production rate was estimated to be 209 l/kkg (25 gal/1,000 lb) of steel.
7. Waste water production rate was estimated to be 209 l/kkg (25 gal/1,000 lb) of steel.
8. In the electric arc furnace (semi-wet) subcategory, a spark box or a spray chamber using slightly more water than can be evaporated conditions the gases for further cleaning in a precipitator or baghouse. A small volume of contaminated waste water may be produced depending on how much excess spray water is used. No waste water flow rates are given because recycle facilitates no discharge.
9. In the electric arc furnace (wet) subcategory, high-energy scrubbers or wet gas washers are used to cool and condition furnace gases, producing much larger volumes of contaminated waste water than in the semi-wet systems. Waste water production rate was estimated to be 209 l/kkg (25 gal/1,000 lb) of steel.
10. Waste water production rate was estimated to be 209 l/kkg (25 gal/1,000 lb) of steel.
11. Waste water production rate was estimated to be 521 l/kkg (6.25 gal/1,000 lb) of steel.

BEST PRACTICABLE CONTROL TECHNOLOGY CURRENTLY AVAILABLE AND ITS COSTS

Technology and Standards Recommended by Rice (1974)

As noted previously, Public Law 92-500 requires that EPA specify the best practicable control technology currently available (BPCTCA), the best available technology economically achievable (BATEA), and the technology for new source performance standards for point source waste water discharges. Levels of critical constituents in effluents that can be achieved by implementation of these technologies and associated costs must also be specified. For this reason Rice (1974) determined a base technology common to the iron and steel industry. Costs* of additional technologies were added to this base level. Rice presents a detailed analysis of the base technology, of the intermediate technology, of the advanced technology and of the estimated impact of each on energy consumption. Space permits the inclusion only of summaries of the technologies and recommended effluent limitations guidelines in this chapter. These summaries and guidelines and the reviews and criticisms of them by industry are the basis for the effluent limitations guidelines promulgated by EPA.† These guidelines are discussed subsequently.

Rice developed cost estimates as follows:

1. National annual production rate data were collected and tabulated along with the number of plants in each subcategory. From this, an "average-size plant" was established.
2. Waste water flow rates determined from the data collected during the study were utilized (see Table 7.9).
3. Treatment models and flow diagrams were developed for each subcategory, based on how the majority of the plants studied handle wastes.
4. Finally, a cost estimate was made on the developed flow diagram. (See Rice, 1974, for details of cost estimation procedure.)

It was noted that the additional energy required to achieve the proposed effluent limitations guidelines amounts to less than 1.5 percent of the 51.6 x 10^9 kwh of electrical energy used by the steel industry in 1972.

Rice concluded that the proposed effluent limitations would have an impact on air quality because waste water recycle systems and stripping columns may increase the loss of volatile substances to the atmosphere. Recycle systems requiring the use of cooling towers were noted to contribute significantly to reductions of effluent loads and only minimally to air pollution. It was noted also that careful operation can avoid or minimize air pollution problems.

Consideration was also given to the solid waste aspects of water pollution controls, especially with respect to sludges produced by treatment. Much of this material was noted to be inert iron oxide which can be reused profitably.

* Costs in August 1971 dollars.

† The guidelines and the reviews and criticisms are published in the *Federal Register*, Vol. 39, No. 126, Part 2, June 28, 1974, pp. 24114-24133.

Other sludges not suitable for reuse must be discarded to landfills. Disposal of solid wastes by landfilling is discussed in Chapter 11. For steel wastes Rice recommends that in cases where geologic conditions may not reasonably preclude pollution of ground water, adequate mechanical precautions (e.g., impervious liners) be employed in landfills to ensure long-term protection of the environment. Periodic sampling and analysis of leachates is advisable.

Table 7.10 presents summaries of (1) the portions of Table 7.9 estimated by Rice to be the best practicable control technology currently available, (2) costs of its implementation, (3) effluent constituent levels achievable by its implementation and (4) various assumptions on which Rice based estimates. Flow rates are those presented in Table 7.9 unless noted otherwise.

Beehive coke production recommendations presented by Rice have been omitted because use of the process will undoubtedly decline in the future.

Regulations Promulgated by EPA

On the basis of the preceding information and recommendations and reviews thereof, the U.S. Environmental Protection Agency promulgated regulations for point source waste discharges by the iron and steel industry (Tables 7.11 through 7.18).* Waste water flow rates are given in Table 7.10.

Effluent limitations specified under the EPA regulations include the maximum allowable discharge for any one day and the average of daily values for 30 consecutive days. These guidelines, representing the quality attainable by application of the best practicable control technology currently available, are given for the following operations:

By-Product Coke Plants: Effluent limitations and variations to be granted by EPA, under certain conditions, for point source discharges from by-product coke plants are shown in Table 7.11.

Sintering Plants: Effluent limitations for point source discharges from sintering plants are shown in Table 7.12.

Blast Furnaces (Iron): Effluent limitations for point source discharges from iron blast furnaces are shown in Table 7.13.

Blast Furnaces (Ferromanganese): Effluent limitations for point source discharges from ferromanganese blast furnaces are shown in Table 7.14.

Steel Making by Basic Oxygen Process and Electric Arc Furnaces: Effluent limitations for point source discharges from basic oxygen and electric arc furnaces using semi-wet air pollution control methods are shown in Table 7.15.

Steel Making by Basic Oxygen Process, Open Hearth Furnaces, and Electric Arc Furnaces: Effluent limitations for point source discharges from basic oxygen and electric arc furnaces using wet air pollution control methods and from open hearth furnaces are shown in Table 7.16.

Vacuum Degassing of Steel: Effluent limitations for point source discharges from vacuum degassing of steel are shown in Table 7.17.

Continuous Casting of Steel: Effluent limitations for point source discharges from continuous casting of steel are shown in Table 7.18.

* These regulations were published in the *Federal Register*, Vol. 39, No. 126, Part 2, June 28, 1974, pp. 24114-24133.

Table 7.10 *Recommended Effluent Limitations Guidelines*

Critical parameters	*BPCTCA limitations (kg/kkg or lb/1,000 lb)*	*BPCTCA limitations (mg/l)*	*Control & treatment technology*	*Estimated total cost ($/kkg)*	*Estimated total cost ($/ton)*
Subcategory: By-product coke					
Cyanide T[1]	0.0219	30	Weak ammonia liquor equalization and storage	0.173	0.157
Phenol	0.0015	2	Free and fixed leg ammonia still operation with lime addition		
Ammonia (as NH_3)	0.0912	125	Dephenolization		
BOD_5	0.1095	150	Sedimentation		
Oil & grease	0.0109	15	Final cooler blowdown to dephenolizer		
Suspended solids	0.0365	50	Benzol wastes blowdown to dephenolizer		
pH	6.0-9.0		Once-through crystallizer effluent to sedimentation		
			pH neutralization		
Flow	Most probable value for tight system is 730 liters effluent per kkg of coke produced (175 gal/ton), excluding all noncontact cooling water				
Subcategory: Sintering					
Suspended solids	0.0104	50	Thickener with chemical flocculation; tight recycle with minimal blowdown to control cycles of concentration	0.0565	0.0513
Oil & grease	0.0021	10	Natural adsorption to settling solids in thickener; provision required for surface skimming		
pH	6.0-9.0		Neutralization		
Flow	Most probable value for tight system is 209 liters effluent per kkg of sinter produced (50 gal/ton), excluding all noncontact cooling water				

Subcategory: Blast furnace (iron)					
Suspended solids	0.0260	50	Thickening with polymer addition	0.271	0.246
Cyanide T[1]	0.0078	15	Vacuum filtration of thickener sludge		
Phenol	0.0021	4	Recycle loop utilizing cooling tower		
Ammonia (as NH_3)	0.0651	125			
pH	6.0-9.0				
Flow	Most probable value for tight system is 522 liters effluent per kkg of iron produced (125 gal/ton), excluding all noncontact cooling water				
Subcategory: Blast furnace (ferromanganese)					
Suspended solids	0.1043	100	Thickener with polymer addition	1.30	1.18
Cyanide T[1]	0.0312	30	Vacuum filtration of thickener underflow		
Phenol	0.0042	4	Scrubber water recycle with evaporative cooling		
Ammonia (as NH_3)	0.2086	200	pH adjustment		
pH	6.0-9.0				
Flow	Most probable value for tight system is 1,043 liters effluent per kkg of ferromanganese produced (250 gal/ton), excluding all noncontact cooling water				

Source: After Rice 1974.

Notes: Recommended for best practical control technology currently available. Available technology listed is not necessarily all inclusive, nor does it reflect all possible combinations or permutations of treatment methods.

Costs may vary, depending on such factors as location, availability of land and chemicals, flow to be treated, treatment technology selected where competing alternatives exist, and extent of preliminary modifications required to accept the indicated control and treatment devices. Estimated total costs shown are only *incremental* costs required above those facilities normally existing within a plant.

1. Total cyanide. Presumably the guidelines for toxic elements will supersede the BPCTCA guidelines. The toxic element guidelines are likely to be more restrictive. The guidelines proposed initially were printed in the *Federal Register*, Volume 38, No. 247, December 27, 1973 (pp 35388-35395) but these guidelines are being revised.

Table 7.10 *(continued)*

Critical parameters	*BPCTCA limitations (kg/kkg or lb/1,000 lb)*	*BPCTCA limitations (mg/l)*	*Control & treatment technology*	*Estimated total cost ($/kkg)*	*Estimated total cost ($/ton)*
Subcategory: Basic oxygen furnace (semi-wet air pollution control methods)					
Suspended solids Fluoride pH Flow	No discharge of process wastewater pollutants to navigable waters (excluding all noncontact cooling water)		Settling tank with chemical and/or magnetic flocculation; complete recycle with no aqueous blowdown – makeup water as required; wet sludge to reuse or landfill	0.0241	0.0219
Subcategory: Basic oxygen furnace (wet air pollution control methods)					
Suspended solids	0.0104	50	Classifier/thickener with chemical and/or magnetic flocculation; tight recycle with minimal blowdown to control cycles of concentration	0.091	0.082
pH	6.0-9.0		Neutralization		
Flow	Most probable value for tight system is 209 liters effluent per kkg of steel produced (50 gal/ton), excluding all noncontact cooling water				
Subcategory: Open hearth furnace					
Suspended solids	0.0104	50	Classifier/thickener with chemical and/or magnetic flocculation; tight recycle with minimal blowdown to control cycles of concentrations	0.0608	0.0552
pH	6.0-9.0		Neutralization		
Flow	Most probable value for tight system is 209 liters effluent per kkg of steel produced (50 gal/ton), excluding all noncontact cooling water				

Subcategory: Electric arc furnace (semi-wet air pollution control methods)					
Suspended solids Fluoride Zinc pH Flow	No discharge of process wastewater pollutants to navigable waters (excluding all noncontact cooling water)		Settling tank with chemical and/or magnetic flocculation; complete recycle with no aqueous blowdown – makeup water as required; or controlled wetting of gases to form sludge only – no recycle or blowdown; wet sludge to reuse or landfill	Zero (0)	
Subcategory: Electric arc furnace (wet air pollution control methods)					
Suspended solids	0.0104	50	Classifier/thickener with chemical and/or magnetic flocculation; tight recycle with minimal blowdown to control cycles of concentration	0.083	0.0753
pH	6.0-9.0		Neutralization		
Flow	Most probable value for tight system is 209 liters effluent per kkg of steel produced (50 gal/ton), excluding all noncontact cooling water				
Subcategory: Vacuum degassing					
Suspended solids	0.0052	50	Settling with coagulation; tight recycle with minimal blowdown; filtration and cooling over a cooling tower for entire recycle flow	0.568	0.156
pH	6.0-9.0				
Flow	Most probable value for tight system is 104 liters effluent per kkg of steel degassed (25 gal/ton), excluding all noncontact cooling water				

Table 7.10 *(continued)*

	BPCTCA limitations			*Estimated total cost*	
Critical parameters	*(kg/kkg or lb/1,000 lb)*	*(mg/l)*	*Control & treatment technology*	*($/kkg)*	*($/ton)*
Subcategory: Continuous casting					
Suspended solids	0.0260	50	Scale pit with dragout conveyor	Zero (0)	
Oil & grease	0.0078	15	Oil skimmer		
			Flatbed filtration		
			Recycle loop with blowdown and cooling tower		
pH	6.0-9.0				
Flow	Most probable value for tight system is 522 liters effluent per kkg of steel cast (125 gal/ton), excluding all noncontact cooling water				

Table 7.11 *Effluent Limitations Promulgated for By-Product Coke Plants*

Effluent characteristics	*Effluent limitations*			
	Maximum for any one day	*Average of daily values for 30 consecutive days shall not exceed*	*Maximum for any one day*	*Average of daily values for 30 consecutive days shall not exceed*
	Metric units (kg/kkg of product)		*English units (lb/1,000 lb of product)*	
Ammonia	0.2736	0.0912	0.2736	0.0912
Total cyanide	0.0657	0.0219	0.0657	0.0219
Oil & grease	0.0327	0.0109	0.0327	0.0109
Phenol	0.0045	0.0015	0.0045	0.0015
TSS	0.1095	0.0365	0.1095	0.0365
pH	Within the range 6.0 to 9.0		Within the range 6.0 to 9.0	

Note: These guidelines represent the effluent quality attainable by the application of the best practicable control technology currently available. Variation from these effluent limitations will be granted for point source discharges from coke plants utilizing desulfurization units. The limitations specified may be exceeded up to 15 percent by those facilities equipped with gas desulfurization units to the extent that such measured discharge is necessary by reason of the increased effluent volume generated by these facilities. Variation will also be granted for any point source discharges from coke plants utilizing the indirect ammonia recovery process. The basic limitations specified may be exceeded up to 30 percent by those facilities recovering ammonia by this technique, to the extent that such measured discharge is necessary by reason of the increased effluent volume generated by this process.

Table 7.12 *Effluent Limitations Guidelines Promulgated for Sintering Plants*

Effluent characteristics	*Effluent limitations*			
	Maximum for any one day	*Average of daily values for 30 consecutive days shall not exceed*	*Maximum for any one day*	*Average of daily values for 30 consecutive days shall not exceed*
	Metric units (kg/kkg of product)		*English units (lb/1,000 lb of product)*	
TSS	0.0312	0.0104	0.0312	0.0104
Oil & grease	0.0063	0.0021	0.0063	0.0021
pH	Within the range 6.0 to 9.0		Within the range 6.0 to 9.0	

Note: These guidelines represent the effluent quality attainable by the application of the best practicable control technology currently available.

Table 7.13 *Effluent Limitations Guidelines Promulgated for Blast Furnaces (Iron)*

Effluent characteristics	*Effluent limitations*			
	Maximum for any one day	*Average of daily values for 30 consecutive days shall not exceed*	*Maximum for any one day*	*Average of daily values for 30 consecutive days shall not exceed*
	Metric units (kg/kkg of product)		*English units (lb/1,000 lb of product)*	
TSS	0.0780	0.0260	0.0780	0.0260
Total cyanide	0.0234	0.0078	0.0234	0.0078
Phenol	0.0063	0.0021	0.0063	0.0021
Ammonia	0.1953	0.0651	0.1953	0.0651
pH	Within the range 6.0 to 9.0		Within the range 6.0 to 9.0	

Note: These guidelines represent the effluent quality attainable by the application of the best practicable control technology currently available.

Table 7.14 *Effluent Limitations Guidelines Promulgated for Blast Furnaces (Ferromanganese)*

Effluent characteristics	*Effluent limitations*			
	Maximum for any one day	*Average of daily values for 30 consecutive days shall not exceed*	*Maximum for any one day*	*Average of daily values for 30 consecutive days shall not exceed*
	Metric units (kg/kkg of product)		*English units (lb/1,000 lb of product)*	
TSS	0.3129	0.1043	0.3129	0.1043
Total cyanide	0.4689	0.1563	0.4689	0.1563
Phenol	0.0624	0.0208	0.0624	0.0208
Ammonia	1.5636	0.5212	1.5636	0.5212
pH	Within the range 6.0 to 9.0		Within the range 6.0 to 9.0	

Note: These guidelines represent the effluent quality attainable by the application of the best practicable control technology currently available.

Table 7.15 *Effluent Limitations Promulgated for Steel Making by Basic Oxygen Process and Electric Arc Furnaces*

Basic oxygen process (semi-wet air pollution control methods)	The following limitations constitute the quantity or quality of pollutants or pollutant properties which may be discharged after application of best practicable control technology currently available by a point source: there shall be no discharge of process waste water pollutants to navigable waters.
Electric arc furnaces (semi-wet air pollution control methods)	

Table 7.16 *Effluent Limitations Guidelines Promulgated for Steel Making by Basic Oxygen Process, Open Hearth Furnaces, and Electric Arc Furnaces*

Effluent characteristics	*Effluent limitations*			
	Maximum for any one day	*Average of daily values for 30 consecutive days shall not exceed*	*Maximum for any one day*	*Average of daily values for 30 consecutive days shall not exceed*
	Metric units (kg/kkg of product)		*English units (lb/1,000 lb of product)*	
TSS	0.0312	0.0104	0.0312	0.0104
pH	Within the range 6.0 to 9.0		Within the range 6.0 to 9.0	

Note: These guidelines represent the effluent quality attainable by application of the best practicable control technology currently available when wet air pollution control methods are used for the basic oxygen process and electric arc furnace.

Table 7.17 *Effluent Limitations Guidelines Promulgated for Vacuum Degassing of Steel*

Effluent characteristics	*Effluent limitations*			
	Maximum for any one day	*Average of daily values for 30 consecutive days shall not exceed*	*Maximum for any one day*	*Average of daily values for 30 consecutive days shall not exceed*
	Metric units (kg/kkg of product)		*English units (lb/1,000 lb of product)*	
TSS	0.0156	0.0052	0.0156	0.0052
pH	Within the range 6.0 to 9.0		Within the range 6.0 to 9.0	

Note: These guidelines represent the effluent quality attainable by application of the best practicable control technology currently available.

Table 7.18 *Effluent Limitations Guidelines Promulgated for Continuous Casting of Steel*

Effluent characteristics	*Effluent limitations*			
	Maximum for any one day	*Average of daily values for 30 consecutive days shall not exceed*	*Maximum for any one day*	*Average of daily values for 30 consecutive days shall not exceed*
	Metric units (kg/kkg of product)		*English units (lb/1,000 lb of product)*	
TSS	0.0780	0.0260	0.0780	0.0260
Oil & grease	0.0234	0.0078	0.0234	0.0078
pH	Within the range 6.0 to 9.0		Within the range 6.0 to 9.0	

Note: These guidelines represent the effluent quality attainable by application of the best practicable control technology currently available.

BEST AVAILABLE TECHNOLOGY ECONOMICALLY ACHIEVABLE AND ITS COSTS

Technology and Standards Recommended by Rice (1974)

Rice (1974) extracted those portions of the control technology summarized in Table 7.9 which it concluded to be the best technology economically achievable (BATEA) for point source discharges and calculated the additional costs of their implementation. Impacts on airborne and solid wastes also are discussed in the Rice report. Table 7.19 presents the summarized information regarding loadings, concentrations, technology identification, and costs.*

Regulations Promulgated by EPA

On the basis of the preceding information and recommendations and reviews thereof, the U.S. Environmental Protection Agency promulgated the following regulations for point source waste water discharges for iron and steel industry operations (see Tables 8.20 through 8.29).† Waste water flow rates are as given in Table 7.19.

Effluent limitations specified under the EPA regulations include the maximum allowable discharge for any one day and the average of the daily values for 30 consecutive days. These guidelines, representing the quality attainable by the application of the best available technology economically achievable, are given for the following operations:

By-Product Coke Plants: Effluent limitations and variations to be granted by EPA, under certain conditions, for point source discharges from by-product coke plants are shown in Table 7.20.

Sintering Plants: Effluent limitations for point source discharges from sintering plants are shown in Table 7.21.

Blast Furnaces (Iron): Effluent limitations for point source discharges from iron blast furnaces are shown in Table 7.22.

Blast Furnaces (Ferromanganese): Effluent limitations for point source discharges from ferromanganese blast furnaces are shown in Table 7.23.

Steel Making by Basic Oxygen Process and Electric Arc Furnaces: Effluent limitations for point source discharges from basic oxygen and electric arc furnaces using semi-wet air pollution control methods are shown in Table 7.24.

Steel Making by Basic Oxygen Process: Effluent limitations for point source discharges from basic oxygen furnaces using wet air pollution control methods are shown in Table 7.25.

Steel Making by Open Hearth Furnaces: Effluent limitations for point source discharges from open hearth furnaces are shown in Table 7.26.

Steel Making by Electric Arc Furnaces: Effluent limitations for point source discharges from electric arc furnaces using wet air pollution control methods are shown in Table 7.27.

* Costs in August 1971 dollars.

† These regulations were published in the *Federal Register*, Vol. 39, No. 126, Part 2, June 28, 1974, pp. 24114-24133.

Vacuum Degassing of Steel: Effluent limitations for point source discharges from vacuum degassing of steel are shown in Table 7.28.

Continuous Casting of Steel: Effluent limitations for point source discharges from continuous casting of steel are shown in Table 7.29.

NEW SOURCE PERFORMANCE STANDARDS

Standards Recommended and Rationale Presented by Rice (1974)

On the basis of data obtained during its study of the steel industry, Rice (1974) recommended the following performance standards for new point source waste water discharges. The rationale for these recommendations is also presented in summary form.

Coke By-Product Production

By-product production from coke produces 80 to 165 l of waste water per kkg (19 to 40 gal/ton) of coke produced. This water constitutes the water originally present in the raw coal fed to the ovens. This water emanates as gas and is condensed at the primary cooler and the final cooler. Approximately 75 percent condenses at the primary cooler in the form of ammonia liquor. The remaining 25 percent condenses in the final cooler as final cooler drains.

Coke plant waste water in excess of this approximately 104 l/kkg (25 gal/ton) is added to aid in processing of the coke or the by-products. Waste water may emanate from coke quenching tower overflow (or blowdown), coke wharf drains, steam condensed in the ammonia stills, cooling tower blowdowns, boiler blowdowns, cooling system leaks, and wash water used in the coke plant area.

Pyrolytic decomposition of coal seemingly cannot avoid 80 to 165 l/kkg (19 to 40 gal/ton) of highly contaminated liquid present in the coal. The coke wharf and quenching water can be eliminated by dry coke quenching as practiced in other countries or by using the wharf drainage as quench tower makeup water, while precluding overflow from the quench tower. Heat and corrosion problems may be generated by this practice, but these problems are considered solvable with conventional designs. But Rice (1974) notes that elimination of discharge of waste water from modern coke plants requires a means of elimination of the 80 to 165 l/kkg (19 to 40 gal/ton) of liquid in the coal. All the wastes in this water, with the possible exception of suspended solids, are pyrolytically decomposable with an expenditure of approximately 126,000 kilogram calories per metric ton of coke. However, pyrolytic decomposition of these wastes can be expected to create serious air pollution problems. The gases emanated from less than optimum incineration would contain oxides of nitrogen and sulfur and some particulate matter. If an incinerator with a wet scrubber were used, the pollutants would simply be transferred back to another waste water stream and not eliminated. Volatilization of the wastes if evaporation of the liquid to dryness were employed would result in much the same problems as incineration.

Rice concluded, therefore, that present technology does not appear to be capable of producing zero discharge of coke by-product waste waters. For this reason zero discharge as a new source performance standard (NSPS) was not recommended. New source performance standards identical to those for best available technology economically achievable were recommended.

Sintering

A sinter agglomerates dust from blast furnaces and other processes for charge into the blast furnace for recovery of iron from the dust. The fines screened from coke serve as fuel for sinter plants.

Stringent air pollution standards have antiquated conventional dry dust collection methods and wet scrubbing of the dust-laden gases has been adopted. The scrubbers produce liquid discharges. Thus, Rice (1974) concluded that so long as air quality standards are such that they can only be met by wet scrubbing methods, there will be a liquid discharge from sinter plants. Therefore, the guidelines recommended for best available technology economically achievable were also recommended for new source performance standards.

Blast Furnace (Iron) and Blast Furnace (Ferromanganese)

Noncontact cooling water and process water from gas cleaning operations for air pollution control constitute the waste waters from blast furnaces. The noncontact cooling water should contain only heat as a pollutant and it must somehow be rejected to the environment in order for the process to operate. It can be rejected to water by water-cooling systems or to the air via cooling towers.

As noted previously, the process water which cleans and cools blast furnace top gases by direct contact becomes contaminated with suspended solids, cyanide, phenol, ammonia, and sulfur compounds. This gas cleaning and cooling water can be recycled after suspended solids removal by settling chambers and after passage through a cooling tower to remove heat. Rice (1974) concluded that attempts have been made to close these systems but that it has not been clearly demonstrated that operation can proceed without some blowdown. Therefore Rice recommended BATEA limitations for NSPS limitations for both ferromanganese and iron-making furnaces.

Steel Making

Direct-contact cleaning of gases for air pollution control produces waste water discharges from open hearths, oxygen processes, and electric furnaces. Higher air quality standards have produced a shift to wet cleaning methods. Rice (1974) noted that, as long as the technology of dry gas cleaning lagged behind the requirements of air quality standards, liquid discharges from steel making must continue. For this reason, Rice recommended no additional reductions in flow or pollutant loads over BATEA.

Table 7.19 *Recommended Effluent Limitations Guidelines*

Critical parameters	BATEA limitations (kg/kkg or lb/1,000 lb)	BATEA limitations (mg/l)	Control & treatment technology	Estimated total cost ($/kkg)	Estimated total cost ($/ton)
Subcategory: By-product coke					
Cyanide A[1]	0.00010	0.25	BPCTCA plus:		
Phenol	0.00021	0.5	Recycle crystallizer effluent to final cooler recycle system		
Ammonia (as NH_3)	0.0042	10	Sulfide oxidation (aeration)		
BOD_5	0.0083	20	Clarification	0.405	0.367
Sulfide	0.00012	0.3	Abandon dephenolization		
Oil & grease	0.0042	10	Multi-stage biological oxidation with methanol addition		
Suspended solids	0.0042	10	Pressure filtration		
pH	6.0-9.0				
Flow	Most probable value for tight system is 417 liters effluent per kkg of coke produced (100 gal/ton), excluding all noncontact cooling water				
Subcategory: Sintering					
Suspended solids	0.0053	25			
Oil & grease	0.0021	10	(Implemented under BPCTCA standards)		
Sulfide	0.00006	0.3			
Fluoride	0.0042	20	Blowdown treatment using lime precipitation of fluorides	0.0694	0.0630
pH	6.0-9.0		Neutralization		
Flow	Most probable value for tight system is 209 liters effluent per kkg of sinter produced (50 gal/ton), excluding all noncontact cooling water				

Subcategory: Blast furnace (iron)					
Suspended solids	0.0052	10	BPCTCA plus:	0.267	0.242
Cyanide A[1]	0.00013	0.25	Treatment of cooling		
Phenol	0.00026	0.5	Tower blowdown via:		
Ammonia	0.0052	10	Alkaline chlorination		
Sulfide	0.00016	0.3	Pressure filtration		
Fluroide	0.0104	20	Carbon adsorption		
pH	6.0-9.0		pH neutralization		
Flow	Most probable value for tight system is 522 liters effluent per kkg of coke produced (125 gal/ton), excluding all noncontact cooling water				

Source: After Rice 1974.

Notes: Recommended for best available technology economically achievable (BATEA). Available technology listed is not necessarily all inclusive nor does it reflect all possible combinations of permutations of treatment methods.

Costs may vary, depending on such factors as location, availability of land and chemicals, flow to be treated, treatment technology selected where competing alternatives exist, and extent of preliminary modifications required to accept the indicated control and treatment devices. Estimated total costs shown are only *incremental* costs required above those facilities normally existing within a plant and/or have been installed as a result of complying with standards for best practicable control technology currently available (BPCTCA).

1. Cyanides amenable to chlorination. Reference ASTM D 2036-72, Method B. Presumably the guidelines for toxic elements will supersede the BPCTCA guidelines. The toxic guidelines are likely to be more restrictive. The guidelines proposed initially were printed in the *Federal Register*, Volume 38, No. 247, December 27, 1973 (pp 35388-35395) but these guidelines are being revised.

Table 7.19 *(continued)*

Critical parameters	BATEA limitations (kg/kkg or lb/1,000 lb)	BATEA limitations (mg/l)	Control & treatment technology	Estimated total cost ($/kkg)	Estimated total cost ($/ton)
Subcategory: Blast furnace (ferromanganese)					
Suspended solids	0.0104	10	BPCTCA plus:	1.927	1.749
Cyanide A[1]	0.00026	0.25	Treatment of system		
Phenol	0.00052	0.5	blowdown via:		
Ammonia (as NH_3)	0.0104	10	Alkaline chlorination		
Sulfide	0.00031	0.3	Pressure filtration		
Manganese	0.0052	5	Carbon adsorption		
pH	6.0-9.0		pH neutralization		
Flow	Most probable value for tight system is 1,043 liters per kkg of ferromanganese produced (250 gal/ton), excluding all noncontact cooling water				
Subcategory: Basic oxygen furnace (semi-wet air pollution control methods)					
Suspended solids; Fluoride; pH; Flow	No discharge of process wastewater pollutants to navigable waters (excluding noncontact cooling water)		Same as BPCTCA	Zero (0)	

Subcategory: Basic oxygen furnace (wet air pollution control methods)					
Suspended solids	0.0052	25	Blowdown treatment with sand filtration or improved settling with coagulation	0.0328	0.0298
Fluoride	0.0042	20	Blowdown treatment using lime precipitation of fluorides		
pH	6.0-9.0		Neutralization		
Flow	Most probable value for tight system is 209 liters effluent per kkg of steel produced (50 gal/ton), excluding all noncontact cooling water				
Subcategory: Open-hearth furnace					
Suspended solids	0.0052	25	Blowdown treatment with sand filtration or improved settling with coagulation	0.126	0.114
Fluoride	0.0042	20	Blowdown treatment using lime precipitation of fluorides		
Nitrate (as NO_3)	0.0094	45	Anaerobic denitrification		
Zinc	0.0010	5	Reduction occurs as a result of improved suspended solids removal		
pH	6.0-9.0		Neutralization		
Flow	Most probable value for tight system is 209 liters effluent per kkg of steel produced (50 gal/ton), excluding all noncontact cooling water				
Subcategory: Electric arc furnace (semi-wet air pollution control methods)					
Suspended solids, Fluoride, Zinc, pH, Flow	No discharge of process wastewater pollutants to navigable waters (excluding all noncontact cooling water)		Same as BPCTCA	Zero (0)	

Table 7.19 *(continued)*

Critical parameters	BATEA limitations (kg/kkg or lb/1,000 lb)	BATEA limitations (mg/l)	Control & treatment technology	Estimated total cost ($/kkg)	Estimated total cost ($/ton)
Subcategory: Electric-arc furnace (wet air pollution control methods)					
Suspended solids	0.0052	25	Blowdown treatment with sand filtration or improved settling with coagulation		
Fluoride	0.0042	20	Blowdown treatment using lime precipitation of fluorides		
Zinc	0.0010	5	Reduction occurs as a result of improved suspended solids removal	0.0988	.0897
pH	6.0-9.0		Neutralization		
Flow	Most probable value for tight system is 209 liters effluent per kkg of steel produced (50 gal/ton), excluding all noncontact cooling water				
Subcategory: Vacuum degassing					
Suspended solids	0.0026	25	Blowdown treatment with sand filtration (suspended solids, zinc, manganese, lead)		
Zinc	0.00052	5			
Manganese	0.00052	5			
Lead	0.00005	0.5		0.492	0.446
Nitrate (as NO_3)	0.0047	45	Blowdown treatment with anaerobic denitrification, (or substitution of another gas for blanketing instead of nitrogen)		
pH	6.0-9.0		Neutralization		
Flow	Most probable value for tight system is 104 liters effluent per kkg of steel degassed (25 gal/ton), excluding all noncontact cooling water				

Subcategory: Continuous casting					
Suspended solids	0.0052	10	BPCTCA plus: Filtration of blowdown	0.0752	0.0682
Oil & grease	0.0052	10			
pH	6.0-9.0				
Flow	Most probable value for tight system is 522 liters effluent per kkg of steel cast (125 gal/ton), excluding all noncontact cooling water				

Table 7.20 *Effluent Limitations Promulgated for By-Product Coke Plants*

Effluent characteristics	*Effluent limitations*			
	Maximum for any one day	*Average of daily values for 30 consecutive days shall not exceed*	*Maximum for any one day*	*Average of daily values for 30 consecutive days shall not exceed*
	Metric units (kg/kkg of product)		*English units (lb/1,000 lb of product)*	
Cyanide A[1]	0.0003	0.0001	0.0003	0.0001
Phenol	0.0006	0.0002	0.0006	0.0002
Ammonia	0.0126	0.0042	0.0126	0.0042
Sulfide	0.0003	0.0001	0.0003	0.0001
Oil & grease	0.0126	0.0042	0.0126	0.0042
TSS	0.0312	0.0104	0.0312	0.0104
pH	Within the range 6.0 to 9.0		Within the range 6.0 to 9.0	

Note: These guidelines represent the effluent quality attainable by the application of the best available technology economically achievable. Variation will be granted for point source discharges from coke plants utilizing desulfurization units. The limitations specified may be exceeded up to 25 percent by those facilities equipped with gas desulfurization units to the extent that such measured discharge is necessary by reason of the increased effluent volume generated by these facilities. Variation will also be granted for point source discharges from coke plants utilizing the indirect ammonia recovery process. The basic limitations specified may be exceeded up to 70 percent by those facilities recovering ammonia by this technique, to the extent that such measured discharge is necessary by reason of the increased effluent volume generated by this process.

1. Cyanide A refers to cyanides amenable to chlorination as described in, "1972 Annual Book of ASTM Standards," Standard D2036-72, Method B, p. 553.

Table 7.21 *Effluent Limitations Promulgated for Sintering Plants*

Effluent characteristics	*Effluent limitations*			
	Maximum for any one day	*Average of daily values for 30 consecutive days shall not exceed*	*Maximum for any one day*	*Average of daily values for 30 consecutive days shall not exceed*
	Metric units (kg/kkg of product)		*English units (lb/1,000 lb of product)*	
Oil & grease	0.0063	0.0021	0.0063	0.0021
Sulfide	0.00018	0.00006	0.00018	0.00006
Fluoride	0.0126	0.0042	0.0126	0.0042
TSS	0.0156	0.0052	0.0156	0.0052
pH	Within the range 6.0 to 9.0		Within the range 6.0 to 9.0	

Note: These guidelines represent the effluent quality attainable by the application of the best available technology economically achievable.

Table 7.22 *Effluent Limitations Promulgated for Blast Furnaces (Iron)*

Effluent characteristics	*Effluent limitations*			
	Maximum for any one day	*Average of daily values for 30 consecutive days shall not exceed*	*Maximum for any one day*	*Average of daily values for 30 consecutive days shall not exceed*
	Metric units (kg/kkg of products)		*English units (lb/1,000 lb of product)*[1]	
TSS	0.0390	0.0130	0.0390	0.0130
Cyanide A[1]	0.0004	0.00013	0.0004	0.00013
Phenol	0.0008	0.00026	0.0008	0.00026
Ammonia	0.0156	0.0052	0.0156	0.0052
Sulfide	0.0005	0.00016	0.0005	0.00016
Fluoride	0.0312	0.0104	0.0312	0.0104
pH	Within the range 6.0 to 9.0		Within the range 6.0 to 9.0	

Note: These guidelines represent the effluent quality attainable by the application of the best available technology economically achievable.

1. Cyanides amenable to chlorination (see note on Table 7.20).

Table 7.23 *Effluent Limitations Promulgated for Blast Furnaces (Ferromanganese)*

Effluent characteristics	*Effluent limitations*			
	Maximum for any one day	*Average of daily values for 30 consecutive days shall not exceed*	*Maximum for any one day*	*Average of daily values for 30 consecutive days shall not exceed*
	Metric units (kg/kkg of product)		*English units (lb/1,000 lb of product)*	
TSS	0.0780	0.0260	0.0780	0.0260
Cyanide A[1]	0.0008	0.00026	0.0008	0.00026
Phenol	0.0016	0.00052	0.0016	0.00052
Ammonia	0.0312	0.0104	0.0312	0.0104
Sulfide	0.0009	0.0003	0.0009	0.0003
Manganese	0.0156	0.0052	0.0156	0.0052
pH	Within the range 6.0 to 9.0		Within the range 6.0 to 9.0	

Note: These guidelines represent the effluent quality attainable by the application of the best available technology economically achievable.

1. Cyanides amenable to chlorination (see note on Table 7.20).

Table 7.24 *Effluent Limitations Promulgated for Steel Making by Basic Oxygen Process and Electric Arc Furnaces*

Basic oxygen process (semi-wet air pollution control methods) Electric arc furnaces (semi-wet air pollution control methods)	The following limitations constitute the quality of pollutants or pollutant properties which may be discharged after application of the best available technology economically achievable by a point source: there shall be no discharge of process waste water pollutants to navigable waters.

Table 7.25 *Effluent Limitations Promulgated for Steel Making by Basic Oxygen Furnaces (wet air pollution methods)*

Effluent characteristics	*Effluent limitations*			
	Maximum for any one day	*Average of daily values for 30 consecutive days shall not exceed*	*Maximum for any one day*	*Average of daily values for 30 consecutive days shall not exceed*
	Metric units (kg/kkg of product)		*English units (lb/1,000 lb of product)*	
TSS	0.0156	0.0052	0.0156	0.0052
Fluoride	0.0126	0.0042	0.0126	0.0042
pH	Within the range 6.0 to 9.0		Within the range 6.0 to 9.0	

Note: These guidelines represent the effluent quality attainable by the application of the best available technology economically achievable.

Table 7.26 *Effluent Limitations Promulgated for Steel Making by Open Hearth Furnaces*

Effluent characteristics	*Effluent limitations*			
	Maximum for any one day	*Average of daily values for 30 consecutive days shall not exceed*	*Maximum for any one day*	*Average of daily values for 30 consecutive days shall not exceed*
	Metric units (kg/kkg of product)		*English units (lb/1,000 lb of product)*	
TSS	0.0156	0.0052	0.0156	0.0052
Fluoride	0.0126	0.0042	0.0126	0.0042
Nitrate	0.0282	0.0094	0.0282	0.0094
Zinc	0.0030	0.0010	0.0030	0.0010
pH	Within the range 6.0 to 9.0		Within the range 6.0 to 9.0	

Note: These guidelines represent the effluent quality attainable by the application of the best available technology economically achievable.

Table 7.27 *Effluent Limitations Promulgated for Steel Making by Electric Arc Furnaces (wet air pollution control methods)*

Effluent characteristics	*Effluent limitations*			
	Maximum for any one day	*Average of daily values for 30 consecutive days shall not exceed*	*Maximum for any one day*	*Average of daily values for 30 consecutive days shall not exceed*
	Metric units (kg/kkg of product)		*English units (lb/1,000 lb of product)*	
TSS	0.0156	0.0052	0.0156	0.0052
Fluoride	0.0126	0.0042	0.0126	0.0042
Zinc	0.0030	0.0010	0.0030	0.0010
pH	Within the range 6.0 to 9.0		Within the range 6.0 to 9.0	

Note: These guidelines represent the effluent quality attainable by the application of the best available technology economically achievable.

Table 7.28 *Effluent Limitations Promulgated for Vacuum Degassing of Steel*

Effluent characteristics	*Effluent limitations*			
	Maximum for any one day	*Average of daily values for 30 consecutive days shall not exceed*	*Maximum for any one day*	*Average of daily values for 30 consecutive days shall not exceed*
	Metric units (kg/kkg of product)		*English units (lb/1,000 lb of product)*	
TSS	0.0078	0.0026	0.0078	0.0026
Zinc	0.0015	0.0005	0.0015	0.0005
Manganese	0.0015	0.0005	0.0015	0.0005
Lead	0.00015	0.00005	0.00015	0.00005
Nitrate	0.0141	0.0047	0.0141	0.0047
pH	Within the range 6.0 to 9.0		Within the range 6.0 to 9.0	

Note: These guidelines represent the effluent quality attainable by the application of the best available technology economically achievable.

Table 7.29 *Effluent Limitations Promulgated for Continuous Casting of Steel*

Effluent characteristics	*Effluent limitations*			
	Maximum for any one day	*Average of daily values for 30 consecutive days shall not exceed*	*Maximum for any one day*	*Average of daily values for 30 consecutive days shall not exceed*
	Metric units (kg/kkg of product)		*English units (lb/1,000 lb of product)*	
TSS	0.0156	0.0052	0.0156	0.0052
Oil & grease	0.0156	0.0052	0.0156	0.0052
pH	Within the range 6.0 to 9.0		Within the range 6.0 to 9.0	

Note: These guidelines represent the effluent quality attainable by the application of the best available technology economically achievable.

Vacuum Degassing

This process removes dissolved gases from the molten metal. The liquid discharge consists of the condensed steam from the high-capacity steam jet ejectors that produce the vacuum.

All of the withdrawn gases, plus particulate matter produced by the violent boiling which occurs when the vacuum is created, contact the water and contribute particulate and dissolved constituents to the condensate. Rice (1974) concluded that substitution of another type of vacuum system does not seem practical at this time and so no further reductions over BATEA limitations were recommended.

Continuous Casting

Considerable contact cooling water is used in the continuous casting process. It becomes contaminated primarily with suspended solids (iron oxide), with small amounts of oil and grease from the lubricants used on the equipment and occasionally with hydraulic fluid if a leak occurs. Rice (1974) concluded that the technology for significantly reducing either the volume or the level of contamination is not available at this time. Thus, no further reductions from BATEA limitations were proposed.

Regulations Promulgated by EPA

On the basis of the preceding information and recommendations and reviews thereof by industrial groups, the U.S. Environmental Protection Agency promulgated regulations for NSPS identical to those for BATEA.

REFERENCES

Battelle Memorial Institute 1970. Final Report on Evaluation of Process Alternatives to Improve Control of Air Pollution from Production of Coke.

Bennett, K. W. 1968. Pollution Control–Is Steel Meeting the Challenge?, *Iron Age*, p. 95.

Bramer, Henry C., and Gadd, William L. 1970. *Magnetic Flocculation of Steel Mill Waste Waters.* Proceedings, 25th Industrial Waste Conference, Purdue Univ., pp. 154-165.

Brough, John R., and Voges, Thomas F. 1969. *Basic Oxygen Process Water Treatment.* Proceedings, 24th Industrial Waste Conference, Purdue Univ., pp. 762-769.

Brough, John R., and Voges, Thomas F. 1970. Water Supply and Wastewater Disposal for a Steel Mill. *Water and Wastes Engineering*, Vol. 7, No. 1, pp. A25-A27.

Catchpole, J. R. 1958. *The Treatment and Disposal of Effluents in the Gas and Coke Industry, Air and Water Pollution in the Iron and Steel Industry.* Iron and Steel Institute Special Report 1961, pp. 219-225.

Cave, R. W. 1970. *Effluent Disposal in an Integrated Works, Management of Water in the Iron and Steel Industry.* Iron and Steel Institute Special Report 128, pp. 124-130.

Cooper, R. L. 1958. *Methods of Approach to Coke Oven Effluent Problems, Air and Water Pollution in the Iron and Steel Industry.* Iron and Steel Institute Special Report 61, pp. 198-202.

Council on Environmental Quality 1972. *A Study of the Economic Impact on the Steel Industry of the Costs of Meeting Federal Air and Water Pollution Abatement Requirements, Parts I, II, and III.* Washington, D.C.

Decaigny, Roger A. 1970. *Blast Furnace Gas Washer Removes Cyanides, Ammonia, Iron, and Phenol.* Proceedings, 25th Industrial Waste Conference, Purdue Univ., pp. 512-517.

Edgar, W. D., and Muller, J. M. 1973. The Status of Coke Oven Pollution Control. Paper presented at April meeting of AIME, Cleveland, Ohio.

EPA 1972. Bibliography of Water Quality Research Reports, Water Pollution Control Research Series, pp. 1-40. U.S. Environmental Protection Agency, Office of Research and Monitoring, Washington, D.C.

Kemmetmueller, R. 1973. Dry Coke Quenching–Proved, Profitable, Pollution Free Quenching Technology. Paper presented at April meeting of AISE, Chicago, Illinois.

Koziorowski, B., and Kucharski, J. 1972. *Industrial Waste Disposal.* Translated from the Polish by J. Bandrowski. New York: Pergamon Press.

McMichael, Francis C., Maruhnich, Edward D., and Samples, William R. 1971. Recycle Water Quality from a Blast Furnace. *Journal of the Water Pollution Control Association*, Vol. 43, pp. 595-603.

Patton, R. S. 1973. Hooded Coke Quenching System for Air Quality Control. Paper presented at April meeting of AISE, Chicago, Illinois.

Rice, 1974. Development Document for *Proposed Effluent Limitations Guidelines and New Source Performance Standards for the Steel Making Segment of the Iron and Steel Manufacturing Point Source Category.* EPA Document 440/1-73-024, Supt. of Documents, U.S. Govt. Printing Office, Washington, D.C.

Stove, Ralph, and Schmidt, Carter 1968. *A Survey of Industrial Waste Treatment Costs and Charges.* Proceedings, 23rd Industrial Waste Conference, Purdue Univ., pp. 49-63.

Talbott, John A. 1971. Building a Pollution-Free Steel Plant. *Mechanical Engineer*, Vol. 93, No. 1, pp. 25-30.

Wallace, De Yarman 1970. Blast Furnace Gas Washer Water Recycle System. *Iron and Steel Engineer Yearbook*, pp. 231-235.

8. *Ferroalloy industry*

SOURCES OF INFORMATION

The sources of information for this chapter are the published literature and the effluent guidelines document prepared under EPA contract no. 68-01-1527, awarded to Datagraphics Incorporated as authorized under Public Law 92-500 (see Datagraphics 1973). The document comprises the most comprehensive available investigation of the origin and disposal of ferroalloy wastes. Information referred to in this chapter as "reported" is information supplied to Datagraphics in response to its extensive study of the industry. Because the Datagraphics report is unique, much of this chapter consists of extractions from it and condensations of certain sections.

FERROALLOYS

Ferroalloys are those elements used primarily for deoxidation, alloying, and graphitization of steel and cast iron. Manganese is the most widely used ferroalloy element, followed by silicon, chromium, and phosphorus. Others include molybdenum, tungsten, titanium, zirconium, vanadium, boron, and columbium.

Datagraphics (1973) reports that 40 plants in the United States produce ferroalloy metals. They cite 34 establishments in Standard Industrial Classifications (SIC) 3313, i.e., those primarily engaged in the production of electrometallurgical products. Twenty of these reported using 75.7 million liters (20 million gallons) or more of water annually. The 20 large water-using plants account for some 88 percent of the value of shipments in SIC 3313. They account for 80 percent of the total value of shipments of ferroalloys.

Four principal methods produce ferroalloy and metallic additives for steel making. These are (1) blast furnace, (2) electric smelting furnace, (3) aluminothermic or silicothermic process, and (4) electrolytic deposition. Ferromanganese is produced in the blast furnace or the electric arc furnace, the latter being predominant. As in the iron and steel industry (Chapter 7), ore, coke, and limestone or dolomite constitute the charge. Although some batch melt furnaces exist, most furnaces are the continuous submerged arc type; thus a large supply of electric power is necessary, and waste production is more or less continuous. The major ferroalloys produced in submerged arc furnaces are:

1. Silicon alloys—ferrosilicon (50 to 90 percent Si) and calcium silicide.
2. Chromium alloys—high-carbon ferrochromium in various grades and ferrochrome-silicon.
3. Manganese alloys—standard ferromanganese and silicomanganese.

Submerged arc electric furnaces may be open, closed, or sealed. Open furnaces have no cover, and air is supplied at the top to burn the CO gas from the charge. Closed furnaces have a top cover through which the electrodes extend down into the charge; the space around the electrodes is filled with the charge to prevent escape of gases generated. The sealed furnace has a similar cover, but with mechanical seals around both the electrodes and mix spouts. Use of the sealed furnace has thus far been restricted to calcium carbide, pig iron, standard ferromanganese, and silicomanganese. Sealed covers are reported to be difficult to adapt to existing furnaces because extensive revisions are required. A modified cover has been developed, incorporating electrode seals, but covering only the reaction zones around the electrodes and leaving the outer rim of the furnace open. These seals, called gas collection sleeves or smoke rings, collect the gas in the region of maximum generation, allow partial stoking of the mix, and are reported to be cheaper than a complete cover.

As in the case of other industries discussed, the largest source of waterborne pollutants in the ferroalloy industry is the effluent from wet scrubbers used for air pollution control. Emissions are from raw material handling, mix delivery, crushing, grinding, sizing, and furnaces. The submerged arc electric furnace utilizes carbon reduction of metallic oxides and continuously generates carbon monoxide (CO). Other sources of gas are charge moisture, reducing agent volatile matter, thermal decomposition products of the ore or smelting mix, and intermediate reaction products. Datagraphics reports that carbon monoxide normally accounts for about 70 percent of the gases by volume. The fumes rising from the furnace entrain the finer constituents of the mix or charge. In an open furnace, all the CO is burned at the top of the charge, resulting in a large volume of gas. In a closed furnace most or all of the CO is withdrawn from the furnace.

Fume emission also occurs at furnace tapholes. Most furnaces are tapped intermittently so taphole fumes emanate only about 10 to 20 percent of the furnace operating time. Melting operations are conducted in open arc furnaces in some plants. Datagraphics observed that no major quantities of gas are generated in this operation, but thermally inducted air flow may result in fume emission.

The water and air pollution features of the electric arc furnace are discussed in considerable detail in Chapter 7. Consequently, the discussion presented here is somewhat limited; the similarities are obvious. Air pollution control devices used include baghouses, wet scrubbers, and electrostatic precipitators. As noted earlier, wet scrubbers produce high-viscosity slurries containing most of the particulates in the off-gases. Spray towers which cool the gases produce slurries containing some of the particulates in the gases. Baghouses generally produce only solid wastes.

Datagraphics concluded that the venturi-type wet scrubber is the only currently feasible type of wet collector for cleaning the large gas volume from the open-type electric furnaces of the ferroalloy industry. Most venturi-type scrubbers allow recirculation of scrubbing liquor so that water consumption is limited to that evaporated into the gas plus the effluents which carry solids. In a ferrosilicon or ferrochrome-silicon operation nearly all the sulfur from the reducing agent is in the gas phase; thus corrosion occurs in liquid recycle systems unless neutralization or special materials are employed. Electrostatic precipitators are being used on open furnaces which produce silicon, ferrosilicon, high-carbon ferrochromium, and silicomanganese. But the electrical resistivity of most ferroalloy fumes at temperatures below 259.7°C (500°F) is too high for use of electrostatic precipitators. Wet electrostatic precipitators reportedly have been used at one installation in Europe. However, all parts of the precipitator exposed to dirty water and to the wet gas were constructed of stainless steel. Reported use of electrostatic precipitators is limited in United States ferroalloy plants.

The venturi-type scrubber has been used to clean CO gas after collection, but the required pressure drops are reported to be high, and an explosion hazard may exist where baghouses or electrostatic precipitators are used to clean CO-rich gases. Datagraphics reported no application of baghouses on ferroalloy furnaces in the United States; only one is reported in the remainder of the world.

Primary sources of waste water from electrolytic processes are considered to be spent electrolytes and washings. Other sources are cooling water, boiler feed water, air conditioning water, and sanitary water. Waste waters also result from slag processing wherein slag is crushed and sized for sale or reuse and/or recovery of metal values.

RAW MATERIALS

The United States is heavily dependent on South America, Africa, Turkey, India, and Russia for manganese and chromium ores. Columbium-bearing concentrates or slags, some titanium concentrates, and zirconium concentrates are also imported; but commercial sources of vanadium and tungsten exist in the United States.

Some ores are concentrated by flotation, but they are less than desirable as a direct furnace charge because they are low in porosity and inhibit the release of reaction gases; thus dust losses are high. Fine ores are processed more efficiently by melt furnaces. Research has been conducted on briquetting fine ores but costs are high. The 10 to 20 percent free moisture in ores is a source of gaseous or liquid effluent.

SUBMERGED ARC ELECTRIC FURNACES

Datagraphics (1973) deduced the following information about the air pollution features of the submerged arc furnace and associated processes. This condensed summary presents the essential information but much greater detail can be obtained from the Datagraphics report.

The molten alloy from the carbon reduction of the ore or concentrate accumulates at the base of the electrodes in the furnace where it is removed through a taphole. Slag moves upward and is skimmed off and salvaged or discarded as a solid waste. The conventional furnace utilizes carbon reduction of the oxide ores, and continuously produces hot carbon monoxide gas which vents from the top and entrains finer-sized constituents of the charge.

In an open furnace, all CO and other combustibles in the gas burn with air induced at the top of the charge, resulting in the discharge of high-temperature gas via stacks. In a covered furnace, most or all of the CO and other gases are captured prior to combustion. The properties and quantity of particulates emitted depend in part on the alloy being produced. Except for ejected mix particles, the fume size is reported to be below two microns and ranges from 0.1 to 1.0 micron with a geometric mean of 0.3 to 0.6; but agglomeration may occur, and the effective particle size may be larger. Loadings and flow rates are dependent on furnace type and hooding. Open submerged arc furnaces have high flow rates and moderate loadings, whereas closed furnaces have moderate flow rates and generally high loadings. In the dry state, the emissions are very light: the bulk density varies from 64.1 to 480.6 kilograms per cubic meter (4 to 30 pounds per cubic foot).

Silicon alloys produce a gray fume containing a high percentage of silicon dioxide (SiO_2). Tars and carbon derive from the coal, coke, or wood chips used in the charge. Chromium furnaces emit SiO_2 similar to a ferrosilicon operation, but with additional chromium oxides. Manganese operations produce a brown emission, which is largely a mixture of SiO_2 and manganese oxides. Composition of the fumes is similar to oxides of the product being produced. Typical chemical analyses obtained by Datagraphics are given in Table 8.1.

Greater-than-normal emissions are characteristic of an electric arc furnace following shutdowns caused by electrode failure. Metallurgical problems, water leaks, furnace hearth failure, major taphole problems, or transformer or electrical system failures also may cause shutdowns. When a cold hearth is restarted after a shutdown, above-normal emissions may occur for as long as a week. The quantity of emissions may reach several times the normal emission level over a period of 1 to 3 percent of the operating time due to furnace interruptions.

Table 8.2 summarizes production and emission characteristics for submerged arc furnaces and Table 8.3 summarizes the types of air pollution control devices used in ferroalloy furnaces in the United States.

EXOTHERMIC PROCESSES

Exothermic processes combine silicon or aluminum with oxygen at temperatures of several thousand degrees in the reaction vessel. The process produces

Table 8.1 *Typical Furnace Fume Characteristics*

Furnace product	*50% FeSi*	*SMZ* [1]	*SiMn* [2]	*SiMn* [2]	*FeMn*	*HC FeCr*	*Chrome ore-lime melt*	*Mn ore-lime melt*
Furnace type fume shape	Open spherical, sometimes in chains	Open spherical, sometimes in chains	Covered spherical	Covered spherical	Open spherical	Covered spherical	Open spherical and irregular	Open spherical and irregular
Fume size characteristics (microns)								
Maximum	0.75	0.8	0.75	0.75	0.75	1.0	0.50	2.0
Most particles	0.05 to 0.3	0.05 to 0.3	0.2 to 0.4	0.2 to 0.4	0.05 to 0.4	0.1 to 0.4	0.05 to 0.2	0.2 to 0.5
X-ray diffraction	All fumes were primarily amorphous							
Trace constituents	FeSi, $FeSi_2$	Fe_3O_4, Fe_2O_3, Quartz, SiC	Mn_3O_4, MnO, Quartz	Quartz, SiMn, Spinel	Mn_3O_4, MnO, Quartz	Spinel, Quartz	Spinel	CaO

Chemical analysis (%)							
SiO_2	61.12	15.68	24.60	25.48	20.96	10.86	3.28
FeO	14.08	6.75	4.60	5.96	10.92	7.48	1.22
MgO	1.08	1.12	3.78	1.03	15.41	7.43	0.96
CaO	1.01	–	1.58	2.24	–	15.06	34.24
MnO	6.12	31.35	31.92	33.60	2.84	–	12.34
Al_2O_3	2.10	5.55	4.48	8.38	7.12	4.88	1.66
TCr as Cr_2O_3	–	–	–	–	29.27	14.69	–
SiC	1.82	–	–	–	–	–	–
ZrO_2	1.26	–	–	–	–	–	–
PbO	–	0.47	–	–	–	–	0.98
Na_2O	–	–	2.12	–	–	1.70	2.05
BaO	–	–	–	–	–	–	1.13
K_2O	–	–	–	–	–	–	13.08

Source: After Datagraphics, 1973.

1. Si, 60 to 65%; Mn, 5 to 7%; Zr, 5 to 7%.
2. Manganese fume analyses in particular are subject to wide variations, depending on the ores used.

Table 8.2 *Production and Emission Data for Ferroalloy Furnaces*

	Uncontrolled particulate emissions				*Electric energy*		
Product	*(kg/kkg alloy)*	*(lb/ton alloy)*	*(kg/mwhr)*	*(lb/mwhr)*	*(mwhr/kkg alloy)*	*(mwhr/ton alloy)*	*Ratio of charge to product weight*
Silvery iron	58	116	20.4	45	2.9	2.6	1.8
50% FeSi	223	446	40.4	89	5.5	5.0	2.5
65-75% FeSi	458	915	47.2	104	9.7	8.8	4.5
Si metal	500-1,000	1,000-2,000	33-65	72-144	15.4	14.0	4.9
SMZ	ND[1]	ND[1]	ND[1]	ND[1]	9.7	8.8	4.5
Mn ore/lime melt	67	133	37.7	83	1.8	1.6	3.5
CaSi	672	1,343	51.7	114	13.0	11.8	3.9
HCFeMn	168	335	28.1	62	2.6	2.4	3.0
SiMn	110	219	22.7	50	4.9	4.4	3.1
FeMnSi	158	315	26.3	58	6.0	5.4	4.3
FeCrSi	416	831	50.8	112	8.2	7.4	3.4
Chg Cr	168	335	28.1	62	4.6	4.2	4.0
HCFeCr	168	335	28.1	62	4.6	4.2	4.0
Cr ore/lime melt	6	11	4.1	9	1.3	1.2	1.2

Source: After EPA 1973.

1. No data.

Table 8.3 *Types of Air Pollution Control Systems Used on U.S. Ferroalloy Furnaces*

Control device	*Products*	*Control device*	*Products*
Wet scrubbers	ferromanganese, 50 to 75% ferrosilicon HC ferrochromium silicomanganese	Wet scrubbers	50 to 85% ferrosilicon silicomanganese HC ferrochromium ferrochrome-silicon
		Cloth-type filters	silicomanganese ferromanganese silicon 75% and higher grades of ferrosilicon silicon metal ferrochrome-silicon
		Electrostatic precipitator	HC ferrochromium ferrochrome-silicon

Source: After Datagraphics 1973.

Note: Covered furnaces with withdrawal and cleaning of gases prior to burning.

high-grade alloys with low-carbon content. In the silicon reduction process, chromium ore and lime are fused in a furnace to form a chromium ore/lime melt, a portion of which is transferred to a reaction ladle along with molten ferrochrome-silicon produced in a second ladle. The reaction in the first ladle reduces the chrome and produces ferrochromium and a calcium silicate slag. The slag from the first ladle contains recoverable chromium oxide, but a second silicon reduction in the ladle with the molten ferrochrome-silicon directly from the submerged arc furnace produces the intermediate ferrochrome-silicon used in the first ladle reaction. The reaction in these ladles produces strong agitation of the molten bath and a rise in temperature. The elevated temperature and agitation produce emissions for about 5 to 15 minutes per heat. They are similar physically to the emissions from the submerged arc furnaces described previously.

The aluminum reduction process is used to produce chromium metal ferrotitanium, ferrovanadium, and ferrocolumbium. The use of aluminum as a reductant is more expensive than using carbon or silicon, but the products are reported to be purer. Mixed aluminothermal-silicothermal processing is used for the production of ferromolybdenum and ferrotungsten. Aluminum reduction produces emissions similar to those of silicon reduction and the electric arc furnace.

According to Datagraphics (1973), the quantity of emissions from the exothermic processes ranges from 9.08 to 18.6 kg (20 to 40 lb) of particulates per ton of ferroalloy produced. The total tonnage of ferroalloys made by the exothermic process amounts to 10 to 15 percent of the total ferroalloy production in the United States.

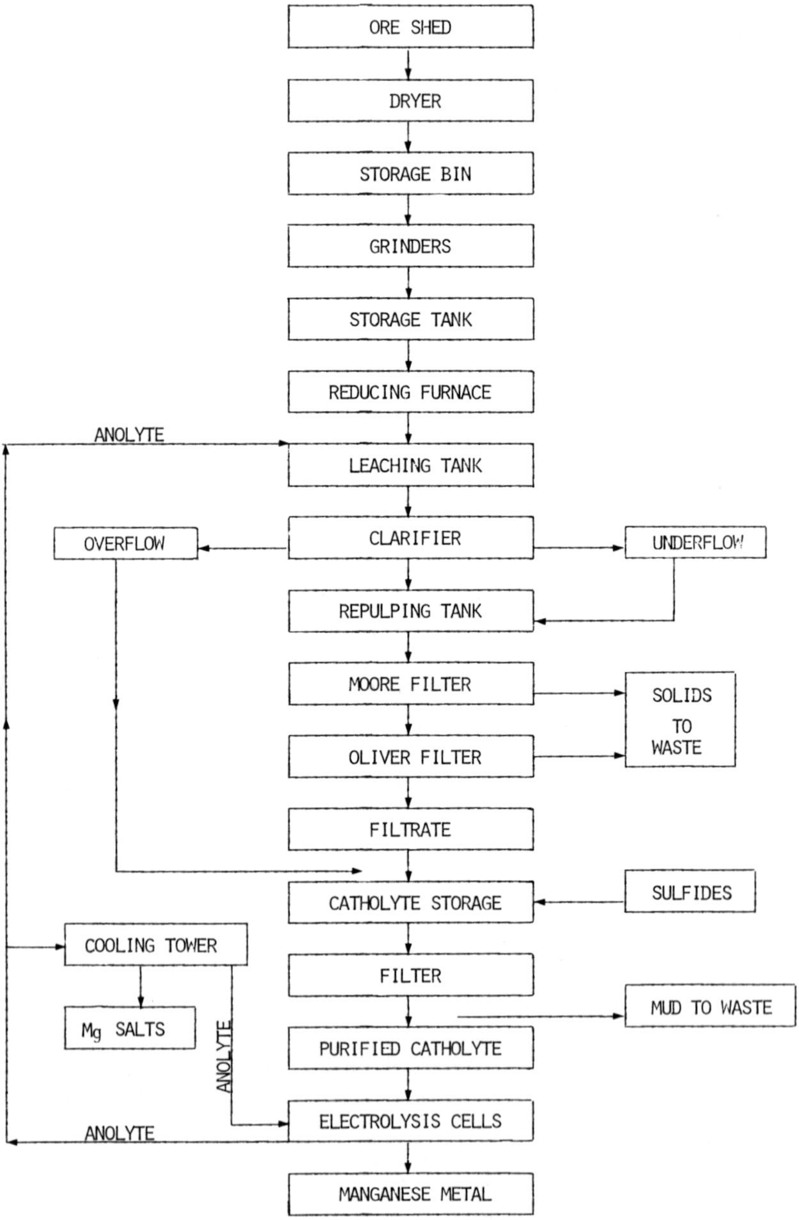

Figure 8.1. *Electrolytic manganese flowsheet (after Datagraphics 1973).*

ELECTROLYTIC PROCESSES

Manganese metal and chromium metal are produced electrolytically by a method wherein ions of the metal contained in an electrolyte are plated on cathodes. After plating, the cathodes are withdrawn from the electrolytic cells and the nearly pure metal is removed. The feed materials are ores, high-metal oxide slags, and ferroalloys produced in submerged arc furnaces.

Electrolytic Manganese

Electrolytic manganese is produced from manganese sulfate in the presence of ammonium salts (see Figure 8.1). The manganese sulfate is prepared from ore mixed with 10 percent coal and 10 percent fuel oil and calcined. The calcined ore is then mixed with a solution of 40 g/l of sulfuric acid and 150 g/l of ammonium sulfate. Manganese dissolves as $MnSO_4$ and the iron in the ore precipitates as the hydroxide. The solution is settled or filtered and sent to electrolysis after further purification. The prepared electrolyte generally contains 33 to 35 g/l Mn and 150 to 160 g/l $(NH_4)_2SO_4$.

The electrolysis baths have stainless steel electrodes. The anodes are placed in frames with diaphragms designed to separate anodic and cathodic regimes in the bath. The spent electrolyte is continuously withdrawn at about 12 to 13 g/l Mn, and fresh electrolyte is added. The baths are cooled by water circulating through stainless steel coils. Manganese settles as a thin, brittle layer at the cathode, where it is removed, crushed, and sold as chips; or it may be melted and cast into bars. The spent electrolyte is reused in leaching and the anodic slime containing manganese dioxide is sent to leaching vats. The cathodes are cleaned with potassium bichromate, washed, and dried. Wastes are a combination of water and solids from the process and fumes from air pollution control.

Electrolytic Chromium

Electrolysis is used to produce high-purity chromium. An aqueous solution of CrO_3 or of the sulfate salts of trivalent chromium may be used as the electrolyte. The metal obtained by electrolysis of CrO_3 is purer than that from the sulfate; sulfate electrolysis, however, reportedly requires one-half of the electrical energy required for CrO_3 electrolysis.

Typically, chromium ores are reduced in an electric furnace to produce high-carbon ferrochromium. The finely ground alloy is leached with sulfuric acid and anolyte solution, which contains sulfuric acid, chromic acid, and ammonium sulfate. Waste product sources are similar to those for the electrolytic manganese process.

RECOVERY OF METALS FROM SLAG

Several electric arc smelting processes produce slag along with the ferroalloy product. These processes are:

- Low-carbon ferrochrome-silicon
- High-carbon ferrochromium
- High-carbon ferromanganese
- Silicomanganese

These slags may contain metal recoverable by crushing and separation by a wet flotation process. The slag fines are also separated from the coarser particles so that a slag usable for road building and similar purposes is produced.

"Shotting" is another method of recovering metal from slag. Afterward the "shots" become the raw material for electrolytic production of the metal. Shotting consists of rapid quenching of the molten slag in flowing water, which produces a small-sized particle (shot) that can be leached with acid to form the electrolyte solution.

PRODUCT SIZING AND HANDLING

Ferroalloys are marketed in particles ranging from 34.1 kg (75 lb) in weight to small granules.

Molten ferroalloys generally are tapped into refractory-lined cooling ladles or molds before cleaning and sizing to market specifications.

Sizing consists of comminution with weights, hammers, and crushers, and screening. Conveyors and elevators generate considerable dust while moving the product between crushers and screens. This dust constitutes a solid waste which may be recyclable.

FACTORS AFFECTING QUANTITIES OF EMISSIONS

As noted previously, emission identity and rates vary with (1) type of alloy, (2) process (continuous or batch), (3) raw materials and pretreatment, (4) operating techniques, (5) furnace size, and (6) maintenance practices. For example, as the silicon in the alloy increases, SiO_2 loss increases; therefore, a silicon-metal furnace emits more SiO_2 than an equivalent-size 50 percent ferrosilicon furnace.

Violent gaseous emissions from open arc batch furnaces occur after addition of a mix that contains reactive constituents (coal, moisture, aluminum). Under these conditions, temperature, dust loading, and gas flow peak simultaneously. Chromium ore-lime melt furnaces, to which few or no gas-releasing constituents are fed, are dissimilar.

Volatiles and fines may promote bridging and nonuniform descent of the charge. The collapse of a bridge causes a momentary burst of gaseous emissions. For some products, economics dictates the use of raw materials with more fines or with more volatile matter than desirable for air pollution control.

Operating techniques influence emissions. The average rate of electric furnace gas production is nearly directly proportional to electrical input. But in some cases, emissions increase at a rate greater than the load increase due to rough operation and inadequate gas withdrawal. Under a fixed load and with gas generation remaining approximately constant, the weight of particulates emitted per hour reportedly can vary by a factor of five.

For example, operating with insufficient electrode immersion promotes increased emissions, and operation under higher voltage promotes higher electrode positions and increases emission rate. In some operations, especially silicon metal production, stoking the charge to break up crusts and assure flow of reaction gases influences emissions. Thus emission is a function of how well and how often the furnace is stoked. Maintenance practices affect emissions on covered furnaces because accumulation of material under the cover and in gas ducts reduces the efficiency of the exhaust system.

WASTE CHARACTERIZATION

Datagraphics (1973) chose to subdivide the ferroalloy industry into four major categories for purposes of waste characterization. These categories are:

I. Cooling water, electric furnace smelting.
II. Water for wet air pollution control devices.
  A. Electric furnace smelting.
    1. Disintegrator-type scrubbers.
    2. High-energy scrubbers.
    3. Electrostatic precipitator spray towers.
    4. Steam/hot water scrubbers.
  B. Exothermic smelting processes.
III. Sanitary, boiler feed and air conditioning water.
IV. Slag processing water.

Data published by Person (1969) and Fetelsdorf et al. (1969) plus in situ observations were the basis for the observations that cooling water use by electric furnaces should be about 48.8 liters (12.9 gallons) per kilowatt-hour of furnace power, with thermal pollution being the major impact. Blowdown was estimated to average about 1 to 2 percent of the recirculation rate, with two cycles of concentration, implying that blowdown should contain about twice the concentrations of constituents in the intake water. Constituents include chromates or phosphates used for corrosion control and other treatment chemicals such as biocides and slime inhibitors.

It was assumed that 556 liters (147 gallons) of water per megawatt-hour of furnace power is evaporated in the gas streams from open furnaces using wet air pollution control devices and that such evaporation in the case of covered furnaces is in proportion to the gas volume. On this basis the effluent volumes were calculated to be: (1) high-energy scrubbers (open furnace), 19,682 l/mwhr (5,200 gal/mwhr); (2) high-energy scrubbers (covered furnace), 609 l/mwhr (161 gal/mwhr); and (3) electrostatic precipitator, 79.5 l/mwhr (21 gal/mwhr).

Datagraphics gave the following reasons for presenting loading data in terms of loading per megawatt-hour:

1. Uncontrolled air emission (upon which the waste loads depend) are more uniform over the various types of products when expressed as kilograms (pounds) per megawatt-hour, rather than as kilograms per metric ton (pounds per ton).
2. Power usage is such a large factor in production costs (about 30 percent) that an increase in power consumption to raise the permissible effluent discharge would be very unlikely.
3. Power usage is carefully monitored at the furnace, usually with a continuous automatic recording device.
4. Furnaces are commonly referred to in the industry as "10 mw" or "35 mw," rather than "50 ton" or "150 ton," as is common practice in the steel industry.
5. The tonnage which may be produced for a given power consumption is fairly wide (a factor of 10) and depends on the product, and numerous products can be produced in a given furnace.

Table 8.4 presents waste loads for a submerged arc furnace with off-gases scrubbed by steam or hot water scrubbers.

Table 8.5 gives loads for an open electric furnace with a venturi scrubber.

Table 8.6 presents raw waste loads for an open submerged arc furnace in which the off-gases are conditioned in a spray tower preceding an electrostatic precipitator.

The data in Table 8.6 for water flow are similar to those Datagraphics predicted (84.0 vs 79.5 l/mwhr) (22.2 vs 21 gal/mwhr). The flow rate from the

Table 8.4 *Raw Waste Loads for Open Chromium Alloy and Ferrosilicon Furnaces with Steam or Hot Water Scrubbers*

Constituents	*(kg/mwhr)*	*(lb/mwhr)*
Suspended solids	8.2	18.1
Phosphate	0.010	0.023
Phenol	0.0004	0.001
Oil	0.002	0.004
Total cyanide	0.0001	0.0002
Free cyanide	0.00005	0.0001
Iron	0.069	0.152
Manganese	0.005	0.010
Zinc	0.068	0.149
Total chromium	0.003	0.007
Hexavalent chromium	0.002	0.004
Lead	0.008	0.018
Aluminum	0.158	0.348
	(l/mwhr)	*(gal/mwhr)*
Flow	2,691	711

Source: After Datagraphics 1973.

Table 8.5 *Raw Waste Load for High Energy Scrubber on Open Electric Furnace*

Constituents	*(kg/mwhr)*	*(lb/mwhr)*
Suspended solids	23.74	52.29
Phosphate	0.0	0.0
Phenol	0.0005	0.001
Oil	0.006	0.014
Total cyanide	0.0	0.0
Total iron	0.041	0.089
Manganese	10.06	22.15
Zinc	0.33	0.72
Total chromium	0.002	0.005
Lead	0.07	0.16
Aluminum	1.13	2.49
	(l/mwhr)	*(gal/mwhr)*
Flow	6,382	1,686

Source: After Datagraphics 1973.

Table 8.6 *Raw Waste Loads for Open Chromium Alloy Furnaces with Electrostatic Precipitators*

Constituents	*(kg/mwhr)*	*(lb/mwhr)*
Suspended solids	0.289	0.836
Phosphate	0.00001	0.00002
Oil	0.0001	0.0002
Iron	0.001	0.002
Manganese	0.0012	0.0026
Zinc	0.001	0.003
Total chromium	0.0016	0.0036
Aluminum	0.0070	0.0155
	(l/mwhr)	*(gal/mwhr)*
Flow	84.0	22.2

Source: After Datagraphics 1973.

stream or hot water scrubbers was considered incomparable with any calculations, and the values for flow from the high-energy scrubber are about one-third of those predicted. However, the flow from the high-energy scrubber does not take into account recirculation of scrubber water which is done prior to clarification. This water may account for the difference.

The data in Table 8.7 describe the waste water from disintegrator scrubbers operating on covered furnaces producing silicon alloys. Raw waste loads of suspended solids and cyanides are given on the basis of the furnace power during 16-hour sampling periods.

The data in Table 8.8 consist of waste loads for a sealed silicomanganese furnace where the off-gases are scrubbed in a spray tower and a disintegrator scrubber. Table 8.9 shows similar loadings for the exothermic process (aluminum) with combination wet scrubbers and baghouses.

The covered furnaces surveyed by Datagraphics displayed water uses higher than those of open furnaces with high-energy scrubbers. Datagraphics considered this to be due to the fact that water use in disintegrator scrubbers is higher, for a particular gas volume, than the water use in high-energy scrubbers. Most of the covered furnaces surveyed used disintegrator, rather than high-energy scrubbers. However, this explanation was not always observed to be valid. No alternative explanation was presented. It is probable that management practices are a significant variable.

Table 8.10 characterizes waste loads from slag processing operations.

Datagraphics was unable to calculate waste loads directly for the slag shotting process because tonnage figures were unavailable. However, estimates from production figures indicated that 2.0 tons of slag and particulates are produced for every ton of alloy. The emission rate for high-carbon (HC) FeMn was reported to be 335 lb/ton of product, so the slag produced would be 2.0 tons minus 335 pounds, or 3,665 lb/ton of alloy. This figure times operating load divided by the electrical energy required per ton of alloy was used to obtain a production figure for slag of 24,452 lb/hr. This number was then divided into the water flow rate to give the following figures: a water use rate of 8,588 gal/ton processed, a suspended solids raw waste load of 15.5 lb/ton, and a manganese load of 3.87 lb/ton.

Table 8.7 *Waste Loads for Covered Furnaces with Disintegrator Scrubbers*

	Suspended solids		*Cyanides*		*Flow*	
Product	*(kg/mwhr)*	*(lb/mwhr)*	*(kg/mwhr)*	*(lb/mwhr)*	*(l/mwhr)*	*(gal/mwhr)*
SiMnZr	20.1	44.3	0.0338	0.0745	8,270	2,185
75% FeSi	39.2	86.3	–	–	8,967	2,369
50% FeSi	5.1	11.3	0.0001	0.0002	8,823	2,331
75% FeSi	6.8	15.0	0.0139	0.0307	7,562	1,998

Source: After Datagraphics 1973.

Table 8.8 *Waste Loads for Sealed Silicomanganese Furnace with Distintegrator Scrubber*

Constituents	*(kg/mwhr)*	*(lb/mwhr)*
Suspended solids	16.6	36.6
Phosphate	0.052	0.114
Phenol	0.009	0.019
Oil	0.038	0.084
Total cyanide	0.044	0.098
Free cyanide	0.011	0.024
Iron	0.056	0.123
Zinc	0.224	0.493
Total chromium	0.0004	0.001
Lead	0.033	0.072
Aluminum	0.413	0.91
Manganese	4.858	10.70
	(l/mwhr)	*(gal/mwhr)*
Flow	10,863	2,870

Source: After Datagraphics 1973.

Table 8.9 *Waste Loads for Aluminothermic Smelting with Combination Wet Scrubbers and Baghouse*

Constituents	*(kg/kkg)*	*(lb/ton)*
Suspended solids	3.6	7.1
Phosphate	0	0
Phenol	0	0
Oil	0.048	0.095
Total cyanide	0	0
Free cyanide	0	0
Iron total	0	0
Manganese	0.0005	0.001
Zinc	0	0
Total chromium	2.98	5.95
Hexavalent chromium	0.95	1.90
Lead	0	0
Aluminum	0	0
	(l/kkg)	*(gal/ton)*
Flow	26,332	6,310

Source: After Datagraphics 1973.

Table 8.10 *Raw Waste Loads for Slag Concentration Process*

Constituents	*(kg/kkg)*	*(lb/ton)*
Suspended solids	46.0	91.9
Phosphate	0	0
Phenol	0	0
Oil	0.064	0.128
Total cyanide	0.0003	0.0007
Total iron	0.543	1.085
Manganese	0.245	0.489
Zinc	0.012	0.023
Aluminum	0.569	1.138
Lead	0	0
Total chromium	0.109	0.217
	(l/kkg)	*(gal/ton)*
Flow	48,259	12,750

Source: After Datagraphics 1973.

Table 8.11 *Waste Loads for Noncontact Cooling Water for Submerged Arc Furnaces*

Parameters	*(kg/mwhr)*	*(lb/mwhr)*
Suspended solids	0.09	0.19
Phosphate	0.001	0.003
Phenol	0.00005	0.0001
Oil	0.002	0.004
Total iron	0.001	0.002
Manganese	0.0005	0.001
Zinc	0.00001	0.00003
Total chromium	0.000005	0.00001
Aluminum	0.002	0.004
	(l/mwhr)	*(gal/mwhr)*
Flow	502.6	132.8

Source: After Datagraphics 1973.

Table 8.11 presents information on the waste water from the recirculation and reuse of cooling water on electric smelting furnaces producing silicon alloys and utilizing no wet air pollution control devices. A softener is used on one cooling water circuit handling about 50 percent of the total cooling water flow. Treatment chemicals are used along with a phosphate reagent rather than a chromate. The raw waste loads are given in Table 8.11 on the basis of the total furnace power during the period of sampling by Datagraphics.

The data in Table 8.12 provide waste loads resulting from the recirculation and reuse of cooling water on electric smelting furnaces utilizing chromate treatment for corrosion inhibition.

Datagraphics also presents constituent concentration data for various effluents from eight United States ferroalloy plants. One set of these data (identified by Datagraphics as plant E) is presented in Tables 8.12 through 8.21 for illustration. The plant consists of seven covered and two open submerged arc furnaces wherein 50 percent ferrosilicon, silicomanganese, standard and medium-carbon ferromanganese, and high-carbon ferrochromium are produced. This operation also reported owning metals refining and slag shotting facilities. The furnaces operated at 82 mw during the Datagraphics survey.

Plant E's nine furnaces were reported to use cooling water on a once-through basis. Sanitary sewage is treated and discharged with the cooling water. The total plant effluent is 1.16 X 10^6 m/day (305.5 mgd), the majority of which is cooling water from the plant's power generating station. The water and waste water systems for the plant are shown in Figure 8.2. The sampling points that Datagraphics used during the survey are also identified.

The fumes from the furnaces are scrubbed with either venturi- or low-energy-type scrubbers. There are 5 venturi- and 12 disintegrator-type scrubbers available for the nine furnaces. The scrubbers use between 22 and 32 l/sec (350 and 500 gpm) of water when operating. The metals refining operation also utilizes a venturi scrubber with effluent flowing via a common line to the first of two 78-acre lagoons (area combined) operated in series. The wash water from the electrolytic operations mixes with the scrubber waste water before entering the lagoons. The acid waste water from the electrolytic operations flows to the second lagoon where a hydrated lime slurry is added as a neutralizing agent. This second lagoon also receives the effluent from a fly ash removal system at the power plant. The effluent from the second lagoon is discharged to the environment. The waste water from the slag concentrator flows to a separate 4.3-acre tailings lagoon and then to the environment.

Summarized analytical data for sampling points designated in Figure 8.2 are shown in Tables 8.12 through 8.21.

CRITICAL CONSTITUENTS

Datagraphics (1973) used the preceding data to select critical pollutants in ferroalloy plant waste waters. They selected those originating in the production processes in significant amounts and for which control and treatment technologies are reasonably available. The pollutants selected are shown by category in Table 8.22. This categorization of the industry with respect to wastes produced

Table 8.12 *Raw Waste Loads for Noncontact Cooling Water for Submerged Arc Furnaces*

Parameters	*(kg/mwhr)*	*(lb/mwhr)*
Suspended solids	2.71	5.98
Total iron	0.011	0.025
Manganese	0.007	0.016
Zinc	0.14	0.30
Oil	0.08	0.18
Total chromium	0.26	0.58
Hexavalent chromium	0.0001	0.0003
Aluminum	0.013	0.03
	(l/mwhr)	*(gal/mwhr)*
Flow	734	194

Source: After Datagraphics 1973.

Note: The average cooling water usage at the plants surveyed by Datagraphics was 53,690 l (14,185 gal) per mwhr at an average temperature rise of 7.6°C (13.7°F).

Table 8.13 *Illustrative Analytical Data for Furnace Scrubber Discharge*

	Concentrations (mg/l)[1]			
Constituents	*Minimum*	*Maximum*	*Average*	*Net average*
Suspended solids	318	426	373	340
Total chromium	0.09	0.09	0.09	0.09
Hexavalent chromium	–	–	–	–
Total cyanide	0.87	0.87	0.87	0.87
Free cyanide	–	–	–	–
Manganese	256	256	256	256
Oil	1.6	1.6	1.6	1.6
Iron	18.0	18.0	18.0	17.4
Zinc	48	48	48	48
Aluminum	13.0	13.0	13.0	12.3
Phenol	0.22	0.22	0.22	0.22
Phosphate	–	–	–	0
Lead	5.6	5.6	5.6	5.6
pH (units)	6.4	6.9	6.7	–

Source: After Datagraphics 1973.

Note: Average flow is 25.2 l/sec (400 gal/min).

1. Milligrams per liter unless otherwise indicated.

Table 8.14 *Illustrative Analytical Data for Metals Refining Scrubber Discharge*

	Concentrations (mg/l)[1]			
Constituents	*Minimum*	*Maximum*	*Average*	*Net average*
Suspended solids	874	1,674	1,204	1,171
Total chromium	0.06	0.06	0.06	0
Hexavalent chromium	–	–	–	–
Total cyanide	–	–	–	0
Free cyanide	–	–	–	0
Manganese	597	597	597	597
Oil	1.2	1.2	1.2	1.2
Iron	1.82	1.82	1.82	1.24
Zinc	0.46	0.46	0.46	0.44
Aluminum	0.87	0.87	0.87	0.20
Phenol	–	–	–	0
Phosphate	–	–	–	0
Lead	–	–	–	0
pH (units)	7.8	8.7	8.2	–

Source: After Datagraphics 1973.

Note: Average flow is 22.1 l/sec (350 gal/min).

1. Milligrams per liter unless otherwise indicated.

Table 8.15 *Illustrative Analytical Data for Slag Shotting Waste Water*

	Concentrations (mg/l)[1]			
Constituents	*Minimum*	*Maximum*	*Average*	*Net average*
Suspended solids	132	302	217	184
Total chromium	0.03	0.03	0.03	0
Hexavalent chromium	–	–	–	–
Total cyanide	–	–	–	0
Free cyanide	–	–	–	0
Manganese	54	54	54	54
Oil	1.2	1.2	1.2	1.2
Iron	0.28	0.28	0.28	0
Zinc	0.13	0.13	0.13	0.11
Aluminum	10.5	10.5	10.5	9.8
Phenol	–	–	–	0
Phosphate	–	–	–	0
Lead	–	–	–	0
pH (units	7.3	7.4	7.5	–

Source: After Datagraphics 1973.

Note: Average flow is 110.3 l/sec (1,750 gal/min).

1. Milligrams per liter unless otherwise indicated.

Table 8.16 *Illustrative Analytical Data for Scrubber Settling Basin Discharge*

Constituents	Concentrations (mg/l)[1] Minimum	Maximum	Average	Net average
Suspended solids	3,348	11,364	6,080	6,047
Total chromium	0.17	0.17	0.17	0.01
Hexavalent chromium	–	–	–	–
Total cyanide	–	–	–	0
Free cyanide	–	–	–	0
Manganese	1,322	1,322	1,322	1,322
Oil	1.0	1.0	1.0	1.0
Iron	7.16	7.16	7.16	6.58
Zinc	89	89	89	89
Aluminum	178	178	178	177
Phenol	–	–	–	0
Phosphate	–	–	–	0
Lead	9.1	9.1	9.1	9.1
pH (units)	8.5	8.6	8.6	–

Source: After Datagraphics 1973.

Note: Average flow is 44.1 l/sec (700 gal/min).

1. Milligrams per liter unless otherwise indicated.

Table 8.17 *Illustrative Analytical Data for Slag Concentrator Waste Water*

Constituents	Concentrations (mg/l)[1] Minimum	Maximum	Average	Net average
Suspended solids	856	872	864	831
Total chromium	2.04	2.04	2.04	1.88
Hexavalent chromium	–	–	–	0
Total cyanide	–	0.013	0.007	0.007
Free cyanide	–	–	–	–
Manganese	4.39	4.81	4.60	4.11
Oil	0.2	2.2	1.2	1.2
Iron	5.8	14.6	10.2	9.6
Zinc	0.22	0.22	0.22	0.20
Aluminum	10.7	10.7	10.7	10.0
Phenol	–	–	–	0
Phosphate	–	–	–	0
Lead	–	–	–	0
pH (units)	6.1	6.2	6.2	–

Source: After Datagraphics 1973.

Note: Average flow is 107.1 l/sec (1,700 gal/min).

1. Milligrams per liter unless otherwise indicated.

Table 8.18 *Illustrative Analytical Data Slag Tailings Pond Discharge*

	Concentrations (mg/l)[1]			
Constituents	*Minimum*	*Maximum*	*Average*	*Net average*
Suspended solids	46	90	62	29
Total chromium	0.02	0.16	0.09	0
Hexavelent chromium	–	–	–	0
Total cyanide	0.006	0.006	0.006	0.006
Free cyanide	–	–	–	–
Manganese	0.95	1.26	1.08	0.59
Oil	–	–	–	0
Iron	1.14	1.54	1.32	0.74
Zinc	0.032	0.058	0.048	0.026
Aluminum	0.72	1.13	0.86	0.19
Phenol	0.25	0.25	0.25	0.25
Phosphate	–	–	–	0
Lead	–	–	–	0
pH (units)	6.2	6.9	6.4	–

Source: After Datagraphics 1973.

Note: Average flow is 107.1 l/sec (1,700 gal/min).

1. Milligrams per liter unless otherwise indicated.

Table 8.19 *Illustrative Analytical Data for Lagoon Influent*

	Concentrations (mg/l)[1]			
Constituents	*Minimum*	*Maximum*	*Average*	*Net average*
Suspended solids	32	972	183	150
Total chromium	0.55	1.2	0.77	0.61
Hexavalent chromium	0.190	0.205	0.198	0.198
Total cyanide	–	–	–	0
Free cyanide	–	–	–	0
Manganese	24.4	26.1	25.4	24.9
Oil	0.4	0.4	0.4	0.4
Iron	0.86	1.49	1.28	0.70
Zinc	4.22	7.90	5.55	5.53
Aluminum	1.44	2.60	2.04	1.37
Phenol	–	–	–	0
Phosphate	–	–	–	0
Lead	–	0.06	0.04	0.04
pH (units)	6.7	7.0	6.8	–

Source: After Datagraphics 1973.

Note: Average flow is 447.3 l/sec (7,100 gal/min).

1. Milligrams per liter unless otherwise indicated.

Table 8.20 *Illustrative Analytical Data for Cooling Water Discharge*

	Concentrations (mg/l)[1]			
Constituents	*Minimum*	*Maximum*	*Average*	*Net average*
Suspended solids	90	176	125	92
Total chromium	6.6	6.6	6.6	6.4
Hexavalent chromium	–	–	–	0
Total cyanide	–	0.014	0.005	0.005
Free cyanide	–	–	–	–
Manganese	0.053	4.61	1.58	1.09
Oil	–	0.4	0.3	0.3
Iron	3.42	32.	15.0	14.4
Zinc	0.045	0.045	0.045	0.023
Aluminum	4.28	4.28	4.28	3.61
Phenol	–	–	–	0
Phosphate	1.98	1.98	1.98	1.98
Lead	–	–	–	0
pH (units)	3.8	7.2	5.4	–

Source: After Datagraphics 1973.

Note: Average flow is 3,571 l/sec (56,680 gal/min).

1. Milligrams per liter unless otherwise indicated.

Table 8.21 *Illustrative Analytical Data for Combined Slag Shotting and Cooling Water Discharge*

	Concentrations (mg/l)[1]			
Constituents	*Minimum*	*Maximum*	*Average*	*Net average*
Suspended solids	148	192	170	137
Total chromium	0.02	0.02	0.02	0.02
Hexavalent chromium	–	–	–	–
Total cyanide	–	–	–	0
Free cyanide	–	–	–	0
Manganese	19.1	19.1	19.1	18.6
Oil	1.0	1.0	1.0	1.0
Iron	3.72	3.72	3.72	3.14
Zinc	0.049	0.049	0.049	0.027
Aluminum	4.95	4.95	4.95	4.28
Phenol	–	–	–	0
Phosphate	–	–	–	0
Lead	–	–	–	0
pH (units)	7.5	7.5	7.5	–

Source: After Datagraphics 1973.

Note: Average flow is 50.4 l/sec (800 gal/min).

1. Milligrams per liter unless otherwise indicated.

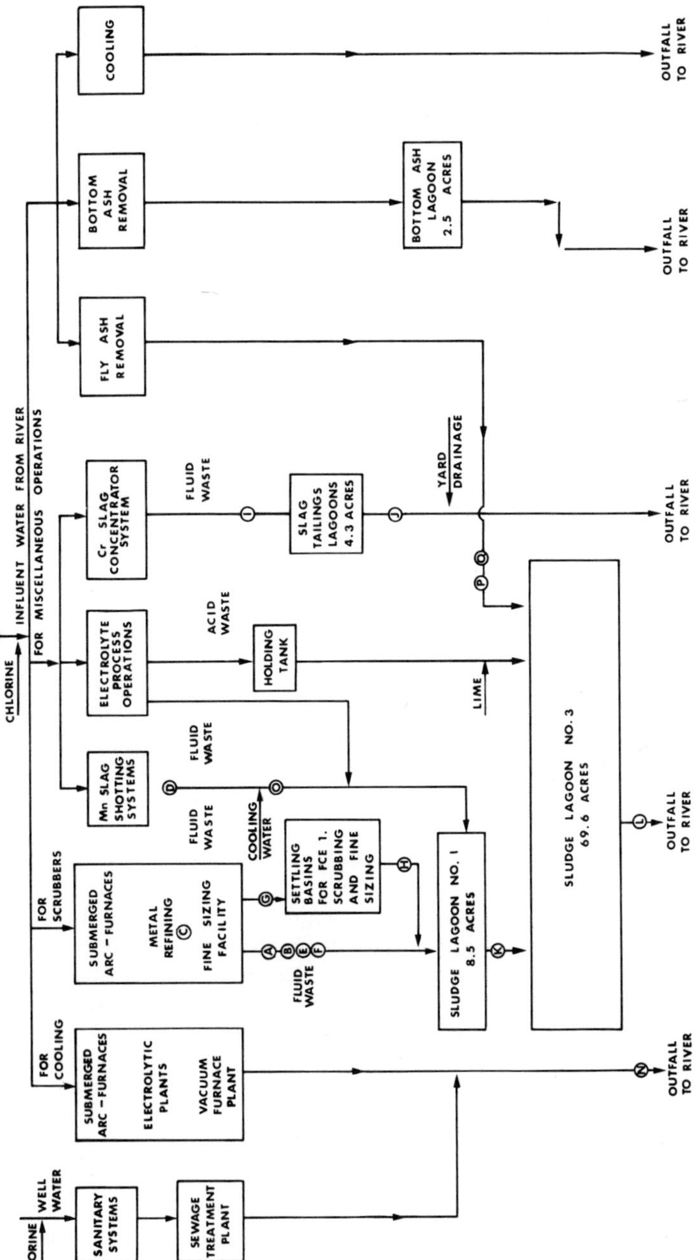

Figure 8.2. *Water and waste water systems for a typical plant (after Datagraphics 1973).*

Table 8.22 *Pollutant Parameters for Industry Categories*

Parameters	*Industry categories*[1] I	II	III	IV
Heat content	–	–	–	x
Suspended solids	x	x	x	x
pH	x	x	x	x
Total chromium	x	x	x	x
Hexavalent chromium	x	x	–	x
Total cyanide	–	x	–	–
Manganese	x	x	x	–
Oil	x	x	x	x
Phenol	x	x	–	–
Phosphate	x	x	–	x

Source: After Datagraphics 1973.

1. Industry categories are:
   I. Open electric furnaces with wet air pollution control devices.
   II. Covered electric furnaces and other smelting operations with wet air pollution control devices.
   III. Slag processing.
   IV. Noncontact cooling water.

also is shown in Table 8.22. The remainder of this section summarizes pertinent aspects of each constituent as revealed during the Datagraphics survey.

Suspended solids result from wet air pollution control devices and slag processing in concentrations up to 7,600 mg/l. They appear to a lesser extent in noncontact cooling water blowdown from recirculating systems. Chromium, manganese, iron, zinc, and aluminum are the principal metals discharged from the production processes. Manganese concentrations observed were as high as 1,576 mg/l, while maximum chromium concentrations observed were 8.36 mg/l from an electrostatic precipitator spray tower and 121 mg/l from an exothermic chromium smelting operation. Hexavalent chromium was considered a pollutant parameter because it may be harmful at low levels.

Cyanide appears in waste water because it is not oxidized in the reducing atmospheres of covered furnaces. Phenols were concluded to derive from electrode binding materials in covered furnaces. They originate principally from covered furnaces, although some may be present in fume from open furnaces.

Other parameters such as dissolved solids, chlorides, and sulfates were observed in effluents in significant concentrations, but largely result from treatment processes such as neutralization, softener regeneration, and water reuse. Many of the metals contained in the raw waste, particularly iron, zinc, aluminum, and lead are part of the solids generated in the smelting furnaces. Datagraphics concluded that they are controlled if suspended solids concentrations are controlled.

CONTROL AND TREATMENT TECHNOLOGY

Sedimentation in lagoons was considered to be the water pollution control and treatment technology traditionally used in the ferroalloy industry.

Air pollution abatement has accounted for the majority of expenditures. According to Datagraphics (1973), air pollution control systems installed, being built, or planned should be capable of meeting existing state regulations; in cases where controls have been installed for five years or more, such controls may be marginal insofar as newer regulations are concerned.

The United States plants surveyed by Datagraphics are classified in Table 8.23 in terms of the industry categorization given previously and in terms of process and treatment of air and water.

The treatment and control technologies available for principal waste parameters common to this industry's waste water were grouped according to Table 8.24. The subsequent discussion summarizes the highlights of each technique.

Cooling ponds are designed to cool water approximately to the average surface water temperature in an area. Cooling ponds can limit the temperature rise in an effluent to 2.78°C (5°F) over that of ambient water. Spray ponds and cooling towers are designed to approximate the wet-bulb temperature. Spray ponds occupy about 10 percent of the area of cooling ponds, at about twice the price. Cooling towers are even more expensive, but occupy comparatively little area and can be operated to recirculate water with blowdown of 1 to 5 percent of the recirculation rate. Treatment chemicals such as chromates or phosphates, added to control scaling and corrosion, contaminate the cooling water.

Datagraphics noted that recirculation can minimize the volume of water requiring suspended solids removal and that lagoons and clarifier-flocculators can achieve effluent concentrations of 25 mg/l, when well operated. Sand

Table 8.23 *Characteristics of Plants Surveyed by Datagraphics*

Plants	*Categories*	*Processes and water uses and air controls*
A	IV	Baghouses being built, recirculated cooling water.
B	II, IV	Disintegrator scrubbers, once-through cooling water use.
C	II, IV	Sealed furnace, disintegrator scrubbers, recirculated cooling water and scrubber water.
D	I, IV	Steam/hot water scrubbers, recirculation of cooling and scrubber water.
E	I, II, III, IV	Disintegrator scrubbers, venturi scrubbers, once-through water use, slag processing.
F	IV	Baghouse/no air controls, recirculated cooling water.
G	I, III, IV	Electrostatic precipitators, with water sprays, recirculated cooling water, slag processing.
H	II, IV	Exothermic process, wet scrubbers and baghouse.

Table 8.24 *Treatment Technologies for Critical Constituents in Ferroalloy Plant Waste Water*

Constituents	*Treatment available*
Heat content	Cooling ponds, spray ponds, cooling towers.
Suspended solids	Water recirculation, lagoons, clarifier-flocculators, sand filters.
pH	Neutralization.
Chromium	Hexavalent chromium reduction, precipitation, sedimentation.
Cyanide	Alkaline chlorination, ozonation.
Oil	Flotation, skimming, air flotation, agglomeration, and filtration.
Manganese	Neutralization of acid salts, precipitation, sedimentation.
Phenol	Biological oxidation, breakpoint chlorination, activated carbon.
Phosphate	Chemical treatment, activated sludge.

Source: After Datagraphics 1973.

filters achieve effluent concentrations of 10 to 15 mg/l but are more expensive than clarifier-flocculators.

Neutralization, as explained in Chapter 3, consists of adding an acid or base to control pH. As indicated by Datagraphics (1973), hexavalent chromium is removed at pH levels below 2.5 by exposure to sulfur dioxide. Raising the pH to about 8.2, then, precipitates the chromium. The ultimate concentration of chromium depends on good pH control and adequate sedimentation.

Cyanide is oxidized rapidly to the less harmful form of cyanite by well-known alkaline chlorination at a pH of 10.5. Cyanate is oxidized to CO_2 and N_2 by continued chlorination at a pH of about 7.0 and a reaction time of about 60 minutes. Ozonation is an alternative method for the destruction of cyanide.

Oil can be removed from waste water by flotation and skimming. Air flotation can also be used with chemicals added to break emulsions and agglomerate the oil.

Manganese and iron can be removed by neutralization to a pH above 9.5 for manganese and above 8.0 for iron, followed by precipitation and sedimentation. Sufficiently high pH, adequate sedimentation, and oxidation are required for low concentrations in the effluent (see Chapter 4).

Phenol can be oxidized by chlorine and chlorine dioxide. Phenol can also be removed by activated carbon. Biological oxidation is possible but may require addition of nutrients.

As has been the case for other industries, the choice of air pollution control technology governs waste water production. Most plants with open furnaces reported to the Datagraphics survey that they utilize dry baghouses and produce no liquid effluent. Datagraphics (1973) reports two known cases in the world of the use of dry dust collectors on sealed or covered furnaces; neither is in the United States. Covered furnaces commonly use wet scrubbers.

Datagraphics utilized the data and treatment technologies discussed up to this point to construct a general waste treatment scheme in which most critical parameter waste constituents can be reduced to minimal concentrations. Not all plant waste streams contain all constituents, and appropriate modifications of this general scheme must be made for compatibility.

In this scheme, initially, the pH is raised to about 11, and sufficient chlorine is added to maintain a free residual, followed by sedimentation. This step oxidizes phenol and cyanide (to cyanite) and phosphates and manganese are precipitated. In the second step, additional chlorine is added and the pH is lowered to 7.0 by a suitable acid. With a reaction time of 60 minutes, the cyanate is oxidized to CO_2 and N_2. In the third step, the pH is lowered to 2.5 and sulfur dioxide is added. After a reaction time of about 30 minutes, the hexavalent chromium is reduced to the trivalent state. The fourth step consists of raising the pH to 8.2, adding a polyelectrolyte and allowing sedimentation. At this point, the trivalent chromium is removed and final clarification is accomplished. With a sufficiently low overflow rate and addition of flocculants in sufficient quantities, a concentration of 25 mg/l suspended solids can be attained and metals can be reduced to low levels. Sand filtration of the final clarifier effluent, with backwash returned to the clarifier, can reduce suspended solids concentrations to 15 mg/l or less. After filtration, the water may be recycled back to the scrubbers.

For a waste stream containing no phenol or cyanide, but with hexavalent chromium, a single initial clarifier operated at a pH of 10.0 and with polyelectrolyte addition is employed by the scheme to reduce phosphate and metals to low concentrations; chlorination is unnecessary. In this case, the first two steps of Figure 8.3 are combined, and the remainder of the treatment is as shown.

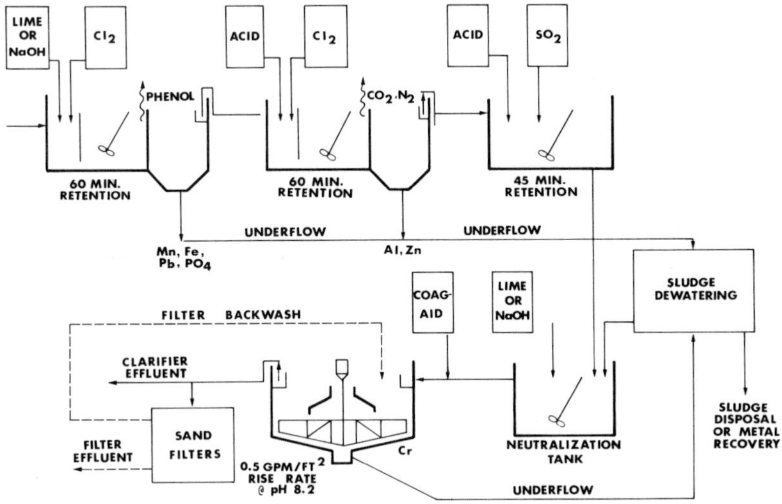

Figure 8.3. *Diagram of waste water treatment system for ferroalloy industry (after Datagraphics 1973).*

For a waste stream containing no hexavalent chromium, but with phenol and cyanide, the chromium reduction step is eliminated from the scheme.

Datagraphics points out that all or part of this treatment scheme is practiced at one or more of the three United States plants studied. Table 8.25 lists those categories to which the scheme should be applicable.

Previous chapters have demonstrated that dry air quality control is preferable to wet air quality control, whenever it can be practiced, because the liquid stage in the usual sequence of gaseous waste, liquid waste, solid waste is omitted. However, the treatment scheme shown in Figure 8.3 handles the waste water problem created by wet air quality control and it reduces the solid waste problem to one of handling dewatered sludge, either by land disposal or recycle for additional metal recovery. Other solid waste disposal problems common to the industry must be handled similarly.

Datagraphics (1973) concluded that utilization of these technologies can reduce waste water constituents to the levels which are indicated in Tables 8.26 through 8.29.

Table 8.25 *Control and Treatment Technologies by Category*

Categories[1]	*Treatment technology level*[2]	*Description*
I	1	Chemical treatment, clarifier-flocculators, recirculation at the scrubber.
	2	Chemical treatment, clarifier-flocculators, sand filters and process water recirculation.
	3	Dry dust collectors for air pollution control.
II	1	Chemical treatment, clarifier-flocculators.
	2	Chemical treatment, clarifier-flocculators sand filters and process water recirculation.
III	1	Clarifier-flocculators, chemical treatment (if necessary).
	2	Total process water recirculation.
IV	1	Once-through water use, cooling ponds, or alternate treatment for control of thermal pollution (chemical treatment of blowdown, if present, recirculating cooling water).
	2	Cooling towers, recirculation, chemical treatment of blowdown.

Source: After Datagraphics 1973.

1. See Table 8.22 for category identification.

2. Level 1: Best practicable control technology currently available recommended by Datagraphics (1973).
   Level 2: Best practicable control technology economically achievable recommended by Datagraphics (1973).
   Level 3: New source performance standards recommended by Datagraphics (1973).

Table 8.26 *Industry Category I: Waste Loads Considered Achievable for Open Furnace with Wet Air Pollution Controls*

				Level 1 effluent						Level 2 effluent					
	Raw waste load			30-day average			24-hr maximum			30-day average			24-hr maximum		
Constituents	(kg/mwhr)	(lb/mwhr)	(mg/l)	(kg/mwhr)	(lb/mwhr)	(mg/l)	(kg/mwhr)	(lb/mwhr)	(mg/l)	(kg/mwhr)	(lb/mwhr)	(mg/l)	(kg/mwhr)	(lb/mwhr)	(mg/l)
Suspended solids	24.0	52.8	1,460	0.160	0.352	25.0	0.319	0.703	50	0.012	0.026	15.0	0.024	0.052	30
Total chromium	0.078	0.172	4.76	0.0032	0.007	0.5	0.006	0.014	1	0.0004	0.0009	0.5	0.0008	0.0017	1.0
Hexavalent chromium	0.005	0.012	0.32	0.0002	0.0004	0.03	0.0006	0.0014	0.1	0.00001	0.00002	0.01	0.00002	0.00004	0.02
Manganese	10.07	22.17	613	0.032	0.070	5.0	0.064	0.141	10	0.0039	0.0086	5.0	0.008	0.017	10
Oil	0.011	0.025	0.7	0.045	0.098	7.0	0.064	0.141	10	0.0055	0.012	7.0	0.008	0.017	10
Phenol				0.0032	0.007	0.5	0.004	0.010	0.7	0.0002	0.0003	0.2	0.0003	0.0007	0.4
Phosphate	0.010	0.023	0.635	0.0064	0.0141	1.0	0.013	0.028	2.0	0.0001	0.0002	0.1	0.0002	0.0003	0.2
	(gal/mwhr)		(l/mwhr)		(gal/mwhr)			(l/mwhr)			(gal/mwhr)			(l/mwhr)	
Flow	4,335		16,410		1,686			6,382			207			783	
		Value					Value					Value			
pH		7.2					6.0-9.0					6.0-9.0			

Source: After Datagraphics 1973.

Table 8.27 *Industry Category II: Waste Loads Considered Achievable for Covered Electric Furnaces and Other Smelting Operations with Wet Air Pollution Control Devices*

	Raw waste load			Level 1 effluent						Level 2 effluent					
				30-day average			24-hr maximum			30-day average			24-hr maximum		
Constituents	(kg/mwhr)	(lb/mwhr)	(mg/l)	(kg/mwhr)	(lb/mwhr)	(mg/l)	(kg/mwhr)	(lb/mwhr)	(mg/l)	(kg/mwhr)	(lb/mwhr)	(mg/l)	(kg/mwhr)	(lb/mwhr)	(mg/l)
Suspended solids	13.02	28.67	1,555	0.209	0.461	25	0.419	0.922	50	0.016	0.035	15	0.032	0.071	30
Total chromium	0.040	0.088	4.76	0.004	0.009	0.5	0.008	0.018	1.0	0.0005	0.0012	0.5	0.001	0.002	1.0
Hexavalent chromium	0.003	0.006	0.32	0.0003	0.0006	0.03	0.0008	0.0018	0.1	0.00001	0.00002	0.01	0.00002	0.00005	0.02
Total cyanide	0.021	0.046	2.49	0.002	0.005	0.25	0.004	0.009	0.5	0.0003	0.0006	0.25	0.0005	0.001	0.5
Manganese	3.74	8.24	447	0.042	0.092	5.0	0.084	0.184	10	0.005	0.012	5	0.011	0.023	10
Oil	0.038	0.083	4.5	0.059	0.129	7.0	0.084	0.184	10	0.007	0.016	7.0	0.011	0.023	10
Phenol	0.061	0.134	7.27	0.004	0.009	0.5	0.006	0.013	0.7	0.0002	0.0005	0.2	0.0004	0.0009	0.4
Phosphate	0.009	0.020	1.11	0.008	0.018	1.0	0.017	0.037	2.0	0.0001	0.0002	0.1	0.0002	0.0005	0.2
	(gal/mwhr)		(l/mwhr)		(gal/mwhr)			(l/mwhr)			(gal/mwhr)			(l/mwhr)	
Flow	2,210		8,365		2,210			8,365			280			1,060	
		Value					Value						Value		
pH		6.0-9.0					6.0-9.0						6.0-9.0		

Source: After Datagraphics 1973.

Table 8.28 *Industry Category III: Waste Loads Considered Achievable for Slag Processing*

				Level 1 effluent					
	Raw waste load			*30-day average*			*24-hr maximum*		
Constituents	*(kg/kkg processed)*	*(lb/ton processed)*	*(mg/l)*	*(kg/kkg processed)*	*(lb/ton processed)*	*(mg/l)*	*(kg/kkg processed)*	*(lb/ton processed)*	*(mg/l)*
Suspended solids	46.0	91.9	864	1.330	2.659	25	2.659	5.319	50
Total chromium	0.109	0.217	2.04	0.026	0.053	0.5	0.053	0.106	1
Manganese	2.87	5.74	54	0.266	0.532	5	0.532	1.064	10
Oil	0.064	0.128	1.2	0.372	0.745	7	0.532	1.064	10
	(gal/ton)	*(l/kkg)*		*(gal/ton)*	*(l/kkg)*		*(gal/ton)*	*(l/kkg)*	
Flow	12,750	53,100		12,750	53,100		12,750	53,100	
	Value			*Value*			*Value*		
pH	6.2			6.0-9.0			6.0-9.0		

Source: After Datagraphics 1973.

Table 8.29 *Industry Category IV: Waste Loads Considered Achievable for Noncontact Cooling Water*

	Raw waste load			Level 1 effluent						Level 2 effluent					
				30-day average			24-hr maximum			30-day average			24-hr maximum		
Constituents	(kg/mwhr)	(lb/mwhr)	(mg/l)	(kg/mwhr)	(lb/mwhr)	(mg/l)	(kg/mwhr)	(lb/mwhr)	(mg/l)	(kg/mwhr)	(lb/mwhr)	(mg/l)	(kg/mwhr)	(lb/mwhr)	(mg/l)
Suspended solids	1.590	1.30	11.0	1.343	2.959	25.0	2.686	5.917	50	0.067	0.148	25	0.134	0.296	50
Total chromium	0.0005	0.001	.01	0.027	0.059	0.5	0.054	0.118	1.0	0.001	0.003	0.5	0.003	0.006	1.0
Hexavalent chromium	0.0005	0.001	.01	0.002	0.004	0.03	0.005	0.012	0.1	0.00003	0.00006	0.01	0.00005	0.0001	0.02
Oil	0.038	0.083	0.7	0.376	0.828	7.0	0.537	1.183	10	0.019	0.041	7.0	0.027	0.059	10.0
Phosphate	0.012	0.026	0.22	0.161	0.355	3.0	0.269	0.592	5.0	0.004	0.009	1.5	0.008	0.018	3.0
	(kg-cal/mwhr)	(BTU/mwhr)		(kg-cal/mwhr)	(BTU/mwhr)		(kg-cal/mwhr)	(BTU/mwhr)		(kg-cal/mwhr)	(BTU/mwhr)		(kg-cal/mwhr)	(BTU/mwhr)	
Heat content	408,000	1,621,000		149,000	592,000		298,000	1,184,000		7,500	30,000		14,900	59,000	
	(gal/mwhr)		(l/mwhr)		(gal/mwhr)			(l/mwhr)			(gal/mwhr)			(l/mwhr)	
Flow	14,185		53,690		14,185			53,690			710			2,685	
		Value					Value						Value		
pH		7.0					6.0-9.0						6.0-9.0		

Source: After Datagraphics 1973.

Category I: Open Electric Furnaces with Wet Air Pollution Control Devices

Datagraphics defined flow as including the total flow in the scrubber rather than the effluent.

Treatment level 1 concentrations shown in Table 8.26 are those Datagraphics considered achievable by the treatment system shown in Figure 8.3, less the sand filter and recirculation portions. Loads are based on concentrations shown and a water use rate of 6,382 l/mwhr (1,686 gal/mwhr).

Treatment level 2 concentrations are based on the entire treatment system as shown in Figure 8.3, including sand filtration and recirculation. These levels would require better operation of the treatment system than was necessary in level 1. Loads are based on a blowdown rate of 783 l/mwhr (207 gal/mwhr).

Treatment level 3 concentrations are based on use of a dry dust collection system for air pollution control which results in no discharge of waterborne pollutants. During the survey this type of dust collection equipment was reported to be more widely used than wet scrubbers on this type of furnace.

Category II: Covered Electric Furnaces and Other Smelting Operations with Wet Air Pollution Control Devices

Treatment level 1 concentrations are the same as for Category I, treatment level 1 (see Table 8.27).

Treatment level 2 concentrations are the same as for Category I, treatment level 2. The assumed blowdown rate is 280 gal/mwhr.

Category III: Slag Processing

Treatment level 1 concentrations are based on the use of clarifier-flocculators (see Table 8.28).

Treatment level 2 concentrations are based on recirculation of all water after precipitation of fine suspended solids in clarifier-flocculators.

Category IV: Noncontact Cooling Water

Treatment level 1 concentrations and heat load (Table 8.29) are based on once-through cooling water use, with cooling ponds to limit temperature rise to 2.78°C (5°F) above that of ambient water (or intake water, if from a surface body).

Treatment level 2 heat load is based on a blowdown rate of 5 percent of recirculation rate in a cooling tower, with a temperature rise of 2.78°C (5°F) above that of ambient water (or intake water, if from a surface body). Other limits are deemed achievable by treatment of blowdown.

The 24-hour maximums are generally twice the 30-day averages and are based on maximum concentrations observed by Datagraphics at exemplary plants, or those which might be attained during system upsets. In the case of oil and phenol, and phosphate in category IV, level 1, the limitations are 1.5 times the 30-day average, and for hexavalent chromium, they are 3 times the 30-day average.

RECOMMENDED EFFLUENT LOADINGS FOR BPCTCA AND BATEA

The data for the four categories (I, II, III, IV) were transformed by Datagraphics (1973) into recommended loadings achievable by best practicable control technology currently available and best available technology economically achievable. These data are presented in Tables 8.30 and 8.31 respectively.

STARTUP AND SHUTDOWN PROBLEMS

Datagraphics (1973) identified no problems of consequence in connection with the startup or shutdown of production facilities insofar as waste water control and treatment is concerned. Cooling water and scrubber water were observed to flow or recirculate during short-term production stoppages.

COSTS

Capital and operating cost information was obtained from each plant surveyed. The following discussion pertains to selected portions of the survey as summarized by the data in Tables 8.32 through 8.34.

The capital costs (per mw capacity) for water treatment systems at the plants surveyed varied from $4,528 (for a cooling water system, including cooling towers) to $27,507 (for a scrubber and cooling water treatment and recirculation system). Operating costs ranged from a low of $0.021/mwhr (for spray tower waste water treatment and cooling tower operation), to a high of $0.770/mwhr (for scrubber and cooling water treatment and recirculation).

Datagraphics (1973) presented capital costs in terms of installed capacity. Operating costs were presented in terms of units of production and also in terms of waste water flows. These costs were based on cost of capital at an interest rate of 8 percent and a depreciation period of 15 years. Capital costs were adjusted to August 1971 dollars using the *Chemical Engineering* "Plant Cost Index." Operating costs were adjusted, when necessary, on the basis of an average of 3.5 percent per year.

Power and lagoon costs were taken from EPA (1973). Power costs were calculated on the basis of flow rates and pumping head, and were assumed to be one cent per kilowatt-hour. This estimate was reportedly confirmed by the Ferroalloys Association as being equal to the average cost in the industry.

The following bases were used for cost calculation according to category and treatment level:

Category I

Treatment level 1 costs were developed as shown in Figure 8.3 on the basis of a 63.1 l/sec (1,000 gpm) flow rate. At a water use rate of 6,362 l/mwhr (1,686 gal/mwhr), this is equivalent to the flow rate from a furnace operating at 35 mw. Costs include expenditures for mechanical equipment, tanks, piping, valves, electrical, engineering, and installation. The basis is the complete system less sand filter and recirculation. Datagraphics (1973) believes that investment

Table 8.30 *Best Practicable Control Technology Currently Available Guidelines and Limitations*

	Category I				*Category II*				*Category III*				*Category IV*			
	30-day average		*24-hr maximum*		*30-day average*		*24-hr maximum*		*30-day average*		*24-hr maximum*		*30-day average*		*24-hr maximum*	
Constituents	*(kg/mwhr)*	*(lb/mwhr)*	*(kg/mwhr)*	*(lb/mwhr)*	*(kg/mwhr)*	*(lb/mwhr)*	*(kg/mwhr)*	*(lb/mwhr)*	*(kg/kkg proc.)*	*(lb/ton proc.)*	*(kg/kkg proc.)*	*(lb/ton proc.)*	*(kg/mwhr)*	*(lb/mwhr)*	*(kg/mwhr)*	*(lb/mwhr)*
Suspended solids	0.160	0.352	0.319	0.703	0.209	0.461	0.419	0.922	1.330	2.659	2.659	5.319	1.343	2.959	2.686	5.917
Total chromium	0.0032	0.007	0.006	0.014	0.004	0.009	0.008	0.018	0.026	0.053	0.053	0.106	0.027	0.059	0.054	0.118
Hexavalent chromium	0.0002	0.0004	0.0006	0.0014	0.0003	0.0006	0.0008	0.0018	–	–	–	–	0.002	0.004	0.005	0.012
Total cyanide	–	–	–	–	0.002	0.005	0.004	0.009	–	–	–	–	–	–	–	–
Manganese	0.032	0.070	0.064	0.141	0.042	0.092	0.084	0.184	0.266	0.532	0.532	1.064	–	–	–	–
Oil	0.045	0.098	0.064	0.141	0.059	0.129	0.084	0.184	0.372	0.745	0.532	1.064	0.376	0.828	0.537	1.183
Phenol	0.0032	0.007	0.004	0.010	0.004	0.009	0.006	0.013	–	–	–	–	–	–	–	–
Phosphate	0.0064	0.0141	0.013	0.028	0.008	0.018	0.017	0.037	–	–	–	–	0.161	0.355	0.269	0.592
pH		6.0-9.0				6.0-9.0				6.0-9.0				6.0-9.0		
													(kg-cal/mwhr)	*(BTU/mwhr)*	*(kg-cal/mwhr)*	*(BTU/mwhr)*
Heat content													149,000	592,000	298,000	1,184,000
Cost items	*($/mw)*		*($/mwhr)*		*($/mw)*		*($/mwhr)*		*($/ton/day)*		*($/ton)*		*($/mw)*		*($/mwhr)*	
Investment	17,143				22,222				2,526				1,266			
Capital costs	–		0.103		–		0.134		–		0.344		–		0.007	
Depreciation	–		0.138		–		0.178		–		0.459		–		0.010	
Operating costs less power	–		0.606		–		0.785		–		0.421		–		0.0	
Power costs	–		0.012		–		0.016		–		0.051		–		0.0	
Total operating costs	–		0.859		–		1,113		–		1.28		–		0.017	

Source: After Datagraphics 1973.

Table 8.31 *Best Available Technology Economically Achievable Guidelines and Limitations*

	Category I				*Category II*				*Category IV*			
	30-day average		*24-hr maximum*		*30-day average*		*24-hr maximum*		*30-day average*		*24-hr maximum*	
Constituents	*(kg/mwhr)*	*(lb/mwhr)*	*(kg/mwhr)*	*(lb/mwhr)*	*(kg/mwhr)*	*(lb/mwhr)*	*(kg/mwhr)*	*(lb/mwhr)*	*(kg/mwhr)*	*(lb/mwhr)*	*(kg/mwhr)*	*(lb/mwhr)*
Suspended solids	0.012	0.026	0.024	0.052	0.016	0.035	0.032	0.071	0.067	0.148	0.134	0.296
Total chromium	0.0004	0.0009	0.0008	0.0017	0.0005	0.0012	0.001	0.002	0.001	0.003	0.003	0.006
Hexavalent chromium	0.00001	0.00002	0.00002	0.00004	0.00001	0.00002	0.00002	0.00005	0.00003	0.00006	0.00005	0.0001
Total cyanide	–	–	–	–	0.0003	0.0006	0.0005	0.001	–	–	–	–
Manganese	0.0039	0.0086	0.008	0.017	0.005	0.012	0.011	0.023	–	–	–	–
Oil	0.0055	0.012	0.008	0.017	0.007	0.016	0.011	0.023	0.019	0.041	0.027	0.059
Phenol	0.0002	0.0003	0.0003	0.0007	0.0002	0.0005	0.0004	0.0009	–	–	–	–
Phosphate	0.0001	0.0002	0.0002	0.0003	0.0001	0.0002	0.0002	0.0005	0.004	0.009	0.008	0.018
pH		6.0-9.0				6.0-9.0				6.0-9.0		
									(kg-cal/ mwhr)	*(BTU/ mwhr)*	*(kg-cal/ mwhr)*	*(BTU/ mwhr)*
Heat content									7,500	30,000	14,900	59,000
Cost item	*($/mw)*		*($/mwhr)*		*($/mw)*		*($/mwhr)*		*($/mw)*		*($/mwhr)*	
Investment	21,063				27,303				8,444			
Capital costs	–		0.127		–		0.165		–		0.049	
Depreciation	–		0.169		–		0.219		–		0.065	
Operating costs less power	–		0.745		–		0.965		–		0.354	
Power costs	–		0.015		–		0.019		–		0.044	
Total operating costs	–		1.056		–		1.368		–		0.512	

Source: After Datagraphics 1973.

Note: Category III: No discharge of waste water pollutants to navigable waters.

Table 8.32 *Treatment Level Costs on Unit of Production Basis*

Industry category and treatment level	*Investment ($/mw or tpd)*	*Annual costs ($/mwhr or ton)* Capital	Depreciation	Operating cost less power	Power	Total
Category I:						
Treatment level 1	$17,143	$0.103	$0.138	$0.606	$0.012	$0.859
Treatment level 2	21,063	0.127	0.169	0.745	0.015	1.056
Category II:						
Treatment level 1	22,222	0.134	0.178	0.785	0.016	1.113
Treatment level 2	27,303	0.165	0.219	0.965	0.019	1.368
Category III:						
Treatment level 1	2,526[a]	0.344[a]	0.459[a]	0.421[a]	0.051[a]	1.28[a]
Treatment level 2	2,604[a]	0.357[a]	0.485[a]	0.421[a]	0.051[a]	1.31[a]
Category IV:						
Treatment level 1	1,266	0.007	0.010	0.0	0.0	0.017
Treatment level 2	8,444	0.049	0.065	0.354	0.044	0.512

Source: After Datagraphics 1973.

Note: Costs on basis of megawatts and megawatt-hours except where tons are indicated; a = tons.

costs may be less (per mw) for a plant larger than the model, and greater for a plant smaller than the model. One treatment system for the entire plant was judged to be more economical in general.

The "Inorganic Chemicals Industry Profile" (59 plants surveyed) indicated operating costs per annual unit of production equal to 28 percent of the capital cost per annual unit of production. Operating costs presented by Datagraphics (1973) were estimated on the basis of 30 percent per year of the estimated capital cost.

Treatment level 2 costs were expanded from level 1 to include recirculation and sand filtration. This resulted in a proportionate increase in annual and operating costs.

For treatment level 3, Datagraphics utilized investment and annual costs which have been published for scrubber systems and for fabric filter systems (all on 30-mw furnaces) for four products (Table 8.34): standard high-carbon (HC) ferromanganese; silicomanganese; HC ferrochromium; and 50 percent ferrosilicon (EPA, 1973). The costs for scrubber systems include a slurry settler and two filters for slurry dewatering. Investment costs for a fabric filter (dry) system are identical to or slightly lower than those for a wet scrubber system, but annual costs are about half. Less power is required for the operation of the fabric filter system.

Table 8.33 *Treatment Level Costs on Waste Water Flow Basis*

Industry category and treatment level	*Investment ($/gpm)*	*Annual costs ($/1,000 gal)*				
		Capital	*Depreciation*	*Operating cost less power*	*Power*	*Total*
Category I:						
Treatment level 1	600	0.057	0.076	0.336	0.007	0.476
Treatment level 2	737	0.070	0.094	0.413	0.008	0.585
Category II:						
Treatment level 1	600	0.057	0.076	0.336	0.007	0.476
Treatment level 2	737	0.070	0.094	0.413	0.008	0.585
Category III:						
Treatment level 1	285.29	0.027	0.036	0.033	0.004	0.100
Treatment level 2	294.12	0.028	0.038	0.033	0.004	0.103
Category IV:						
Treatment level 1	16.54	0.0016	0.0021	0	0	0.004
Treatment level 2	119	0.013	0.017	0.240	0.011	0.281

Source: After Datagraphics 1973.

Table 8.34 *Scrubber Costs vs Fabric Filter Costs for Air Pollution Control on 30-mw Open Electric Submerged Arc Furnaces*

	Scrubber system				
	Investment		*Annual costs*		
	($)	*($/mw)*	*($)*	*($/mwhr)*	*($/ton)*
HC ferromanganese	1,640,000	54,667	826,000	3.49	8.37
Silicomanganese	1,640,000	54,667	826,000	4.29	18.88
HC ferrochromium	1,472,000	49,067	699,000	2.83	11.87
50% ferrosilicon	3,180,000	106,000	1,986,000	7.94	39.72
	Fabric filter system				
	Investment		*Annual costs*		
	($)	*($/mw)*	*($)*	*($/mwhr)*	*($/ton)*
HC ferromanganese	1,640,000	54,667	445,000	1.88	4.51
Silicomanganese	1,640,000	54,667	445,000	2.31	10.17
HC ferrochromium	1,190,000	39,667	385,000	1.56	6.54
50% ferrosilicon	2,340,000	78,000	734,000	2.94	14.68

Source: After Datagraphics 1973.

Category II

Treatment level 1 costs were developed as for Category I, level 1 at a flow of 63.1 l/sec (1,000 gpm or 8,365 l/mwhr), which is equivalent to the flow from a furnace operating at 27 mw. Expected investment cost per mw was noted to be somewhat higher for small plants and less for large plants.

Treatment levels 2 and 3 were identical to category I, level 2, but were based on a 27-mw furnace.

Category III

Treatment level 1 costs were calculated for two clarifier flocculators, with the necessary piping and pumps and other appurtenances. Costs were based on the use of 53,148 l/kkg (12,750 gal/ton) processed.

Treatment level 2 costs were noted to be greater than for level 1 by the addition of pumps and pipes necessary for recirculation.

Category IV

Treatment level 1 cooling pond costs were calculated assuming 315.15 l/sec (5,000 gpm) and 2.2°C (4°F) approach to equilibrium temperature. For treatment level 2, Datagraphics (1973) selected flow rates as the average of selected plants surveyed. Hexavalent chromium reduction and precipitation costs were included in the costs shown in Tables 8.30 and 8.31. The scheme is complex and the reader should see the Datagraphics report for details.

INCREMENTAL COSTS OF ACHIEVING LEVELS OF TREATMENT TECHNOLOGY

It was concluded that the cost of implementing the various levels of treatment technology in the industry will vary from plant to plant, depending on the treatment currently in use. The best estimates of these costs were presented by category, based on the assumed levels of present technology in typical plants surveyed. Summaries of the pertinent points follow.

Category I

A lagoon in which the scrubber waste water is treated by sedimentation prior to discharge was concluded to be typical of United States plants. Such a lagoon may be usable as part of a new treatment scheme, but almost all of the costs shown in Table 8.32 probably would be incurred to bring the effluent down to the suggested limitation. Only an additional sand filter and recirculation system would be required for a plant to go to level 2 treatment from this base. Thus, the incremental cost of reaching level 1 would be $17,143/mwhr in investment, and $0.859/mwhr in annual costs, and the incremental cost of reaching level 2 would be $3,920/mw for investment, and $0.197/mwhr in annual costs.

Category II

The typical plant was considered to have a lagoon for cyanide oxidation and scrubber waste water treatment by sedimentation prior to discharge. It was assumed that the cost of reaching level 1 is 100 percent of that shown in Table 8.32 and the cost of reaching level 2 from level 1 is the arithmetic difference. Therefore, the incremental cost of reaching level 1 was calculated to be $22,222/mw in investment cost and $1.113/mwhr in annual costs with an additional $5,081/mw in investment cost, and $0.255/mwhr in annual costs to reach level 2.

Category III

The typical plant for this category was assumed to have a small lagoon which would require expenditure of 100 percent of the costs shown in the table. Thus, the incremental cost for level 1 would be $2,526/tpd of slag in investment cost and $1.28/ton of slag in annual operating costs. Reaching level 2 would require an additional investment of $78/tpd and increase annual costs by $0.03.

Category IV

The typical plant was assumed to employ once-through cooling water at no cost. The incremental cost to reach level 1 treatment technology was calculated to be $1,266/mw for investment cost and $0.017/mwhr for annual costs. A cooling tower would be a wise alternative to a cooling pond. The incremental cost for a cooling tower system (a recommended alternative for level 2 treatment) would be $8,444/mw for investment cost and $0.512/mwhr for annual operating costs.

AIR AND SOLID WASTES

Datagraphics (1973) concluded that the solid waste produced by treatment of waste waters in the ferroalloys industry derives principally from the smelting operation as waste from air pollution control devices. The solid waste from air pollution control is produced whether a dry or wet system is utilized and varies only in that the former produces a slurry or sludge and the latter a fine dust. The slurry or sludge generally is accumulated in sludge lagoons (a form of landfilling) while the dry dust must be bagged and landfilled. It was recommended that greater attention be paid to the disposal of these potentially harmful materials. Landfilling in a sealed site or encapsulation in concrete or polymers were recommended as possible improvements. It was reported that little success has been realized by efforts to agglomerate these solids for recharging to smelting furnaces, and that dry dust probably could be utilized more easily than wet sludge.

REGULATIONS PROMULGATED BY EPA

On the basis of the data and information presented by Datagraphics (1973) and of comments received by affected parties, EPA promulgated the following rules and regulations for the iron and steel industry (see Tables 8.35 through 8.40).* Justifiable exceptions can be granted by a regional administrator.

Open Electric Furnaces with Wet Air Pollution Control Devices

Effluent limitations representing the degree of effluent reduction attainable by application of the best practicable control technology currently available (level 1) are shown in Table 8.35. For regulations representing the degree of effluent reduction attainable by application of the best available technology economically achievable (level 2) and standards of performance for new point sources (level 3), see Table 8.36.

Covered Electric Furnaces and Other Smelting Operations with Wet Air Pollution Control Devices

Effluent limitations representing the degree of effluent reduction attainable by application of the best practicable control technology currently available (level 1) are shown in Table 8.37. For regulations representing the degree of effluent reduction attainable by application of the best available technology

Table 8.35 *Effluent Limitations Promulgated for Open Electric Furnaces with Wet Air Pollution Control Devices (level 1)*

Effluent characteristics	*Effluent limitations*			
	Maximum for any one day	*Average of daily values for 30 consecutive days shall not exceed*	*Maximum for any one day*	*Average of daily values for 30 consecutive days shall not exceed*
	Metric units (kg/mwhr)[1]		*English units (lb/mwhr)*[1]	
TSS	0.319	0.160	0.703	0.352
Total chromium	0.006	0.0032	0.014	0.007
Chromium VI	0.0006	0.0003	0.0014	0.0007
Total manganese	0.064	0.032	0.141	0.070
pH	Within the range 6.0 to 9.0		Within the range 6.0 to 9.0	

Note: Level 1 represents the degree of effluent reduction attainable by the application of the best practicable control technology currently available.

1. Mwhr means megawatt hour(s) of electrical energy consumed in the smelting *process* (furnace power consumption).

* These rules and regulations were published in the *Federal Register*, February 22, 1974, Vol. 39, No. 37, Part 2, pp. 6806-6814.

Table 8.36 *Effluent Limitations Promulgated for Open Electric Furnaces with Wet Air Pollution Control Devices (levels 2 and 3)*

Effluent characteristics	*Effluent limitations*			
	Maximum for any one day	*Average of daily values for 30 consecutive days shall not exceed*	*Maximum for any one day*	*Average of daily values for 30 consecutive days shall not exceed*
	Metric units (kg/mwhr)		*English units (lb/mwhr)*	
TSS	0.024	0.012	0.052	0.026
Total chromium	0.0008	0.0004	0.0017	0.0009
Chromium VI	0.00008	0.00004	0.0002	0.0001
Total manganese	0.008	0.0039	0.017	0.0086
pH	Within the range 6.0 to 9.0		Within the range 6.0 to 9.0	

Note: Level 2 represents the degree of effluent reduction attainable by the application of the best available technology economically achievable. Level 3 represents standards of performance for new sources.

Table 8.37 *Effluent Limitations Promulgated for Covered Electric Furnaces and Other Smelting Operations with Wet Air Pollution Control Devices (level 1)*

Effluent characteristics	*Effluent limitations*			
	Maximum for any one day	*Average of daily values for 30 consecutive days shall not exceed*	*Maximum for any one day*	*Average of daily values for 30 consecutive days shall not exceed*
	Metric units (kg/mwhr)[1]		*English units (lb/mwhr)*[1]	
TSS	0.419	0.209	0.922	0.461
Total chromium	0.008	0.004	0.018	0.009
Chromium VI	0.0008	0.0004	0.0018	0.0009
Total manganese	0.084	0.042	0.184	0.092
Total cyanide	0.004	0.002	0.009	0.005
Phenols	0.006	0.004	0.013	0.009
pH	Within the range 6.0 to 9.0		Within the range 6.0 to 9.0	

Note: Level 1 represents the degree of effluent reduction attainable by the application of the best practicable control technology currently available.

1. Provided, however, that for nonelectric furnace smelting processes, the units of the effluent limitations set forth in this section shall be read as "kg/kkg of product (lb/ton of product)," rather than "kg/mwhr (lb/mwhr)" and the limitations (except for pH) shall be three (3) times those listed in the table in this section.

Table 8.38 *Effluent Limitations Promulgated for Covered Electric Furnaces and Other Smelting Operations with Wet Air Pollution Control Devices (levels 2 and 3)*

Effluent characteristics	*Effluent limitations*			
	Maximum for any one day	*Average of daily values for 30 consecutive days shall not exceed*	*Maximum for any one day*	*Average of daily values for 30 consecutive days shall not exceed*
	Metric units (kg/mwhr)[1]		*English units (lb/mwhr)*[1]	
TSS	0.032	0.016	0.071	0.035
Total chromium	0.001	0.0005	0.002	0.0012
Chromium VI	0.0001	0.00005	0.0002	0.0001
Total manganese	0.011	0.005	0.023	0.012
Total cyanide	0.0005	0.0003	0.001	0.0006
Phenols	0.0004	0.0002	0.0009	0.0005
pH	Within the range 6.0 to 9.0		Within the range 6.0 to 9.0	

Note: Level 2 represents the degree of effluent reduction attainable by the application of the best available technology economically achievable. Level 3 represents standards of performance for new sources.

1. Provided, however, that for nonelectric furnace smelting processes, the units of the effluent limitations set forth in this section shall be read as "kg/kkg of product (lb/ton of product)," rather than "kg/mwhr (lb/mwhr)" and the limitations (except for pH) shall be three (3) times those listed in the table in this section.

Table 8.39 *Effluent Limitations Promulgated for Slag Processing (level 1)*

Effluent characteristics	*Effluent limitations*			
	Maximum for any one day	*Average of daily values for 30 consecutive days shall not exceed*	*Maximum for any one day*	*Average of daily values for 30 consecutive days shall not exceed*
	Metric units (kg/kg processed)		*English units (lb/ton processed)*	
TSS	2.659	1.330	5.319	2.659
Total chromium	0.053	0.026	0.106	0.053
Total manganese	0.532	0.266	1.064	0.532
pH	Within the range 6.0 to 9.0		Within the range 6.0 to 9.0	

Note: Level 1 represents the degree of effluent reduction attainable by the application of the best practicable control technology currently available.

Table 8.40 *Effluent Limitations Promulgated for Slag Processing (levels 2 and 3)*

Effluent characteristics	*Effluent limitations*			
	Maximum for any one day	*Average of daily values for 30 consecutive days shall not exceed*	*Maximum for any one day*	*Average of daily values for 30 consecutive days shall not exceed*
	Metric units (kg/kkg processed)		*English units (lb/ton processed)*	
TSS	0.271	0.136	0.542	0.271
Total chromium	0.0054	0.0027	0.011	0.0054
Total manganese	0.054	0.027	0.108	0.054
pH	Within the range 6.0 to 9.0		Within the range 6.0 to 9.0	

Note: Level 2 represents the degree of effluent reduction attainable by the application of the best available technology economically achievable. Level 3 represents standards of performance for new sources.

economically achievable (level 2) and standards of performance for new point sources (level 3), see Table 8.38.

Slag Processing

Effluent limitations representing the degree of effluent reduction attainable by application of the best practicable control technology currently available (level 1) are shown in Table 8.39. For guidelines representing the degree of effluent reduction attainable by application of the best available technology economically achievable (level 2) and standards of performance for new point sources (level 3), see Table 8.40.

REFERENCES

Datagraphics Inc. 1973. Development Document for *Proposed Effluent Limitations Guidelines and New Source Performance Standards for the Smelting and Slag Processing Segment of the Ferroalloy Manufacturing Point Source Category.* Document EPA-440/1-73-008, U.S. Environmental Protection Agency, Washington, D.C.

Eckenfelder, W. W. 1970. *Water Quality Engineering.* New York: Barnes and Noble.

EPA 1973. *Air Pollution Control Engineering and Cost Study of the Ferroalloy Industry.* (Draft Report), U.S. Environmental Protection Agency, Office of Air and Water Programs, Washington, D.C.

Fetelsdorf, H. J., Hodapp, E., and Endell, N. 1969. *Experiences with an Electric Filter Dust Collecting System in Connection with a 20-MW Silicochromium Furnace.* Proceedings, AIME Electric Furnace Conference, Vol. 27, Detroit, pp. 109-114.

Person, R. A. 1969. *Control of Emissions from Ferroalloy Furnace Processing.* Proceedings, AIME Electric Furnace Conference, Vol. 27, Detroit, pp. 81-92.

9. *Recovery of copper, lead, and zinc*

This chapter deals with the primary recovery of copper, lead, and zinc. The primary recovery of copper includes smelting, electrolytic refining, and hydrometallurgy (leaching).* The primary recovery of lead consists of roasting the mineral galena and smelting the oxide. The primary recovery of zinc consists of roasting the mineral sphalerite and subsequent pyrometallurgical and/or leaching, electrolytic recovery and casting operations. All these operations are associated with the production of sulfuric acid by contact plants.

SOURCES OF INFORMATION

The sources of information for this chapter are the published literature and the effluent guidelines document prepared by Battelle Research Institute's Columbus Laboratory under authorization by Public Law 92-500 (see Battelle 1974). This document constitutes the most comprehensive available treatment of these industries with respect to sources of wastes. The published literature in general treats the waste production and treatment characteristics of the copper, lead, and zinc industry in a piecemeal fashion. Consequently, a considerable portion of this chapter is condensed or extracted from the Battelle report.

Neither the Battelle document nor the promulgated effluent guidelines are available in final form at this writing; consequently, the draft was used as a source of information. The drafts and final versions of these development documents generally are very similar except for the sections on proposed best practicable and best available economically achievable treatment technologies and the sections pertaining to effluent limitations guidelines. Therefore, it is extremely important to keep in mind that the analogous sections contained herein report approximate values rather than the final values on which promulgation will eventually be based. They were considered sufficiently important to

* Since the completion of the manuscript for this book, hydrometallurgy has been transferred by EPA to the ore and mineral dressing category for purposes of effluent guidelines development.

include in this volume because, even though the numbers are approximate, the technology discussed is available and the sources and the nature of effluents are valid. The validity of "practicable" and "economically achievable" is the factor still under debate. Disagreements relative to the meaning of these terms will be resolved only after the lengthy review with industry and other groups that is now in progress. Since these negotiations will be completed after this volume goes to press, results could not be incorporated in this edition.

The third major source of information is the author's experience in the Coeur d'Alene district of North Idaho, where primary recovery of lead and zinc is practiced.

The best practicable waste water control technology and the best available economically achievable waste water control technology proposed by Battelle (1974) for these industries is presented here. Again the author wishes to emphasize that these are not necessarily the exact procedures EPA will adopt as a basis for promulgation of waste discharge guidelines for these industries, but the guidelines can be expected to be based on variations of them.

PRIMARY COPPER

The Industry

According to the American Bureau of Metal Statistics (1973), the production of copper by United States refineries totaled 1,618,000 short tons in 1971. Demand had slackened to some extent from previous years in 1971, because of competition for the electrical market by aluminum. Historically much of the United States production has been from mines in Michigan, Montana, New Mexico, Arizona, and Utah. Currently, Arizona and Utah hold first and second positions, respectively, in copper production.

Since mine production is in the Western United States, copper smelters are located in the West. Fifteen smelters were operating in the United States in 1972. Arizona has seven smelters and eight other states have one each.

Electrolytic copper refining, historically, has been concentrated in New York, New Jersey, and Maryland; but refineries are moving westward toward the smelters. Of the 14 refineries in the United States, 3 are pyrometallurgical and 11 are electrolytic.

Sulfuric acid plants have been built at 11 of the 15 U.S. copper smelters. The impetus has been the need to collect sulfur dioxide as a means of air pollution control.

The Process

Most copper is obtained from one or more copper sulfide minerals, although native copper and copper oxides are mined also. Most mining (80 percent) is done by the open pit method. Sulfide ore extraction is followed by comminution and then by flotation.

Concentrators consist of flotation cells in which copper sulfides rise to the top of a reagent and water slurry. This floating material is skimmed off and sent to thickeners and the waste water and solids are sent to tailings ponds.

The copper concentrate then goes to a smelter where roasting, smelting, and conversion are accomplished. Roasting (if utilized) dries the concentrates and removes sulfur and other impurities. The smelting process consists of feeding copper concentrates and fluxes (limestone) into a heated reverberatory furnace wherein two immiscible liquid phases form. The bottom phase is a copper-iron sulfide known as matte and the top phase is liquid iron silicate slag. The matte and slag flow from the furnace, whereupon the slag is discarded and the matte enters a converter.

Conversion consists of the oxidation of the iron-copper sulfide matte to blister copper by blowing air through the molten material. The iron sulfide in the matte oxidizes to sulfur dioxide and iron oxide. Sulfur dioxide leaving the converter is collected for a sulfuric acid plant. The iron oxide reacts with a silica flux added to the converter to form an iron silicate slag which is recirculated to the matte smelter for recovery of its copper content.

After removal of the iron silicate slag, additional air is introduced to the converter to oxidize the copper sulfide. Sulfur dioxide is evolved and collected. The converted copper contains arsenic, some sulfur, and precious metals. It is moved to a refining furnace (molten state) or cast into blister copper pigs.

Refining is accomplished electrolytically or pyrometallurgically. Gold, silver, nickel, selenium, and tellurium are not removed pyrometallurgically so the product is often further refined electrolytically. Electrolytic refining produces high-purity copper. It is conducted in tanks containing copper anodes. The cathodes are thin copper sheets. The electrolyte is a dilute solution of sulfuric acid and copper sulfate. Copper and impurities such as nickel, arsenic, and iron are electrolytically oxidized and dissolved at the anode. Gold, silver, selenium, and tellurium occur in the electrolyte as finely divided solids. Copper deposits on the cathode.

During the mining process copper oxides, which are not concentrated with the sulfides, are placed in waste dumps. These dumps are leached with the sulfuric acid produced by SO_2 recovery at the smelter. Leaching can also be accomplished in containers called vats.

Leaching solutions are treated by cementation or by solvent extraction. In cementation, copper is replaced in solution by iron (usually old tin cans). Solvent extraction employs organic solvents which extract copper in the organic phase. The copper is stripped from the organic molecules by mixing with sulfuric acid followed by electrowinning.

New processes, most notably the Arbiter process (see Arbiter 1973), in the future may replace one or more of the above procedures. However, Battelle (1974) concluded that there are not sufficient data available to evaluate reliably their waste production characteristics.

Sources and Characteristics of Wastes

Battelle (1974) tentatively recommended subdividing the primary copper industry in two ways for purposes of analysis of origin and control of wastes, particularly waste waters.

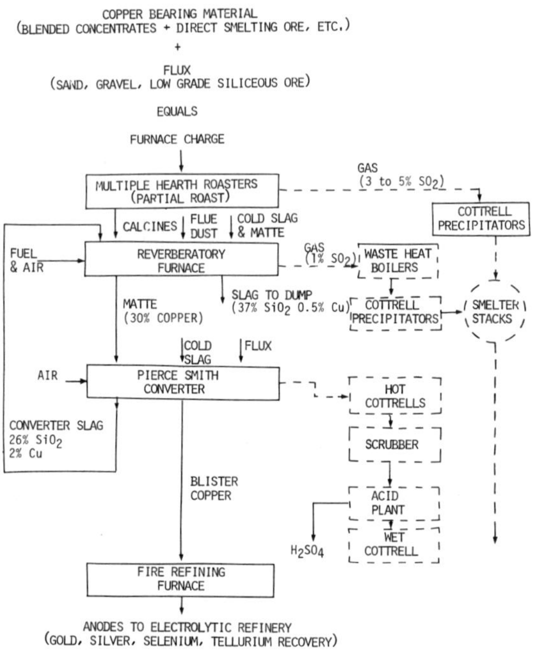

Figure 9.1. *Generalized diagram of conventional smelting flow sheet (after Battelle 1974).*

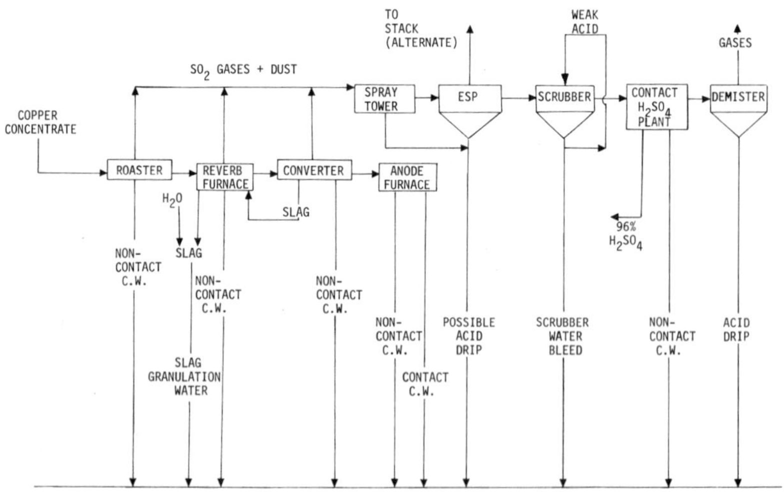

Figure 9.2. *Schematic showing waste effluents from the copper smelting process (after Battelle 1974).*

The two subcategories are based on the type of climate and on the process employed. The climatic categorization is due to the geographic variation in the evaporation rates in the United States. The areas wherein annual evaporation exceeds annual precipitation are designated as Category I. Battelle tentatively recommended total impoundment with no discharge of waste water as both the best practicable control technology currently available and the best available control technology economically achievable for plants in Category I areas. In areas where annual precipitation exceeds annual evaporation (Category II), evaporation from ponds is less practical and various waste water treatment technologies must be applied.

Smelting, electrolytic refining and leaching, and related recovery of copper from leach solutions are the three additional subcategories Battelle tentatively recommended. Although milling was not considered for purposes of the Battelle study, it is an important part of smelter operations because the wastes from mills (a water-suspended solids slurry) have a pH of at least 10. These waste may constitute a ready-made treatment mechanism for acid smelter wastes if mixing is employed.

Smelting

A smelter separates copper from the iron, sulfur, and other gangue in the concentrate produced by the mill. The process is schematically illustrated in Figure 9.1.

The concentrates may or may not be roasted prior to smelting and conversion. Roasting removes moisture, some sulfur, and volatile arsenic, antimony, and selenium. Roasting is not used if the concentrate is to be smelted as "green feed" to a reverberatory. Arsenic and antimony may not even be present. Drying can be achieved with forced air.

Sulfur dioxide gas is produced by roasting. The roaster-produced gas has enough SO_2 for sulfuric acid manufacture and is captured for that purpose.

Roasting occurs in multiple-hearth furnaces or in fluidized-bed roasters. Heat is provided by igniting the concentrates and injecting air.

The wastes from the roasting process are noncontact cooling water, fume, SO_2, and particulate matter as shown by Figure 9.2.

Subsequent to roasting, matte smelting in a reverberatory furnace collects most of the copper in a molten copper-iron-sulfide material called matte. One or more of the following constitutes the feed material to the furnace:

1. Copper concentrates, roasted or unroasted.
2. High-grade copper ore.
3. Cement copper.
4. Converter slag, recycled for additional recovery of copper.
5. Reagglomerated flue dust from smelting operations.
6. Fluxes, such as limestone, which convert silicates to fluid slag.

Heat is supplied by powdered coal, oil, or natural gas. Temperatures ranging from about 1,100°C to 1,650°C (2,000°F to 3,000°F) are achieved. These temperatures melt the charge, and a molten matte containing almost all of the recoverable copper as a sulfide forms on the furnace bottom. Liquid slag composed of silicates of iron and other metals floats atop the matte prior to discharge of the matte into ladles and disposal of the slag.

Gases containing sulfur dioxide are evolved from the matte-smelting operation but the SO_2 concentration may be less than 1 percent of the total gas volume. Acid-plant technology requires 4 percent SO_2, so the weak reverberatory off-gas cannot be sent to an acid plant. Battelle (1974) points out that conversion of the SO_2 to sulfuric acid requires either replacement of the reverberatory furnaces with furnaces such as the Outokumpu Oy flash-smelting furnace or the electric furnace. The SO_2 in the reverberatory gases can also be concentrated by scrubbing the weak gases with dimethyl aniline (DMA). Each of these alternatives is being adopted by one or more smelters in the United States. Smelting also produces noncontact cooling water and water used in many smelters to granulate the slag as it emanates from the furnace.

Conversion follows matte smelting. The process converts the matte copper to a relatively impure form called blister copper. The operation is done in two stages by blowing air through the molten material. First, the iron sulfide is oxidized to sulfur dioxide and iron oxide. Sulfur dioxide leaves the converter as a gas which goes to a sulfuric acid plant. Fumes and dusts are collected by baghouses or precipitators and sent to lead smelters. The iron oxide reacts with silica flux which is added to the converter and forms molten iron silicate slag. The approximate reaction in the first stage of converting is:

$CuFeS_2$	+	O_2	+	SiO_2	$\longrightarrow$	SO_2	+	$FeSiO_3$	+	Cu_2S
Matte		Added as air		Silica added for fluxing		Sulfur dioxide to sulfuric acid manufacture		Iron silicate slag removed and recirculated		Cuprous sulfide (white metal)

When this reaction is completed, the slag is removed from the converter. If it contains enough copper, it is recirculated to matte-smelting. After slag removal, air injection continues until the remaining sulfur is oxidized and removed as sulfur dioxide. The converted copper contains heavy metals, arsenic, sulfur, and the precious metals from the concentrates.

Battelle (1974) points out that converters produce more than half of copper-smelter SO_2 emissions in concentrations above 6 percent SO_2. Most sulfuric acid from smelters is manufactured from converter gas.

As indicated by the previous discussion, smelters produce few waste water streams other than contact and noncontact cooling water. The other streams are from flue-gas cleaning, sulfuric acid production, and slag granulation. Gas cleaning waste streams are due to the fact that dust and SO_2 evolve from roasters, reverberatory furnaces, and converters. Dusts include condensed substances (fume) volatilized in the furnace. Effective removal of particulate matter from the gases is achieved by electrostatic precipitators called

"Cottrells." Water sprays cool the hot gases prior to entering the Cottrells. Most of this water is evaporated but some of it collects in sumps of spray chambers or the electrostatic precipitators. If SO_2 is not recovered, the gases are discharged through a tall stack so as to disperse them as much as possible.

Battelle notes that in recent years the tendency has been to abandon roasters, thereby eliminating one source of air pollution. The gases from reverberatory furnaces are too dilute for acid production, but the gases from converters are sufficiently concentrated to permit manufacture of sulfuric acid in a contact plant. In order for sulfuric acid to be recovered from the converter gases the off-gases must be further cleaned and cooled by passing the gases from the Cottrell electrostatic precipitators through a 2 to 20 percent acid scrubber. Sulfates, chlorides, and other soluble salts build up in the scrubbers and chloride concentration may control the minimum necessary bleed rate. If the particulate removal devices are not performing effectively, a buildup of solids in the scrubber-water circuit may occur unless there is either a high percentage of bleed or a settling pond in the circuit. Recycle may be limited by the solids in the scrubber water as well as by the acidity and corrosiveness. Bleed-stream volumes reported to Battelle range from 3 l/sec (50 gpm) to as high as 35 l/sec (550 gpm). Rates as high as 250 l/sec (4,000 gpm) were reported at some smelters where water is used on a once-through basis.

The results of the analyses of samples taken at three smelters during the Battelle investigation are presented in Table 9.1. The table shows that the

Table 9.1 *Examples of Effluent Concentrations from Selected Copper Smelting Operations*

	Effluents from acid plant scrubbers			*Effluents from slag granulation*		
Parameters	*Plant 102*	*Plant 103*	*Plant 110*	*Plant 103*	*Plant 110*	*Plant 102*
pH	2.0-2.5	1.8	<2	7.7	810	6.4-7.6
TDS	0.1	5,000.	15,500.	140.	3,800.	30.4
TSS	1.4	6.5	70.2	6.8	151.	45.2
$SO_4^=$	–	490.	2,200.	62.	310.	–
CN^-	0.01	<0.005	0.077	<0.005	0.050	0.030
As	19.0	8.2	9.3	3.7	0.048	5.70
Cd	0.09	0.09	0.81	<0.001	<0.001	0.042
Cu	0.058	0.12	7,600.	0.12	0.05	0.604
Fe	0.65	0.10	3.81	0.04	0.30	340.
Pb	2.24	0.91	8.97	0.04	0.070	7.4
Hg	0.028	<0.0001	0.010	<0.0001	<0.0001	<0.0001
Ni	0.015	<0.001	0.100	<0.001	0.06	0.16
Se	0.040	<0.001	1.08	<0.001	0.54	0.040
Te	0.017	<0.001	–	<0.001	–	0.100
Zn	0.75	13.7	11.1	0.44	0.023	36.
Oil & grease	0.04	–	0.0	–	0.0	0.02

Source: After Battelle 1974.

Note: In concentration units of milligrams per liter.

scrubber effluents have a low pH and high sulfate content. Arsenic, zinc, and lead are also prominent pollutants. Bismuth and antimony would be expected also.

The scrubbed gases pass to the sulfuric acid plant for conversion to sulfuric acid. A sulfuric acid effluent may emanate from a mist separator located downstream from the absorption tower in the acid plant.

Slag from the reverberatory furnace is the most significant solid waste produced by copper smelting. Converter slag is usually recycled to the reverberatory furnaces for copper recovery. When slag is poured onto a slag dump and allowed to solidify in place, negligible waste water results because the solidified massive slag is considered essentially unleachable. But if the slag is granulated with water and pumped to a disposal site, solubilization may occur during quenching, and the water may become contaminated, particularly with suspended solids and some metals.

Refining

The blister copper from the converter ordinarily is cast into anodes for electrolytic refining. An anode furnace degasses it to provide satisfactory anodes. Anode molds are filled with blister copper and sprayed with large quantities of water to quench and cool them. The solidified anodes may be removed from the molds and suspended in cooling tanks for further cooling. The contact cooling water should be in a closed circuit with a cooling pond or tower. If buildup of sludge occurs, it is removed. Analyses of field samples acquired at two plants during the Battelle (1974) investigation are presented in Table 9.2. The pollutants in the water are copper and iron, and to a lesser extent lead, all in low concentrations.

The copper anodes are purified to essentially pure copper by electrolytic dissolution at the anode and deposition as the pure metal at the cathode in an electrolyte of copper sulfate and sulfuric acid. Electrolytic refining produces a very high-purity copper and recovers the valuable metal and metalloid impurities that are present in the copper from the anode furnace. The impurities dissolve in the electrolyte or fall to the bottom of the cells as slime. A schematic diagram of an electrolytic refinery is shown in Figure 9.3.

Copper oxide (Cu_2O) is the major impurity in the anode. It reacts with the sulfuric acid, and one-half precipitates and one-half dissolves according to:

$$Cu_2O + H_2SO_4 \longrightarrow CuSO_4 + Cu\downarrow + H_2O$$

The precipitated copper joins the slime and is later recovered. The dissolved copper causes gradual buildup of copper in the electrolyte.

Electrolytic refining is conducted in rectangular tanks usually constructed of concrete lined with lead or antimonial lead. The impure anodes are hung in the tanks. The cathodes are thin sheets of electrolytically refined copper. These "starting" sheets and the anodes are immersed in the electrolyte. A refining circuit may consist of a thousand or more tanks.

Table 9.2 *Analysis of Water Used to Cool Cast Copper Anodes and Refinery Shapes*

	Plant 110				*Plant 102*				
						Anode casting		*Wirebar cooling*	
Parameters	*Inlet water*	*Anode bosh tank overflow*	*Wirebar cooling*	*Semi-continuous cake casting*	*Inlet water*	*Recycle*	*Bleed*	*Recycle*	*Bleed*
pH	7.6	7.5	7-8	8.	7.1-7.6	7.4-8.4	7.4-8.4	8.0-8.4	8.0-8.4
TDS	1,430.	3,700.	1,250.	1,400.	0.2	0.2	0.3	0.1	0.1
TSS	0.0	43.0	12.5	0.0	0.5	0.6	0.3	0.4	0.1
SO_4	240	270.	240.	270.	–	–	–	–	–
As	<0.001	0.11	<0.001	<0.001	<0.001	<0.001	<0.001	<0.001	<0.001
Cd	<0.001	<0.001	<0.001	<0.001	0.0008	0.0001	0.0011	0.0021	0.0024
Cu	0.30	3.1	0.69	0.18	0.021	0.42	0.81	3.5	2.8
Fe	0.02	0.15	0.13	0.04	1.2	0.80	1.1	1.7	2.8
Pb	0.007	0.65	0.007	0.003	0.078	0.19	0.35	0.068	0.073
Hg	0.00350	0.00385	0.00425	<0.0001	0.00004	0.00004	0.00002	0.00004	0.00002
Se	<0.001	0.23	<0.001	<0.001	0.040	0.040	0.040	0.040	0.040
Zn	<0.001	0.01	0.067	<0.001	0.35	0.21	0.19	0.088	0.14
Oil & grease	0.0	2.0	0.0	0.0	0.14	0.02	0.02	0.1	0.1

Source: After Battelle 1974.

Note: In concentration units of milligrams per liter.

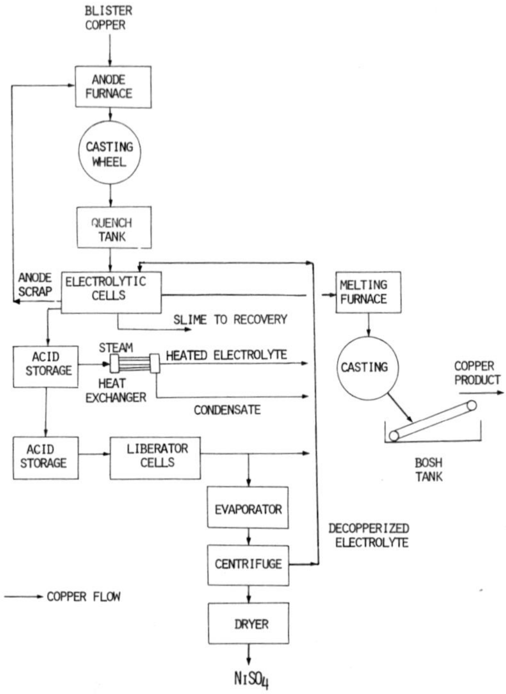

Figure 9.3. *Generalized flow diagram of electrolytic copper refinery (after Lanier 1965).*

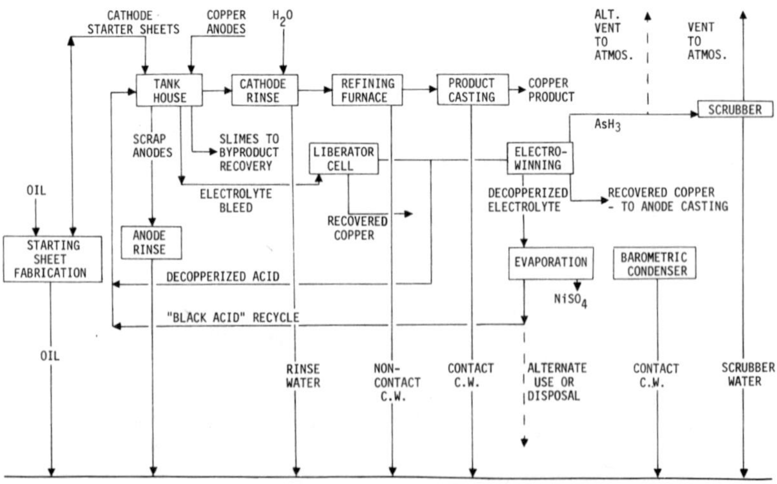

Figure 9.4. *Generalized schematic of waste effluents from the electrolytic refining of copper (after Battelle 1974).*

The current first traverses the anodes, where copper and impurities such as nickel, arsenic, and iron are electrolytically oxidized and dissolved. Gold, silver, selenium, and tellurium enter the electrolyte as finely divided solids and settle as slimes.

The current then traverses the electrolyte to the cathode where reduction and deposition of the copper occurs. Nickel, arsenic, bismuth, and antimony are not deposited with the copper if their concentrations are maintained below certain levels.

This process is repeated from tank to tank in series. After about two weeks, the cathodes are removed and washed and new starting sheets are inserted on each side of the anodes. The process is repeated until the anode is 80 to 90 percent consumed, then it is removed from the tank, washed, and sent back to the anode-refining plant to be remelted and recast into full-size anodes.

Washing or rinsing of the copper electrodes constitutes a source of waste water (see Figure 9.4). Historically, rinsing has tended to be nonconserving of water; but management has improved, and discharges have decreased. According to Battelle (1974), the cathode rinse circuit can now be closed with conservation measures and a moderate amount of concentration by evaporation. The scrap anodes may also be rinsed before return to the anode furnace, whereupon similar types of effluents result. The copper cathodes are melted in a furnace and cast into products which may be cooled with noncontact and/or contact cooling water. Contact cooling water must be of fairly high quality to avoid staining the product. It is generally recirculated to a cooling pond, often with some bleed. Because of the purity of the copper, pickup of contaminants other than copper is minimal. The slimes in the tanks are removed periodically and their precious content is recovered.

As noted above, soluble nickel, arsenic, antimony, bismuth, and copper concentrations build up in the electrolyte. Control of these concentrations within tolerable limits is achieved by withdrawal of a portion of the electrolyte and replacement with fresh solution. Copper is commonly recovered from the bleed solution by electrolytic stripping, by passing the solution through liberator cells similar to the refining cells, but with insoluble iron or lead anodes in place of copper anodes. Electrolysis drives the copper out of solution as copper metal onto a cathode. In the final stages volatile, poisonous AsH_3 is driven off at the cathode. The fumes should be collected and scrubbed before release. Effluent from such a scrubber constitutes a source of waste water. The decopperized solution is evaporated for further concentration and crystallization of nickel sulfate. The residual "black-acid" liquid may be returned to the refining circuit, used for leaching operations in associated plants, or used for fertilizer manufacture.

The slimes containing the insoluble metals and compounds that settle to the bottom of the electrolytic cells periodically are pumped to a slimes treatment plant. Passage across a trommel screen removes particulate copper, and subsequent sulfuric acid leaching effects more complete decopperization. After filtering and washing, the cake is fused in a reverberatory furnace. This furnace may constitute a source of air pollution and, if scrubbers are used to clean the gases, a waste water effluent is produced. The first slag from the furnace contains the lead and it is sent to a lead smelter. Soda ash and sodium nitrate are then added to flux the selenium and tellurium into a "soda" slag. The soda slag is dissolved in hot water and filtered; the residue is returned to the furnace. The filtrate is acidified to pH 5.5, which precipitates TeO_2. It is filtered out before additional refining. The filtrate, containing selenium, is acidified with H_2SO_4, and the selenium is reduced to metal with sulfur dioxide gas. The selenium is filtered, dried, and packaged at a minimum selenium content of 99.5 percent.

The tellurium dioxide in the TeO_2 filter cake is dissolved in a sodium hydroxide-sodium sulfide solution. Precious metals remain in the insoluble residue, which is filtered off and returned to the furnace. The filtrate is reacidified to a pH of 5.5, which reprecipitates tellurium dioxide. It may be dried, ground, and sold, or it may be reduced to metallic tellurium by pyrometallurgical methods. Alternatively, tellurium may be reduced to metal form in HCl solution with gaseous sulfur dioxide, or it may be dissolved in caustic and electrolyzed to produce commercial-grade tellurium. If HCl is used, a scrubber must be employed to collect volatilized HCl for air pollution control. This scrubber produces waste water. The liquor from which the tellurium dioxide precipitates is also a waste water.

After removal of the soda slag from the slimes-smelting furnace, air is bubbled through the remaining metal to oxidize copper, and the oxide is skimmed off. The remainder is cast into anodes and sent to an electrolytic refinery for silver removal. Gold and platinum collect as slimes and are melted and cast into anodes for electrolytic refining.

Hydrometallurgy

Hydrometallurgical methods* can recover copper from oxide ores or from ores too low in grade for treatment in a concentrator. In the process copper is dissolved by sulfuric acid from which it is recovered in elemental form, and the stripped leach solution is recycled for further leaching. Ore can be leached in situ, collected in piles and leached (heap leaching), or leached in large concrete tanks called vats. If vat leaching is employed, a solid waste transport problem is encountered because leaching leaves a residue in the vats. Other solid wastes are left in place. In heap leaching, piles of predominantly oxide ore are leached with circulating, dilute acidic solutions. The heap should be carefully prepared to insure that the leach solution is recovered and will not infiltrate into the ground and be lost or contaminate ground water.

The leach solution is distributed evenly over the top of the pile and allowed to trickle through the pile. The leach solution drains to reservoirs at the base of the pile, from where it may be pumped back to the top of the heap for more leaching or it may be sent to the solvent extraction plant.

In heap leaching, the copper oxides dissolve in the leach solution according to the reaction:

$$CuO + H_2SO_4 \longrightarrow CuSO_4 + H_2O$$

* See footnote on the first page of this chapter.

As indicated by the following reaction, the dissolution of sulfide minerals is slower and indirect, and requires oxidation by the entry of air into the heap during leaching:

$$CuS + 2O_2 \longrightarrow CuSO_4$$

Ferric sulfate in the leach solution also promotes the oxidation of copper sulfides, by a reaction similar to:

$$4Fe_2(SO_4)_3 + CuS + 4H_2O \longrightarrow 8FeSO_4 + CuSO_4 + 4H_2SO_4$$

Bacteria as leaching aids apparently produce both ferric sulfate and sulfuric acid.

The pregnant leach solutions, containing up to several grams per liter of copper as copper sulfate, are then treated by cementation or solvent extraction.

In situ leaching consists of allowing the leach solution to percolate through an in-place ore body to which the necessary permeability has been imparted by blasting. Two thousand short tons (1,800 metric tons) of explosive were detonated recently at the Old Reliable Mine of Ranchers Exploration Company in Arizona for this purpose. It is expected that about 14 million kilograms (30 million pounds) of copper can be recovered from this deposit by leaching.

Vat leaching consists of passing the leaching solution through vats of ore crushed to about 1 to 1¼ centimeters (3/8 to 1/2 in.). These vats hold up to 20,000 tons of ore. The floors of leach vats are filters which facilitate the upflow or downflow of leach solutions and wash solutions.

After the copper oxides are converted to soluble copper sulfate, the ore is washed free of soluble copper by a countercurrent process in which fresh wash water is added to the vat. The pregnant solution may comprise both the leach and wash solutions. The wash solutions generated in these cases are considerably diluted. They are treated by cementation or solvent extraction because they are too dilute for the recovery of copper by electrolysis. Care must be taken to assure that these processes do not produce waste waters during washing.

Cementation precipitates copper from solution as a finely divided metallic product by replacement with some more active metal (iron). Scrap tin cans are the cementing material commonly used. They are placed in channels called launders and the copper-rich solution flows over them. Copper is washed out of a launder through screens into settling basins by high-pressure water sprays. After recovery of copper precipitate (called cement copper), the stripped solution containing ferrous sulfate and some free acid is returned to leaching operations with or without the addition of more sulfuric acid. In some cases it may become an effluent. Some solid waste is also produced by cementation because of impure sources of iron. These are discarded in landfills, tailings ponds, or along with overburden from open pit operations. Most cement copper is sent to reverberatory matte furnaces.

Solvent extraction of pregnant leach solutions consists of stripping the dilute leach solutions by contact with an organic solvent which selectively extracts copper into the organic molecule. The organic solution is then exposed to an acid "stripping" solution which causes the copper to leave the organic phase and enter the aqueous solution. The result is treated by electrowinning to produce cathode copper. Electrolytes and solvents are recycled.

Air Pollution Control Equipment and Sulfuric Acid Plants

During roasting, smelting, converting and refining, dust, fume, and sulfur dioxide emissions are produced. Air pollution control regulations now require major reductions in emissions. Thus, smelters are being equipped with more particulate removal equipment and plants are being installed to convert sulfur dioxide to sulfuric acid. Aqueous effluents result from such air pollution control devices. Smelter gases containing sulfur dioxide can be used as feed material for a sulfuric acid plant; however, the gases must be cleaned before use in the contact process. Each acid plant must be designed for the particular smelter-gas feed that is available. Battelle (1974) characterized the SO_2 emissions from roasters as containing between 12 and 14 percent SO_2 by volume, while those from converters might be about 6 percent. The gas from a reverberatory furnace may have an SO_2 concentration of only 1 or 2 percent, which is below the 3 to 4 percent required for a contact acid plant. Thus, it is not feasible simply to add an acid plant to control all of the SO_2 emissions from a copper smelter. The SO_2 from the reverberatory furnace must be concentrated or the furnace replaced by electric furnaces or the flash-smelting process, both of which produce higher concentrations of SO_2 than does the reverberatory smelter. Concentrations of SO_2 from flash smelting are between 12 and 14 percent, which is suitable for acid manufacture. Additional concentration of the gas from the reverberatory furnaces or converters can be accomplished by the anhydrous dimethylaniline (DMA) process. After cleaning by electrostatic precipitators, the gases are scrubbed, cooled, and dried. Then, in an adsorption tower, the descending DMA removes the SO_2 from the ascending gas. Additional scrubbing of the gas occurs in the tower before the gas, having only about 0.05 percent SO_2, is exhausted to the atmosphere. Steam strips the SO_2 from the pregnant DMA solution in another adsorbing tower. Then the wet concentrated SO_2 is dried with 98 percent H_2SO_4 and compressed to liquid sulfur dioxide. Some DMA is lost to the water circuit as blowdown; it can be removed by treatment with activated carbon, Battelle reports.

Scrubbers are needed to prepare the gases for the contact plant after passage through flues and Cottrell precipitators. Blowdowns result even though the scrubbers operate on a recycle basis. They contain the pollutants associated with the off-gas streams. Table 9.1 presents analyses of three scrubber effluents.

Acid plant effluents and effluents from slime treatment operations are the two principal wastes from air pollution control in the primary copper industry. Cooling and condensation of volatile fumes necessitate use of water.

Dusts are produced in various operations. Cyclones, precipitators, scrubbers, and/or baghouses can be satisfactorily applied but improved baghouses may provide the optimum solution because they produce no liquid effluent. However, they cannot be used on some hygroscopic emissions.

Waste Characterization

Battelle (1974) presents analytical data on various waste waters from primary copper recovery operations. The data were obtained from the permit applications submitted under the U.S. Army Corps of Engineers waste discharge permit system. Table 9.1 presents these analyses for acid plant scrubber effluents and effluents from slag granulation. Tables 9.3 through 9.5 present

Table 9.3 *Waste Effluents from Plant No. 102*

Contributing Operations: Reverberatory furnace, converter, anode furnace

Parameters	Intake (mg/l)	Discharge (mg/l)	Net change (mg/l)	Net loading (kg/day)	Net loading (kg/kkg)	Net loading (lb/ton)
Alkalinity	80	10.2	neg	–	–	–
COD	5.2	27.7	22.5	313	1.00	2.00
Total solids	150	243	93	1,293	4.15	8.30
Dissolved solids	150	241	91	1,266	4.07	8.14
Suspended solids	–	2	2	27.8	0.09	0.18
Oil & grease	1.0	1.0	0	–	–	–
Sulfate (as S)	36	24	neg	–	–	–
Chloride	3.7	46	42	584	1.88	3.76
Aluminum	0.10	0.10	0	–	–	–
Arsenic	0.015	0.415	0.40	0.56	0.018	0.036
Cadmium	0.010	0.030	0.02	0.28	0.009	0.018
Calcium	4.8	66	1.8	25.0	0.080	0.160
Chromium	0.01	0.01	0	–	–	–
Copper	0.02	1.37	0	18.8	0.060	0.120
Iron	0.08	0.76	0.68	9.5	0.031	0.062
Lead	0.10	0.10	0	–	–	–
Mercury	0.001	0.001	0	–	–	–
Nickel	0.01	0.01	–	–	–	–
Selenium	0.02	0.02	0	–	–	–
Silver	0.01	0.01	0	–	–	–
Sodium	3.4	34	30.6	425	1.37	2.74
Zinc	0.22	1.23	1.01	14.0	0.045	0.090
pH	7.0	6.4	–	–	–	–
Flow, 10^6						
(l/day)	–	13.91	–			
(gal/day)	–	3.675	–			

Source: After Battelle 1974.

Note: Production is 311 metric tons/day (343 short tons/day).

these analyses for reverberatory furnace, converter, and anode furnace; electrolytic refinery, electrolytic purification, and recovery of heavy metals; and electrolytic refinery, electrolytic purification, and casting of copper.

On the basis of the data in these tables, other permit data not presented here, analyses of samples collected in the field, and plant visits, Battelle tentatively defines the following list of significant pollutants in primary copper recovery plant effluents:

- Total suspended solids (TSS)
- Copper
- Metals associated with copper

Arsenic	Iron	Mercury	Selenium	Zinc
Cadmium	Lead	Nickel	Tellurium	Oil and grease

Table 9.4 *Waste Effluents from Plant No. 102*

Contributing Operations: Electrolytic refinery, electrolyte purification, and recovery of precious metals

Parameters	Intake (mg/l)	Discharge (mg/l)	Net change (mg/l)	Net loading (kg/day)	Net loading (kg/kkg)	Net loading (lb/ton)
Alkalinity	80	–	–	–	–	–
COD	5.2	35.4	30.2	36.8	0.119	0.238
Total solids	150	229	79	96.4	0.310	0.620
Dissolved solids	150	194	44	53.7	0.173	0.346
Suspended solids	–	35	35	42.7	0.137	0.274
Oil & grease	1.0	1.0	0	–	–	–
Sulfate (as S)	36	56	20	24.5	0.080	0.160
Chloride	3.7	13	9.3	11.3	0.037	0.074
Aluminum	0.10	0.10	0	–	–	–
Arsenic	0.001	0.67	0.67	0.82	0.003	0.006
Cadmium	0.01	0.01	0	–	–	–
Chromium	0.01	0.01	0	–	–	–
Copper	0.02	0.006	neg	–	–	–
Iron	0.08	6.90	6.82	–	–	–
Lead	0.10	0.10	0	8.32	0.027	0.054
Mercury	0.001	0.001	0	–	–	–
Nickel	0.01	1.58	1.57	1.91	0.006	0.012
Selenium	0.02	0.14	0.12	0.14	0.0004	0.0008
Zinc	0.03	0.83	0.80	0.98	0.003	0.006
pH	7.0	4.6	–	–	–	–
Flow, 10^6						
(l/day)	–	1.22	–			
(gal/day)	–	0.32	–			

Source: After Battelle 1974.

Note: Production is 311 metric tons/day (343 short tons/day).

Table 9.5 *Waste Effluents from Plant No. 102*

Contributing Operations: Electrolytic refinery, electrolyte purification, casting of copper

Parameters	Intake (mg/l)	Discharge (mg/l)	Net change (mg/l)	Net loading (kg/day)	Net loading (kg/kkg)	Net loading (lb/ton)
Alkalinity	80	12.9	neg	–	–	–
COD	5.2	10.5	5.3	1.78	0.006	0.012
Total solids	150	218	68	22.9	0.074	0.148
Dissolved solids	150	213	63	21.2	0.068	0.136
Suspended solids	–	5	5	1.7	0.006	0.012
Oil & grease	1.0	1.0	0	–	–	–
Sulfate (as S)	36	46	10	3.37	0.011	0.022
Calcium	4.8	42.6	37.8	12.7	0.040	0.080
Chromium	0.01	0.01	0	–	–	–
Copper	0.02	1.02	1.0	0.34	0.001	0.002
Iron	0.08	0.87	0.79	0.27	0.0008	0.0016
Lead	0.10	0.30	0.20	0.07	0.0002	0.0004
Mercury	0.001	0.001	0	–	–	–
Nickel	0.01	0.39	0.38	0.13	0.0004	0.0008
Selenium	0.02	0.20	0.18	0.06	0.0002	0.0004
Silver	0.01	0.01	0	–	–	–
Zinc	0.03	0.20	0.17	0.06	0.0002	0.0004
pH	7.0	6.7	–	–	–	–
Flow, 10^6						
(l/day)	–	0.337	–			
(gal/day)	–	0.089	–			

Source: After Battelle 1974.

Note: Production is 311 metric tons/day (343 short tons/day).

From this list TSS, oil and grease, arsenic, cadmium, copper, lead, mercury, selenium, and zinc were selected as pollutant parameters for the primary copper industry; pH was added subsequently. Battelle concluded that arsenic appears to arise from acid plants and from by-product operations recovering selenium and tellurium from refinery slimes.

Proposed Best Practicable Control Technology Currently Available, Effluent Limitations Guidelines, and Costs*

Battelle (1974) based tentative judgment of the best practicable control technology currently available to the primary copper industry on the average of best existing performance by plants of various sizes, ages, and unit processes investigated during its study. Consideration was also given to:

1. Total cost of application relative to benefits to be achieved.

* In 1971 dollars.

2. Size and age of facilities.
3. Process employed and changes anticipated.
4. Nonwater-quality environmental impact (including energy requirements).

All guideline values were (and are) expressed as consecutive 30-day averages and as maximum daily values in units of kilograms of pollutant per metric ton (and pound of pollutant per short ton) of product produced in order that the limitations can be applied to plants with differing production rates. In order to determine compliance, the information needed from the plant would be:

1. Net concentration of pollutants, in terms of weight per unit volume, e.g., mg/l, resulting from the process: mg/l (out)–mg/l (in) = mg/l (net).
2. Flow rate in liters/min (gal/min) and duration of flow, e.g., continuous, determined at the exit of the treatment process.
3. Production rate as daily, weekly, or monthly averages in terms of mass, e.g., kkg (tons).

With this information, the amount of each pollutant discharged per day can be calculated as:

$$\text{Discharge} = \frac{\text{mg/l(net)} \times \text{l/hr} \times \text{hr/day}}{10^{-6}} = \text{kg/day}$$

$$\text{Unit discharge} = \frac{\text{kg/day}}{\text{Production, kkg/day}} = \text{kg/kkg}$$

Comparison of the latter figure with the effluent limitation for that pollutant shows whether the discharge is in compliance with the guidelines.

Smelting

The tentatively recommended effluent limitations were subdivided according to climate. The Category I subdivision includes those smelters located in areas where annual evaporation exceeds annual precipitation. Category II covers those smelters located in areas where the reverse is true.

Category I: Battelle (1974) tentatively recommended that the effluent limitation attainable through application of the best practicable control technology currently available is total impoundment with no discharge of process waste water pollutants.

Category II: For smelters in Category II, the tentatively recommended effluent limitations attainable through the application of the best practicable control technology currently available are presented in Table 9.6.

Battelle based these recommendations on information from 15 smelters, of which 12 are in net-rainfall-deficiency areas (Category I), and 3 are in net-excess-rainfall areas (Category II). Most of the 12 reported no discharge of either pollutants or water and thus provided no basis for evaluation of treatment technologies. Of the smelters currently discharging waste water, most are operated on a water circuit with a concentrator or refinery or both. Thus, the smelter contributions were not determinable from the data available.

Table 9.6 *Proposed Effluent Guidelines for BPCTCA, Copper Smelting*

	Effluent limitations[1,2]			
	30-day average		*Daily maximum*	
Parameters	*(kg/kkg Cu)*	*(lb/ton Cu)*	*(kg/kkg Cu)*	*(lb/ton Cu)*
TSS (25 mg/l)	0.50	1.0	1.0	2.0
As	0.20	0.40	0.4	0.8
Zn	0.50	1.0	1.0	2.0
Se	0.10	0.20	0.2	0.4
Total other metals[3]	0.20	0.40	0.4	0.8
Oil & grease	0.10	0.20	0.2	0.4
pH	>7.0	<10.0	>7.0	<10.5

Source: After Battelle 1974.

Note: These are tentative recommendations by Battelle and are subject to change based on comments received and on further internal review by EPA. At this writing they are the only such data available.

1. Effluent limitations for metals include both dissolved and suspended fractions.
2. Flow is 20,000 l/kkg (4,800 gal/ton).
3. Total other metals (alkalies & alk. earths excepted), e.g., Fe, Cu, Ni, Al, Cr.

One concentrator-smelter operation reported having an exemplary liming and settling treatment, but it was concluded to be impossible to transfer the effluent concentrations to more typical operations because the inlet concentrations before treatment were below guidelines. This operation was observed to demonstrate that natural atmospheric carbonation after liming can produce an effluent with pH below 10 when residence times are sufficient.

It was concluded that: (1) the practical approaches to water pollution abatement from smelter process waste water discharges include recycle, reuse, segregation and, finally, neutralization and settling (liming and settling); (2) constituents in effluents from a smelter (when adequately neutralized with lime and when precipitated solids are effectively removed by settling or filtration) can be reduced in concentration to the levels of the constituents shown in Table 9.7.

In order to achieve these concentrations, Battelle recommended pH levels in the range of 10 to 11. Natural recarbonation from the atmosphere is expected to gradually lower the pH as settling proceeds in the pond and, in a pond designed to minimize suspended solids, the pH is expected to be below 10 by the time the effluent is discharged from the settling pond.

Because arsenic, selenium, tellurium, and zinc are not effectively removed at pH's above 7, absorption and coprecipitation were relied on for their removal. Battelle concluded that there is not a sufficient degree of confidence in the engineering and economic practicability of the alternative control and treatment technologies to warrant their recommendation at this time. These alternative technologies consist of (1) lime and sedimentation followed by or

Table 9.7 *Constituents and Concentrations Considered Achievable by Best Practicable Control Technology*

Constituents	*Concentrations (mg/l)*	*Constituents*	*Concentrations (mg/l)*
As	10	Ni	0.25
Cd[1]	0.25	Se	5
Cu	0.25	Zn[2]	25
Fe	0.25	Oil & grease	5
Pb	0.5	TSS	25

Source: After Battelle 1974.

Note: See general note on Table 9.6.

1. Cadmium and mercury were proposed as toxic elements and may be subject to stronger restrictions than BPCTCA concentrations (see *Federal Register*, Vol. 38, No. 247, December 27, 1973, pp. 35388-35395).
2. In the opinion of the author, the zinc concentration will be much less than 25 mg/l if the pH levels suggested subsequently are achieved. The level recommended for zinc plant wastes is 5 mg/l (Table 9.39). Even 5 mg/l is below the capability of the technology as shown in Appendices E and F.

combined with sulfide precipitation; (2) evaporation accomplished by spraying the mother liquor (discussed previously) into the flue gases of the ore furnace where they might become an air pollution problem; and (3) precipitation with sulfide in alkaline solutions.

Battelle concluded that the consumptive use of water by smelter operations employing some degree of water conservation ranges from 23,000 to 35,000 l/kkg of copper produced (6,100 to 9,200 gal/short ton), exclusive of noncontact cooling water. It was concluded that 20,000 l/kkg (5,300 gal/ton) may be a reasonable average for all plants.

Noncontact cooling water is used to cool furnace doors, tapholes, converter hoods, and other equipment similar to that described in Chapter 7. It is contaminated only by heat and not included in the above figure.

Battelle proposed the water-flow circuit shown in Figure 9.5 for controlling the major pollution sources. The slag-granulation circuit is closed but the waste water from slag granulation could be eliminated by allowing molten flow of slag to a solid waste disposal site. The higher quality contact and noncontact cooling waters are recycled through a process-water pond for settling and cooling. The pond also supplies makeup water for the gas scrubber which might eliminate the need for a bleed stream from this circuit. Otherwise bleed water could be treated by liming and settling.

The effluent from gas scrubbers cleaning the flue gases upgradient from the acid plant is recycled. Bleed from the scrubber is limed and settled in a pond sufficiently large to achieve a 25 mg/l total suspended solids level before discharge. Other acid effluents from the acid plant join the scrubber bleed for treatment, as does any bleed from the process-water pond. All process water enters this pond prior to discharge. The sediments in the ponds in Figure 9.5 reflect the final solid waste product produced by the conversion of gaseous wastes to liquid wastes, which in turn are converted to solid wastes.

Costs for handling the waste streams generated primarily by air pollution control devices were obtained by Battelle in the following manner. The waste streams were assumed to be combined. The combined waste water was characterized by a low pH (<2) and assumed to contain suspended solids from the dust picked up in the gas stream as well as dissolved solids obtained from the dust. The following problems were noted with respect to cost estimates:

1. The solids content of the water can vary from 200 ppm to 3 percent.
2. The amount of scrubbing necessary is a function of the efficiency of the gas cleaning device (usually electrostatic precipitator).
3. The quantity of acid produced presently has only an indirect relationship to copper production. It is a function of the amount of off-gas treated and the SO_2 concentration in this off-gas.

Battelle handled these variables by the following assumptions:

1. Only the SO_2 in the converter gases is involved in sulfuric acid manufacture.
2. The quantity of SO_2 generated in the converting operation is about 2 tons per ton of copper produced.
3. Ninety percent of the SO_2 is converted to acid.

Using the above assumptions, it was calculated that a smelter producing 90,700 kkg (100,000 tons) of copper per year (259 kkg [285 tons] per day) produces about 466 kkg (513 tons) of sulfuric acid per day. On the basis of data supplied by smelters on volumes of waste water produced by acid plants, an average value of 1.25 l/min/kkg (0.3 gpm/ton) was calculated. For Battelle's representative case, this quantity represents an effluent volume of approximately 3.240 l/kkg of copper (780 gal/ton).

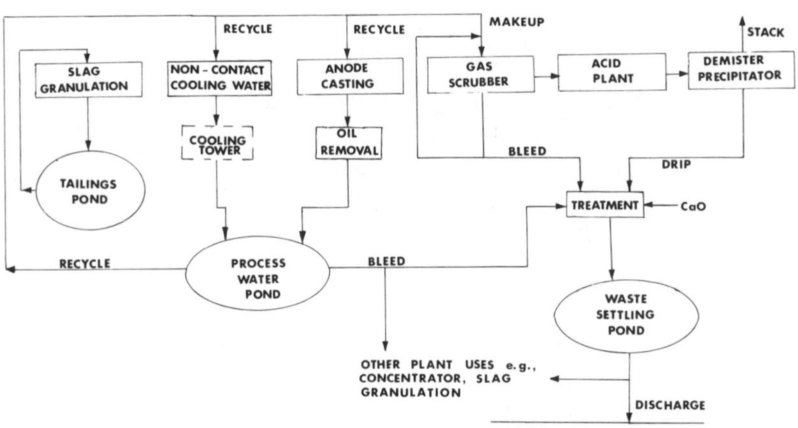

Figure 9.5. *Integrated water pollution control circuit for process wastes from a copper smelter (after Battelle 1974).*

Table 9.8 *Cost Effectiveness of Treatment for Waste Water from Example Acid Plant in a Copper Smelter*

	Annual copper costs				Effluent levels (mg/l)								
	Capital		Operating										
Treatment	*($/kkg)*	*($/ton)*	*($/kkg)*	*($/ton)*	*TSS*	*Cu*	*As*	*Cd*	*Pb*	*Hg*	*Se*	*Te*	*Zn*
None	0		0		200-15,000	7,500	10-20	0.1-1.0	1-10	0.001-0.03	0.001-1.0	0.001-0.02	1-20
Primary clarifier	1.28	1.16	0.29	0.26	15	1-7,500	10-20	0.1-1.0	1-10	0.001-0.03	0.001-1.0	0.001-0.02	1-20
Lime and settle[1]	2.65	2.40	0.83	0.75	15	0.1	10-20	0.1	<1	0.001-0.03	0.001-1.0	0.001-0.02	1-20
Reverse osmosis	5.40	4.90	2.06	1.87	0	<0.1	<0.1	<0.1	<0.1	<0.1	<0.1	<0.1	<0.1
Evaporation (multiple effect VTE)	6.51	5.90	1.95	1.77	0	0	0	0	0	0	0	0	0
Evaporation pond	3.31	3.00	0.62	0.56	0	0	0	0	0	0	0	0	0

Source: After Battelle 1974.

1. Costs obtained from a plant currently implementing this technology.

The costs associated with several methods of treating this volume of waste water effluent levels are shown in Table 9.8 for a plant producing 90,700 kkg (100,000 tons) of copper annually. Since liming and settling were selected as the tentatively recommended best practicable control technology for smelter wastes, it was concluded that capital costs would consist of \$2.66 per annual metric ton (\$2.40 per annual short ton) and that operating costs would consist of \$0.83 per metric ton (\$0.75 per short ton).

Refining

Battelle (1974) concluded that the best practicable control technology for Category I is total impoundment with no discharge of pollutants. Best practicable control technology for Category II would produce an effluent subject to the guidelines recommended in Table 9.9.

The study on which these recommendations were based revealed that there are no exemplary primary electrolytic copper refineries performing all the unit operations associated with refining, including the processing of by-products, because the discharge of untreated effluents has been practiced.

However, on the basis of the similarity in effluent compositions to those of smelters, it was concluded that the same approaches to water pollution abatement are applicable, and that liming and settling constitute the best practicable control technology presently available. The reservations concerning removal of selenium and tellurium in smelter wastes also apply to refinery wastes.

Battelle's study also revealed that, discounting noncontact cooling water, the minimum discharge quantities reported by refineries were in the 2,000 to

Table 9.9 *Proposed Effluent Guidelines for BPCTCA, Copper Refining*

	Effluent limitations[1]			
	30-day average		*Daily maximum*	
Parameters	*(kg/kkg Cu)*	*(lb/ton Cu)*	*(kg/kkg Cu)*	*(lb/ton Cu)*
TSS (25 mg/l)	0.05	0.10	0.10	0.20
As	0.02	0.04	0.04	0.08
Zn	0.05	0.10	0.10	0.20
Se	0.01	0.02	0.02	0.04
Total other metals[2]	0.02	0.04	0.04	0.08
Oil & grease	0.01	0.02	0.02	0.04
pH	>7	<10	>7	<10.5

Source: After Battelle 1975.

Notes: See general note on Table 9.6.

Flow is 2,000 l/kkg (480 gal/ton).

1. Effluent limitations for metals include both dissolved and suspended fractions.
2. Total other metals (alkalies & alk. earths excepted), e.g., Fe, Cu, Pb, Ni, Al, Cr.

4,000 l/kkg (480 to 960 gal/ton) range. But analysis of the operations revealed that a volume less than the average of the best represents a realistic consumptive use of water and, more specifically, that flow of 2,000 l/kkg (480 gal/ton) should be attainable by refineries through the application of reasonable water conservation measures.

In order to minimize water use, Battelle recommended measures illustrated in Figure 9.6. These measures include a normally closed circuit around the tank house and electrolyte purification; any necessary bleed from this circuit would be routed to the liming and settling treatment facilities. The higher quality water in the process water pond would serve the product casting line or enter a closed circuit with a cooling tower, with the bleed diverted to the process water pond as necessary.

However, under current conditions, the major streams are the bleed from copper electrowinning, from wet scrubbers on various furnaces, and the waste stream from the selenium and tellurium recovery operations if present.

Data Battelle obtained from refineries led to the conclusion that a reasonable value for the volume of waste water arising from the various treatment and recovery operations is approximately 15,000 l/day (4,000 gal/day) for the 90,700 kkg/yr (100,000 ton/yr) case, with most of it emanating from the Dore furnace and the selenium-tellurium recovery operations. Assuming that this stream can be treated by liming and settling (which is conjectural), or that one of several proprietary solidification schemes will work, estimated treatment costs were figured (Table 9.10). The costs for lime treatment were calculated as for lime treatment of smelter waste water. Lime treatment of refinery waste water was ultimately recommended at a capital cost of $2.19 per annual metric ton ($1.92 per annual short ton) and an operating cost of $1.01 per metric ton ($0.92 per short ton).

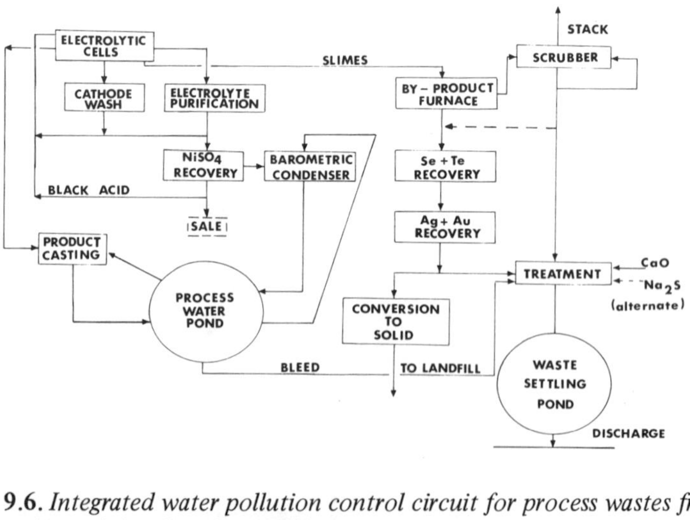

Figure 9.6. *Integrated water pollution control circuit for process wastes from a copper refinery (after Battelle 1974).*

Table 9.10 *Costs of Treating Process Waste Water from Electrolytic Copper Refinery*

	Costs			
	Capital		Operating	
Treatment	*($/annual kkg)*	*($/ton)*	*($/kkg)*	*($/ton)*
Lime and settle	$2.19	$1.92	$1.01	$0.92
Evaporation (multiple effect VTE)	4.63	4.20	1.18	1.07
Chem fix	0.73	0.66	0.83[a]	0.75[a]
Evaporation ponds	0.50	0.45	0.88	0.80

Source: After Battelle 1974.

Note: Example case: 90,700 kkg/yr (100,000 ton/yr) refinery recovering selenium and tellurium.

a. Based on treatment cost of approximately $0.011/l ($0.04/gal) treated.

Hydrometallurgy

Battelle (1974) concluded tentatively that the best practicable control technology for Category I is total impoundment with no discharge of pollutants. No copper leaching operations* were reported to be located in Category II areas; therefore, no other effluent limitations guidelines were recommended.

Cooling Water

Noncontact Cooling: Noncontact cooling water is used to cool reverberatory furnace doors and converter hoods. The Battelle study revealed that an average of about 5,300 cu m/day (1.4 mgd) is used on the basis of 90,700 kkg (100,000 tons) of copper produced annually. Few pollutants other than heat are contained in this waste stream if care is taken to preclude its exposure to other sources of waste water.

The two techniques available for control of noncontact cooling water were concluded to be recirculation through a cooling tower or spray pond, and reuse in other portions of the operations such as slag granulation, anode-casting cooling, or concentrator. The costs of these two alternatives were calculated on the basis of the following assumptions:

1. A total length of piping to a cooling tower of 610 m (2000 ft); pumps were rated at 53,000 cu m/day (1.4 mgd) at 7.5 m (23 ft) head.
2. For reuse, the water was assumed to be pumped a distance of 610 m (2,000 ft), with pumps rated at 15 m (46 ft) head.

The calculated costs of the two cases are presented in Table 9.11.

* See footnote on the first page of this chapter.

Table 9.11 *Costs of Decreasing the Temperature of Noncontact Cooling Water*

	Capital		*Operating*	
	($/annual kkg)	*($/annual ton)*	*($/kkg)*	*($/ton)*
Recycle	$1.07	$0.97	$0.23	$0.21
Reuse	0.67	0.61	0.15	0.14

Source: After Battelle 1974.

Contact Cooling for Anode and Product Casting: Battelle (1974) concluded that, for a copper production rate of 90,700 kkg/yr (100,000 tons/yr), water usage for cooling anodes after casting averages about 2,840 cu m/day (0.75 mgd). The major method of control is either recycle or reuse. However, evidence revealed that a slight pickup of dissolved and suspended solids occurs, although the amount is relatively small. It was concluded that these solids should have no effect on the reuse requirements for either slag granulation or return to the concentrator circuit. However, for operations where dissolved-solids content is critical, reuse of the contact cooling water may not be advisable. In addition, the employment of a cooling tower or spray tower would necessitate a blowdown to keep the dissolved and suspended solids below a tolerable level. One copper smelter reported to Battelle a blowdown of about 10 percent of the water use rate. Therefore, Battelle made cost calculations in a fashion similar to that for noncontact cooling, with the exception that a blowdown of 10 percent from the recycle operation was assumed to require treatment in a lime and settling scheme. Capital and operating costs amounted to $2.04/annual kkg ($1.85/annual ton) and $0.47/kkg ($0.43/ton) of copper, respectively.

Analogous cooling in copper refineries involves cooling the product castings, billets, cakes, ingots, and wirebars. The above figures were recommended as good approximations of these costs. No effluent limitations for cooling water were recommended by Battelle.

Sludge Production and Disposal

Sludge-disposal costs depend primarily on the physical location of the plant. Most primary copper smelters were observed to have adequate room for their sludge in vacant land, slag-granulation ponds, or concentrator tailings ponds. Battelle (1974) calculated that a pond 8 hectares (20 acres) in area would fill up at a maximum rate of about 0.3 m (1 ft)/yr (suspended solids of 3 percent in the waste water). Removal of the sludge to a sanitary landfill site was estimated to cost $3.86/kkg ($3.50/ton) of copper, at $11/kkg ($10/ton) cost for sludge collection and disposal. However, most operations probably would not be forced to go this route because of available solid waste disposal sites from which removal would be unnecessary.

Proposed Best Available Technology Economically Achievable, Effluent Limitations Guidelines, and Costs*

Battelle (1974) based its tentative judgment of best available economically achievable control technology on the best control and treatment technology employed by a point source in the industrial category or on a technology that is readily transferable from one industrial process to another.

The best economically achievable technology currently available assesses the availability of in-process controls as well as control or additional treatment techniques employed at the end of a production process.

Smelting

Battelle (1974) recommended that, for smelters located in Category I areas, the best available technology economically achievable is total impoundment with no discharge of process waste water pollutants. For smelters located in Category II areas, the recommended effluent limitations attainable through the application of the best available technology economically achievable are presented in Table 9.12.

Battelle (1974) points out explicitly that, "the identification of the best available technology economically achievable for smelting operations in Category II areas is complicated by the fact that the advanced technologies potentially most completely satisfying this requirement have not yet been demonstrated in practice. What are presumed to be the very best control and

Table 9.12 *Proposed Effluent Guidelines for BATEA, Copper Smelters*

	Effluent limitations[1]			
	30-day average		*Daily maximum*	
Parameters	*(kg/kkg Cu)*	*(lb/ton Cu)*	*(kg/kkg Cu)*	*(lb/ton Cu)*
TSS (25 mg/l)	0.0625	0.125	0.125	0.250
As	0.025	0.050	0.050	0.100
Zn	0.0625	0.125	0.125	0.250
Se	0.0125	0.025	0.025	0.050
Total other metals[2]	0.025	0.050	0.050	0.100
Oil & grease	0.125	0.025	0.025	0.050
pH	7	10	7.0	10.5

Source: After Battelle 1974.

Notes: See general note on Table 9.6.

Flow is 2,500 l/kkg (600 gal/ton).

1. Effluent limitations for metals include both dissolved and suspended fractions.
2. Total other metals (alkalies & alk. earths excepted), e.g., Fe, Cu, Pb, Ni, Al, Cr.

* In 1971 dollars.

treatment technologies are not employed by any point source in this industrial category, nor is evidence available to suggest that they can be readily transferred from one industrial process to another."

As noted previously, neutralization cannot be relied upon to precipitate arsenic and other pollutants common to copper processing. But Battelle notes that precipitation with sulfide, especially in neutral or alkaline solutions, would be expected to reduce dissolved arsenic concentration to 1 mg/l or less and that it would be expected to reduce soluble concentrations of lead, mercury, copper, and cadmium, to tenths of milligrams per liter. However, the behavior of selenium and tellurium through this processing scheme was noted to be uncertain and the sulfide process is undemonstrated for smelter wastes. Thus, the sulfide process was not recommended as the best available technology economically achievable.

On this basis, Battelle concluded that the treatment technology of liming and settling, coupled with the control technology of minimizing effluent water flows, is the best available technology economically achievable.

In terms of conservation of water, which is considered essential, Battelle noted that anode casting and gas cleaning scrubbers preceding acid plants are the two main water users in a copper smelter. It was calculated that 1.6×10^8 calories (0.63×10^6 BTU) are released in freezing and cooling a metric ton of copper to ambient conditions, which is enough to evaporate approximately 300 1 (79 gal) of water. A bleed of no more than 300 1/kkg (79 gal/ton) from the anode casting circuit was estimated to be needed to maintain dissolved-solids concentrations at satisfactory levels; it was recommended that this bleed be diverted to the acid plant scrubbers for ultimate rejection.

Battelle calculated the effluent loadings in Table 9.12 by assuming that the average S:Cu ratio in concentrates does not exceed 1.5:1, and that it is possible to recover up to 4.5 tons H_2SO_4/ton of copper. At the 90 percent recovery-level target of air pollution control this was observed to represent approximately four tons of acid per ton of copper. Battelle reported that the average of the best effluent discharges from gas cleaning scrubbers in the industry is about 600 l/kkg (145 gal/ton) of acid, equivalent to about 2,400 l/kkg (580 gal/ton) of copper. This figure was rounded to 2,500 l/kkg (600 gal/ton) for the Table 9.12 value, which is an 87.5 percent reduction from the effluent volumes for best practicable technology currently available.

Refining

Battelle (1974) tentatively recommended that, for copper refineries located in Category I areas, the best available technology economically achievable is total impoundment with no discharge of pollutants. For refineries in Category II, the effluent limitations shown in Table 9.13 were recommended.

Battelle concluded that the appropriate treatment and control technology for this category is neutralization with lime and minimization of effluent production rate. It was noted that effluents from by-product recovery, the principal pollutant source, can be held to about 60 l/kkg (14 gal/ton) of copper. The bleed from the casting cooling-water circuit can be minimized

Table 9.13 *Proposed Effluent Guidelines for BATEA, Copper Refineries*

	Effluent limitations[1]			
	30-day average		*Daily maximum*	
Parameters	*(kg/kkg Cu)*	*(lb/ton Cu)*	*(kg/kkg Cu)*	*(lb/ton Cu)*
TSS (25 mg)	0.005	0.010	0.010	0.020
As	0.002	0.004	0.004	0.008
Zn	0.005	0.010	0.010	0.020
Se	0.001	0.002	0.004	0.008
Total other metals[2]	0.002	0.004	0.008	0.006
Oil & grease	0.001	0.002	0.004	0.008
pH	>7	<10	>7	<10.5

Source: After Battelle 1974.

Notes: See general note on Table 9.6.

Flow is 200 l/kkg (48 gal/ton).

1. Effluent limitations for metals include both dissolved and suspended fractions.
2. Total other metals (alkalies & alk. earths excepted), e.g., Fe, Cu, Pb, Ni, Al, Cr.

according to the calculation that the heat evolved in cooling a metric ton of molten copper is enough to evaporate approximately 300 l (79 gal) of water. It was assumed that the casting cooling-water circuit can accept a threefold concentration, so that a bleed of about 100 l/kkg (24 gal/ton) is indicated. Other sources, including bleeds from scrubber circuits, electrolyte purification, and $NiSO_4$ recovery, plus other uncategorized sources, were estimated to contribute another 40 l/kkg (10 gal/ton) for a total of 200 l/kkg (48 gal/ton) which is a 90 percent reduction from best practicable technology currently available. If effluent limitations are not achieved by the application of liming and settling technology, additional control and treatment technologies will have to be investigated.

Hydrometallurgy

Battelle (1974) tentatively recommended total impoundment with no discharge as the best available technology economically achievable for hydrometallurgical operations. No recommendations were made for leaching operations* in Category II because no plants are located in Category II areas.

New Source Performance Standards

The term *new source* is defined by PL 92-500 as "any source the construction of which is commenced after the publication of proposed regulations prescribing a standard of performance."

* See footnote on first page of this chapter.

Battelle (1974) tentatively recommended new source performance standards be the same as the guidelines for the best available technology economically achievable.

This technology for Category I areas consists of total impoundment with no discharge of pollutants. For Category II areas, it is liming and settling.

A number of entirely new copper process technologies, including the Arbiter, Cymet, and CLEAR processes, were found to be in various stages of development and it was admitted that some of these may eliminate some effluents and/or produce new effluents. However, Battelle concluded that none of these processes appears to be far enough along for evaluation. The author is not in agreement with respect to the Arbiter process. A plant has been constructed at Anaconda, Montana.

PRIMARY LEAD

The Industry

The mineral galena (PbS) is the principal source of lead.* Galena frequently occurs in association with the mineral sphalerite (ZnS), a principal source of zinc; therefore, lead and zinc frequently are produced together. Minerals containing gold, silver, cadmium, mercury, copper, and antimony commonly occur with galena and sphalerite and are produced concurrently.

Most of the lead in the United States is produced by 48 mines and 31 concentrators (mills and flotation facilities). There are 6 smelters and 5 refineries in the United States.

In 1971, approximately 74 percent of the lead in the United States was mined in Missouri; the 10 leading mines produced 85 percent, and the 25 leading mines produced 98 percent. Locations and pertinent data for the six lead smelters and five lead refineries are presented in Table 9.14. Ninety-seven percent of the lead is produced by the following states: southeastern Missouri, 74 percent; Idaho, 12 percent; Utah, 7 percent; and Colorado, 4 percent.

Processes and Sources of Wastes and Waste Waters

The production of lead consists of mining, concentrating, smelting and refining. Underground mining is heavily predominant. Methods employed include block-caving, cut and fill, and room and pillar. The major waste produced by underground mining is mine water which frequently is acid and contains dissolved iron and heavy metals. Disposal and preventive measures are discussed in Chapter 4.

Concentration consists of crushing, grinding, and separation of galena and other ore minerals from the gangue and from each other. Froth flotation is the predominant process. After crushing and grinding, the mineral slurry is exposed to chemical reagents which render the minerals wettable or nonwettable. The sulfide minerals are then made aerophilic, one by one, by reagent addition.

* See section entitled "Sources of Information" at beginning of chapter.

They then attach themselves to rising bubbles in the slurry and rise to the top where they are skimmed off and dried to become concentrates. The remainder of the slurry (tails or tailings) is discarded in a tailings pond. Flotation reagents may add dissolved constituents to effluents from these ponds. Tailings ponds also frequently receive wastes from primary lead production processes; therefore they are important to the subject of this chapter. Other aspects of tailings disposal are discussed in Chapter 11.

Smelting

Battelle (1974) prepared the flow sheet in Figure 9.7 to illustrate the processes involved in lead smelting and refining, along with sources of some wastes. The processes include charge preparation (blending of the concentrate with flux and other products, sintering, blast furnace smelting, and refining to remove and, wherever possible, to recover metallic impurities. Figure 9.8 portrays the source of waste water. Sources of waste water also identify points of production of airborne wastes because wet scrubbers are frequently used to control them.

Charge Preparation: Charge preparation commonly blends lead concentrates with fluxes and recycle products such as collected dust, fume, and slag. This blend is pelletized and becomes the product which subsequently is sintered. Dust is the only significant waste produced by charge preparation. Dust may also be produced by conveyor transfer points prior to charge preparation. If produced in sufficient quantities, it can be collected by cyclones or baghouses.

Sintering: Sintering of the charge is accomplished on a movable grate furnace. During sintering, pressurized air is supplied from below the movable grate of the sintering machine. A layer of pelletized concentrates on the bed of the sintering machine is ignited by burners above the bed. The balance of the 16-inch-deep charge is spread on the burning layer, and the grate is moved into the windbox where the updraft makes the bed burn from the bottom up. The practice relies on the sintering reaction to remove nearly all sulfur from the concentrates. The sulfur emanates as SO_2 in the first portion of the sintering operation. The SO_2 gas is usually divided into two portions. The gases from the initial sintering stage, containing approximately 4 percent SO_2, are passed through a gas cleaning system consisting of baghouses and wet scrubbers, then to a contact sulfuric acid plant if an acid plant is employed. The weaker SO_2 gas from the latter portion of the sintering operation is usually cooled, passed through a baghouse to remove dust, then vented to the atmosphere. If wet gas scrubbers are used on the weak gas, they operate with closed-circuit scrubbing-water systems, in which the dust- and fume-laden scrubber water is clarified by settling and reused in the scrubbers. The weak gas may be recycled back through the sinter machine and the bleed stream treated. These measures are to minimize air pollution by the sinter gases.

The objectives of sintering are: (1) to remove sulfur as SO_2 and SO_3; (2) to eliminate, by volatilization, undesirable impurities such as arsenic, antimony and cadmium; and (3) to produce a material of suitable size and strength for melting in a blast furnace.

Table 9.14 *Lead Smelters and Refineries*

Company	*Location*	*Annual tons of lead-containing material treated 1972*	*First year of operation*	*Raw materials used*	*Slag treatment*	*Acid plant*	*Products*
American Smelting and Refining Company	Glover, Mo.	159,600	1968	Lead conc. Waelz residue Lead residues	None	None	Refined lead Copper dross Retort bullion
	East Helena, Mont.	192,000	1888	Lead conc. Siliceous ores Zinc residue	Slag fuming furnace	None	Lead bullion Soda ash matte Soda ash speiss Lead baghouse dust Zinc fume
	El Paso, Texas	NR[1]	1887		Slag fuming	None	Lead bullion Zinc fume
	Omaha, Neb.	136,000	1870	Lead bullion Secondary lead	None	None	Refined lead Antimonal lead Solder Misc. lead alloys Bismuth Copper matte Sodium telluride slag

Bunker Hill Co.	Kellogg, Idaho	550,000	1918	Lead conc.	Slag fuming	300 TPD	Refined lead Gold, silver Antimony
St. Joe Minerals	Herculaneum, Mo.	336,000	1892	Lead conc.	None	300 TPD	Refined lead Silver bullion Copper matte
Missouri Lead Operating Co.	Boss, Mo.	192,500	1968	Lead conc.	None	225 TPD	Refined lead Copper matte Dross Silver bullion

Source: After Battelle 1974.

1. NR = not reported.

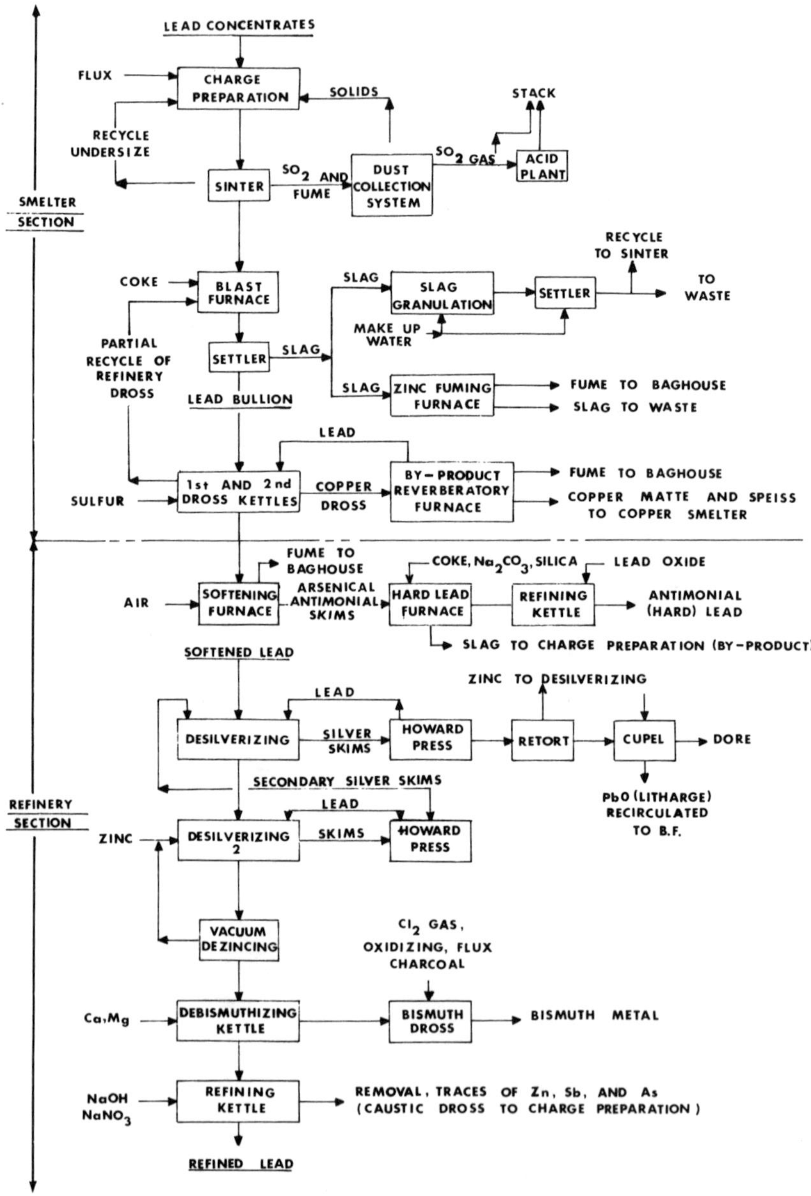

Figure 9.7. *Generalized flow sheet of processes involved in lead smelting and refining (after Battelle 1974).*

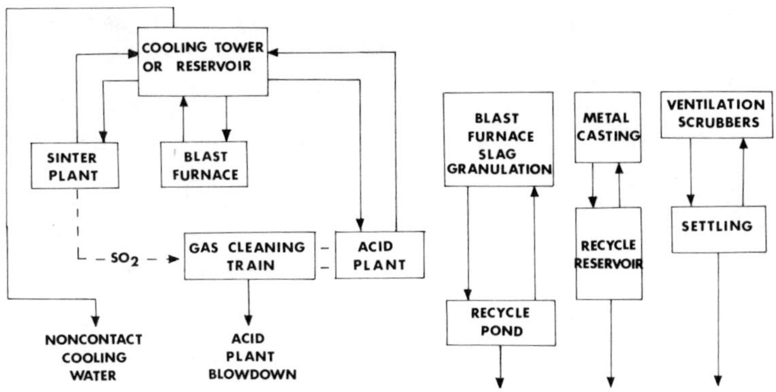

Figure 9.8. *Generalized diagram of water uses and waste water sources in primary lead plants (after Battelle 1974).*

After sintering is completed, the material is broken up by rollers and sized for the blast furnace. Sizing produces dust which is captured by wet scrubbers or baghouses and recycled back to the sinter operation. Effluent from these scrubbers usually enters a larger waste stream.

Blast Furnace: The blast furnace in a lead smelter is similar to that in a steel plant in that it utilizes a combination of heat and gases to separate the charge produced by the sinter machine into molten metal and slag. The easily reduced metals including lead, copper, silver, gold, bismuth, antimony, and arsenic enter the metal phase, whereas other metals and the nonmetals become part of the slag phase.

The lead blast furnace is water cooled and it is charged on both sides at the top. Associated facilities include materials handling equipment, an exhaust gas handling system, a conveyor system for the charge, and a refractory-crucible structure at the bottom, designed for continuous or intermittent tapping of molten metal and slag.

In addition to sinter, the charge includes coke and fluxing materials such as silica and limestone, recycled slag, cadmium plant residues, and refinery dross.

The many chemical reactions that occur in the blast furnace have in common the reduction of metal oxides or sulfides to the elementary metal by reaction with oxidizing agents such as carbon monoxide or iron. Wastes produced include gaseous SO_2, CO_2, H_2O and dust.

The molten metal from the blast furnace generally contains more than 95 percent lead with impurities consisting of:

- Copper: up to 2.5 percent
- Zinc: negligible
- Precious metals: variable
- Antimony: up to 2 percent
- Arsenic: up to 1 percent
- Bismuth: up to 0.03 percent

The slag consists of iron, calcium, and magnesium silicates, small quantities of arsenic and antimony, and variable amounts of lead (1.5 to about 4 percent). The slag is treated in a fuming furnace to recover zinc if enough is present. The charge is heated and air is blown in when the slag is molten. The zinc is boiled off and oxidized to zinc-oxide dust particles which are collected in dust collecting equipment such as cyclones, precipitators, dust chambers, and baghouses. Slag after fuming, or that which is discarded without fuming, is usually granulated by impacting it with a high-pressure water jet. The granulated slag may be either recycled to the winter process, or discarded onto a slag pile, frequently in the form of a slurry. The slag granulating water circuit should be designed so as to permit water recirculation. Water losses by evaporation during granulating should be made up with other waste waters.

In some cases, in addition to molten lead and slag an operation may utilize a matte phase consisting of copper and iron sulfides with precious metals. The matte phase forms as a discrete liquid layer between lead and slag. It is usually separated and sent to outside processors for metal recovery.

Battelle (1974) observed during its study of United States lead smelters that in some plants water circuits for cooling of hot gases from blast furnaces and sintering operations or for air pollution control in wet scrubbers operate in closed-loop fashion without a discharge. The characteristics of combined waste effluents from the slag granulation process, blast furnace cooling water, and miscellaneous blowdowns are presented in Table 9.15. Table 9.16 shows the Battelle data for the characteristics of water from a slag settling pond.

Refining

Drossing is the first step in the refining of the lead melt if the melt contains copper. The purpose of drossing is to remove copper from the melt.

Drossing is conducted in large steel kettles holding up to 227 kkg (250 tons). Copper is separated by lowering the temperature of the metal close to, but still above, the melting point of lead. At this temperature, the solubility of copper in lead is minimal, and the lighter copper forms a surficial crust, which is skimmed off. Sulfur may be added to the melt to enhance the removal of copper as copper sulfide, which also appears in the crust. The crust also contains lead oxide, which is skimmed off with the copper and copper sulfide.

Prior to the process known as softening, the lead melt may be transferred to a second kettle, where a second decoppering cycle may be performed. After granulation with water under pressure, the dross is treated in a by-product reverberatory furnace to recover lead as bullion and to produce copper matte, which is handled in the fashion described in the preceding section. The dross kettles may be hooded and produce fume which may in some plants be scrubbed with wet scrubbers that produce an effluent. Dross granulation may also produce an effluent that contains arsenic. If so, it must be handled in its own closed circuit because it is difficult to remove from water.

After copper removal, the lead melt is softened by removal of antimony. Lead alloyed with antimony is called "hard lead" or antimonial lead. Softening may be done by oxidation via the introduction of air into a reverberatory

Table 9.15 *Characteristics of Combined Waste Effluents from Slag Granulation, Blast Furnace Cooling Water and Miscellaneous Blowdowns*

Parameters	*Total plant intake (mg/l)*	*Total plant discharge (mg/l)*	*Net change (mg/l)*	*Net loading (kg/day)*	*(kg/kkg)*	*(lb/ton)*
Alkalinity	203	186	−17	NLC[1]	–	–
COD	8	8	0		0	0
Dissolved solids	408	500	92	551.54	0.97	1.94
Suspended solids	3	36	33	197.84	0.35	0.70
Sulfate (as S)	145	215	70	419.65	0.73	1.46
Chloride	18	–	–	–	–	–
Calcium	70	–	–	–	–	–
Copper	0.02	0.02	0	0	0	0
Iron	1.70	–	–	–	–	–
Lead	0.12	0.30	0.18	1.08	0.002	0.004
Magnesium	0.31	–	–	–	–	–
Nickel	0.03	0.04	0.02	0.12	0.0002	0.0004
Zinc	0.12	0.50	0.38	2.28	0.004	0.008
pH	7.6	8.3	–	–	–	–
Flow						
(l/day)	–	5,995,440	–			
(gal/day)	2,592,000	1,584,000	–			

Source: After Battelle 1974.

Note: Production is 571.4 kkg/day (630 tons/day).

1. NLC = no load calculable.

furnace or by oxidative slagging with a sodium hydroxide and sodium nitrate flux. The air-oxidation method oxidizes copper, antimony, arsenic, and tin which become insoluble and are skimmed off in a primary slag. Aeration is continued and a final slag forms. The slag is further treated to recover antimony, antimony oxide, antimonial (hard) lead, tin, and sodium arsenate, which is discarded. The lead melt is then drained from beneath the slag and treated further by pyrolytic methods.

Oxidative slagging utilizes a sodium hydroxide-sodium nitrate mixture to oxidize arsenic, antimony, and tin to form a slag of arsenates, antimonates, and stannates of sodium. The metals in the slag may be recovered or discarded.

After softening, the gold and silver are extracted from the lead melt by the Parkes process wherein zinc is added to the melt. The zinc combines with gold and silver to form alloys which are lighter than, and insoluble in, lead. They accumulate on the surface and are skimmed off for subsequent separation.

The lead melt is subsequently relieved of its bismuth content via the Betterton process and of its calcium and magnesium by the addition of sodium hydroxide and sodium nitrate. It is then cast.

Table 9.16 *Waste Effluents from a Settling Pond Receiving Granulated Slag*

Parameters	Total plant intake (mg/l)	Total plant discharge (mg/l)	Net change (mg/l)	Net loading (kg/day)	Net loading (kg/kkg)	Net loading (lb/ton)
Alkalinity	0	8	8	11.82	0.04	0.08
COD	<4	4.4	0.4	0.59	0.002	0.004
Dissolved solids	198	301	103	152.14	0.57	1.14
Suspended solids	18	15	−3	NLC[1]	–	–
Sulfate (as S)	2	112	110	162.48	0.61	1.22
Chloride	14	14	0	0	0	0
Cyanide	0	0	0	0	0	0
Arsenic	–	<0.02	–	–	–	–
Cadmium	–	0.08	0.08	0.12	0.0004	0.0008
Calcium	44	48	4	5.91	0.02	0.04
Chromium	–	<0.01	–	–	–	–
Copper	0.055	0.009	−0.046	NLC[1]	–	–
Lead	0	0.85	0.85	1.26	0.005	0.01
Magnesium	21	23	2	2.95	0.01	0.02
Mercury	–	0.0024	–	–	–	–
Nickel	<0.05	0.05	–	–	–	–
Sodium	1.2	12	10.8	15.95	0.06	0.12
Zinc	0.03	1.2	1.17	1.73	0.006	0.012
pH	8.0	7.6	–	–	–	–
Flow						
(l/day)	1,506,430	1,477,058	–			
(gal/day)	398,900	390,240	–			

Source: After Battelle 1974. Compiled from producer's data.

Note: Production is 268.5 kkg/day (296 tons/day).

1. NLC = no load calculable.

Battelle (1974) concluded on the basis of its study that none of these refinery procedures, including by-product processing, require water for other than indirect furnace cooling. Thus, the process should produce no waste water other than that which may be required for air pollution control.

Summary of Waste Water Production

Battelle (1974) observed during its study that one U.S. lead smelting plant discharges no waste water by virtue of climatic conditions at its location; one other plant was observed to discharge only noncontact cooling water, with no discharge of process waste waters. All other U.S. lead smelters discharge some waste water.

Of those lead smelters and refineries discharging waste water, Battelle identified the significant process waste water streams that are components of the discharged waters to be acid plant blowdown, slag-granulation waste water, and wet scrubber waste water. The first and last effluents result from air pollution control measures. Battelle presents data for one smelter showing discharges totaling about 8,175,600 l/day (1,500 gpm) carrying suspended and dissolved solids (mostly sulfates) and identifiable amounts of cadmium, lead, and zinc. The data in Table 9.17 characterize the present discharge, and the data in Table 9.18 characterize the discharge anticipated after completion of certain revisions of the water circuits at the plant. The total flow of effluent will be reduced from 8,175,600 l/day (1,500 gal/min) to 1,635,120 l/day (300 gal/min), with lime treatment of the acid plant blowdown, to achieve the projected waste loads listed in Table 9.18. The calculated discharge rate would be reduced from about 14,000 l/kkg (3,400 gal/ton) to 2,900 l/kkg (690 gal/ton). The latter figure would include 408,780 l/day (75 gpm) of treated acid plant blowdown, equivalent to a calculated minimum rate of blowdown of 1,500 l/kkg (360 gal/ton) of acid or 713 l/kkg (170 gal/ton) of lead produced.

Table 9.17 *Characteristics of Effluent from Casting, Cooling, Acid Plant Blowdown, Dross Granulation, and Noncontact Cooling Water*

Parameters	*Total plant intake (mg/l)*	*Total plant discharge (mg/l)*	*Net change (mg/l)*	*Net loading*		
				(kg/day)	*(kg/kkg)*	*(lb/ton)*
Alkalinity	203	260	57	124.26	0.22	0.44
COD	8	200	192	418.56	0.73	1.46
Dissolved solids	408	980	572	1,246.96	2.18	4.36
Suspended solids	3	500	497	1,083.46	1.90	3.80
Sulfate (as S)	145	500	355	773.90	1.35	2.70
Chloride	18	–	–	–	–	–
Cadmium	–	1.9	1.9	4.13	0.007	0.014
Calcium	70	40	–30	NLC[1]	–	–
Copper	0.02	0.10	0.08	0.17	0.0003	0.0006
Iron	1.70	1.0	–0.7	NLC[1]	–	–
Lead	0.12	0.50	0.38	0.83	0.001	0.002
Magnesium	0.31	–	–	–	–	–
Nickel	0.03	–	–	–	–	–
Sodium	–	200	–	–	–	–
Zinc	0.12	10	9.88	21.54	0.038	0.076
pH	7.6	6.8	–	–	–	–
Flow						
(l/day)	–	2,180,160	–			
(gal/day)	2,592,000	576,000	–			

Source: Battelle 1974. Based on data from Rapp 1971.

Note: Production is 571.4 kkg/day (630 tons/day).

1. NLC = no load calculable.

Table 9.18 *Characteristics of Waste Water Described in Table 9.17 After Completion of Certain Water Circuit Revisions*

Parameters	Total plant intake (mg/l)	Total plant discharge (mg/l)	Net change (mg/l)	Net loading		
				(kg/day)	(kg/kkg)	(lb/ton)
Alkalinity	203	–	–	–	–	–
COD	8	–	–	–	–	–
Dissolved solids	408	550	142	232.17	0.41	0.82
Suspended solids	3	30	27	44.15	0.08	0.16
Sulfate (as S)	145	215	50	81.75	0.14	0.28
Chloride	18	–	–	–	–	–
Calcium	70	–	–	–	–	–
Copper	0.02	0.02	0	0	0	0
Iron	1.70	0.6	–1.1	NLC[1]	–	–
Lead	0.12	0.30	0.18	0.29	0.0005	0.001
Magnesium	31	–	–	–	–	–
Nickel	0.03	0.08	0.05	0.08	0.0001	0.0002
Zinc	0.16	0.50	0.34	0.56	0.001	0.002
pH	7.6	8.0	–	–	–	–
Flow						
(l/day)	9,810,720	1,635,120	–			
(gal/day)	2,592,000	432,000	–			

Source: Battelle 1974.

Note: Production is 571.4 kkg/day (630 tons/day).

1. NLC = no load calculable.

In another smelter, the effluents from a number of operations (including lead smelting) are mixed, and Battelle presents data for the projected characteristics of the effluent from a central treatment plant (Table 9.19) which treats all the wastes combined. The discharge concentrations and unit waste loads are those expected from lime-and-settle treatment. The unit waste loads were calculated on the basis of the input flow (to the treatment plant) attributable to the lead smelting portion of the operation. This operation is the one discussed in detail in Appendix E.

Battelle's study led to the conclusion that in lead smelters, sulfuric acid plant blowdown is the major producer of waste water loadings. It was noted that not all primary lead smelters now have acid plants but it was also noted that there is a need for all smelters to have acid plants.

On the basis of the information it obtained, Battelle tentatively recommended that the entire U.S. primary lead industry be considered as one category for purposes of waste characterization. It tentatively recommended that the critical pollutant parameters in lead smelter waste waters be pH, total

Table 9.19 *Estimated Composition of Discharge from Projected Central Treatment Plant Treating Smelter Wastes, Zinc Plant Wastes and Acid Mine Drainage*

Parameters	*Total plant intake (mg/l)*	*Total plant discharge (mg/l)*	*Net change (mg/l)*	*Net loading*		
				(kg/day)	*(kg/kkg)*	*(lb/ton)*
Dissolved solids	135	–	–	–	–	–
Suspended solids	3.0	30	27	–	0.145	0.290
Antimony	0.03	–	–	–	–	–
Arsenic	0.03	–	–	–	–	–
Cadmium	0.002	0.07	0.068	–	0.00036	0.00072
Copper	0.012	0.006	−0.006	–	NLC[1]	–
Iron	0.50	–	–	–	–	–
Lead	0.06	0.10	0.04	–	0.00021	0.00042
Manganese	0.57	–	–	–	–	–
Mercury	0.0032	0.001	−0.0022	–	NLC[1]	–
Silver	0.01	–	–	–	–	–
Zinc	5.8	0.1	−5.7	–	NLC[1]	–
pH	6.3	10	–			
Flow						
(l/day)	7,766,800	1,771,400	–			
(gal/day)	2,057,000	486,000	–			

Source: After Battelle 1974.

Note: Production is 331 kkg/day (365 tons/day).

1. NLC = no load calculable.

suspended solids (TSS) lead, zinc, cadmium,* arsenic, and mercury.*

Mercury occurs in lead-smelter effluents when they are combined with waste waters from adjoining zinc plant and acid plant units. The mercury originates in the zinc concentrates and it accumulates as a sludge in roaster gas scrubbers. It is now held to less than the local limit of 0.277 kg/day (0.5 lb/day) in the effluent from both the lead and zinc plants observed by Battelle.

Available Control and Treatment Technologies

Battelle (1974) conducted a state of the art analysis of available control and treatment technologies for lead smelter wastes. The basic components, mainly neutralization and sedimentation of waste water, are discussed in Chapters 3 and 4. Battelle recommended that the control of lead smelter and refinery waste waters be subdivided into six categories of effort:

* Cadmium and mercury have been proposed as toxic elements and presumably the toxic element standards will supersede when they are finally specified.

1. Segregation of water streams into: (a) noncontact cooling water; (b) process water; and (c) smaller, sometimes intermittent, streams such as cooling tower or furnace jacket blowdowns.
2. Water conservation techniques (recycling).
3. Housekeeping provisions (for spills, leaks, storm water runoff, pond failures, blowdowns of cooling towers, furnace jackets, and auxiliary equipment).
4. Special in-plant abatement measures.
5. Clarification (settling) techniques.
6. Monitoring techniques.

All United States smelters currently rely mainly on removal of heavy-metal pollutants by precipitation of the hydroxides.

Control of contact cooling water can be accomplished by separation of it from process water used in the granulation of blast furnace slags, zinc fuming furnace slags, and other slags. This requires handling direct contact water in separate recycle systems using separate dewatering bins and settling basins. Bleed-off from such settling basins accounts for most of the waste water from a lead smelter. Acid plant waste water bleed-off is often combined with other process water into a single settling basin which, in modern plants, is checked for pH at regular intervals. Lime additions are made on the basis of these checks. Continuous monitoring of pH with automatic lime feeders would be more desirable.

Battelle recommended that noncontact cooling water be cooled and recycled because once-through systems are inefficient users of water and constitute an unnecessary addition of water to the waste stream.

Sludge and underflow from thickeners and settling basins were considered recyclable. Eventually, the pollutant being recycled builds up to a point where it must be bled off. For example, in some lead smelters cadmium builds up slowly; when it reaches a level deemed to be economic, it is bled off and shipped out for cadmium recovery. Heavy-metal pollutants when concentrated by recycle, can be sold. This does not apply to arsenic because supply exceeds demand. Conservation of arsenic contained in the baghouse dust of the nonferrous smelters could provide a domestic supply that approaches demand, but under present technology it is uneconomic. Only one firm presently recovers white arsenic, As_2O_3, from baghouse dust. It is separated and purified by volatilization. Thus, much of the arsenic removed must be discarded which continues to be a problem. The Boliden Company of Sweden reported a method of mixing arsenic with portland cement in blocks which resist leaching out in storage. Battelle found no evidence that this fixation system is in use in the United States lead industry.

Additional available treatment methods for further reduction of pollutants from process-water discharges include hydrogen sulfide treatment, reverse osmosis, evaporation, and chemical fixation. All were determined to be unproven for the primary lead industry, or uneconomical, by Battelle. It should be noted that one U.S. primary zinc recovery plant reports plans to implement recovery of metals, with hydrogen sulfide, from a waste water not dissimilar to the typical lead smelter waste water.

Proposed Best Practicable Control Technology Currently Available, Effluent Limitations Guidelines, and Costs*

Battelle (1974) based tentative judgment of the best practicable control technology currently available† on the factors listed for primary copper smelters in the preceding section. The procedure for determining compliance with guidelines is identical to that for copper.

Battelle recommended the guidelines listed in Table 9.20 as the effluent limitations attainable through the application of the best practicable control technology currently available (BPCTCA). The values in the table are based on a technology which would produce a minimum flow of effluent with constituent concentrations achievable by lime neutralization and settling. The data in Table 9.20 were concluded to represent the complete control or recycle of all process waste water streams except a blowdown from acid plant operation at a rate of 825 l/kkg (200 gal/ton) of lead produced (the maximum value observed by Battelle in existing plants). The concentrations applied to this flow are as shown in Table 9.21.

The BPCTCA which is the basis for these recommendations assumes that discharges from primary lead smelters consist of waste waters from slag granulation and/or acid plant operation. Two of five existing plants reportedly discharge no water from the slag granulation operation; discharges of this waste water are intermittent in other cases. Slag granulation reportedly does not require water of any special level of quality. The quality of the water used appears to be limited only by the quality required for pumping, which is

Table 9.20 *Recommended Effluent Limitations for the Primary Lead Industry Based on BPCTCA.*

	Effluent limitations			
	30-day average		*Daily maximum*	
Parameters	*(kg/kkg Pb)*	*(lb/ton Pb)*	*(kg/kkg Pb)*	*(lb/ton Pb)*
Total suspended solids	0.025	0.05	0.05	0.1
Lead	0.0004	0.0008	0.0008	0.0016
Zinc	0.002	0.004	0.004	0.008
Cadmium	0.0008	0.0016	0.0016	0.0032
Mercury	8×10^{-7}	1.6×10^{-6}	1.6×10^{-6}	3.2×10^{-6}
Arsenic	2.5×10^{-5}	5×10^{-5}	5×10^{-5}	1×10^{-4}
pH	5-10		5-10	

Source: After Battelle 1974.

Note: See general note on Table 9.6.

* In 1971 dollars.

† These are tentative recommendations by Battelle (1974) and are subject to change based upon comments received and further internal review by EPA.

Table 9.21 *Concentrations of Constituents Obtainable by BPCTCA*

Constituents	*Concentrations (mg/l)*
Total suspended solids	30
Lead	0.5
Zinc	3
Cadmium[1]	1
Mercury[1]	0.001
Arsenic[2]	0.03

Source: After Battelle 1974.

Note: See general note on Table 9.6.

1. Cadmium and mercury were proposed as toxic elements and are subject to stronger restrictions than BPCTCA concentrations. (See *Federal Register*, Vol 38, No. 247, December 27, 1973, pp. 35388-35395).
2. The arsenic level recommended for neutralized copper smelter wastes shown in Table 9.7 is 10 mg/l. Because 0.03 mg/l is not realistic, this figure is undoubtedly an error and probably should read 10 mg/l, as shown in Table 9.7.

achievable by settling. The slag granulation operation also represents an evaporative sink usable for the disposal of waste water from other processes. Thus, Battelle concluded that one element of the best practicable control technology currently available is the complete recycle and reuse of water in slag granulation, with no discharge of waste water from the circuit. Alternatively, blast furnace slag could be discarded molten in some cases, which would eliminate the need for granulation water.

In the case of sulfuric acid plant blowdown two smelters reportedly treat and discharge this stream; and a third smelter treats the stream and reuses it for slag granulation, with a subsequent intermittent discharge from that operation. Air pollution control requirements make acid plants necessary at smelters so this waste treatment process is necessary. Acid plant blowdown appears to be a necessity of operation because of accumulated dissolved solids in the gas cleaning water circuit and the need to prevent corrosion of acid plant equipment, notably by high levels of chlorides.

Battelle recommended that this blowdown be treated by lime neutralization and clarification to achieve the best practicable technology currently available. Neutralization with lime to a pH in the 8 to 10 range should reduce the concentrations of those metals precipitable as hydroxides and, with properly designed retention facilities, will also reduce total suspended solids to below effluent limitation guidelines. As noted in Chapters 3 and 4, lime offers the advantage that it, unlike sodium-based alkalies, forms relatively insoluble sulfate, $CaSO_4$, which will tend also to reduce the concentrations of dissolved sulfate in the effluent.

Battelle concluded that the effluent reduction benefits balance the costs of the technology and that those plants not presently achieving the recommended limitations (Table 9.20) would require an estimated maximum capital investment of about \$2.78/annual kkg (\$2.52/annual ton) and an increased operating cost of about \$1.02/kkg (\$0.92/ton) to achieve the effluent limitations.

The costs of existing control and treatment technologies were determined by collecting data on capital costs and on operating costs for present control practices from selected lead smelters. Battelle modified these data as follows to put all costs on a common basis as shown in the two tables:

1. The capital costs reported were changed to 1971 dollars by the use of the Marshall and Swift Index.*
2. The annual costs were recalculated to reflect common capitalized charges. To do this, the annual costs were calculated by using a factor method as follows:
   a. Operating and maintenace: as reported by the lead smelters.
   b. Depreciation: 5 percent of the 1971 capital.
   c. Administrative overhead: 4 percent of operating and maintenance.
   d. Property tax and insurance: 0.8 percent of the 1971 capital.
   e. Interest: 8 percent of the 1971 capital
   f. Other: as reported by the smelters.

The costs of waste water treatment additional to current practices were because either the pertinent cost data were not available from existing smelters or the reported plant cost data showed substantial variation due to differences among smelters with respect to water usage and treatment and cost-reporting procedures. The cost data for the present waste water practices in the primary lead industry are summarized in Table 9.22. Variations reflect differences among the smelters in the presence or absence of an acid plant, extent of plant water circuit modification, water usage, waste water treatment practices, and cost-reporting procedures employed. Costs of additional treatment alternatives determined by Battelle are presented in Table 9.23.

Table 9.22 *Capital and Operating Costs of Present Waste Water Treatment Practices in Primary Lead Industry*

Plant designation	Annual lead production (kkg)	Capital costs (Total \$)	Capital costs (\$/annual kkg)	Operating costs (Total \$/yr)	Operating costs (\$/kkg)
A	208,610	\$256,900	\$1.23	\$ 85,800	\$0.41
B	120,963	493,900	4.08	212,400	1.76
C	97,956	135,900	1.39	56,200	0.57
D	120,948	595,300	4.92	198,200	1.64
E	58,048	171,400	2.95	63,000	1.09

Source: After Battelle 1974.

* Quarterly values of this index appear in *Chemical Engineering.*

Table 9.23 *Additional Treatment Alternatives and Costs for Process Water Discharges in Primary Lead Industry*

Treatment alternatives and costs	*Plant A*	*Plant B*	*Plant C*	*Plant D*[1]
Lime and settle[2]				
Total flow (cu m/day) [gpm]	758 [139]	409 [75]	381 [70]	381 [70]
Total capital costs (1971 $)	150,000	104,000	100,000	336,000
($/annual kkg) [$/annual ton]	0.72 [0.65]	0.86 [0.78]	1.05 [0.93]	2.78 [2.52]
Total operating costs ($/year)	37,000	23,000	22,000	123,000
($/kkg) [$/ton]	0.18 [0.16]	0.19 [0.17]	0.22 [0.20]	1.02 [0.92]
Multiple effect evaporation/distillation				
Total flow (cu m/day) [gpm]	1,166 [214]	660 [121]	381 [70]	762 [140]
Total capital costs (1971 $)	660,000	425,000	285,000	765,000
($/annual kkg) [$/annual ton]	3.16 [2.87]	3.51 [3.19]	2.91 [2.64]	6.33 [5.74]
Total operating costs ($/year)[3]	318,000	195,000	125,000	318,000
($/kkg) [$/ton][3]	1.52 [1.38]	1.61 [1.46]	1.28 [1.16]	2.63 [2.38]

Source: After Battelle 1974.

1. Treatment costs for lime and settle include: capital costs of $232,000 and operating costs of $100,000/year for slag granulation recycle system, estimated from costs data supplied by Plant B. Treatment costs for evaporation include additional costs for sinter plant scrubber water recycle system estimated at $53,000 capital and $17,000/year operating costs based on cost data supplied by Plant A.
2. Costs based on $SO_4^=$ concentration in the waste water at 1,000 mg/l.
3. Including costs for the disposal of concentrated liquid waste produced from evaporator by the Chemfix process.

Proposed Best Available Technology Economically Achievable, Effluent Limitations Guidelines, and Costs*

Battelle (1974) based tentative judgment of BATEA guidelines† on what it determined to be the best control and treatment technology planned by an existing facility. However, it was noted that the best available technology economically achievable may necessitate some industrially sponsored development work prior to its application.

The tentatively recommended effluent limitations attainable through the application of the best available technology economically achievable are presented in Table 9.24.

The guidelines are based on the assumption of complete control or recycle of all process waste water streams except a blowdown from acid plant operation at a rate of 825 l/kkg (200 gal/ton) of lead produced (the maximum observed in existing plants). The concentrations on which the limitations in Table 9.24 are based are presented in Table 9.25.

The technology which Battelle (1974) identified as achieving the tentatively recommended effluent limitation guideline is similar to that discussed under BPCTCA. The principal elements of the technology are:

1. The reuse and recycle of water or use of a nonwater slag granulation process to achieve a level of performance of no discharge of process waste water from the slag granulation process. (See discussion for BPCTCA).
2. The application of a lime and settling treatment process to control the selected pollutant parameters in acid plant blowdown.

Table 9.24 *Recommended Effluent Limitations for the Primary Lead Industry, based on BATEA*

	Effluent limitations			
	30-day average		*Daily maximum*[1]	
Parameters	*(kg/kkg Pb)*	*(lb/ton Pb)*	*(kg/kkg Pb)*	*(lb/ton Pb)*
Total suspended solids	0.025	0.05	–	–
Lead	8.3×10^{-5}	1.7×10^{-4}	–	–
Zinc	8.3×10^{-5}	1.7×10^{-4}	–	–
Cadmium[2]	5.8×10^{-5}	1.2×10^{-4}	–	–
Mercury[2]	8×10^{-7}	1.6×10^{-6}	–	–
Arsenic	2.5×10^{-5}	5×10^{-5}	–	–

Source: After Battelle 1974.

Note: See general note on Table 9.6.

1. Insufficient data available to establish daily variations.
2. Cadmium and mercury were proposed as toxic elements and are subject to stronger restrictions than BPCTCA concentrations (see *Federal Register*, Vol. 38, No. 247, December 27, 1973, pp. 35388-35395).

* In 1971 dollars.

† These are tentative recommendations by Battelle (1974) and are subject to change based upon comments received and further internal review by EPA.

Table 9.25 *Concentrations of Constituents Obtainable by BATEA*

Constituents	*Concentrations (mg/l)*
Total suspended solids	30
Lead	0.1
Zinc	0.1
Cadmium[1]	0.07
Mercury[1]	0.001
Arsenic	0.03

Source: After Battelle 1974.

Note: See general note on Table 9.6.

1. Cadmium and mercury were proposed as toxic elements and are subject to stronger restrictions than BPCTCA concentrations (see *Federal Register*, Vol. 38, No. 247, December 27, 1973, pp. 35388-35395).

The principal difference between the guidelines Battelle recommended for BPCTCA and for BCTEA lies in the degree of removal of the critical waste water pollutants. The degree of reduction identified is derived from the expected levels to be achieved in a planned treatment plant to be installed at an existing lead smelter. Reportedly, the levels of constituents predicted are based on test work already completed.

The technology consists of carefully controlled lime and settle treatment, with appropriate provisions for optimum mixing, increased retention time, the use of flocculant aids, and, as an alternative to settling, the use of higher performance vacuum filtration techniques. However, as noted previously, neutralization cannot reliably precipitate completely pollutants such as arsenic.

Precipitation with hydrogen sulfide would be expected to reduce dissolved arsenic, as well as lead, cadmium, and mercury, to tenths of milligrams per liter. Battelle went so far as to observe that economics of sulfide precipitation appear attractive and that some reduction in lime costs might be possible because a pH in excess of 8 is unlikely to be required. However, Battelle did not recommend sulfide precipitation as BCTEA because it has not been demonstrated in the industry. Similar conclusions were reached for chemical fixation which conceivably could convert waste solutions to a solid for disposal to a landfill or tailings pond.

Proposed New Source Performance Standards and Costs*

Battelle (1974) tentatively recommended that New Source Performance Standard (NSPS)† be identical to that for BCTEA. Essentially no rationale for

* In 1971 dollars.

† These are tentative recommendations by Battelle (1974) and are subject to change based upon comments received and further internal review by EPA.

the recommendation was presented. However, it is clear that a number of the technologies alluded to previously might be considered.

PRIMARY ZINC

The Industry

Primary zinc and primary lead plants are often located together because ores frequently contain both sphalerite and galena.

In 1971, 51 percent of the zinc produced in the United States came from zinc ores, 30 percent from lead-zinc ores, 10 percent from lead ores, and 7 percent from copper ores. The 25 largest zinc producing mines in the United States are listed in Table 9.26.

Fifteen mines that produced 26,000 tons of zinc in 1970 and 20,000 tons in 1971 were shut down by the end of 1971, largely because of closure of smelters due to air pollution controls. More mines and smelters closed in 1972.

United States consumption of primary zinc exceeds mine production by over 100 percent. U.S. zinc plants are listed in Table 9.27. Commercial zinc is produced at zinc reduction plants; thus refining is not similar to that of copper and lead. Higher quality zinc products are produced by distillation. Because of diminishing smelter capacity, a larger proportion of imported slab zinc relative to concentrates can be anticipated.

According to U.S. Geological Survey estimates, the United States has about 18 percent of the world's "recoverable" zinc and 7 percent of the subeconomic zinc resources.

Processes and Sources of Wastes and Waste Waters

A zinc smelter employs practices consisting of blending concentrates, removing magnesium (if necessary), roasting and/or sintering to convert zinc sulfide to zinc oxide, leaching with sulfuric acid to dissolve the oxide in the case of electrolytic recovery, or pyrometallurgical procedures in lieu of electrolytic recovery. A flow sheet for a pyrolytic plant is shown in Figure 9.9. The waste waters from certain processes are extremely contaminated with respect to pH and dissolved solids. Sources of waste water are shown in Figure 9.10.

Concentrate Preparation

The initial step consists of storing, drying, and blending concentrates. If electrolytic reduction is employed (as opposed to pyrolytic reduction), the concentrates may require "pretreatment" if they contain sufficient magnesium. The necessity for removing magnesium is that if it is not removed prior to the electrolysis process, the magnesium displaces zinc in the zinc plant electrolytic solution circuit. Losses of 15 percent of the zinc have been observed when the magnesium content of the electrolyte is allowed to reach equilibrium. Electric current efficiency decreases concomitantly (6 percent), and more purification cycles are required.

Table 9.26 *Leading Zinc-Producing Mines in the United States in 1971 in Order of Output*

Rank		*City and State*	*Operator*	*Source of zinc*
1	Balmat	St. Lawrence, N.Y.	St. Joe Minerals Corp.	Lead-zinc ore
2	Eagle	Eagle, Colo.	The New Jersey Zinc Co.	Zinc ore
3	Sterling Hill	Sussex, N.J.	The New Jersey Zinc Co.	Zinc ore
4	Friedensville	Lehigh, Pa.	The New Jersey Zinc Co.	Zinc ore
5	Bunker Hill	Shoshone, Idaho	The Bunker Hill Co.	Lead-zinc ore, lead-zinc tailings
6	Buick	Iron, Mo.	Missouri Lead Operating Co.	Lead ore
7	Young	Jefferson, Tenn.	American Zinc Co.[1]	Zinc ore
8	New Market	Jefferson, Tenn.	New Market Zinc Co.[1]	Zinc ore
9	Zinc Mine Works	Jefferson, Tenn.	United States Steel Corp.	Zinc ore
10	Immel	Knox, Tenn.	American Zinc Co.[1]	Zinc ore
11	Austinville and Ivanhoe	Wythe, Va.	The New Jersey Zinc Co.	Zinc ore
12	Edwards	St. Lawrence, N.Y.	St. Joe Minerals Corp.	Zinc ore
13	Star Unit	Shoshone, Idaho	Bunker Hill Co. and Hecla Mining Co.	Lead-zinc ore
14	Ground Hog	Grant, N. Mex.	American Smelting and Refining Co.	Lead-zinc ore
15	Jefferson City	Jefferson, Tenn.	The New Jersey Zinc Co.	Zinc ore
16	Burgin	Eureka, Utah	Kennecott Copper Corp.	Lead, lead-zinc ore
17	Idarado	Ouray and San Miguel, Colo.	Idarado Mining Co.	Copper-lead-zinc ore
18	U.S. and Lark	Salt Lake, Utah	United States Smelting, Refining, Mining Co.	Lead, lead-zinc ores
19	Flat Gap	Hancock, Tenn.	The New Jersey Zinc Co.	Zinc ore
20	Bruce	Yavapai, Ariz.	Cyprus Mines Corp.	Copper-zinc ore
21	Shullsburg	Lafayette, Wis.	Eagle-Picher Industries, Inc.	Zinc ore
22	Leadville	Lake, Colo.	American Smelting and Refining Co.	Lead-zinc ore
23	Coy	Jefferson, Tenn.	American Zinc Co.[1]	Zinc ore
24	Magmont	Iron, Mo.	Cominco American, Inc.	Lead ore
25	Ozark	Reynolds, Mo.	Ozark Lead Co.	Lead ore

Source: After Battelle 1974.

1. Purchased by the American Smelting and Refining Co., November 29, 1971.

Elimination of the poor-quality effluent from the acid leaching process requires the elimination of the need for acid leaching of concentrates. Research directed at this objective can take two forms: (1) mineral beneficiation processes can be studied in order to develop methods of removing the magnesium-bearing minerals during concentration, and (2) methods can be studied for preventing the displacement of zinc by magnesium in the electrolyte cells.

Table 9.27 *Primary Zinc Plants in the United States*

Company	*Location*	*Type of operation*	*Annual zinc producing capacity* (*kkg/yr*)	*Annual zinc producing capacity* (*tons/yr*)	*Acid plant operation*	*Products produced at plant site*
American Smelting and Refining Co.	Amarillo, Texas	Horizontal retort	42,303	46,641	No	Prime western zinc
American Smelting and Refining Co.	Corpus Christi, Texas	Electrolytic	97,956	108,000	Yes	Slab zinc, zinc alloys, zinc sulfate, cadmium, sulfuric acid
Amax	Sauget, Ill.	Electrolytic	64,149	70,727	Yes	Slab zinc, cadmium, zinc sulfate, sulfuric acid
Bunker Hill (Gulf Resources and Chemical Co.)	Wallace, Idaho	Electrolytic	110,312	121,624	Yes	Slab zinc, zinc alloys, cadmium, sulfuric acid
New Jersey Zinc (Gulf and Western Industries)	Palmerton, Pa.	Vertical retort	103,398	114,000	Yes	Slab zinc, zinc alloys, zinc dust and pellets, zinc oxide, ferroalloy, sulfuric acid
St. Joe Minerals Corp.	Monaca, Pa.	Electrothermic	226,750	250,000	Yes	Slab zinc, zinc alloys, zinc oxide, cadmium, ferrosilican, mercury, sulfuric acid
National Zinc Co.	Bartlesville, Okla.	Horizontal retort	45,350	50,000	Yes	
Amax	Blackwell, Okla.	Horizontal retort	72,832	80,300		Zinc

Source: After Battelle 1974.

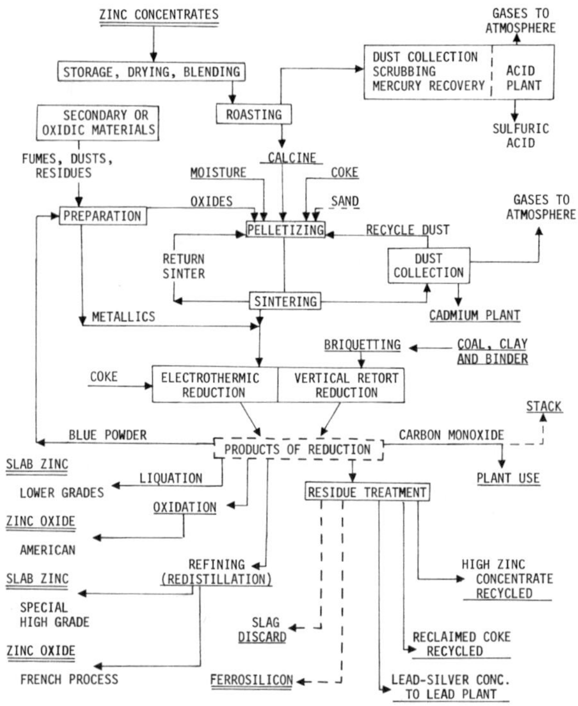

Figure 9.9. *Generalized flow sheet of pyrolytic zinc plants (after Battelle 1974).*

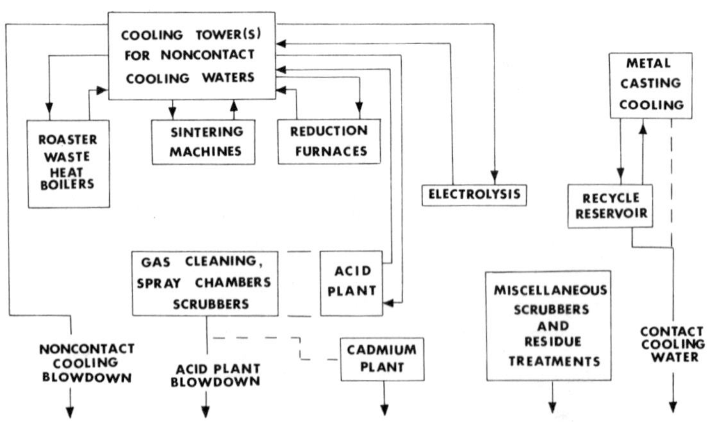

Figure 9.10. *Generalized diagram of waste water streams in primary zinc operation (after Battelle 1974).*

Pretreatment consists of acid leaching followed by filtering with the filtrate going through a float cell and then to disposal. The filter cake is stored in storage bins. This material and the concentrates not requiring leaching are loaded by fixed proportions into blending hoppers and fed into a rotary dryer. The pretreatment float cells produce decanted water from the acid leach process. Water is also derived from the drum filters which filter the acid leached concentrate. The leaching solution is essentially sulfuric acid, which accounts for the low pH and high concentrations of dissolved metals in the pretreatment float cell effluent, as shown in Table 9.28. Not all zinc plants treat high-magnesium concentrates, therefore not all zinc plants produce this waste water.

Roasting

The concentrates must be roasted to convert zinc sulfides to zinc oxides for subsequent leaching and electrolysis. Roasting is carried out as completely as is economically feasible in the electrolytic plants since residual zinc sulfide would be essentially insoluble in the subsequent acid dissolution step. In pyrolytic or retort plants, roasting is followed by sintering; therefore complete removal of sulfur is not so necessary during roasting. Battelle (1974) reports that, with a typical feed of 54 percent zinc and 31 percent sulfur, some 96 percent of the sulfur is removed by roasting in one of the U.S. pyrolytic plants studied.

Various types of roasters are employed by U.S. zinc plants. Multiple-hearth roasters permit different conditions on different hearths and are used effectively in deleading. Fluid bed roasters are employed to give stable operating conditions. Most roasters provide a 5.5 to 7.0 percent SO_2 feed to sulfuric acid plants which are the principal means of precluding air pollution. Because of the comparatively high temperatures generated in roasters, waste heat boilers are commonly used to cool the emitted gases down to about 400°C (750°F) prior to cleaning.

Table 9.28 *Sources, Flow Rates, and Quality of Effluents in the Zinc Recovery Process*

Source	*Average flow (gpm)*	*pH*	*Suspended solids (mg/l)*	*mn (ppm)*	*Fe (ppm)*	*Pb (ppm)*	*Zn (ppm)*	*Cu (ppm)*	*Cd (ppm)*	*Hg (ppm)*
Pretreatment float cells	200	2.5	3,500	45	268	630	1,275	1.7	11.0	.02
Peabody scrubber	450	2.0	180	3.5	56	86	450	2.6	9.0	5.4
Recycle overflow	530	1.0	312	88	5.8	1.8	5,050	0.4	0.1	0
Scrubbers for zinc melting & casting	80	6.0	22	4	0.1	0.15	42	0.1	25	0
Filter and dryer overflow	30	6.0	360	7.5	81	79	200	4.7	1.5	0.02

Source: Data obtained from author's research.

After passage through a waste heat boiler, the roaster gas is sent to a baghouse and wet scrubber(s) gas cleaning system prior to entering the sulfuric acid plant, where the SO_2 is converted to sulfuric acid. The wet scrubber system produces an effluent identified in Table 9.28 as "Peabody scrubber."

Roasting eliminates some of the impurities from the zinc other than sulfur and lead. Any mercury present is volatilized and goes with the gas stream to the scrubbers which precede the acid plant. Some cadmium may be eliminated also, but most of it remains with the oxide to be recovered in solution purification in electrolytic plants or eliminated to a greater extent by sintering in the case of pyrometallurgical plants. The roasted concentrate is called calcine. It goes to the sintering operation if additional reduction is done pyrolytically and to the leaching operation if electrolysis is employed.

To recapitulate, the purpose and operation of the roaster gas cleaning system is as follows: After leaving the pretreatment stage, the dry, finely ground zinc sulfide concentrate is fed into the roaster and oxidized. The SO_2 gas (and a small amount of SO_3 gas) that is produced, along with considerable fine dust, is blown out of the roaster and into a waste heat boiler where the heat is captured and used as an energy source. The SO_2 gas, still containing some dust, is then blown to a baghouse, and possibly through a cyclone, which removes most of the dust. The SO_2 gas leaves the baghouse and enters a humidifier and scrubber unit, one type of which is a Peabody scrubber. The "liquor" consists of partially recycled waste water and fresh water which has been injected into the system, along with dissolved SO_2 gas and dust particles. There is sufficient SO_3 in the gas entering the scrubber to assist in making the solution acid, and soluble metal ions available in the dust are dissolved in the liquor. As the liquor becomes too poor in quality and temperature to wash the gas, some of it is bled off and fresh water is added. The bled-off liquor is scrubbed of SO_2 gas and the remaining liquor constitutes the waste water of the bleed.

Electrostatic precipitators, called mist precipitators, are also employed in acid plants. The waste water from these precipitators joins the waste stream from the scrubber. The acid waste is poor in quality and hot; therefore, it cannot be reused in some other part of the plant without further treatment.

Battelle (1974) considers the identifiable waste water streams from this sequence of operations to be the stream from the humidification chamber and the bleed stream from the wet scrubber. Battelle refers to them collectively as "acid plant blowdown" and has determined three bases for their existence:

1. Control of temperature of the gases and, implicitly, the recirculating stream, involving, for example, the use of cool makeup water as the means of temperature control.
2. The prevention of buildup of chloride concentration in the recirculating stream to a level which would cause significantly accelerated corrosion of the materials of construction. A level of 0.002 percent chloride by weight in the recirculating stream was considered tolerable in terms of stainless steel equipment, whereas a chloride level of 0.2 percent by weight would be considered a level to justify an increased bleed rate.

3. The maintenance of a tolerable level of suspended solids in the recirculating system. The tolerable level was defined as the level at which the system functioned continuously, and was noted to vary with pipe sizes and pump characteristics. It was characterized as having a maximum ranging from 2 to 3 percent solids by weight. The origin of the suspended solids is the particulates in the gas stream entering the electrostatic precipitator and subsequently the wet scrubber. This particulate level depends on the number and performance level of the preceding dust control devices. Performance levels of these devices may vary with time, maintenance, and charge material, or ambient atmospheric conditions.

Acid plant blowdown was identified by Battelle as a component of discharge streams in all but one of the United States plants studied. In the exception it was routed to a cadmium by-product operation. But a subsequent chloride bleed stream was necessary.

Characteristics and flow rates of acid plant blowdowns from zinc plants observed by Battelle are presented in Tables 9.29 and 9.30, respectively.

Sintering and Briquetting (Pyrolytic Process)

In order to achieve a more compact feed for pyrolytic reduction of the oxide, as well as to eliminate more of the impurities, the calcine from roasting is treated by sintering (in pyrolytic plants). However, since the SO_2 has been substantially removed during roasting, the gases from sintering are weaker in SO_2 and frequently are vented to a stack. Whether this practice will be allowed to continue is as yet undetermined.

Table 9.29 *Characteristics of Waste Water from an Acid Plant Blowdown*

		Loading		
Parameters	*Discharge (mg/l)*	*(kg/day)*	*(kg/kkg)*	*(lb/ton)*
Dissolved solids	5,400	7,063	21.98	43.95
Suspended solids	200	261.6	0.814	1.628
Cadmium	33	43.16	0.134	0.268
Lead	48	62.8	0.195	0.391
Tellurium	–	–	–	–
Zinc	1,500	1,962	6.105	12.2
CrO_4	0.10	0.1308	0.0004	0.0008
pH	2.8	–	–	–
Flow				
(l/day)	1,308,096			
(gal/day)	345,600			

Source: After Battelle 1974.

Note: Production is 321.4 kkg/day (360 tons/day).

Table 9.30 *Rates of Flow of Acid Plant Blowdown Streams*

	Acid plant blowdown flow rates		*Flow per unit of zinc produced*		*Flow per unit of sulfuric acid produced*	
Plant	*(l/day)*	*(gal/ton)*	*(l/kkg)*	*(gal/ton)*	*(l/day)*	*(gal/ton)*
B	850,300	224,600	3,170	760	4,900	1,180
C	741,300	195,800	4,220	1,010	1,960	470
D	1,363,000	360,000	4,500	1,080	4,290	1,030
E	2,400,000	633,600	7,300	1,760	5,300	1,270
G	817,600	216,000	6,590	1,580	3,280	785

Note: As observed by Battelle 1974.

The feed for sintering includes, in addition to calcined concentrates, baghouse dusts or scrubber sludges from dust collection units, and various residues from other processes. These materials are pelletized prior to sintering.

Most of the cadmium and lead are driven off during sintering. Consequently the dust is collected and recycled until cadmium and lead are economically recoverable (elsewhere).

Pyrolytic Reduction

If pyrolytic reduction is employed (as opposed to electrolysis) the sintered calcine is reduced in a vertical or horizontal gas heated retort. In the case of the vertical retort, the gases from the process (principally zinc vapor and carbon monoxide) are passed through a splash-type zinc condenser and then through a venturi wet scrubber. The zinc vapor condenser is about 95 percent efficient. The vertical retort utilizes the cleaned carbon monoxide from the zinc condensation chamber and subsequent scrubber to partially heat the furnace. Blue powder, a mixture of zinc and zinc oxide, is collected as a residue from the scrubbing system and recycled each time the condenser is cleaned.

In the process of electrothermic reduction of zinc, a larger, internally heated fire brick retort or furnace is used. Cooling water flows in steel jackets surrounding the reduction portion of the furnace. Preheated coke and sinter, along with miscellaneous zinc-bearing products, are fed into the top of the furnace. This charge moves slowly downward while the coke serves as reductant and conductor for the electric current emanating from graphite electrodes. Zinc vapor generated by the reduction passes from the furnace to a vapor ring and then to a water-cooled condenser where it is condensed by bubbling through molten zinc. The remaining gas, mostly carbon monoxide, emanating from the condenser is scrubbed by impinge jets and used as fuel elsewhere in the plant. Some zinc and zinc oxide sludge is recovered by settling the scrubber slurry. This sludge can be dried, briquetted, and recycled to the furnace feed. Furnace residue after zinc distillation is reclaimed of residual coke and some unreacted zinc. These are recovered and recycled. If sand is added to make a hard sinter, sufficient ferrosilicon may be present in the furnace residue to warrant recovery of it as a by-product.

The gas washing water produced by the gas scrubbers referred to above is characterized by Battelle (1974) as involving high volume and as containing zinc and metal oxides, possibly hydrocarbons, and various particulates (as suspended solids) and the corresponding products of hydrolysis.

The characteristics of one such stream are presented in Table 9.31. The gases from reduction furnaces are referred to as internal process streams, meaning they are not discharged as characterized, but pass through other production processes, waste treatment, or mixing before becoming a component of discharge. The characteristics listed are as reported to Battelle by producers, and the constituents are those considered significant by the producers. The characteristics apply to settled wastes and the metal concentrations are reported in terms of total metals. Waste water production rate was calculated to be 461 l/kkg (182 gal/ton). The rate is minimal because recycle is employed with the bleed rate (waste water production rate) representing 1 percent of the total recirculated flow rate.

Zinc Production by Electrolysis

If electrolysis is employed to reduce zinc oxide to the metal, the thoroughly roasted (oxidized) concentrates (calcine) are sized and prepared for attack by sulfuric acid.

Leaching: The acid used for leaching consists of spent electrolyte which is a strong sulfuric acid solution (about 200 gms/1) containing residual zinc sulfate. Continuous leaching has become the trend in the United States because less equipment, space, and labor are required.

Leaching occurs in a series of stainless steel leach tanks of approximately 100 volume tons each. The zinc oxide is dissolved, as is copper, cadmium, arsenic, antimony, cobalt, and nickel. Such materials as lead, silver, gold, iron, silica, and calcium remain undissolved.

Table 9.31 *Characteristics of Gas Scrubbing Waste Water from Pyrolytic Recovery of Zinc*

Constituents	Concentrations (mg/l)[1]	Calculated raw waste load (kg/day)	(lb/day)	(kg/kkg)	(lb/ton)
Dissolved solids	7.2	45.5	–	0.170	0.34
Suspended solids	220	10.4	–	0.039	0.077
Cadmium	0.15	0.031	–	0.00012	0.00023
Lead	58	12	–	0.045	1.5×10^{-4}
Cyanide	12	2.48	–	0.009	0.019

Source: After Battelle 1974.

Note: This waste water is a scrubber-bleed stream, not a discharge stream; average flow value of 207,100 l/day (54,720 gal/day), equivalent to 761 l/kkg (182 gal/ton) of zinc produced.

1. Metals concentrations based on total metals.

Purification: The leach slurry, made up of the insoluble materials and the zinc-rich solution, is drawn from the last tank of the leaching system, as needed, into batch pressure filters for solids separation. The residue from these filters, which contains the undissolved material mentioned above, is washed, then passed through a hydroclone (or through a float cell for sulfide recovery) before being put into thickeners. When the proper specific gravity is attained, this residue is filtered on drum filters and dried to approximately 25 percent moisture. This material is then sent to a lead smelter where it is processed for the metals it contains. The liquid discharge from the filters (sometimes Oliver, but usually Burt filters) is pumped to a storage tank.

Subsequently the impure zinc sulfate solution (sometimes called Burt solution) is drawn from storage into a series of tanks for first-stage purification. In this stage zinc dust, copper sulfate, and arsenic are added to precipitate copper, cadmium, nickel, and cobalt-rich cakes from the zinc solution. The discharge from the final tank is filtered and the filtrate is pumped to another storage tank. The residue (or residues) is sent for processing to a cadmium plant, and/or other plants if the properties of the residues require.

The zinc-rich solution from this purification process is passed through another tank circuit (called second-stage purification). Here antimony and 95 percent of the cadmium are precipitated from the solution with the aid of additional zinc dust and copper sulfate, and again the solution is put through a filter and sent to another (second-stage) storage tank. The residue from this filter is also sent to the cadmium plant (or other appropriate plant) for processing. From the second-stage storage tank, the solution goes to the third-stage purification, a final tank series which removes the remaining cadmium and antimony. This is also accomplished with zinc dust and copper sulfate. The resulting solution and precipitate are passed through a hydroclone with the overflow going to a filter. The residue from this filter and the underflow of the hydroclone are both returned to the first-stage purification for recycle. The filtrate from the last step goes into a batch step called cleanup purification where the solution is treated a final time with zinc dust to insure purity. The residue goes back to the first-stage purification while the filtrate, which is purified zinc sulfate solution (neutral solution) is pumped to storage tanks to await electrolysis.

All solids and liquids produced by leaching and purification contain recoverable metals. Thus, every effort must be made not to divert any of it to the waste water stream. However, some losses from the filters may occur, as indicated in Table 9.28.

Zinc Electrolysis and Stripping: In order to obtain pure zinc metal, the zinc contained in the purified zinc sulfate solution is deposited on aluminum cathode plates by electrolysis. The purified solution is drawn as needed to maintain the zinc content of the feed to the electrolytic cells. The heat generated by the high amperage through the cells is dissipated by passing the cell feed through balance tanks and an evaporation cooler. Cell discharge is recycled to the balance tanks. At intervals of 8 or 16 hours, the zinc metal plated on the cathodes is stripped off and sent to a casting plant. Spent electrolyte is either sent to the leach department or the cadmium plant, or it is purified and recycled to the electrolytic cells.

The cathode wash water is essentially the only waste water produced by electrolysis. It is contained in the water listed as "recycle overflow" in Table 9.28. This water also contains the effluent from manganese scavenging and certain cell room wash waters, hence the relatively high flow rate.

The manganese scavenger system recovers manganese dioxide which collects on the anode of electrolytic cells. It is subsequently added to the slurry produced by the first stage of leaching described above. The reason for the addition is that manganese dioxide oxidizes undesirable ferrous iron to ferric iron which precipitates and is removed by the Burt filters.

Cadmium Recovery: As mentioned previously, certain residues from the purification process are sent to the cadmium plant. The first-stage purification residue (the copper-cobalt residue) is leached with spent electrolyte to dissolve any cadmium or zinc present. This slurry is then filtered and the copper-cobalt residue is stockpiled for shipment. The filtrate is pumped to a cadmium leach tank where it is mixed with the second-stage purification or cadmium residue, which is also leached with spent electrolyte. This leach solution is filtered and the residue is sent back to the copper residue leach tank. The filtrate is pumped to another tank where zinc dust is added to precipitate the cadmium. The solution from this tank is filtered and the filtrate is sent to the continuous leach tanks described previously. The cadmium sponge residue along with any other cadmium-rich residues (as from a lead smelter or recycled cadmium-rich roaster flue dusts) is put into solution with the aid of spent electrolyte from the cadmium electrolytic cells. The cadmium-rich solution is filtered and the residue returned to the cadmium leach tank. The filtrate is fed to the cadmium electrolytic cells. After a varying length of time, the cathodes in these cells are stripped of their cadmium. The cadmium metal is then melted and cast into balls, sticks, or mossy cadmium. Battelle (1974) reports that all but one of the United States cadmium plants operates with closed aqueous circuits.

Refining, Melting, and Casting

Cathode zinc sheets from electrolytic zinc plants are dried, melted, and cast into various forms of slab zinc. Alloys are also prepared and cast. At times, dependent upon marketing conditions, lead and other desired constituents are purposely added to a relatively high grade of electrolytic zinc to make a select grade for galvanizing. Zinc dust also is made at these plants for use in the purification of solutions and for cadmium recovery, as described previously.

Pyrolytic zinc is normally less pure than that produced electrolytically, but this may be rectified (1) by careful selection of raw material and charge so as to minimize impurities, and (2) by refining via a liquation treatment wherein the melt is allowed to cool to slightly above the melting point of zinc. The lead and iron precipitate are removed mechanically. The partially purified zinc is then cast as slab zinc. By redistillation of this still somewhat impure zinc, the highest purity commercial grades of zinc can be produced.

Melting and casting zinc into commercial forms is a relatively simple operation as the temperature required is only moderate. Water is used in some plants to cool the molds rapidly. Ammonium chloride flux is commonly used on the molten zinc in the melting furnace to retard oxidation at the surface and to collect any oxides formed. Air pollution control devices on gases from the zinc melting operation collect solids containing both zinc oxide and chloride compounds. Any processing to recover the zinc from this air pollution control residue must deal with the chloride content of the residue. This is especially true in electrolytic zinc plants, because the electrolysis reaction is very sensitive to deleterious effects of chloride in the electrolyte.

The nature of the effluent from scrubbers applied to melting and casting gases is listed in Table 9-28 as "scrubbers for zinc melting and casting." Chlorides were not monitored but their concentration would be high.

Waste Waters Produced by Auxiliary Operations

In some zinc plants, waste waters are produced by auxiliary operations which process residues from previously described air pollution control devices and other residues to reclaim their zinc values. Battelle (1974) reports the principal characteristic of these waste waters to be relatively high levels of dissolved solids (sulfates and chlorides) along with varying but significant concentrations of arsenic, cadmium, copper, zinc, and lead. One such stream is characterized in Table 9.32. These waste waters are internal process streams and are subject to mixing and treatment before discharge. Battelle (1974) reported the average flow rate associated with this stream to be 174 l/kkg (42 gal/ton) calculated on the basis of zinc production for the plant.

Acid Plants

Battelle (1974) reports that all but one United States zinc plant produce sulfuric acid as a by-product. The exception has a variance to operate only until June 30, 1975.

The feed for the plants is derived from gases relatively high in SO_2 produced by roasting (oxidizing) zinc concentrates (sulfides). The gases are first passed through a dust collection system (baghouse or electrostatic precipitator, scrubbers, cyclones and flues). In some plants, part of the heat obtained by cooling the gases is utilized in waste-heat boilers. Residual solids in the gas flow are removed and final cooling is done commonly in a cooling or scrubbing tower, where water flows countercurrent to the gas flow. Clean SO_2 gas is converted to SO_3 by catalytic action and this is absorbed into water to form commercial grades of sulfuric acid. After passing through a mist precipitator, the residual gases are vented to the stack. Water used in cooling and scrubbing absorbs sulfuric acid from SO_3 in the roaster gases as described under the discussion of roasting.

The scrubber sludge contains almost all of the mercury present in the concentrates. It is recovered or accumulates as a by-product of the plant. The fate of the sludge is to some extent determined by its mercury content. Recycle is difficult if the mercury content is high.

Table 9.32 *Characteristics of Waste Water from an Auxiliary Metal Recovery Operation*

Parameters	*Intake (mg/l)*	*Discharge (mg/l)*	*Net change (mg/l)*	*Net loading (kg/day)*	*(kg/kkg)*	*(lb/ton)*
Alkalinity	110	–	–	–	–	–
COD	–	56	56 (G)	53.8	0.198	0.396
Dissolved solids	575	146,130	145,555 (N)	139,878	514.1	1,028.2
Suspended solids	0	1,338	1,338 (N)	1,286	4.73	9.46
Oil & grease	–	10	10 (G)	9.61	0.0353	0.0706
Sulfate (as S)	80	9,259	9,179 (N)	8,821	32.42	64.84
Chloride	106	160	54 (N)	51.9	0.191	0.382
Cyanide	–	<0.1	<0.1 (G)	0.096	0.0004	0.0008
Arsenic	–	3.0	3.0 (G)	2.88	0.011	0.022
Cadmium	–	<0.02	<0.02 (G)	<0.019	0.00007	0.00014
Calcium	64	–	−64 (G)	NLC	NLC	NLC
Copper	–	8.0	8.0 (G)	7.69	2.83	5.66
Iron	3.4	33.0	29.6 (N)	28.45	0.105	0.210
Lead	–	300	300 (G)	2.88	1.058	2.116
Magnesium	17	–	−17 (G)	NLC	NLC	NLC
Mercury	–	0.003	0.003 (G)	0.0029	0.00001	0.00002
Nickel	–	1.0	1.0 (G)	0.961	0.0035	0.0070
Selenium	–	<0.01 (G)	<0.0096	0.0096	0.00004	0.0008
Zinc	–	48,000	48,000 (G)	46,128	169.53	339.06
Phenols	–	<0.1	<0.1 (G)	<0.096	–	–
pH	7.9	5.2				
Flow						
(l/day)	–	–	96,139			
(gal/day)	–	–	25,400			

Source: After Battelle 1974.

Notes: N = net; G = gross; NLC = no load calculable.
Production is 272.1 kkg/day (300 short tons/day).

Summary of Characteristics of Primary Zinc Recovery Wastes

On the basis of information similar to that in the preceding sections, Battelle (1974) concluded that the following waste streams and points of derivation are characteristic of the primary recovery of zinc.

1. Roasting–noncontact cooling water.
2. Roaster gas cleaning train–bleed streams from gas cooling spray chambers or wet scrubbers.
3. Sulfuric acid plant–noncontact cooling water.
4. Reduction furnaces–noncontact cooling water.

5. Reduction furnace gas cleaning systems—bleed streams or once-through water streams.
6. Electrolysis (of zinc)—noncontact cooling water.
7. Metal casting cooling—direct contact cooling water waste streams.
8. Cadmium production—spent process liquor.
9. Rectifier cooling—noncontact cooling water.
10. Boiler operations—boiler blowdown. Miscellaneous cooling waters from pump seals, bearing cooling vacuum pumps.
11. Auxiliary air pollution control operations—including dust control and/or wet processing of air pollution control residues to reclaim metal values.

Cathode wash water and manganese dioxide recovery process water could be added to this list.

With respect to general characteristics of current industry practice, Battelle (1974) observed that the volumes of discharges vary from zero at one plant in the Southwestern United States, which is able to practice complete solar evaporation, to 44,663,000 l (11,800,000 gal) per day. Treatment practices at plants discharging waste waters vary from simple settling to lime-and-settle treatments applied to either component streams or to the total plant effluent. These data are summarized in Table 9.33.

In terms of gross concentrations of constituents of combined source streams, the ranges reported by Battelle (1974) for some constituents are:

□ Dissolved solids	455-4,485 mg/l
□ Suspended solids	25-249 mg/l
□ Sulfates	175-2,221 mg/l
□ Chlorides	60-620 mg/l
□ Arsenic	0.1-0.68 mg/l
□ Cadmium	0.02-2.4 mg/l
□ Copper	0.01-0.34 mg/l
□ Iron	0.02-1.93 mg/l
□ Lead	0.02-1.35 mg/l
□ Selenium	0.007-1.8 mg/l
□ Zinc	5-243 mg/l

Thus, wastes from primary zinc plants can be characterized as acidic and containing zinc and sulfates, accompanied by cadmium, lead, and, less significantly, arsenic and selenium. Acid plant blowdown, metal casting cooling water and streams from auxiliary operations such as recovery of metal values from residues of air pollution control devices were specified as the most ubiquitous and significant sources of waste water. The critical pollutants in these waste waters were identified as pH, suspended solids, zinc, cadmium, mercury, arsenic, and selenium.

Available Control and Treatment Technology

Battelle (1974) tentatively concluded that liming and settling is the most appropriate treatment technology for zinc plant waste waters produced by air

Table 9.33 *General, Current, Waste Production and Handling Practices in the Primary Zinc Industry*

Plant	*Type of operation*	*Volume of discharge (l/day)*	*(gal/day)*	*Degree of treatment*[1]
A	Pyrolytic	0	0	Solar evaporation
B	Electrolytic	2,763,050	430,000	Complete lime and settle
C	Electrolytic	18,186,000	4,800,000	Settle
D	Electrolytic	5,450,400	1,440,000	Mix and settle
E	Pyrolytic	2,910,670	769,000	Partial lime and settle
F	Pyrolytic	44,663,000	11,800,000	Partial lime and settle
G	Pyrolytic	1,078,725	285,000	Complete lime and settle
H	Pyrolytic	69,220	18,288	(Lime and settle)

Sources: After Battelle 1974. Verbal and written information submitted by producers under Battelle contract and existing RAPP data.

1. Complete lime and settle indicates that all waste streams are treated; partial lime and settle indicates that one or more component streams are so treated.

pollution control devices and product washing and purification. Other technologies were rejected for reasons similar to those presented for lead smelters. Battelle's rationale for recommending the treatment of zinc plant waste waters with lime is presented here in summary form.

The choice of lime as a reagent for treating zinc plant waste waters is based principally on economics. Caustic soda (NaOH) and soda ash (Na_2CO_3) are both costly alkalies, and both are currently in short supply. Furthermore, they do not form insoluble sulfates, thus neutralization with these alkalies does not limit sulfate concentrations. Ammonia, NH_3, is easy and convenient to handle but it does not reliably precipitate metals because it forms soluble complexes with them.

Addition of a lime slurry to a low pH zinc plant waste water containing heavy metals will precipitate the hydroxides of most of the metals and will reduce dissolved sulfate concentrations through the formation of gypsum, $CaSO_4 \cdot 2H_2O$. Sufficient lime must be added to the zinc plant effluent to raise the pH to between 10.0 and 11.5, and the closed stream conveyed to a settling pond to settle out suspended solids. It was noted that organic polyelectrolytes are available which, though costly per pound, are effective at low concentrations in facilitating flocculation and clarification.

Data on concentrations of metals in treated zinc plant waste water were supplied to Battelle by a zinc plant operating a lime-and-settle treatment facility for mixed plant wastes. The data, as shown in Figure 9.11, represent daily averages of pH and soluble-metal analyses in 24-hour composite samples. Battelle reported that the treatment facility operates with effluent pH ranging from 6.5 to 10.5 and a level of suspended solids ranging from nil to a daily maximum of 64 mg/l. The data represent 29 daily determinations over a period of two months of continuous operation with an average flow rate of 2,650 cu m/day (0.7 mg/l). They fall into the bands delineated in Figure 9.11. The

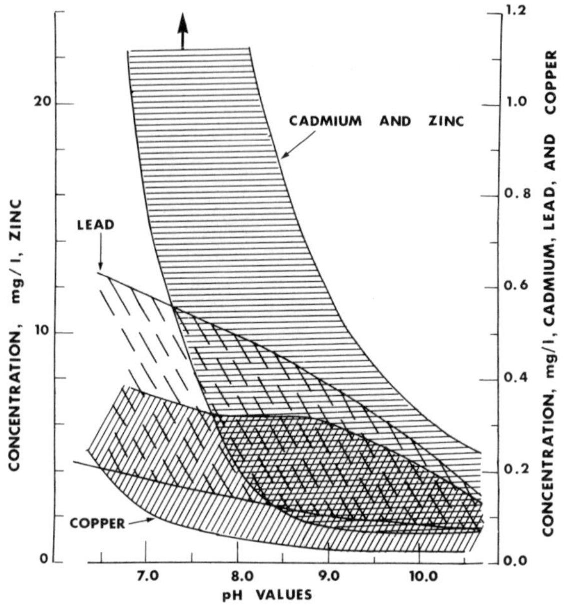

Figure 9.11. *Concentration of metals in the effluent from a lime-and-settle treatment operation for zinc plant waste waters (after Battelle 1974).*

lowest concentration band is for copper. The band descends to the maximum pH included, giving levels of 0.05 to 0.15 mg/l of copper at pH 10.5. The band for lead shows a similar pattern, covering a range of values from 0.6 mg/l at pH 6.5 to 0.05 to 0.2 mg/l at pH 10.5. Zinc and cadmium show somewhat different patterns with a band descending from concentrations including single values on the order of 30 to 100 mg/l at pH 6.5 to 7.0 to 0.1 to 0.3 at pH 10.5. The plant data were observed to show a general trend in agreement with theory, but concentration limits appear somewhat changed in complex coprecipitation compared with theoretical values, which may be attributable to the differences between static (equilibrium) and dynamic systems. It was noted that, in dynamic systems, precipitation may be accompanied by instantaneous variations in pH and continuing separation of precipitate and supernatant so that solid-liquid equilibrium is not possible.

An indication of the effectiveness and performance of the treatment plant that Battelle used as a model for BPCTCA is given in Tables 9.34 and 9.35. The data in these tables are based on composite grab samples taken over a two-day period. The concentrations of constituents of two input streams to the treatment, and the concentrations in the effluent are given in Table 9.34. The effectiveness of the treatment in removing various constituents is shown in Table 9.35, which also shows the mass-balance data Battelle used to derive the effectiveness levels. The data in Table 9.35 are based on an average operating pH of 8.2.

The positive values given for a few of the constituents were explained by Battelle as follows: The increase in suspended solids is atypical of plant operation according to prior operational data, and represents only the reported sampling period. The increase in oil and grease is considered an artifact of sampling and calculation; all input and output concentrations are near the practical minimums of 1 to 50 mg/l. The increase in dissolved solids may be a result of grab sampling or it may be real.

Both sulfates and chlorides were observed to show net decreases due to the treatment. All other constituents were removed to some degree by the lime and settle treatment. Over 90 percent removal was observed for iron, zinc, and cadmium.

Battelle also presented the data in Table 9.36 pertaining to the removal of selected constituents from acid plant blowdown by a lime-and-settle treatment. It was noted that a pH of 11.8 is maintained in the effluent from this plant. While no reduction in suspended solids is indicated by the data, considerable reductions in cadmium, lead, and zinc concentrations are achieved.

Table 9.34 *Analysis of Input and Effluent Streams for a Zinc Plant Waste Water Treatment Plant*

	Inputs to treatment plant		
Parameters	*Stream 1 (mg/l)*	*Stream 2 (mg/l)*	*Treatment plant discharge (mg/l)*
COD	23.0	15	17
Dissolved solids	135	5,650	4,485
Suspended solids	8	13	249
Oil & grease	5.0	<1 → 3	4
Sulfate (as S)	4,572	2,240	2,221
Chloride	105	242	171
Cyanide	–	<0.1	<0.1
Arsenic	1	0.1	<0.1
Cadmium	0.6	0.16	<0.02
Copper	0.2	0.03	<0.02
Iron	19	~0.17	0.06
Lead	2	0.23	0.15
Mercury	0.006	0.004	0.004
Nickel	0.26	0.31	0.13
Selenium	13	0.54	1.8
Zinc	25	1,063	50
pH	2.5	6.5	8.2
Flow			
(l/day)	847,840	1,915,210	2,763,050
(gal/day)	224,000	506,000	730,000

Source: Reported by Battelle 1974.

Table 9.35 *Effectiveness of Removal of Various Constituents from Zinc Plant Wastes*

Parameters	*Input stream 1 (kg/day)*	*Input stream 2 (kg/day)*	*Total input (kg/day)*	*Total output (kg/day)*	*Net change[1] (kg/day)*	*Net change[1] (percent)*
COD	19.50	28.73	48.23	46.97	−1.26	−2.6
Dissolved solids	114.46	10,820.94	10,935.40	12,392.28	+1,456.88	+13.3
Suspended solids	6.78	24.90	31.68	688.00	+657.00	+2,070.
Oil & grease	4.24	1.92-5.75	6.16-9.99	11.05	1.06-+4.89	+49-+17
Sulfate (as S)	3,876.	4,290.07	8,166.07	6,136.73	−2,029.34	−24.8
Chloride	89.02	463.48	552.50	472.48	−80.02	−14.5
Cyanide	–	<0.19	0.19	0.28	–	–
Arsenic	0.85	0.19	1.04	0.28	−0.76	−73.1
Cadmium	0.51	0.31	0.82	0.06	−0.76	−92.7
Copper	0.17	0.06	0.23	0.06	−0.17	−73.9
Iron	16.11	0.33	16.44	0.17	−16.27	−98.97
Lead	1.70	0.44	2.14	0.41	−1.73	−80.8
Mercury	0.005	0.008	0.013	0.010	−0.003	−23.1
Nickel	0.22	0.59	0.81	0.36	−0.45	−55.6
Selenium	11.02	1.03	12.05	4.97	−7.08	−58.8
Zinc	21.20	2,035.87	2,056.89	138.15	−1,918.74	−93.3
pH	2.5	6.7	–	–	–	–
Flow						
(l/day)	847,840	1,915,210	2,763,050	2,763,050	–	–
(gal/day)	224,000	506,000	730,000	730,000	–	–

Note: Calculated by Battelle 1974.

Calculations for each entry $= \frac{mg}{l} \times \frac{1}{day} \times 1 \times 10^{-6} \frac{kg}{mg}$ = kg/day; see Table 9.34 for mg/l values.

1. Negative numbers indicate removal by treatment.

Battelle reported that during its study one zinc plant supplied the input, output, percent removal, and resulting unit waste loads for its *proposed* treatment plant. These data are shown in Table 9.37. The treatment technology was described as including a separate, proprietary process step for the control and removal of mercury, followed by precipitation of heavy metals as sulfides by hydrogen sulfide treatment, with subsequent liming and settling using flocculant aids to achieve a level of less than 15 mg/l suspended solids. The hydrogen sulfide precipitation treatment reportedly offers the advantage of producing relatively dense, recyclable precipitates, and of a decrease in the volume of sludge produced by the lime treatment step, along with a decrease in the cost of sludge disposal. It was noted that the treatment process is anticipated to reduce waste water production from the current flow rate of 18,168,000 l/day (4.8 mgd) to a flow rate of about 1,000,000 l/day (275,000 gal/day) by various changes in water use practices within the plant. The calculated

Table 9.36 *Effectiveness of Treatment of Acid Plant Blowdown by Lime and Settle*

Parameters	Intake (mg/l)	Discharge (mg/l)	Net change[1] (mg/l)	Net change[1] (percent)	Net loading (kg/kkg)	Net loading (lb/short ton)
Dissolved solids	5,400	2,500	−2,900	−53.7	–	–
Suspended solids	200	200	0	0	–	–
Cadmium	33	0.03	−32.97	−99.9	–	–
Lead	48	6.9	−41.1	−85.6	–	–
Zinc	1,500	2.6	−1,497.4	−99.8	–	–
pH	2.8	11.8				
Flow						
(l/day)	1,308,096	–	–			
(gal/day)	345,600	–	–			

Source: After Battelle 1974.

1. Negative numbers indicate removal by treatment.

discharge rates would be reduced from the current value of 104,900 l/kkg (25,150 gal/ton) to a projected 5,800 l/kkg (1,400 gal/ton) of zinc produced.

Proposed Best Practicable Control Technology Currently Available, Effluent Limitations Guidelines, and Costs*

Battelle (1974) tentatively recommended the best practicable control technology currently available (BPCTCA)† for zinc plant waste waters generated by air pollution control devices and direct contact process waters on the basis of the factors discussed previously for copper. Calculations for determining compliance are identical to those presented for copper.

The effluent limitations guidelines presented in Table 9.38 were tentatively recommended as those that can be achieved through the application of the best practicable control technology currently available. Battelle identified the best practicable control technology currently available as the minimization of discharge or process waste waters by practices of recycle, reuse, segregation, and, finally, lime neutralization to achieve controlled precipitation, followed by sedimentation. On the basis of its study of the entire primary zinc industry, Battelle concluded that at present, in some cases, process waters are used on a once-through basis, whereas in other plants, the discharge from the same process is considerably lower on a unit-product basis by virtue of recycle.

* In 1971 dollars.

† These are tentative recommendations by Battelle (1974) and are subject to change based on comments received and further internal review by EPA.

Table 9.37 *Effectiveness of Planned Treatment Plant*

	Input to treatment (mg/l)	*Treated output (mg/l)*	*Treated removal*	*Net loading (kg/day)*	*(kg/kkg)*	*(lb/ton)*
Dissolved solids	45,000	3,400	92.4	3,539	20.43	40.86
Suspended solids	1,000	<15	98.5	15.6	0.032	0.064
Sulfate (as S)	33,700	1,600	95.3	1,666	9.62	19.24
Chloride	397	100	74.8	104	0.60	1.20
Arsenic	74	0.1	99.9	0.104	0.0006	0.0012
Cadmium	28	0.04	99.9	0.04	0.0002	0.0004
Calcium	354	588	NLC[1]	612	3.533	7.066
Chromium	5	0.02	99.6	0.02	0.0001	0.0002
Copper	41	0.15	99.0	0.16	0.0009	0.0018
Iron	99	0.2	99.8	0.21	0.0012	0.0024
Lead	7	0.13	98.1	0.14	0.0008	0.0016
Magnesium	4,850	40	99.2	41.6	0.2402	0.4804
Mercury	0.12	0.0005	99.6	0.0005	0.000003	0.000006
Nickel	30	0.09	99.7	0.09	0.0005	0.0010
Sodium (and potassium)	513	538	NLC[1]	560	3.233	6.466
Zinc	3,990	⩽0.3	⩾99.99	0.3	0.0017	0.0034
pH	–	9.5	–	–	–	–
Flow						
(l/day)	1,040,875	–	–			
(gal/day)	275,000	–	–			

Source: Reported by Battelle 1974.

Note: Production is 176 kkg/day (194 tons/day).

1. NLC = no load calculable.

Those measures that were identified in the Battelle report as having the potential for minimizing the production of waste water by the primary zinc industry include the following:

1. The minimization of acid plant blowdown by careful operation of prescrubber gas cleaning facilities to minimize particulate loadings in the wet scrubbers, by provisions for settling in the scrubber-liquor recycle circuit, and possibly by reusing the scrubber bleed stream elsewhere in the plant. These measures have been discussed previously.
2. The minimization of metal-casting cooling water discharge by recycle, possibly including provisions in the circuit for removal of suspended solids, oil and grease, and thermal load.
3. The exploitation of the evaporative capacity of hot gases or hot metal for in-plant disposal of waste water by evaporation if additional air pollution can be avoided, as discussed previously.

Table 9.38 *Recommended Effluent Limitations for the Primary Zinc Industry, Based on BPCTCA*

	Effluent limitations			
	30-day average		*Daily maximum*	
Parameters	*(kg/kkg Zn)*	*(lb/ton Zn)*	*(kg/kkg Zn)*	*(lb/ton Zn)*
Total suspended solids	0.2	0.4	0.4	0.8
Zinc	0.04	0.08	0.08	0.16
Cadmium[1]	4×10^{-3}	8×10^{-3}	8×10^{-3}	1.6×10^{-2}
Mercury[1]	8×10^{-6}	1.6×10^{-5}	1.6×10^{-5}	3.2×10^{-5}
Selenium	0.2	0.4	0.4	0.8
Arsenic	8×10^{-4}	1.6×10^{-3}	1.6×10^{-3}	3.2×10^{-3}
pH	5-11		5-11	

Source: After Battelle 1974.

Note: See general note on Table 9.6.

1. Cadmium and mercury were proposed as toxic elements and may be subject to stronger restrictions than BPCTCA concentrations (see *Federal Register*, Vol. 38, No. 247, December 27, 1973, pp. 35388-35395).

Battelle noted that lime neutralization of waste waters to a pH in the 10 to 11 range will reduce the concentrations of those metals precipitable as hydroxides and, with properly designed retention facilities, will also reduce total suspended solids to below the recommended effluent limitation guidelines.

It was also pointed out that arsenic and selenium are not effectively removed at pH's above 7, and their removal depends on adsorption and coprecipitation.

The basis for the selection of waste water flow rates for calculations consists of the flow rates of contact process waste waters discussed previously. These rates range from 899 cu m/day (165 gpm) to a maximum of 2,452 cu m/day (450 gpm) and they reflect the recycle practices utilized in one plant in particular. Battelle converted average flow rates of process waste water discharge (see Table 9.33) were converted to discharge rates per unit of product to obtain the following ranges of values:

Plant	*Discharge rates (averages)* *(l/kkg Zn)*	*(gal/ton Zn)*
B	4,880	1,170
C	6,440	1,550
D	7,220	1,730
E	8,700	2,090
F	7,100	1,700
G	13,500	3,240

Table 9.39 *Constituents and Concentrations Considered Achievable by BPCTCA*

Constituents	*Concentrations (mg/l)*
Total suspended solids	25
Zinc	5
Cadmium [1]	0.5
Mercury [1]	0.001
Selenium	5
Arsenic [2]	0.1

Source: After Battelle 1974.

Note: See general note on Table 9.6.

1. See note 1, Table 9.38.
2. The arsenic level recommended for neutralized copper smelter wastes shown in Table 9.7 is 10 mg/l. Because 0.1 mg/l is not realistic, this figure is undoubtedly an error and probably should read 10 mg/l, as shown in Table 9.7.

Plant G was ignored because it is scheduled for closure or conversion. Therefore, a rate of discharge of 8,000 l/kkg (1,920 gal/ton) was selected for use in recommending effluent limitations.

Battelle tentatively concluded that current lime-and-settle treatment operations for which reasonable amounts of data are available show concentrations of constituents to be achievable in effluents as given in Table 9.39.

These demonstrated levels of concentrations were applied, together with the flow rate given above, to derive the recommended effluent limitations shown in Table 9.38.

Battelle tentatively concluded that the effluent reduction benefits balance the costs of the technology and that those plants not presently achieving the recommended limitations (Table 9.38) would require an estimated maximum capital investment of about \$3.78/annual kkg (\$3.43/annual ton) and a maximum increased operating cost of about \$0.98/kkg (\$0.89/ton) to comply with the effluent limitations.

Proposed Best Available Technology Economically Achievable, Effluent Limitations Guidelines, and Costs*

Battelle (1974) tentatively recommended the effluent limitations guidelines shown in Table 9.40† as those attainable by the application of the best available technology economically achievable to primary zinc plant waste waters produced by air pollution control measures or direct contact processes. The best available technology economically achievable includes the BPCTCA

* In 1971 dollars.

† These are tentative recommendations by Battelle (1974) and are subject to change based on comments received and further internal review by EPA.

control measures which segregate and minimize the volume of process waste water streams, as well as the application of additional control, and treatment processes to further decrease waste loads.

The additional treatment and control measures identified are those discussed previously, which reportedly are planned for installation at an existing plant. The additional measures identified include:

1. Rearrangement of water circuits within the plant to reduce the amount of process waste water discharged to 6,540 l/kkg (1,570 gal/ton) of zinc produced.
2. Treatment of process waste water by precipitation of metals with hydrogen sulfide to produce a relatively dense, recyclable material.
3. The further treatment of the process waste water by a lime-and-settle method utilizing increased retention time and flocculant aids.
4. The utilization of special treatment specifically for the control and recovery of mercury and selenium found in the process.

The concentrations listed in Table 9.41 were assumed achievable by these measures and they were applied to the process waste water discharge rate of 6,500 l/kkg (1,570 gal/ton) to obtain the recommendations given in Table 9.40.

Battelle based its selection of these recommendations on what it concluded to be the technology identified as the best that has been demonstrated to be capable of being designed for plant-scale operation. "Demonstration" consists of the discussion presented in the foregoing section on available control and treatment technology.

Table 9.40 *Recommended Effluent Limitations for the Primary Zinc Industry, based on BATEA*

	Effluent limitations			
	30-day average		*Daily maximum*[1]	
Parameters	*(kg/kkg Zn)*	*(lb/ton Zn)*	*(kg/kkg Zn)*	*(lb/ton Zn)*
Total suspended solids	0.098	0.196	–	–
Zinc	0.002	0.004	–	–
Cadmium[2]	0.0003	0.0006	–	–
Mercury[2]	3×10^{-5}	6×10^{-5}	–	–
Selenium	0.001	0.002	–	–
Arsenic	0.0007	0.0014	–	–
pH	7-9.5			

Source: After Battelle 1974.

Note: See general note on Table 9.6.

1. Insufficient data to establish values.
2. See note 1, Table 9.38.

Table 9.41 *Constituents and Concentrations Considered Achievable by BCTEA*

Constituents	*Concentrations*[1]
Total suspended solids	15
Zinc	0.3
Cadmium[2]	0.04
Mercury[2]	0.0005
Selenium	0.2
Arsenic	0.1
pH	9.5

Source: After Battelle 1974.

Note: See general note on Table 9.6.

1. In milligrams per liter (except pH).
2. See note 1 on Table 9.38.

Proposed New Source Performance Standards

Battelle (1974) recommended that new source performance standards for waste waters from zinc plants be the same as the BCTEA effluent guidelines presented in Table 9.40.

The control and treatment technology applicable to the recommended standards of performance is the same as that for BCTEA. No rationale was presented for the recommendation of no additional standards.

ADDITIONAL APPROACHES TO TREATING SMELTER WASTES

As indicated by Battelle (1974), recovery of SO_2 for purposes of air pollution control undoubtedly will be a future necessity for all smelters. Therefore, all smelters can be expected ultimately to produce a low pH waste water as a result of cleaning gases in preparation for single (and probably double) absorption sulfuric acid plants. These low-pH waste waters will dominate other types of direct contact waste waters in volume; thus the net waste water can be expected to be consistently acid. Battelle has relied upon classical treatment and control technologies for its recommendations to EPA regarding guidelines for implementation of Public Law 92-500.

In the case of some plants in some locations, it should be feasible to mix the acid smelter waste waters with basic waste waters from other processes carried out at the same or nearby locations in order to achieve partial neutrality. The process most ubiquitously associated with smelters is flotation. Flotation is conducted at high pH values in order to keep certain ions from going into solution, in order to prevent the deterioration of sodium cyanide used to suppress sphalerite during the flotation of galena, and in order to optimize the flotation and depression of certain minerals. The net result is a high-pH slurry which is discarded in a tailings pond. If the volume of smelter wastes is low compared to flotation wastes, it may be feasible to mix the two wastes to produce a waste which is either easier to treat or for which treatment other than settling is unnecessary. However, mixing may also have a negative effect in that the addition of acid water to tailings may strip ions from the tailings, thereby creating a larger volume of waste water requiring treatment. Clearly, any moves in the direction of mixing should be thoroughly researched.

Certain metallurgical processes also produce high-pH waste waters which may be amenable to mixing with smelter waste waters to achieve partial neutrality. The recovery of antimony is a prominent example. In the Coeur d'Alene district of north Idaho an antimony plant produces a high-pH waste water (pH-11) a few miles up-valley from a lead-zinc smelter. Agreements between such industries to mix wastes may prove economically beneficial under the constraints of Public Law 92-500.

Research on the interaction of the components of such wastes prior to mixing would be advisable. Preliminary information of this type is discussed by Williams and Wallace (1973) but many needed experiments are yet to be conducted.

It is clear also that still other alternatives must be researched if smelter wastes are to be handled and discarded with maximum efficiency. Two areas that show promise are:

1. The development of new processes that either produce no airborne wastes or produce wastes that can be handled by dry technique.
2. The development of more efficient and more reliable dry air pollution control technologies which will facilitate the conversion of airborne wastes directly to solid wastes so that the expense of waste water treatment is precluded. Battelle alludes to such a process as discussed in Chapter 5 under the title of the "Teller process." However, it is proprietary and not much detail is available.

The Arbiter process constitutes one of the more promising new technologies for the primary copper industry (see Arbiter, 1973). A plant for its utilization is currently under construction by the Anaconda Company near Anaconda, Montana. The plant will produce 35,000 tons of metal per year. The technology also shows promise for nickel and zinc; however details are proprietary.

The Arbiter process utilizes an oxygenated ammonia solution as a leach medium for extraction of copper from sulfide concentrates. Leaching is followed by solvent extraction and electrowinning according to the general principles discussed previously under the section entitled hydrometallurgy.

The process features elimination of sulfur oxide gas and dust control problems, without equipment not essential to metallurgy. This is in contrast to pyro-plants for which gas cleaning and collecting equipment, and acid or sulfur-making plants are now essentially required.

Because of the use of ammonia solutions as solvents, the process is selective with respect to metal dissolution. The soluble metals—copper, zinc, nickel and cobalt—can be separated from one another by the use of solvent extraction, chemical precipitation and/or electrowinning. Thus, zinc or nickel in copper concentrates, copper or cobalt in nickel ores or concentrates, or copper, lead, and zinc in complex concentrates of these metals, can be separated and recovered to a degree not generally possible by selective flotation and smelting.

Disposal of the sulfur may be an economic burden of the process. All the sulfide that reacts with leach solutions is converted ultimately to ammonium sulfate which must be eliminated from the leach circuit by a bleed-off. In a location where ammonium sulfate has a market, crystallization of this by-product and its sale is the most economical route for sulfur disposal. It involves less energy; no other raw materials; and no waste disposal.

If ammonium sulfate cannot be sold, then treatment of the bleed stream with lime and steam to recover ammonia and prepare artificial gypsum for disposal is the alternative. It is more expensive in terms of energy, raw materials, and impoundment of gypsum.

A third alternative now under study would convert the ammonium sulfate solution directly to ammonia, and to elemental sulfur. If the sulfur can be sold or used locally, this alternative might be economical.

Precious metals are not recoverable directly with the base metals, but must be reclaimed from the leach residue. Flotation will be used to recover silver along with unleached copper at the Anaconda plant in Montana.

REFERENCES

American Bureau of Metal Statistics 1973. *Yearbook for 1972.* New York.

Arbiter, N. 1973. Anaconda's Ammonia Leach Process. Paper presented at the 1973 annual meeting of the American Mining Congress, Denver; available from American Mining Congress, Washington, D.C.

Battelle 1974. Development Document for Proposed Effluent Limitations Guidelines and New Source Performance Standards for the Primary Copper, Secondary Copper, Primary Lead, and Primary Zinc Subcategories of the Copper, Lead, Zinc Segment of the Nonferrous Metals Manufacturing Point Source Category. Draft report prepared by Battelle's Columbus Laboratories for the Environmental Protection Agency, Contract No. 68-01-1518.

Lanier, H. 1965. *Encyclopedia of Chemical Technology.* 2nd ed. New York: John Wiley and Sons, Inc.

Williams, R. E., and Wallace, A. T. 1973. *The Role of Mine Tailings Ponds in Reducing the Discharge of Heavy Metals to the Environment.* Documents PB-224 730/AS and PB-224 731/AS, National Technical Information Service, U.S. Dept. of Commerce, Springfield, Virginia.

10. *Mill wastes*

SOURCES OF INFORMATION

The sources of information for this chapter are the published literature and the research conducted by the writer and his associates (see Williams and Wallace 1973). Unfortunately no industry-wide study of the type cited in Chapters 5 through 9 has yet been conducted on mill wastes. Such a study has been scheduled but at this writing it has not been initiated nor has a contract been written. Consequently the availability of comprehensive data and information on mill wastes is much more limited than was the case for the subjects discussed in chapters 5 through 9. Similarly no permanent effluent limitations guidelines have been proposed or promulgated for point source waste water discharges for this industry. Waste discharge permits are being issued on the basis of the interim guidelines discussed in Chapter 1.

THE PROCESS AND THE WASTES

Milling consists of the comminution of ore into particles that can be subjected to a recovery process which separates valuable minerals from gangue. Although originally restricted to grinding, the term *milling* is now used in a broader sense to include the concentration or recovery process. Flotation usually constitutes the recovery process, although gravitational methods also are used if the specific gravity of the mineral is sufficiently greater than that of the gangue. Historically gravitational separation methods have left many sulfide minerals unrecovered, and they have been discarded with the gangue. These solid wastes have been discarded in various locations and sometimes constitute a source of water pollution (see Mink, Williams, and Wallace 1972).

Much of the available information on mill wastes pertains to the copper, lead, and zinc industry, although, as indicated in Chapter 6, a considerable amount of information is also available for phosphate wastes. In spite of the latter information, the most crucial problem of phosphate slimes, the high water content, remains unsolved.

Milling is accomplished customarily by crushing and grinding. The details of the procedure depend on the properties of the mineral and the gangue. The most important properties are mineral identity, specific gravity, hardness, mineral particle size, and mineral dissemination. In many cases, it has been established that, following crushing, each ore fraction requires a particular grind to liberate mineral particles most effectively and efficiently. The fraction having the lowest specific gravity frequently also has the greatest hardness, greatest mineral dissemination, and finest particle size. Also, the ore fraction having the highest specific gravity frequently has the lowest hardness, the least mineral dissemination, and the largest mineral particle size. This is the basis for the procedure for primary grinding. The objective is to grind and classify each rock particle to the size required for the best metal recovery.

Summaries of two possible procedures for a lead-zinc mill follow. Comparison of these two examples illustrates the degree to which minor changes in milling procedures can affect the production of wastes, especially waste waters. Both procedures have been utilized on a production basis.

One practice is to implement flotation of both lead and zinc sulphides after primary grind. The zinc rougher tailings are classified; the slime fraction is discarded as tailings; and the sand fraction is reground and returned to the lead rougher circuit. A portion of the lead rougher concentrate and zinc rougher concentrate is also reground and retreated by flotation in this circuit.

The main flotation circuits are controlled so as to put into the two products a particular fraction of the ore. The purpose of the procedure is to keep all the ore particles at a size controllable with respect to classification and flotation and to segregate them into fractions having the same specific gravity and hardness range for regrinding and classification.

An alternative practice after primary grinding is to classify, regrind, and refloat the lead rougher flotation tailings rather than the zinc rougher flotation tailings. In this case, a hydroclone overflow constitutes the zinc rougher flotation feed, and the zinc rougher flotation tailings are the final tailings. All lead and zinc flotation scavenge concentrates, in addition to the first zinc cleaner return and first and second lead cleaner return, are classified in hydroclones and reground; the regrind ball mill discharge, in all cases, goes to lead roughers. This alternative procedure, in some cases, has been demonstrated to minimize the amount of galena in the final zinc concentrate. As indicated in Chapter 9, this facilitates efficiency in the primary zinc recovery process.

The author worked with one company that has utilized both of these alternatives. With respect to efficiency and waste production and control, the second alternative offers several advantages over the first. These advantages are as follows:

1. Overgrinding is reduced.
2. Recovery of lead and zinc is increased, thereby increasing yield and decreasing the probability of their becoming a pollution hazard in the tailings pile after abandonment.

3. The grade of the zinc concentrate is improved, and a greater percentage of lead reports to the lead concentrate rather than to the zinc concentrate.
4. Grinding circuit capacity is increased.
5. Water consumption is reduced, which is of particular importance with respect to cost of waste water treatment.
6. Pulp tonnages handled are reduced, facilitating disposal of wastes.
7. Flotation pulp density is greater. This requires less energy consumption for water removal.
8. Reagent consumption is reduced. This is also important with respect to waste water treatment costs, as will be shown subsequently.

Both the lead and zinc concentrates from the flotation process are thickened and drum-filtered to be sent to the primary recovery processes discussed in Chapter 9. Thickener underflows can be reused at several points in the process for purposes of water conservation and minimum production of waste water.

Various reagents are added to the slurry during the flotation process. In the case of a mill with lead and zinc (galena and sphalerite) flotation circuits, the common flotation reagents are soda ash, sodium cyanide, zinc sulfate, Xanthate (Z-11), and copper sulfate.

Sodium cyanide and zinc sulfate are added to the lead cleaners and to regrind ball mills in a ratio of about 1:3 to depress sphalerite while the galena is being floated. Soda ash is used to maintain a pH of approximately 7.2 in the lead rougher so that the sodium cyanide does not decompose, as it is unstable in an acid environment.

Lime slurry is added to the zinc circuit ahead of the zinc conditioner. Sufficient lime is added to maintain a pH of 10.6 in order to depress any pyrite present. As noted in Chapter 9, iron interferes with the electrolytic recovery of zinc and other metallurgical processes. Copper sulfate is added to activate or refloat the sphalerite, i.e., to offset the effect of depression by zinc sulfate and sodium cyanide. Xanthate (Z-11) is added to both the lead and zinc circuits as a collector. Methyl isobutyl carbinol (MIBC) is used as the frother throughout both circuits and is added to lead rougher feed pumps and to zinc conditioners.

Reagent consumption for two operations is shown in Table 10.1; however, it is clear that reagent consumption can vary greatly from one mill to another. For instance, the amount of sodium cyanide and zinc sulfate required to depress sphalerite during the flotation of lead is a function of the amount of copper present in the ore.

In one of these mills (A), all reagents except soda ash are added to the flotation circuits through flow waters via a centrally located reagent control panel. In other mills, reagent addition is accomplished with a great deal of human subjectivity and a great deal less automation.

The products of these processes are lead and zinc concentrates, which contain the by-products discussed in Chapter 9, and a slurry consisting of water, the gangue minerals, and the reagents added during the flotation process. All three of these wastes are discarded in a tailings pond, which can be constructed according to the methods described in Chapter 11.

Table 10.1 *Reagent Consumption in Lead and Zinc Flotation Operation (A) and Lead Operation Only (B)*

	Lead and zinc flotation operation		
Reagent	*Lead flotation (lb/ton, average)*	*Zinc flotation (lb/ton, average)*	*Total (lb/ton, average)*
Sodium cyanide	0.116	–	0.116
Zinc sulfate	0.245	–	0.245
Xanthate (Z-11)	0.040	0.074	0.114
MIBC	0.065	0.080	0.145
Copper sulfate	–	0.343	0.343
Quicklime	–	1.416	1.416
Separan	–	0.001	0.001
Soda ash	0.312	–	0.312

Lead operation only

	Pounds per ton of ore by month in 1972									
Reagent	*Jan.*	*Feb.*	*Mar.*	*April*	*May*	*June*	*July*	*Aug.*	*Sept.*	*Oct.*
Copper sulfate	0.81	0.99	1.07	0.99	1.02	1.06	1.09	0.95	0.86	0.72
Zinc sulfate	0.48	0.41	0.50	0.55	0.56	0.51	0.52	0.51	0.43	0.41
Xanthate	0.48	0.43	0.43	0.57	0.56	0.69	0.70	0.58	0.60	0.60
Sodium cyanide	0.18	0.20	0.20	0.19	0.19	0.19	0.21	0.20	0.19	0.20
Lime	0.50	0.42	0.66	1.18	1.05	1.11	1.22	1.19	1.17	1.09

WASTE CHARACTERISTICS AND DISPOSAL

In humid regions it may be impossible to recycle all the effluent from a mill waste tailings pond; thus the quality of the effluent becomes important with respect to Public Law 92-500. Selected characteristics of the waste slurries from a lead-zinc concentrator and a lead concentrator are shown in Table 10.2.

Solid Wastes

Suspended solids are the most critical pollutant in the effluent from concentrators and, in fact, the principal purpose of a tailings pond is to reduce the suspended solids load. Given space to construct a sufficiently large pond, in an area with minor winds, it has been demonstrated that this can be accomplished so as to achieve compliance with the interim guidelines discussed in Chapter 1. Tailings ponds must be designed with retention time and grain-size distribution of the tailings in mind, as outlined in Chapter 11. Use can be made of Stokes' law in the manner discussed in Chapter 3. In special cases, it may be necessary to design a pond with baffles so as to minimize the effect of wind on suspended solids removal. It is essential that the grain-size distribution-pond design relationship be established prior to the construction of the pond rather than afterward; otherwise guidelines ultimately will be violated. Wind direction is particularly critical with respect to the design and location of the decant system. Such preventive design measures are considerably less expensive than the alternative use of flocculating agents.

Ultimately, essentially all milling operations must cope with the common problem of production of a waste slurry of suspended solids. The solids differ among ore bodies with respect to grain-size distribution and mineralogy. If the gangue contains a high percentage of clay minerals, the equilibrium water content may approach 90 percent, as is the case with phosphate slimes. If the gangue contains sufficient pyrite, the runoff from the tailings pile can be expected to be acid, and corrective measures such as liming are in order. If, on the other hand, the gangue is composed of carbonates or silicates, this is not likely to be a problem. Lead-zinc tailings average about 0.3 percent lead and 0.3 percent zinc, which may or may not be soluble if low pH runoff develops. Consequently, although solid wastes are common to all concentrator operations, the technology for handling each case may vary somewhat depending on the characteristics of the specific waste. Thus, while the general rules discussed in Chapter 11 are applicable, it is essential that tailings be studied thoroughly before a disposal facility is designed and constructed.

Recent experience has revealed that unexpected subtle factors can cause problems with the disposal of tailings. Among the most publicized has been the presence of asbestiform minerals, which may cause cancer, in the gangue of certain ore bodies. This fact makes their disposal near water supplies inadvisable. Other factors of this type are likely to be found as research continues.

In the case of underground mines, the coarse fraction of the concentrator tailings is used for backfilling. Separation is achieved by cyclones. Backfilling accomplishes several objectives and, as noted in Chapter 11, modern research is revealing that a larger and larger percentage of the grain-size distribution can be backfilled advantageously. From a waste disposal point of view, backfilling is especially important where space limitations are severe, as, for example, in the Coeur d'Alene district of North Idaho.

Table 10.2 *Characteristics of Wastes from a Lead-Zinc Mill (A) and from a Lead Mill (B)*

Lead-zinc mill

Suspended solids (mg/l)	*pH*	*Electrical conductivity*	*Chloride (ppm)*	*F (ppm)*	*Cd (ppm)*	*Cu (ppm)*	*Mn (ppm)*	*Pb (ppm)*	*Sb (ppm)*	*Zn (ppm)*
277,206	10.05	919	4	2.4	<0.02	1.87	0.52	0.1	<5.0	0.2

Lead mill

Suspended solids	*pH*	*Cd (ppm)*	*Cu (ppm)*	*Mn (ppm)*	*Pb (ppm)*	*Zn (ppm)*	*Cn (ppm)*
41,563 mg/l	8.14	<0.02	1.88	1.07	<0.1	0.4	0.52

As increasingly lower grade deposits of all types of minerals are mined, the rate of production of solid wastes by concentrators must increase. It is abundantly clear that a sociologically and environmentally acceptable scheme must be devised for disposing of these solid wastes before any extractive operation is initiated or expanded. If these wastes are permitted to accumulate and constitute a source of dust and contaminated runoff, limited use areas will continue to be proposed wherever new potential ore bodies are discovered.

Liquid Wastes

Table 10.2 lists the concentrations of dissolved constituents in the slurry emanating from lead-zinc and lead concentrators. Some of the constituents, most notably the copper and cyanide, are attributable to the reagents listed in Table 10.1. A more complete summary of mill reagents and their toxicity is presented by Hawley (1972). If expensive treatment of the effluent from a tailings pond is to be avoided, it is essential that the use of flotation reagents be carefully controlled and limited to absolutely minimal amounts. Examination of the data in Table 10.1 for the lead concentrators producing the waste characterized in Table 10.2 reveals that one of the operations used almost three times as much copper sulfate per ton of ore as did the other. Since these reagents are apparently the source of copper and cyanide, and perhaps a partial source of zinc, the use rates are reflected in the effluent data in Table 10.2. Subsequent to the investigation which produced these data, the reagent use rate for the operation in Table 10.1 (B) was reduced and the effluent quality came into compliance with the waste discharge permit shown in Appendix D without further treatment.

Ultimately, intensified research should be directed at the development of flotation reagents which do not add to pollutant loads. Under present conditions, the amount of reagent required is in part a function of the grain size or grind of the ore. The need for fine grind is, in part, determined by the degree of galena and sphalerite "interlock" in the ore. The greater the interlock, the finer the grind needed for separation but, concomitantly, the greater the expense incurred for grinding. Therefore reagent use is an economic juggle with the cost of grinding. This being the case, research on alternatives to high-cost mechanical grinding may help solve the water pollution problem caused by the addition of reagents to suppress, float and reactivate galena and sphalerite. An example of one such possible process is the Snyder process developed at Lone Star Industries, Inc., Brookfield, Connecticut. It is apparently unique in that it splits particles along galena-sphalerite surfaces by a process similar to implosion.

Additional efforts that might be undertaken to minimize waste handling costs in the milling industry include optimal water use rate studies in the various portions of the milling and concentration process. Clearly, water treatment and handling costs require that only minimal amounts of water be utilized, regardless of the mineral extracted. However, the author has been unable to obtain data on optimal water use rates. Apparently such studies have not been published for milling processes.

Other pertinent information undoubtedly will be revealed by the forthcoming effluent guidelines development document for the ore milling industry, mentioned previously. The investigation on which the document will be based will cover the entire ore milling industry, including those ores which produce the minerals treated in Chapters 5 through 9.

REFERENCES

Hawley, John R. 1972. *The Use, Characteristics, and Toxicity of Mine-Mill Reagents in the Province of Ontario.* Special Projects Section, Mining Industrial Wastes Branch, Ontario Ministry of the Environment, West Toronto, Ontario, Canada.

Mink, L. L., Williams, R. E., and Wallace, A. T. 1972. Effect of Early Day Mining Operations on Present Day Water Quality. *Ground Water Journal*, Vol. 10, No. 1.

Williams, R. E., and Wallace, A. T. 1973. *The Role of Mine Tailings Ponds in Reducing the Discharge of Heavy Metals to the Environment.* Documents PB-224 730/AS and PB-224 731/AS, National Technical Information Service, U.S. Dept. of Commerce, Springfield, Virginia.

11. *Production and disposal of mineral resource solid wastes*

INTRODUCTION

Both air and water pollution control technologies produce solid wastes. Solid wastes per se are also produced by mineral resource industries.

Air pollution control frequently employs wet scrubbers to remove particulate matter and certain gases from gaseous emissions. The resulting waste waters are treated by some combination of pH adjustment, coagulation, aeration, thickening, and sedimentation. The final step almost inevitably involves a solid waste, usually a sludge in some combination with water. Even in those cases where dry air pollution control methodologies are employed, the final product is a solid waste which may or may not be recyclable. If it cannot be recycled, it must be discarded by some solid waste disposal technique such as landfilling or a tailings pond.

Water pollution control measures also are implemented for many purposes other than air pollution control. These processes normally produce solid wastes in the form of sludges of varying water content. The treatment and disposal of mill tailings in the form of slurries from which a portion of the water may or may not be removed constitute one of the most important examples of this type of process.

Solid wastes which are produced as solid wastes include such materials as gangue, slag from certain types of furnaces and from coke by-product plants, and overburden which must be removed and relocated during open pit mining.

Ultimately all solid wastes that cannot be recycled within a process, be it strip mining or sludge from waste water treatment, are discarded by accumulation on the earth's crust (mine rock, overburden, or tailings piles) or by the method known as landfilling. However, the form of landfilling or disposal on the surface can vary tremendously; and the consequences of either process can be acceptable or disastrous, depending on the procedure selected and the management practiced.

Much scientific work has been conducted and published on the subject of disposing of solid wastes by landfills, which may be used for sludge disposal. Much of this work was made possible by the Solid Waste Disposal Act of 1965 (Public Law 89-272). Among the more important contributions is that of Lambie (1969) which develops construction and use criteria for sanitary landfills. Hughes and Cartwright (1972) and Hughes, Landon and Farvolden (1971) also have made significant contributions to site evaluation and to the process of waste stabilization and prevention of ground-water pollution. The extensive literature already available for this field has led the author to treat it briefly here. But, as the production of sludges increases because of increasing waste water treatment by mineral resource industries, the need for landfill space will increase concomitantly.

This chapter presents discussions of the major categories of solid wastes that are peculiar to mineral resource industries. The general topic of site selection is covered separately as it applies to all types of solid wastes.

SITE SELECTION

Initial investigations of the feasibility of a mining, milling, or metallurgical facility must include considerations of the fate of the final waste product. Generally, this final product constitutes a solid waste. Historically, such investigations have been concerned primarily with locating one or more areas of acceptable size and proximity to the production site. More recently, however, environmental regulations, particularly Public Law 92-500, and greater concern for public safety have introduced new variables to the site selection process. These variables are influential in:

1. Protecting the quality of ground water from degradation by leachates emanating from and passing through waste piles.
2. Protecting surface water from silt loads and dissolved solid loads generated by erosion and corrosion of waste piles.
3. Prevention of eolian erosion with concomitant downgradient dust problems.
4. Protecting human life from catastrophic failure caused by floods or seismic events.

The additional variables that the site investigation must cover to guarantee this type of protection include topography, hydrogeologic environment, availability of appropriate construction materials, hydrology, seismic conditions, direction and velocity of prevailing winds, and frequency and intensity of precipitation events (atmospheric conditions). Clearly, proper evaluation of this number of variables requires an interdisciplinary approach to the selection of sites for disposal of large volumes of solid mineral resource wastes, even though some wastes will not require the evaluation of all variables. Under modern legal and social constraints, it is not possible for a single engineer or geologist to optimize the site selection process and assure preclusion of unanticipated, post-construction expenditures for corrective measures and legal defenses.

Succeeding paragraphs discuss the above variables in terms of information that an interdisciplinary team must generate prior to selection of a solid waste disposal site. It is assumed that the solid waste to be discarded at the site is subject to lixiviation, damage by wind and water erosion, and failure due to seismic events. The absence of any one of these characteristics, say capacity of eolian erosion, simply means that one variable can be omitted from the investigation. Similarly, if no suitable site is available to handle one of the variables, then transportation and/or design and construction costs will reflect that deficiency.

Topography

Examination of the topography of a potential solid waste disposal site is facilitated by on-site inspection, topographic maps, and aerial and satellite photographs. The latter are particularly useful for identifying potentially dangerous geomorphic features such as old landslide scars which may be reactivated if disturbed. Both maps and photographs are useful in the identification of flood or floodplain features which may indicate infrequent but potentially hazardous hydrologic events. In some areas topographic maps with superimposed geologic and other hazards are available. Topographic maps are also essential to the calculation of storage available at a potential site. Similarly, the natural topography of a potential site provides partial insight into the slope at which the implaced solid wastes can be expected to be stable under existing conditions of wind and water erosion. This observation is especially valid in rugged terrain.

Topography in combination with geology determines the most economical route for pipelines which frequently transport solid waste slurries to disposal sites. Topography in combination with meteorologic and hydrologic data determine the necessity for and route of stream channel diversions if disposal of solid wastes in arroyos and draws is unavoidable. Careful planning and design can turn topographically adverse conditions into safe solid waste disposal sites. Examples of the use of sites such as arroyos for mill tailings disposal are presented by Dahlberg and Laird (1973) and Dopson and McGregor (1973).

Topography and hydrology determine selection of optimum routes for roads for construction and maintenance of waste disposal facilities such as tailings ponds. The U.S. Environmental Protection Agency has published guidelines for the selection of routes for such roads. They have been adapted to mineral resource industries by Sceva (1973). Included among the guidelines are the following points pertinent to topography and, in some cases, to hydrology and geology:

1. Locate all roads so they avoid, or design roads to counteract, unstable soil areas. Select the topographic gradient that will provide for a stabilized road prism with proper drainage. Fit road locations to the topography so that minimum alterations of natural conditions will be necessary.
2. Locate roads on natural benches, ridge tops, and flatter slopes to minimize

disturbances of the terrain and to enhance the stability of the roads. Use all available topographic surveys, soil-type maps, and other soils and geologic information to select the route of minimal erosion.

3. Locate roads on stable areas well away from streams. Avoid routes through the bottoms of steep, narrow canyons; through slide areas; through steep, naturally dissected terrain; through slumps; through marshes or wet meadows; through ponds; or along natural drainage channels. Where alternative locations through stable areas are not available, incorporate corrective stabilization measures into the road design. Cross streams where channel and bank disturbance will be minimal. Avoid excessive sidehill cuts and fills, especially near stream channels. Plan for retaining walls or rip-rap where needed to increase the stability of fill embankments and cuts and to protect fill embankments from water erosion under conditions of adverse topography.
4. In critical situations, design for full bench construction rather than part bench and part compacted fill. Haul all excavated waste material to safe bench or cove locations.
5. In steep topography, reduce backslope sloughing by rounding the tops of cut slopes.
6. Design fills at an angle less than the angle of repose of the natural topography.
7. Clear trees and other vegetation for only the minimum essential width required for construction and maintenance and choose the design alignment and minimum area necessary to serve needs.
8. Alter topography to divert all drainage so that it does not pass over or collect in new cuts, fills, barrow areas, or waste dumps. Do this by providing bridges or adequate culverts at all natural water courses both for permanent and temporary roads. Minimize channel changes and stabilize the fill material of approaches with rip-rap or abutments.
9. Design culverts or bridges large enough to carry at least the 25-year frequency flood for the drainage in question. If runoff records are not available, use geomorphic evidence of previous runoff events for design calculations.
10. To reduce fish passage problems (where appropriate) and pipe abrasion, use bridges or "true-arch" (bottomless) culverts on steep slopes. In locations where bridges or bottomless culverts are impractical, use culverts with baffle plates to provide for fish passage through the culvert. Orient culverts with natural stream channels and extend them beyond the fill slopes. For anadromous fish passage, provide an entrance pool at least 3 feet deep and 12 feet long at the outflow end of the culvert. Stabilize the pool with barrier logs to prevent erosion. Place the logs so there is no impassable drop between the culvert and the stream.
11. Provide an adequate drainage system that will minimize both runoff concentration in the roadway and saturation of poorly drained soils. Use roadside ditches and culverts. Slope each relief culvert at a minimum grade of one percent. Use downspouts and other slope protection measures to prevent erosion of fill areas.

12. Provide dips, water bars, and cross drains to prevent eroding and gullying by fast-moving water. Put such structures close enough together so that collected water volume will be small and easily divertible off roads into safe spreading areas.
13. Design temporary roads to drain by outsloping wherever possible. On long slopes, space dips in the road to assure diversion of runoff from the road surface. For fills with culverts, place a dip at the downgrade approach so that, in the case of a flood or plugged culvert, excess water may flow over the road at that point. Use fords or other low-level crossings for small streams that have a high upstream debris hazard.
14. Prevent muddy and turbid waters from draining off the roadbed and into streams. Where necessary, make provisions to build up the surface of dirt roads with rock. Pave, or otherwise stabilize, road surfaces to minimize subgrade failures, road surface erosion, and road maintenance grading.

Clearly not all these guidelines will be applicable to all areas. The correct approach to their utilization consists of the selection of the appropriate techniques by the interdisciplinary team planning the waste disposal project.

Hydrogeology

The general geologic character of a potential solid waste disposal site should be among the first of the factors examined. Depth to and topography of buried bedrock should be determined by boreholes and geophysical methods for structural and drainage purposes. Klohn (1972) recommends that borings generally penetrate the subsoil profile to a depth of 1.5 times the width of the area to be loaded, if the discarded material is mill tailings and if bedrock is not encountered. Using this approach, the unconsolidated materials should be mapped and their grain-size distribution, permeability, and engineering properties determined so that potential settlement, leakage, and other failure problems can be anticipated and incorporated into the design of the disposal facility. Evidence of faults and landslides (including scars, vegetation, and springs) should be identified and mapped so that failure-prone areas can be avoided if accidental release of the material to be discarded constitutes a hazard to life or to the environment.

It is essential that the ground-water flow system be defined in the vicinity of waste disposal sites from which dissolved solids may be leached or otherwise released to ground water. Table 4.2 Illustrates the nature of water that can emanate from some tailings piles as seepage. Such water should not be allowed to enter a good-quality ground-water flow system. However, a tremendous amount of judgment must be exercised prior to construction if undesirable damage to a ground-water flow system is to be precluded. Hughes and Cartwright (1972) state that at present many rules and regulations developed to control disposal of waste in (or on) the ground are arbitrary and questionably derived rules of thumb. They conclude that others are based on outmoded hydrogeologic concepts; some are based on work done in one hydrogeologic environment, then transferred to another area where the hydrogeology is different. Hughes and Cartwright go on to state that "in humid climates nearly all subsurface [or surface] waste disposal results in some contamination; therefore, if such disposal is to be allowed at all, it is necessary to determine what is acceptable in each case. In so doing, an understanding of the fate of constituents released to the ground-water flow system is essential."

Data published by Mink, Williams, and Wallace (1972), Williams, Kealy, and Mink (1973), and Galbraith, Williams, and Siems (1972) show clearly that ions such as lead, zinc, and cadmium from mineral resource waste disposal sites can enter and travel with ground water under some hydrogeologic conditions. Mink, Williams, and Wallace, in fact, have shown that the source of the water recharging the ground-water flow system need not be the source of the contaminants in the ground water. Other surficial materials may contribute ions to the recharge waters by the process of lixiviation. These studies demonstrate the importance of understanding the ground-water flow system at a mineral resource waste disposal site prior to construction so that protective measures can be designed into the disposal facility.

The ground-water flow system in an area is a result of the fact that part of the precipitation falling on the ground surface infiltrates through the zone of aeration toward the zone of saturation, the top of which is the water table; the principal direction of movement in the zone of aeration is downward. Upon reaching the top of the zone of saturation, water enters the ground-water flow system and, although it is still driven by gravity, its movement is influenced also by a number of other factors. These factors include the geometry of the top of the zone of saturation, the ground-water gradient, the permeability of the earth materials and the effects of any pumping or artificial recharge of ground water in the area. A typical flow system is composed of small local systems superimposed upon larger systems (see Figure 11.1). According to

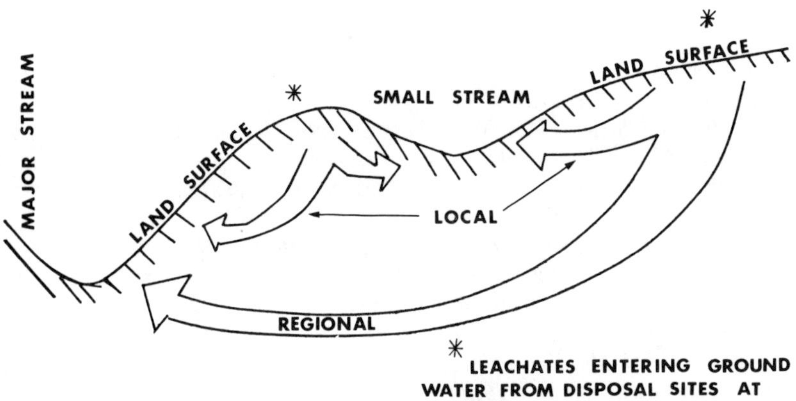

Figure 11.1. *Regional and local ground water flow systems.*

Hughes and Cartwright, local flow systems in Illinois, for example, range from a few yards in length to perhaps a quarter of a mile (0.4 km). In a local flow system, recharge occurs at the high areas on the local land surface. In larger, regional ground-water flow systems, recharge occurs from the topographically high areas of the region whereupon the water flows beneath one or more local systems and discharges into a major stream, lake, or river of the region.

Based on a three-year study, Hughes and Cartwright concluded that the major mechanisms involved in decreasing concentrations of dissolved solids in a leachate introduced into the subsurface are sorption, dilution, dispersion, and bacterial action. At and near its source, bacteria use any organic material present in a leachate for food. As the leachate continues to move in the subsurface along with the ground water, the concentration of the inorganic portion of some of its dissolved solids is reduced primarily by sorption on the earth materials and, to a lesser degree, by dilution and dispersion. Hughes and Cartwright point out that sorption of dissolved solids on the earth materials generally is considered a means of permanent disposal but that in fact, in many cases, it is not. Where all of the ion exchange positions of the clay materials in the matrix are occupied, sorption of new ions from a waste may merely result in the release of the previously sorbed ions, with little net reduction in the quantity of dissolved solids in the moving fluid. As these soils return towards their natural state, the sorbed ions from the effluent may be released and replaced by naturally occurring ions. Under these circumstances, sorption is only a means of extending the travel time of the ions that originated in the waste. However, although permanent disposal is not achieved, this process may increase the effective dilution in the receiving water sufficiently to make permanent disposal unnecessary, depending on the period of use of a site and the material in it.

Fractured rocks and permeable sands and gravels are inefficient in removing dissolved solids. In such materials, ground-water movement is also more rapid. At the other extreme, earth materials with a high percentage of clay-sized particles are the most efficient in attenuating dissolved solids from wastes. The surface area available for sorption is great, and ground-water movement through this type of material is slow.

The character of leachates from classical landfills is similar to the character of leachates of many mineral resource wastes except for organic constituents which are lower in the latter. (The solid wastes discarded in landfills are commonly called refuse.)

Hughes and Cartwright point out that the considerable data available on landfills elucidate the mechanisms involved in the production and migration of dissolved solids from landfills. Regulations set up to control ground-water pollution from landfills in California generally require that (1) wastes be placed above the water table, (2) wastes be placed in materials with a low permeability, and (3) landfills be a specified distance from the nearest point of water use (well or body of water) or aquifer.

Research leading to the requirement that refuse be placed above the zone of saturation was carried out in southern California (California State Water Pollution Control Board, 1954). The study concluded that sites in which refuse does not intersect the water table will not produce a leachate if the rainfall in the area is not adequate to infiltrate the refuse. (This reasoning, of course, would not apply to mineral resource wastes discarded as slurries.)

In humid areas, up to 40 percent or more of the annual precipitation may infiltrate the surface of a landfill if it is sufficiently permeable. This water will produce a leachate containing dissolved solids if the solid waste is capable of being leached. Where such water is available, placing the waste above the top of the zone of saturation will not protect the ground water unless an impermeable liner prevents the leachate from reaching ground water.

In humid areas, sites in which the zone of saturation is deepest are generally in upland recharge areas and in permeable materials. Contamination of ground water is difficult to control under these circumstances, in spite of the depth of the water table. Another problem with disposal above the water table is that leachate collection is difficult because it cannot be accomplished hydraulically. Impermeable liners must be utilized. An additional problem is that the depth of the top of the zone of saturation may be difficult to determine, particularly in materials with low permeability. Wells or piezometers finished in materials of low permeability may be installed for weeks or months before the water level stabilizes. The depth of the top of the zone of saturation in these materials generally is not in accord with water levels in nearby wells. The level of water in the wells reflects the fluid potential in the aquifer in which the well is completed, and this potential depends on the position of the well in the ground-water flow system, the type of materials present above the aquifer, and the amount of ground-water pumpage. Unless a well is shallow and the materials between the surface and the aquifer from which it produces are relatively permeable, it is unlikely that the water level in a well will coincide with that of the top of the zone of saturation.

Hughes and Cartwright point out that low-permeability materials are the safest environment for use as waste disposal sites. Nevertheless, placing wastes in these materials does not imply that the effect of infiltrating precipitation or slurry water can be ignored. If the underlying materials have low permeability, the waste may become saturated and, if the surface of the waste is higher than its surroundings, water will escape around the margin of the wastes in springs or seeps. Disposal of waste in permeable materials must be accompanied by protective measures to prevent leakage of leachates to ground water. Facilities for the collection and handling of leachate may be necessary for leachable wastes at some sites.

In areas where precipitation plus slurry water and collected runoff equals or exceeds potential evaporation, a waste disposal site such as a tailings pond may serve as an infiltration pit. As pointed out by Hughes and Cartwright, earth liners for lagoons are often considered impermeable; however, a well-compacted earth liner with a permeability of 10^{-6} cm/sec, which is realistically low, may permit approximately 450 gal (1,700 l) of leachate per day to leak into the earth materials underlying an area the size of a football field. It is possible, by using soil cements or other additives, to construct earth liners with permeabilities several orders of magnitude lower than this, and liners constructed of plastics or rubber are, for all practical purposes, impermeable.

Williams, Kealy and Mink describe methods of using mill tailings to minimize leakage from lagoons. However, seepage is expensive and difficult to eliminate completely and it is clearly advisable to select sites where hydrogeologic conditions minimize the potential for damage.

The most appropriate rules for site selection are:

1. Select the site underlain by maximum possible thickness of unsaturated material.
2. Select the site underlain by material with lowest possible permeability.
3. Select the site not underlain by shallow aquifers without confining layers.
4. Avoid sites that are adjacent to streams where ground water and the contaminated seepage or leachate which it collects will discharge immediately to the stream.
5. If high and dry low-permeability sites are not available, use extreme caution in designing to minimize seepage. Cutoff zones beneath dams, "impermeable" liners, and segregation of mill tailings are examples of such design parameters.
6. Provide seepage collection facilities where seepage is anticipated to be poor in quality and high in volume or water is needed for industrial purposes.

Construction Materials

Some solid waste disposal facilities require utilization of large volumes of materials for structures such as dams. It is critical that these materials be selected carefully if failures such as that at Buffalo Creek, West Virginia, are to be precluded (see Wahler and Associates 1973). It is not advisable to use less than optimal materials simply because they are available. Thus, the site selection process should include weighing the cost of transport of appropriate construction materials to the site, whether the materials be mine rock or naturally occurring materials.

Exploratory procedures for construction materials consist of pitting, trenching, and drilling. The exploration program should identify alternative sites as to quantities of the various classes of construction materials available. The materials at the alternative sites should be classed according to volume available, water content, grain-size distribution, compaction characteristics, and shear strengths. Availability of materials with the standard appropriate engineering properties should be ranked with the other variables in the site selection process. If waste materials from a mining operation are to be utilized, they should also be subjected to standard engineering analyses. They should be considered competitively with naturally occurring materials only if their physical properties are appropriate. Assuming that all other factors are equal, the process of choosing among competing materials has been summarized by Wahler and Associates (1970) as follows: Construct the embankment of the coarsest, strongest, least compressible and most permeable material available. These should be placed in proper zones at proper densities. Drainage blankets, cores, and filter beds should be utilized as needed and the slimes should be kept as far from the outer embankment as possible.

Hydrology

Hydrologic studies consist primarily of the acquisition of data on precipitation and runoff. These data are often available in U.S. Weather Bureau records and in U.S. Geological Survey publications, respectively. However, runoff records often will not be available and estimates of stream flow rates will have to be made on the basis of channel size and gradient, on geomorphological features such as flood plain size and gradient, and on nearby records if available. Gauging stations should be installed as early in the process as possible. In lieu of these procedures the U.S. Bureau of Reclamation (1965) presents a simplified method of estimating magnitudes of floods. Additional methods are presented by Chow (1964).

Runoff estimates or data affect the selection and design of mineral resource solid waste disposal sites in two ways. First, they indicate the rate at which non-process water can be expected to enter the site. The ideal site collects as little water as possible in addition to direct precipitation. All other things being ideal, the runoff onto or through a selected site should be zero. Since this situation is usually impossible to achieve, the runoff data also indicate the flow volumes and peak flows the site can be expected to experience. Thus, the runoff data, in combination with the direct precipitation data and the rate at which waste water (if any) enters the site along with the solid waste, determine many of the design characteristics of the waste disposal site. These data in combination determine:

1. The amount of freeboard that must be maintained on a dam if the waste disposal site is in a draw.
2. The size of the diversion system required to handle peak flows if they are so great that the dam must be bypassed, or if the quality of the water behind the dam is so poor that it cannot be allowed to enter a stream.
3. Design of the spillway (where a spillway is included in construction plans).
4. Design of the dam itself, especially if the waste is emplaced in a draw as opposed to a gently sloping sidehill or horizontal plateau or large valley bottom.

Kealy (1973) recommends designing for a peak flow equivalent to a 50-year flood and, if possible, for a 100-year flood.

Seismic Conditions

Some solid waste disposal structures must be designed so they will withstand seismic events. Such structures ordinarily enclose fine-grained solid wastes which contain sufficient water to produce low shear strength and a concomitantly low factor of safety unless properly impounded. If such wastes are disposed in earthquake-prone areas, the design of the structures should reflect these conditions.

Blume (1973) delineates zones that are prone to earthquakes of specific intensities on a worldwide basis. He also delineates those localities in the

United States where seismic investigations are being conducted as a consequence of potential nuclear power plant construction.* Insight into the probability of damaging earthquakes in any specific locale can be gained from data supplied by various institutions such as the National Oceanographic and Atmospheric Administration (NOAA) and by geologic evidence such as faults, offset ridges, and linear features interpreted from aerial photographs and satellite data.

Design of mineral resource waste disposal structures (such as tailings dams) must utilize estimates of several aspects of earthquake phenomena in addition to intensity. These include anticipated epicentral distance, focal depth amplitude, period of particle motion, and peak ground acceleration. With these data, one can arrive at joint probabilities for various levels of motion at a specific site and design accordingly, taking into account the foundation material present at each alternative site. Design procedures for liquid-solid waste disposal structures are a subject in themselves and beyond the scope of this book.

For purposes of this work, let it suffice to say that the site selection and evaluation process should include an investigation of the potential for seismic activity and of its probable location so that alternative sites can be compared and appropriate design procedures incorporated in the plan.

Meteorological Conditions

Wind direction and velocity and precipitation and its intensity are the major meteorological data needed to evaluate potential mineral resource solid waste disposal sites. Wind direction and velocity are critical because of dust problems which can be caused by eolian erosion and because persistent high-velocity winds can complicate the task of revegetating abandoned waste piles.

As noted previously, precipitation data for most localities can be obtained from the U.S. Weather Bureau. However, data on wind velocity and direction generally must be obtained at individual locations because conditions vary greatly with locale. Data gathering stations should be established as soon as potential sites are identified.

Precipitation data are collected for three principal reasons. From these data:

1. Waste disposal structures can be designed to handle anticipated direct precipitation and diversions can be incorporated into the project as needed.
2. Erosion control measures such as slope angle and their costs can be predicted during the site evaluation process.
3. A revegetation program can be planned for each alternative site and costs compared among sites before a specific site is selected.

Wind direction and velocity data are needed so that:

1. Sites upgradient from developments such as towns, recreation areas and farm lands can be avoided if there is to be dust at the waste disposal site.

* Information on such studies and on other aspects of seismic events can be obtained from the National Earthquake Information Center, National Oceanographic and Atmospheric Administration, National Ocean Survey, Rockville, Maryland.

2. Areas with unusually high winds can be avoided in order to minimize eolian erosion. If high-wind sites are utilized, revegetation programs are likely to meet with failure.

Selection of sites with high-intensity precipitation and/or high-velocity winds will usually reflect these conditions in the form of high costs for wind barriers and erosion control devices. Thus, ideally, sites with minimum winds and lowest intensity precipitation should be favored.

TAILINGS DISPOSAL

Introduction

Tailings are the solid waste products of milling and possible subsequent processing of ore; however, in some mineral resource industries tailings have some other name. Notable examples are the "slimes" produced by the phosphate industry (see Chapter 6) and the "red muds" of the aluminum refining industry (see Chapter 5). Considerable differences exist among tailings from different industries but considerable similarities exist also. Tailings may range

Although tailings may vary from acidic to basic, all are transported to their final destination as slurry. Drop boxes like the ones in this photograph are used to keep grades below one percent and minimize pipe abrasion.

U.S Bureau of Mines photo

from acidic to basic but all are transported to their final resting place as a slurry. Drop boxes are used ubiquitously to keep grades below one percent to minimize pipe abrasion. Final water content ranges from approximately 15 percent for oil shale tailings to as much as 85 percent for the red muds of the aluminum industry, but all contain some water at equilibrium with gravity. Some contain a percentage of material with respectable shear strength but most contain a percentage of material that is essentially liquid. All contain some chemically inert material but many contain sufficient pyrite to produce the equivalent of an acid mine drainage problem upon exposure to the atmosphere. (See Chapter 4).

The most significant characteristic and difference among tailings is grain-size distribution (see Figure 3.1). This characteristic is most significant because it determines the method that must be used for safe containment if the tailings are discarded terrestrially. If the tailings contain a sufficient percentage of coarse materials (usually called "sands"), the tailings can be segregated during disposal and the coarse sands used as a confining embankment for the finer materials that will not drain and have negligible shear strength. The coarse materials can be separated by gravity in a method called "upstream construction" or they can be segregated by centrifuging ("cycloning") in the method called downstream construction.

If the tailings do not contain sufficient sands to form a well-drained embankment of acceptable density, permeability, shear strength, and factor of safety, or if the sands are used to backfill underground openings, then the embankment must be constructed with imported materials, which may be either natural rock or mine waste rock. Wahler and Associates (1974) conclude that the average gradation distribution of tailings from a 60,000 ton-per-day open pit copper mine and mill in Arizona shows that approximately 52 percent of the tailings are finer than 0.074 millimeters (200 sieve) and that 11 percent average finer than 2 microns. Similarly, Figure 3.2 shows that 50 percent of the tailings from a lead-zinc mill in northern Washington average finer than 0.07 millimeters. Wahler and Associates (1973) conclude that the engineering properties of such tailings are similar to any silt. Thus, it is essential that the minor sandy portion of tailings be segregated and restricted to the embankment area, regardless of the disposal method utilized.

General Design Considerations

The foundation requirements for all three types of tailings embankments (also called "dams" or "dikes") are identical to those for any other kind of dam or embankment. The foundation material must have sufficient strength to support the weight of the structure without shear failure or movement under the load. The latter factor is particularly significant to disposal on steep slopes or steep draws because the weight of the embankment plus the effect of the water provided by the slurry may produce landslide conditions. This is particularly true if the area is subject to seismic events. Design must include consideration of the engineering properties of the foundation material and the probability of seismic events at alternative sites.

The foundation materials should also be sufficiently impermeable to prevent leakage from the tailings pond to the ground-water flow system beneath the pond and embankment and to springs alongside the embankment. If such materials are not present, the pond should be lined (see Miners 1973 for an example of control of seepage with liners), or the foundation material should be treated chemically, or the tailings should be carefully segregated and the water level maintained over the slime zone in the manner described by Williams, Kealy, and Mink (1973) and Webb and Smith (1971). Cutoff walls and blankets and cores may be necessary in order to prevent leakage through and beneath the embankment if existing materials are not appropriate. These concepts are illustrated in Figure 11.2.

The slope of the embankment must be determined by the climatic conditions as revealed by the site selection investigation and by the physical properties of the material used in the embankment (angle of internal friction, density, permeability, shear strength, angle of repose). Model studies are helpful in determining stable slope angles once these variables have been identified. The large variety of climatic conditions and materials common to

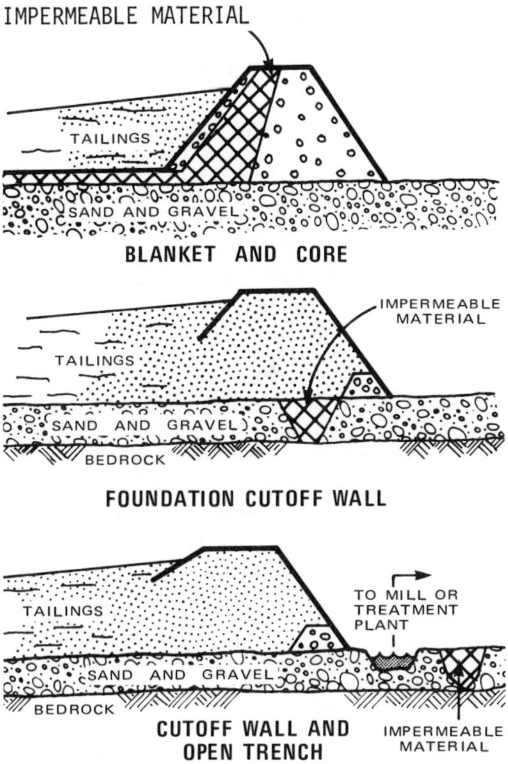

Figure 11.2. *Methods used to minimize seepage outflow (after Swaisgood and Toland 1973).*

Table 11.1 *Suggested Minimum Design Safety Factors*

Parameters on which Design is Based	*Case 1* [a]	*Case 2* [b]
1. Design based on peak shear strength parameters.	1.5	1.3
2. Design based on residual shear strength parameters.	1.3	1.2
3. Analysis that includes predicted return period accelerations applied to the potential failure mass.	1.2	1.1
4. For horizontal sliding on the base of embankments retaining tailings in seismic areas assuming shear strength of the tailings reduced to zero	1.3	1.3

Source: After Canadian Department of Mines, Energy and Resources, Mines Bureau.

Note: See Blume 1973 for seismic considerations.

a. Where persons or property would be damaged by failure.
b. Where persons or property would not be damaged by failure.

mineral resource solid waste disposal sites make general rules difficult. However some are useful. The Mines Branch of the Canadian Department of Energy Mines and Resources (1972) recommends the factors of safety itemized in Table 11.1. The factor of safety is obtainable through standard engineering procedures once the slope, physical properties, and the configuration of the phreatic line are known (see Kealy and Busch 1971 and Kealy and Williams 1971).

Toland (1971) recommends the factors of safety shown in Table 11.2 as guidelines for design of slopes for tailings pond embankments, provided other governmental requirements do not exist.

The phreatic line must be controlled within tailings pond embankments so that a reasonable factor of safety can be achieved. The influence of the phreatic line on factor of safety is illustrated by Figure 11.3. As indicated in

Table 11.2 *Guidelines for Design of Tailings Pond Embankments*

1. Starter dams and continuing embankments that have water adjacent to the embankments will have a safety factor of 1.50 or greater.
2. Embankments that will have flat upstream tailings slopes and maintain ponded water a minimum of 200 feet from the upstream crest will have a safety factor of 1.20 or greater.
3. Dynamic loading resulting from the highest predictable earthquake in the area must be evaluated and will have a design safety factor greater than 1.00.
4. Density standards to prevent liquefaction under the design earthquake conditions must be maintained.

Source: After Toland 1971.

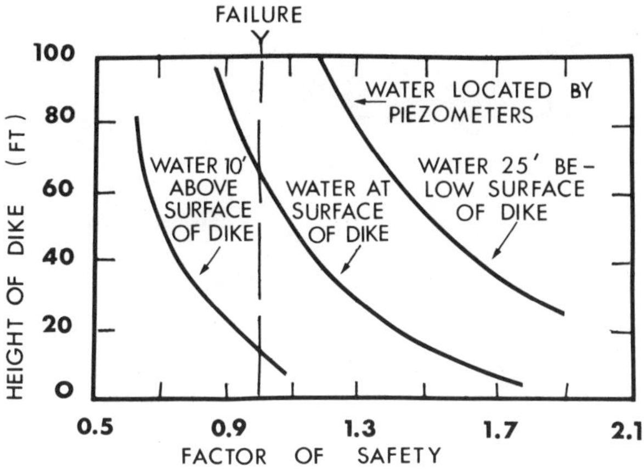

Figure 11.3. *Relationship of factor of safety to location of phreatic line (after Kealy and Busch 1971).*

the figure, the lower the phreatic line, the higher the factor of safety, all other factors being equal. The phreatic line can be lowered by utilization of internal drains, including well points, if necessary, or by the utilization of free draining materials for the embankment. The effect of drains is illustrated in Figure 11.4. Methods of determining appropriate spacing of drains can be obtained from Cedergren (1967).

Density of tailings is a major factor in the analysis for factor of safety of a tailings embankment. If tailings density is too low, other factors cannot compensate for it in the factor of safety analysis and the deposit may liquefy if disturbed. Guerra (1973) concludes that in situ relative densities of $\geqslant 60$ percent are marginally safe against liquefaction. If model studies predict this to be the case, then either external materials must be brought in for embankment construction or, in some cases, cycloned tailings can be compacted and used in the embankment, provided the moisture content is not too far from optimal. Brawner and Campbell (1973) suggest that cohesionless sands (tailings are cohesionless) should be compacted in 12-inch layers with vibrating drum or pneumatic rollers and that cohesive clays and silts be compacted in 9-inch layers using grid or sheep's foot rollers. Guerra reports that tailings have been used as backfill material after compaction with vibratory rollers in 12-inch layers to 95 percent of maximum modified Proctor at a minimum moisture content of 4 percent.

The published literature contains considerable information on the density and other physical properties of tailings from different types of mills. Guerra provides considerable data for the iron industry; Pettibone and Kealy (1971) provide data on the lead-zinc industry and Wahler and Associates (1974) provide data on tailings from the copper industry. These sources can serve as a basis for design of tailings structures if there is no method of predicting what the mill "not-yet-built" will produce.

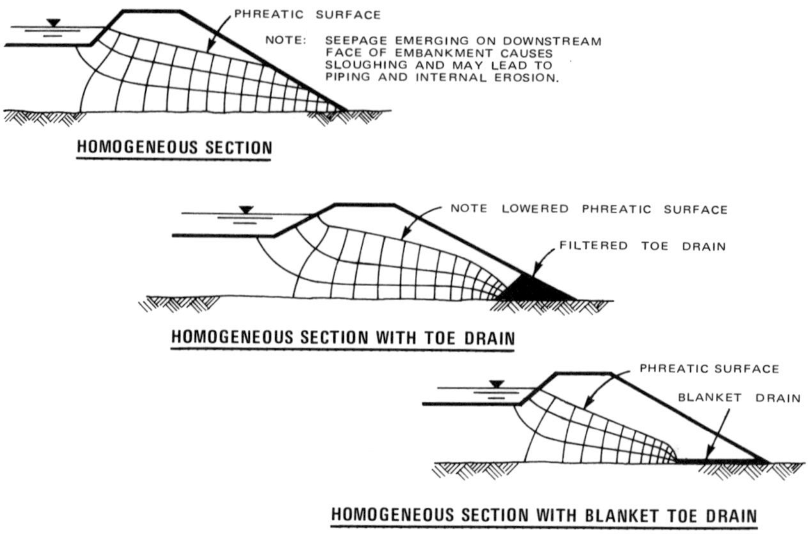

Figure 11.4. *Effects of drains on phreatic line (after Canadian Department of Energy, Mines and Resources-Mines Branch 1971).*

Tailings ponds ordinarily are drained by decant towers, siphons, or pumps. But regardless of the normal method of removing tailings pond water, great care must be taken to design sufficient freeboard to handle peak runoff, particularly if tailings embankments are to be located in draws. Alternatively, diversion structures may be constructed to handle runoff as predicted by the hydrologic analysis conducted during the site selection process. Diversion structures should also be employed if the quality of the water within the pond is sufficiently poor that it cannot be discharged to a stream. Considerable insight into the production of such water can be obtained from Williams (1973) and from Mink, Williams, Wallace, and McNay (1973). Spillways or diversion structures must be adequate to handle peak flows because overtopping is likely to cause failure of the embankment with subsequent disastrous results downgradient. Kealy (1973) recommends designing for a 50-year flood and, if possible, for a 100-year flood.

The design of all embankments should include monitoring devices to ascertain that the structure performs adequately. Instrumentation should include piezometers to monitor the configuration of the phreatic line and subembankment pressures as well as quantity and quality of leakage. Slope indicators should be installed to ascertain that creep does not occur and to detect it before failure if it does occur.

All mineral resource solid waste disposal sites, including tailings piles, should be selected and designed with a plan for abandonment. As will be shown in Chapter 12, revegetation is the most generally accepted, and generally the most permanent and aesthetically pleasing, method of stabilization upon abandonment. If revegetation is not initiated and nurtured until the cover is self sustaining, mineral resource solid waste disposal sites will constitute areas of aeolian and aquatic erosion until the revegetation process occurs naturally. This period may tend toward centuries.

Upstream Construction

The initial step in the construction of an embankment by the upstream method is the raising of a starter dam composed of permeable materials that can also serve as a toe drain. This dike should be accompanied by facilities to collect the drainage and return it to the pond, especially if the drainage does not comply with the waste water discharge permits required by PL92-500 (see Appendix D). The size of materials for such drains (or any other drain) should be coarse enough to permit drainage but fine enough to prevent passage of tailings. If the starter dike (toe drain) is too coarse, a filter blanket should be emplaced behind it. The Mines Branch of the Canadian Department of Energy, Mines, and Resources (1972) suggests the following relationships between filter and protected soil:

1. $\dfrac{\text{15\% size of filter}}{\text{85\% size of protected soil}}$ should be $\leqslant 5$

2. $\dfrac{\text{50\% size of filter}}{\text{50\% size of protected soil}}$ should be $\leqslant 25$

3. The filter materials should be smoothly graded; gap graded materials should be avoided.

4. $\dfrac{\text{15\% size of filter}}{\text{15\% size of protected soil}}$ should be $\geqslant 5$

5. The filter should not contain more than 5% of particles, by weight, finer than the No. 200 sieve, and the fines should be cohesionless.

6. The coefficient of uniformity of the filter should be $\leqslant 20$.

Once the starter dam, appropriate drains, filters, and instrumentation have been installed, the remainder of the dam is constructed by spigotting the tailings in an upstream direction off the starter dam. Spigots should be spaced sufficiently close together that the delta of the spigots coalesce to form a linear sand beach with no slime-filled channels extending back toward the embankment between spigots. Dopson and McGregor (1973) report success with a spigot spacing of 26 feet. If spigots are too far apart, part of the embankment will be composed of low-shear-strength slimes. The outermost limit (toward the pond from the embankment) of the sand portion of the delta should be noted at the end of the first level of spigotting, and the inner portion of the embankment at its maximum height should be limited to this location. Otherwise the upper portion of the embankment will rest on slimes and a low factor of safety will

A cyclone is being used here during upstream construction of a tailings dam to segregate course tailings and divert them to the embankment's inner face. Tailings are spigotted off a peripheral discharge line.

American Smelting and Refining Company photo

result. This line will be a function of the grain-size distribution of the tailings, the original slope of the topography, and the method used to raise dikes. Wahler and Associates (1974) conclude that, at one major upstream disposal project investigated in detail, the percent of fines in the deposited tailings increased at a rate of about 0.1 percent per linear foot from the embankment out to a distance of 250 to 350 feet toward the center of the pond where a discontinuity occurred. Beyond this discontinuity, the relationship of increasing fines versus distance from the embankment was re-established. At a distance of about 450 feet from the embankment, the materials ran about 80 percent finer than the 200 sieve. Noteworthy, however, is the fact that about 30 percent of the tailings were finer than the 200 sieve at the point of discharge (the embankment). Shear strength and density values were compatible with this grain-size distribution.

It is not possible to make generalized recommendations on maximum embankment height and width on the basis of these data alone. If additional data are obtained and compiled, such generalizations could be made. For the present, these data should be collected as tailings ponds are constructed so that the maximum embankment height and width can be determined for each tailings disposal operation individually. If this is not done, tailings embankments may be extended to heights that would permit them to extend out over the slime zone, which would produce an unsatisfactory factor of safety.

The embankment normally is raised by dragging or dozing up material from the previous lift and a new spigotting level is located. The process is repeated until the maximum height is reached. The final product is shown in Figure 11.5.

Dredging and dozing should be conducted in such a way that deep trenches are not left along the inside edges of the embankments. Such trenches are subject to filling with slimes, which may result in an unstable embankment.

Wahler and Associates (1970) report that embankment slopes varying from 1.5 to 1 to 2.5 to 1 are commonly achieved with draglines. Observations by the author indicate that slopes on many existing tailings embankments are too steep to withstand erosion under the meteorological conditions present in the area. It is critical that this factor be taken into account during the site selection and design process. Otherwise embankments may become unstable in the long run and downstream siltation problems will be unavoidable.

Williams, Kealy, and Mink (1973) and Kealy and Busch (1971) have examined the upstream method in some detail. One pond investigated is located in northern Washington. It is a lead-zinc mill tailings pond in which the grain-size distribution is that shown in Figure 3.2. The pond has a peripheral discharge system and the water is decanted from the clean portion of the free-water area.

Permeability tests of the pond revealed a lower permeability at depth, as well as a lateral variation in permeability. The permeability distribution considered as being most representative is shown in Figure 11.6. Piezometers were installed so that field potential values and the configuration of the free-water surface could be measured.

The permeability distribution and boundary conditions shown in Figure 11.6 were used as a basis for the construction of a finite-element mathematical model as shown in Figure 11.7. Details of the application of the mathematical

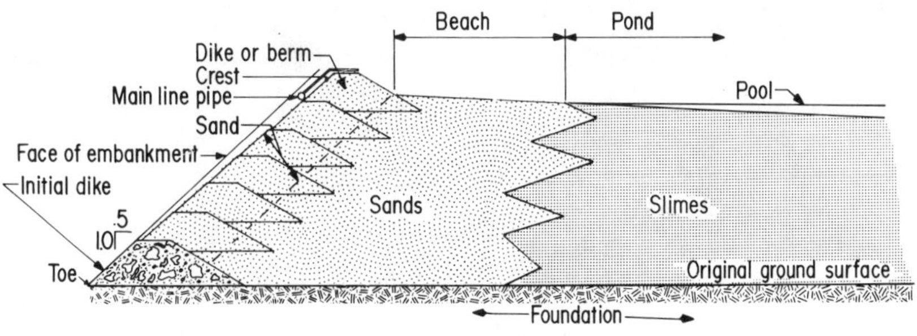

Figure 11.5. *Upstream construction of tailings pond embankment (after Mink, Williams, Wallace, and McNay 1973).*

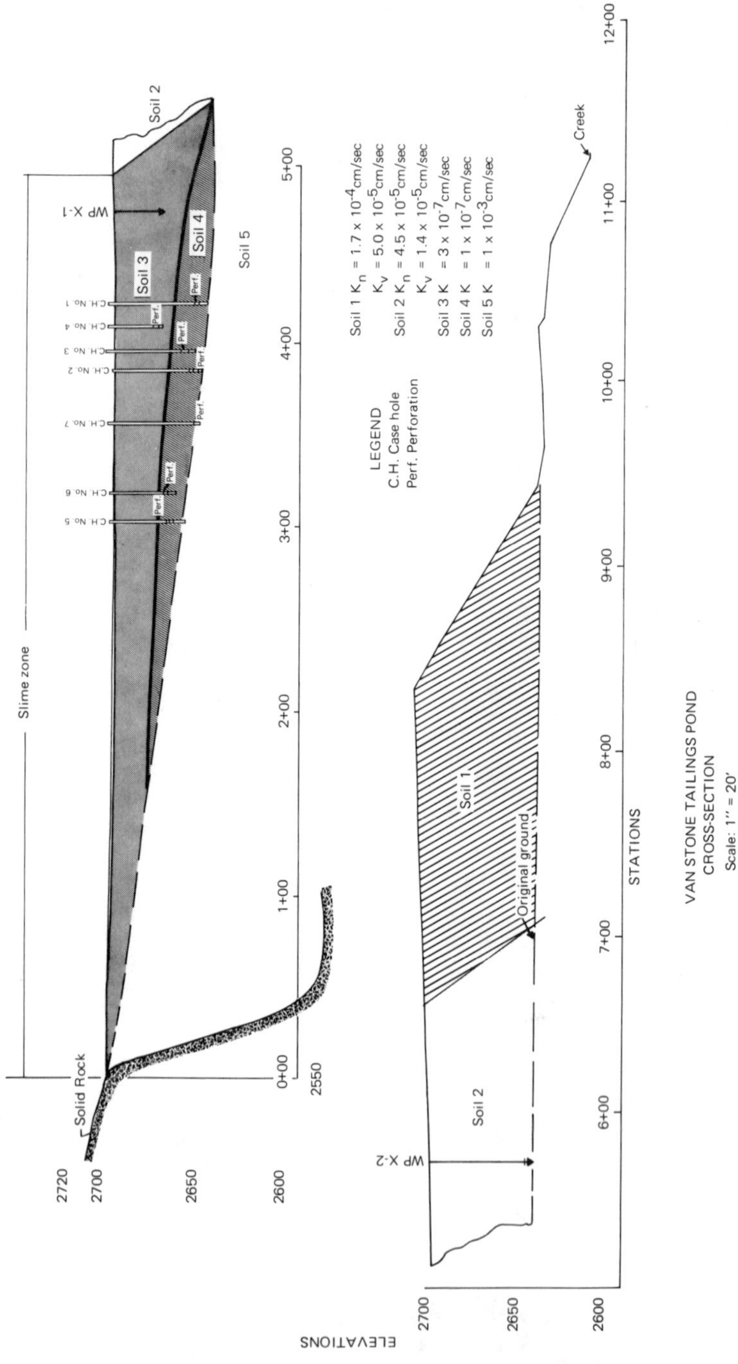

Figure 11.6. *Cross section of tailings pond modeled, showing permeability distribution considered most representative.*

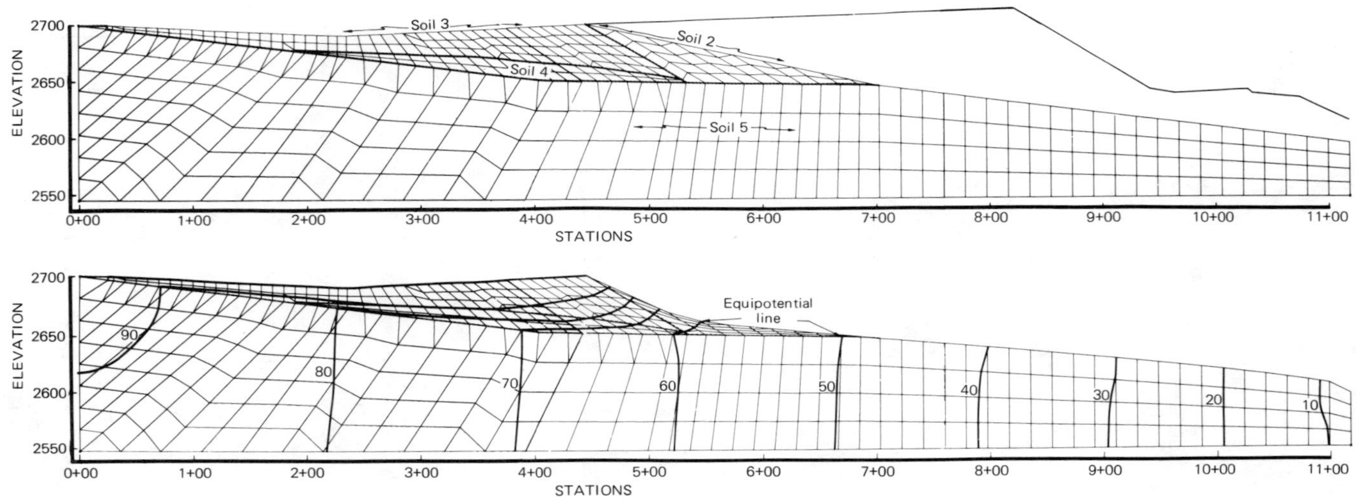

Figure 11.7. *Finite-element mathematical model of tailings pond embankments, showing permeability distributioon (top) and resulting fluid potential distribution for the system (bottom).*

A dragline is used here to raise the embankment height during upstream construction of a starter dam. Until more data are obtained, maximum embankment height and width are determined individually for each tailings operation. *U.S. Bureau of Mines photo*

model to problems such as this are presented by Taylor and Brown (1967), by Kealy and Williams (1971), and by Kealy and Busch (1971). The objective of the application of the model to this particular tailings pond was to examine the structural and environmental characteristics of the pond and to investigate ways of improving the design of this and other tailings ponds when necessary. The fluid potential distribution for the system is shown in Figure 11.7. The discharge from the bottom of the two-dimensional model (unit width in the third dimension) is portrayed in Figure 11.8.

Manipulation of the free-water body in the finite-element mathematical model of the tailings pond revealed that the most significant ground-water outflow beneath the pond emanates from the natural ground-tailings contact and from the soil 3 (slimes)-soil 2 interface (see Figure 11.9). Therefore, by controlling the pond water level, the leakage along the soil 3-soil 2 contact could be minimized, if not eliminated. In addition, proper preparation of the

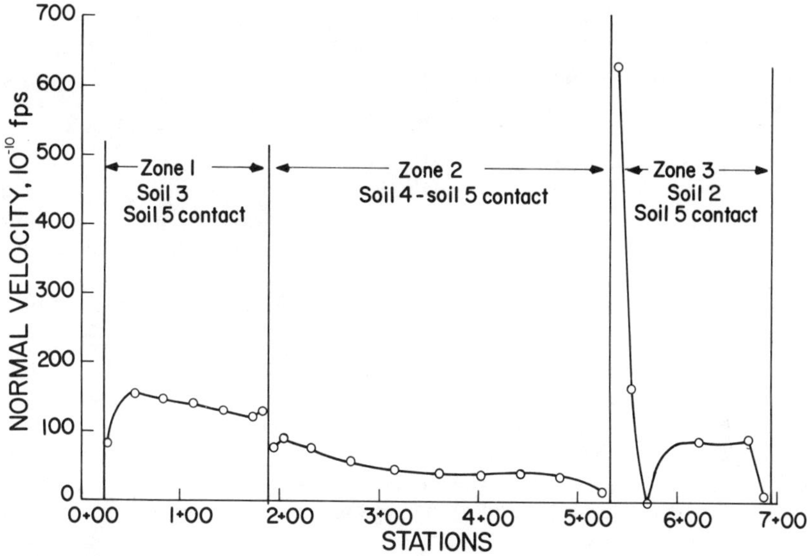

Figure 11.8. *Velocity profile for pond losses as predicted by mathematical model. Station numbers are in hundreds of feet.*

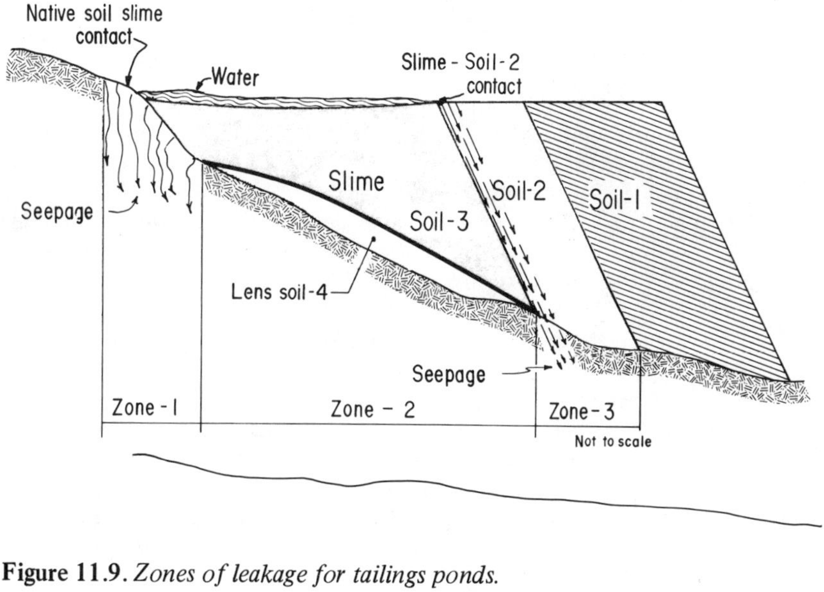

Figure 11.9. *Zones of leakage for tailings ponds.*

natural ground-pond contact would eliminate leakage along this contact. If the flows from each of the elements at the bottom of the slime zone (soils 3 and 4) are summed and deletions are made for the soil 3-soil 5 contact and for the soil 3-soil 2 interface, and if the unit width of the two-dimensional cross section is extended to cover one acre, then the daily rate of leakage from the bottom of the pond, as predicted by the model, is only 110 gallons per acre. However, such a calculation assumes flawless operation and management of the pond. Nevertheless, the transmission from the pond of water or water containing detrimental substances, in theory at least, can be minimized using tailings. It is critical that this condition results only if the pond water level and permeability (slime distribution) at the two contact areas are carefully controlled. Control can be maintained with a peripheral discharge system and with a decant system managed to maintain the free-water surface essentially over the slime zone. A system so designed is also more stable structurally because the phreatic line in the embankment is near the base of the embankment beneath soils 1 and 2, thereby providing for a maximum factor of safety. The relation between the phreatic line as delineated by the mathematical model and slope stability is presented by Kealy and Williams (1971) and summarized in Figure 11.3. In this case it is slightly above 1.

The field piezometer water levels were in close agreement to pressures produced by the mathematical model, thus confirming that a pond "yet-to-be-built" can be designed by use of models and a knowledge of the characteristics of the tailings and information supplied by the site evaluation.

This study indicates that upstream construction can be stable and minimize leakage, provided the tailings have the proper grain-size distribution and provided that the height of the embankment is limited so that it does not extend inward over the slime zone. Maximum success will be realized by:

1. Employing a closely spaced peripheral discharge system to inject the tailings into the tailings pond.
2. Ensuring that the tailings supply contains a coarse fraction sufficient to provide an outer embankment with a permeability three to four orders of magnitude greater than that of the inner slime zone (approximately 10^{-3} vs 10^{-7} cm per sec, respectively). This permeability distribution will result in a phreatic line that will provide acceptable slope stability. If mine backfilling utilizes too much of the coarse fraction, the outer portion of the tailings pond embankment may have to be constructed of borrow material.
3. Designing the decant system in such a way that the free-water body over the slime (low-permeability) zone is maintained so that leakage through the outer coarser portion of the embankment is prevented.
4. Lowering the permeability of the natural ground-tailings contact with chemicals or by other means, minimizing early leakage along this contact.
5. Designing the tailings pond free-water area so that sufficient retention time for removal of all suspended solids is provided throughout the life of the pond. Settling velocities can be predicted from knowledge of the tailings to be produced. The addition of flocculating agents, and possibly pH control agents, may be required at this stage.

6. Ensuring that no wastes other than mill wastes (or similar wastes) enter the pond. In particular, waste waters with low pH and high concentrations of heavy metals should not be allowed to enter the pond unless further treatment of the effluent and seepage is anticipated.
7. Establishing a revegetation program on the slopes as early as possible. Tailings are highly erodable and stream siltation problems are likely to develop if slopes are not revegetated quickly.

Failure to employ steps 3 and 4 will result in unnecessarily high leakage rates.

Downstream Construction

As the name implies, the embankment of a tailings pond constructed by the downstream method rises off the starter dam in the downstream direction (Figure 11.10). Thus there is little concern about slimes underlying the embankment provided the tailings have sufficient coarse material to permit the use of the method. Most of the downstream construction techniques use some variation of a cyclone to segregate the coarse tailings and divert them to the outer face of the embankment. Several methods are illustrated in Figures 11.11 through 11.13. Klohn (1972) lists the major advantages of the downstream method as:

1. None of the embankment is built on previously deposited, loose, fine tailings.
2. Placement and compaction control can be exercised as required over the fill operation.
3. Underdrainage systems can be installed, as required, as the dam is built. The underdrainage permits control of the line of saturation through the dam and, hence, increases its stability.
4. The dam can be designed, and subsequently constructed, to whatever degree of competency that may be required, including resistance to earthquake forces.

Klohn (1972) also points out that the major disadvantage of all methods of downstream dam building is the large volume of sand required to raise the dam. In the early stages of operation, it may not be possible to produce sufficient volumes of sand to maintain the crest of the tailings dam above the rising pond levels. If this is the case, then either a higher starter dam is required or the sand supply must be augmented with borrow fill. Both procedures add to the cost of the initial tailings facility. Insufficient sand would, of course, preclude the use of the upstream method also.

In Kohn's opinion, downstream construction is the only procedure that permits design and construction of tailings dams to conventionally acceptable engineering standards (presumably excluding imported materials). Details of design and construction can vary widely to suit the requirements of any project and produce an acceptable end product at a minimum cost. According to Klohn, all tailings dams located in seismic areas—and all major tailings dams,

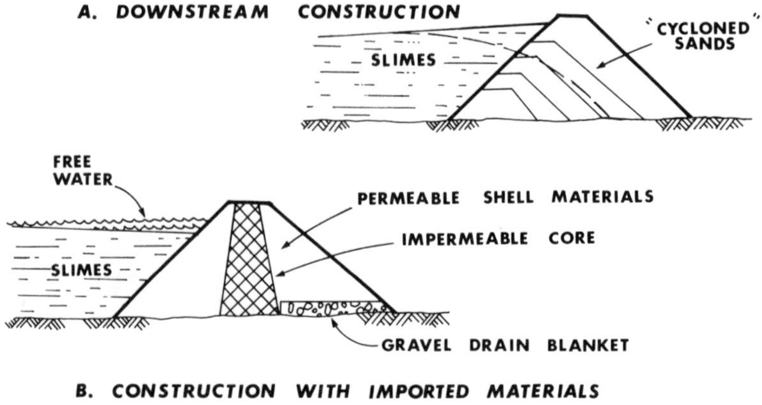

Figure 11.10. *Two methods of tailings embankment construction (after Swaisgood and Toland 1973).*

regardless of their location–should be constructed using some form of the downstream method. It is clear that these comments should be qualified according to the height of the embankment, the grain-size distribution of the tailings, and the slope of the original ground. These factors determine whether the upper portion of the embankment in upstream construction will extend out over the slime zone, which would create an unstable condition.

An embankment constructed by the downstream method normally expands away from the center of the sand as the embankment rises. This is accomplished by moving the cyclones outward with each rise. However, the cyclones can be raised vertically, whereupon Klohn terms the process the centerline downstream method. In this method, slimes and sands interfinger on the inside of the embankment, and the embankment rises more rapidly with a fixed sand fraction which may be critical during the early days of a pond when area is minimal. The centerline method presents some danger of sloughing into the pond if slopes are allowed to rise too steeply. However, harmful effects should be negligible if the embankment is not raised too rapidly.

Construction of Tailings Pond Embankments with Imported Materials

Tailings ponds are constructed with imported materials (Figure 11.10) if the tailings do not contain sufficient coarse-grained material to form a stable embankment. This condition may be a result of the milling procedure or of the reintroduction of the coarse fraction back into abandoned underground openings. This practice is known as sandfilling and it is becoming prevalent in underground mines because abandoned openings constitute health and safety hazards. Consequently, more and more tailings sand embankments near underground mines are being constructed of materials other than tailings. Clearly this adds to the cost of tailings disposal. However, other problems, such as erodibility of embankments, may be reduced.

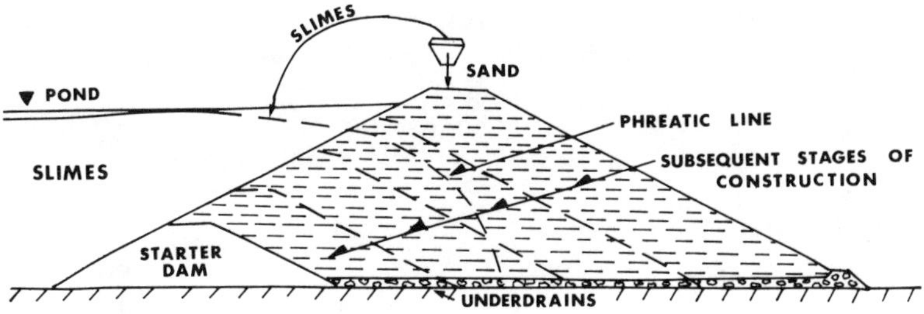

Figure 11.11. *Downstream construction using heavily compacted cycloned sand (after Klohn 1972). Sand is placed in layers and compacted, fill slopes are relatively steep, and there is a good underdrainage system.*

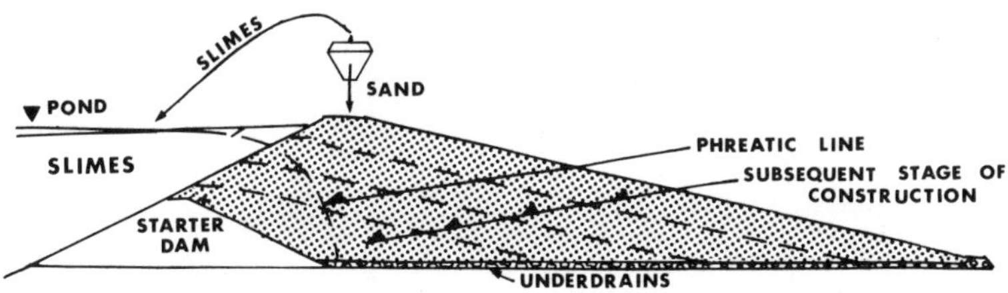

Figure 11.12. *Downstream construction using low- to medium-dense, free-draining, cycloned sand (after Klohn 1972). Sand is clean and free-draining, fill slopes are relatively flat, and there is a good underdrainage system.*

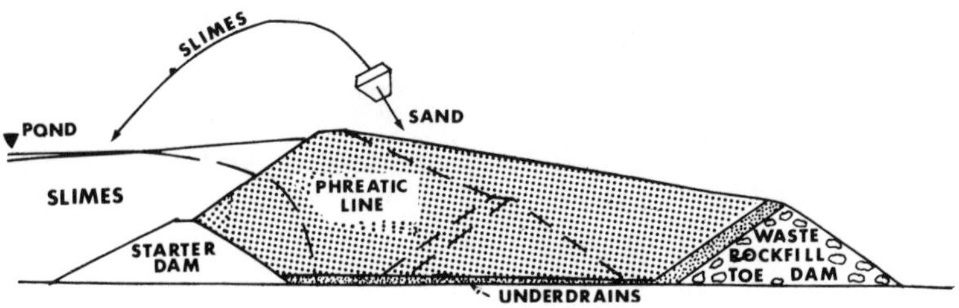

Figure 11.13. *Downstream construction using cycloned sand and waste rock toe (after Klohn 1972). Solid lines represent section if cycloned sand is placed in low- to medium-dense state. Dashed lines represent section if the cycloned sand is heavily compacted.*

In order to keep costs as low as possible, it is common practice to use mine waste rock as the imported material. The Bunker Hill mine, mill, and metallurgical complex waste disposal pond near Kellogg, Idaho, is an excellent example. The effluent from this pond constitutes the water treated in the treatment plant described in Appendix E. This operation combines the cost of discarding the mine rock with the cost of tailings embankment construction, thereby eliminating one set of costs.

The use of imported materials facilitates the design and construction of water storage dams. This technique is advantageous in seismic areas because greater control can be exercised since there is minimal risk of having to use materials that might liquefy upon disturbance.

The technology for constructing water storage dams is well developed and readily available in a variety of published papers and texts. Thus, discussion of it here is considered unnecessary.

Construction and Operational Problems

Klohn (1972) lists six factors that can cause or contribute to construction problems with tailings ponds. These factors are:

1. Hydrological conditions that differ from those originally assumed on the basis of incomplete data.
2. Changes in mill water balance, resulting in either more or less water than being added to the pond.
3. Changes in mill grind, resulting in either more or less sand available for tailings dam construction.
4. Unforeseen cycloning problems, caused by a high clay mineral content in the ore, which may drastically reduce recovery of suitable sand and may require two-stage cycloning.
5. Unexpected and prolonged shutdown of the mill—dam building ceases but pond levels continue to rise owing to continued runoff.
6. Climatic conditions that might combine an unusually long winter season (shortening the dam construction period) with an unusually heavy runoff (requiring maximum freeboard to store the flood surcharge).

Klohn considers overtopping to be the most serious threat to a tailings disposal facility. He also considers excess seepage to be a dangerous threat because of the possible piping it can create. In the opinion of the author, this problem can be minimized by controlling the water level over the slime zone and not allowing careless management to permit it to rise up onto the sand beach. Such an incident would not only cause seepage but would allow slimes to be deposited within the embankment, thereby causing it to be unstable.

It is clear that water quality considerations should be added to the above list of operational problems. The waste discharge permits described in Chapter 1 and in Appendix D are legally binding documents. Thus, continual vigilance will be required to ascertain that those ponds with discharges are managed so that suspended solids regulations, in particular, are not violated. Williams,

Wallace, and Mink (1971) have shown that one or more clarification ponds in series with a tailings pond can be very effective in this regard. This implies very careful location and control of decants or siphons. Seepage can be troublesome in this regard also; some of the waste discharge permits include requirements that ground-water quality monitoring wells be installed around the pond. Much less seepage is required to violate discharge permits than to cause piping.

Subaqueous Disposal of Mill Tailings

Subaqueous disposal of mill tailings has been practiced by a few operations in the United States and Canada and by a greater number of operations on a worldwide basis. Evans, Ellis, and Pelletier (1973) provide elaborate detail on the disposal of copper tailings in Rupert Inlet at the north end of Vancouver Island, British Columbia, Canada. Salazar and Gonzales (1973) describe a subaqueous disposal system with unique tailings transport problems in the Philippines. Perhaps the most famous subaqueous tailings disposal project is that of Reserve Mining Company's taconite operation on Lake Superior. This operation has gained much notoriety, primarily because of asbestos associated with the tailings (see Weston 1971, for details).

The author is of the opinion that waste discharge permits under Public Law 92-500 will be difficult, if not impossible, to obtain for proposed subaqueous disposal projects. The position currently being taken by the courts and the United States Government on the Lake Superior taconite tailings disposal project is part of the evidence on which this opinion is based. Other evidence consists of the fact that most company executives do not elect to go the subaqueous disposal route except as a last resort. For this reason, subaqueous tailings disposal is not treated in detail here.

Uses of Tailings

The use of tailings as a material with acceptable engineering properties is clearly the best method of disposal. Generally, only the coarse fractions are utilized. Guerra (1973) reports such uses but Pettibone and Kealy (1971) provide the most detailed example known to the author. Their paper leads one to conclude that tailings may be more useful than has been believed. The example cited consists of a highway embankment constructed by the Idaho Department of Highways along a four-mile section of U.S. Interstate 90 near Kellogg, Idaho. The case also is discussed by O'Dell (1964). Over 95 percent of the project consists of embankment with the fills ranging from a few feet to over 20 feet. The only major cut is a rock cut on the east end of the project. Thus, the required 1,200 cubic yards of embankment material had to be obtained from a local borrow pit. A large tailings pond owned by the Bunker Hill Company was located adjacent to the Interstate right-of-way, about midway in the project. Its embankments are of mine rock. O'Dell (1964) reports that the Bunker Hill borrow source was chosen because (1) it was conveniently located in the center of the project, (2) it met specifications, and (3) there was no other borrow source of this size near the project.

Tailings had been discharged into the south side of the pond, and the large particles had settled closest to the point of discharge; their size decreases toward the center. The tailings are silty sand, with the minus No. 200 sieve material ranging from 14 to 89 percent, the average being about 50 percent, as shown in Table 11.3. The maximum size for most samples is a No. 30 sieve size; only one sample contained an appreciable amount of plus No. 4 sieve material. The tailings displayed a specific gravity of about 2.91 and a maximum dry density of 115 pounds per cubic foot at an optimum moisture content of 14 percent (AASHO, T-99). The symbols for both the unified soil-classification and the American Association of State High Officials (AASHO) systems are given in Table 11.3. The AASHO classification of the tailings varied from A-2 (near the point of discharge) to A-4 at the outer edges of the pond. To prevent wind erosion, Bunker Hill had covered the pond with about 0.5 foot of black slag. Pettibone and Kealy (1971) report that, although the slag by itself is not readily compacted, a mixture of 95 percent slag and 5 percent tailings can be compacted to coincide with Idaho Department of Highway's specifications (100 percent of laboratory maximum density).

According to O'Dell's report, several contractors looked over the source before bidding and decided that the area would be too hard to work because of the nature of the material. But the successful bidder used a belt loader pulled by a D-9 Caterpillar at approximately two to three miles per hour to scrape a one-foot layer from a vertical face about eight feet high. Its average production was between 1,200 and 1,500 cubic yards per hour.

The material in the embankment never reached the optimum moisture content of 14 percent. However, it was said to have been compacted with a 50-ton, pneumatic-tired roller. Required density was reported with two to three passes of the roller. In areas where the moisture content was near optimum, one roller pass produced at least maximum density. O'Dell (1964) summarizes by stating that the tailings served the purpose well. He termed them easy to work with the belt loader and pointed out that they were compacted readily, and too much water was not a problem, contrary to the prediction of many.

Tailings as Backfill (Sandfill) for Abandoned Stopes

Underground metal mines recently have almost unanimously adopted the practice of refilling worked out stopes with the coarser fraction of mill tailings. This practice reduces the surface area needed for tailings disposal, but it usually leads to the importation of material for the tailings pond embankment. Mine waste rock is frequently used.

Stewart (1958) states that hydraulically emplaced backfill was first practiced in 1864 in an anthracite coal mine at Shenandoah, Pennsylvania, where breaker waste was flushed into abandoned workings for subsidence control. Pariseau and Kealy (1972) cite some 37 references dealing with the subject.

As pointed out by Pariseau and Kealy (1972), hydraulic fill in conventional cut-and-fill stopes provides ground support and a working platform at a lower cost than if the stope were timbered. Even if the stope is timbered first and

Table 11.3 *Summary of Standard Properties, Embankment Material for 190, Near Kellogg, Idaho*

Identification		*Grain-size fractions, as a percentage*			*Consistency limits*					*Moisture density data (AASHO, T-99)*		
Location	*Depth of strata (ft)*	*Silt and clay minus No. 200*	*Sand No. 200 to No. 4 size*	*Gravel No. 4 size to 3 in.*	*Liquid limit, as a percentage*	*Plasticity index, as a percentage*	*Linear shrinkage as a percentage*	*Laboratory classification symbol (unified system)*	*Specific gravity*	*Maximum dry density (lb/cu ft)*	*Optimum moisture as a percentage*	*Classification symbol (AASHO system)*
Test pit 1	0.5-11	21	79	0	NV[3]	NP[4]	–	SM	2.97	116.3	13.9	A-2-4
Test pit 4	0.5-11	39	61	0	NV	NP	–	SM	2.92	115.9	13.7	A-4 (1)
Test pits 5, 6, 7, 8 and 9	0.5-11	71	29	0	NV	NP	–	ML	2.88	117.8	13.4	A-4 (7)
Test pits 10, 11, 16 and 17	–	60	40	0	NV	NP	–	ML	2.89	115.7	14.3	A-4 (5)
Test holes 18, 19, 20 and 21	–	47	53	0	NV	NP	–	SM	2.93	116.6	13.2	A-4 (2)
Test holes 22, 23 and 24	0.5-11	77	23	0	NV	NP	–	ML	2.89	112.1	16.7	A-4 (8)
Test holes 25 and 26	–	89	11	0	NV	NP	–	ML	2.90	111.2	17.6	A-4 (8)
Back slag	–	4	96	0	NV	NP	–	SP	3.42	NA	NA	A-1-b
Tailings	–	28	72	0	NV	NP	–	SM	3.00	118.0	13.4	A-2-4
Test holes 12, 13, 14 and 15	–	66	34	0	NV	NP	–	ML	2.86	110.8	16.2	A-4 (6)
s 1/2 Sec 36 T49N, R2 EBN	0-25	72	27	1	NV	NP	1.6	ML	2.96	122.6	12.5	A-4 (7)
Se 1/4 Sec 36 T49N, R2 E[1]	–	14	86	0	NV	NP	–	SM	–	–	–	–
Same[2]	–	20	80	0	NV	NP	–	SM	–	137.8	13.0	–
Auger hole 9	32-52.6	20	40	40	NV	NP	0.4	GM-SM	–	NA	NA	A-2-4
Edge of pond	–	45	55	0	NV	NP	0.4	SM	–	–	–	–
Undetermined	0-21	93	7	0	NV	NP	2.4	ML	2.90	115.0	15.2	A-4 (8)
Auger hole 3	0-5	45	55	0	NV	NP	0.8	SM	–	–	–	A-4 (2)

Source: After Pettibone and Kealy 1971.

1. Before proctor.
2. After 8th proctor point.
3. No value.
4. Nonplastic.

then filled, the fill provides additional ground control for the mine workings and, in some instances, timber may be recovered.

The cost advantage of hydraulic sandfill compared with dry fill is that it reduces the nonproductive time of the mining cycle. Hydraulic fill also reduces ventilation short circuits through stopes, reduces timber fire hazards, and relieves the underground transportation system. Stopes are made safer through improved support and ground control. For purposes of this book, it reduces the area needed for tailings disposal. The major disadvantages of hydraulic fill are the introduction of additional water into the mine, the occasional spills that occur, and the fact that importation of material for tailings embankments is required if too much of the coarse fraction is removed from the tailings.

In 1961, the U.S. Bureau of Mines initiated a program of applied research concerning hydraulic backfill. Field tests extending over periods of several years were conducted at several mines in the Coeur d'Alene district of north Idaho, where mine depths exceed 6,000 feet. Publications cited by Pariseau and Kealy (1972) describe the important physical characteristics of mine backfills, the influence of various factors on percolation rate, the improvement of backfill quality by vibratory compaction, the effects of cement and dispersant additives on backfill, the stress-strain relationships of backfill undergoing uniaxial consolidation, and the results of field measurements of stope wall convergence and backfill stress changes at two operating mines.

Pariseau and Kealy (1972) also point out that, despite the sizable accumulation of engineering experience and research data pertaining to hydraulic fill, there is a need for a quantitative method for establishing a figure of merit with regard to the support potential of hydraulic fills. Predictions of fill quality requirements and artificial support requirements in general have been based largely on experience and trial and error.

Pariseau and Kealy attempted to fill this need for a quantitative method of evaluating the support potential of backfill by formulating an idealized cut-and-fill stoping procedure in which the final stope configuration was obtained incrementally through an alternating series of cuts and fills in an initially stressed medium, as is done in actual mining practice. They examined four stoping sequences. They idealized the essential physical aspects of stoping operations into a manageable mathematical problem whose solution results in the prediction of quantities such as stope convergence and fill stresses that can be compared to similar quantities measured in the field. Iterative solutions to the differential equations developed for conditions of plasticity (as opposed to elasticity) and constant and variable boundary conditions were obtained by the finite-element method. The finite-element method used consisted of a plane strain, elastic-plastic program which utilized a Mohr-Conlomb yield condition.

The analysis indicated that hydraulic mine fills have a significant potential for stope support and ground control. Relatively low-quality fill (meaning a high percentage of fines) reportedly reduced stope convergence by approximately 50 percent from the open stope case. The extent of the failed zone around the stope walls was approximately 35 percent less than that for an open stope. In the case of the action of the high-quality (coarse) fill, maximum stope convergence was one-third that obtained with the low-quality fill and only one-fifth that of an open stope. The failed zone vanished with the emplacement of the last high-quality fill. In short, the analysis indicates that hydraulic mine fill is an effective method of ground support and control. Perhaps more importantly, the analysis suggests that even fine-grained backfill provides considerable support. This observation suggests that it should be possible to dispose of a greater percentage of mill tailings by filling underground openings than has been practiced previously. More specifically, it should be possible to sandfill with tailings as fine as No. 200 sieve. On the basis of data presented elsewhere in this chapter and in Figure 3.2, it should be advantageous to backfill approximately 50 percent of the average copper, lead, or zinc mill tailings.

The primary objection to backfilling with finer tailings is the risk of poor drainage of the backfill material, which could introduce the possibility of bulkhead failures. This problem has been attacked by Sprute and Kelsh (1973) using electrokinetics as a means of dewatering the finer fractions of tailings. These investigators believe that essentially all tailings can be used for backfilling if electrokinetic dewatering is utilized to assure drainage.

COAL MINE WASTES

Introduction

Interest in the disposal of coal mine wastes, including coal waste banks, has intensified since the failure of a coal refuse dam upstream from Saunders, West Virginia, in 1972. The disaster left 118 persons dead and 7 persons missing, and destroyed over 500 homes. Saunders, located in the valley of the Middle Fork of Buffalo Creek near its confluence with Buffalo Creek, was destroyed. The failure has since become known as the Buffalo Creek disaster or simply as "Buffalo Creek."

Prior to Buffalo Creek, interest in coal refuse disposal was directed primarily at controlling burning of the refuse banks. Consequently, only limited data were published on the disposal methods utilized at the banks, or on their stability. The major exception is the work of W. E. Davies, a geologist with the U.S. Geological Survey. Davies began work on the engineering geology associated with coal waste banks in 1965. In 1966 he stated in a little noticed article:

> Geologic and soil investigations of sites of [coal] waste banks can provide data for proper drainage and foundation control and can assure that banks are not placed where natural landslides are likely to occur. In the case of banks acting as dams for settling ponds, maintenance of internal spillways by careful control of the slopes of waste-bank faces and segregation of materials in the bank, along with the provision for protected surface spillways should be of prime importance.

According to Wahler and Associates (1973), the failure at Buffalo Creek was caused in part by overtopping of a coal mine refuse embankment which impounded a settling pond in a small valley. The pond had only four feet of freeboard. Wahler and Associates state that "apparently none of the embankments were constructed on the basis of engineered plans, and neither were they constructed on the basis of studies of the engineering properties of the coal refuse of which they were built."

Origin of Wastes

Davies (1966) describes coal mine waste banks as consisting largely of bony coal and slabby shale partings, with some blocky sandstone and shale waste from excavation of shafts and extensive rock tunnels. Wahler and Associates (1973) would add construction materials to this list. Davies states that in Appalachia 15 to 40 percent of a waste bank, with extremes of 10 and 80 percent, is low-grade coal and related carbonaceous material. Sandstone comprises less than 10 percent, and 30 to 70 percent may be shale. Davis describes the wastes as follows:

> At the time of dumping, sandstone and shale consist of angular fragments of a few inches to more than two feet. The shale weathers into chips and flakes within a few months. When dumped, the coal fraction is one-half inch to one foot; but, within a few weeks, slaking forms large quantities of fine fragments After a short period of slaking and weathering, the material less than one-eighth inch in grain size generally makes up 30 to 50 percent (with extremes of 5 to 80 percent); fragments in the one-eighth to two inch size generally make up 30 to 50 percent (extremes of 5 to 70 percent), and more than two inch material generally makes up 20 to 30 percent (extremes of 0 to 90 percent).

Wahler and Associates selected the engineering properties shown in Table 11.4 for engineering analysis of the coal refuse in the waste embankment (actually dam) at Buffalo Creek. They note that these materials do not exhibit exceptionally unusual characteristics when compared to soil-like materials conventionally used for earth dam construction. This observation suggests that (with proper site selection, design, construction and management procedures)

Table 11.4 *Engineering Properties of Coal Mine Waste Embankment Materials Selected for Buffalo Creek Engineering Analysis*

Engineering property		*Values used in analysis*
Dry density		90 lb/cu ft
Wet density		106 lb/cu ft
Coefficient of permeability		2×10^{-4} cm/sec
Permeability ratio (horizontal to vertical)		1:1
Specific gravity		1.95
Shear strength parameters effective stress		
	For normal stress less than 16 psi	$\phi' = 41°$
		$c' = 0$
	For normal stress greater than 16 psi	$\phi' = 34°$
		$c' = 500$ psi
Total stress		$\phi = 17°$
		$c = 700$ lb/sq ft

Source: After Wahler and Associates 1973.

the material in coal refuse could be used safely as embankment material if ignition could be eliminated.

Davies states, however, that in general:

> Compaction is low in all [existing] coal waste banks. In truck dumps graded by bulldozers, the void spaces are at least 30 percent; in tramline and hillside dumps, 50 percent. Within a few days after dumping, settlement cracks develop in a zone 10 feet from the face in truck and hillside banks and slumps often develop later along these cracks. On tramline dumps, longitudinal cracks are common along the lower flanks of the bank; in rockfill dams, settlement cracks occur along the crest.

Davies describes waste banks in Appalachia existing in 1966 as:

> . . . hillside dumps–flat-topped benches on valley walls and other declivities with natural slopes averaging 36° but ranging from 25° to 45°. Also, truck dumps–flat-topped fills extending along and across small valleys–are common . . . and often fill a valley 100 feet deep, or more. Aerial tramline dumps are used at some large preparation plants, and the resulting banks are steep-sided ridges and cones, most of which extend across narrow valleys. Many banks consist of several similar or different types of dumps.

The second type of material associated with coal refuse banks is the sludge produced by coal preparation plants. Davies characterizes modern preparation plants as centrally located and serving several mines. Preparation plants include washing operations which utilize vibrating screens, shaker screens, jigs, picking tables, and dryers. The water from these washing operations is sometimes referred to as "black water." Wahler and Associates report that the black water emanating from the preparation plant at Saunders, West Virginia, contained 18 percent solids by weight.

Historically, these waste waters have been discarded by introduction into creeks or settling ponds; frequently the ponds are located in draws dammed by coal refuse embankments. More modern trends are toward the use of classifiers to remove marketable coal fines and thickeners to remove coal sludge so that the wash water can be recycled. Fine unstable sludge from thickeners can thus be discarded in smaller retaining basins or in landfills as are sludges from treatment plants which neutralize effluents from air pollution control scrubbers as discussed in other chapters of this book.

According to Wahler and Associates, the sludge under the refuse embankments at Buffalo Creek would be sandy silt to silty sand under the Unified Soil Classification System. The higher the coal content, the lower the specific gravity (1.34 to 1.66). The important engineering properties of the sludge materials beneath the refuse embankments at Buffalo Creek are summarized in Table 11.5. Wahler and Associates state that, although the shear strength parameters are not unusually low, unless sufficient supplementary loading is superimposed to increase intergranular stress and provide resistance to lateral deformation, the full benefit of its ability to resist shear cannot be realized because of its unusually light unit weight. In short, coal preparation plant sludge should not be used as a foundation material unless rigorous engineering procedures are utilized during construction to increase its density. Examples of

Table 11.5 *Engineering Properties of Coal Preparation Plant Sludge Materials Beneath the Coal Refuse Embankment at Buffalo Creek Failure in West Virginia*

Engineering property	*Values used in analysis*
Dry density	54 lb/cu ft
Wet density	78 lb/cu ft
Equivalent coefficient of permeability ($\sqrt{K_h \times K_v}$)	2×10^{-5} cm/sec
Permeability ratio (horizontal to vertical)	25 : 1 and 100 : 1
Specific gravity	1.43
Shear strength parameters	$\phi' = 37°$
	$c' = 0$
Total stress	$\phi' = 16.5°$
	$c' = 1{,}100$ lb/sq ft

Source: After Wahler and Associates 1973.

such practices would be compaction and electrokinetics. Wahler and Associates concluded that the waste dump failure on Middle Fork, Buffalo Creek, was caused by the construction of a coal refuse dam on preparation plant sludge that had accumulated in a pool behind dams farther downstream. The failure occurred by shearing of the foundation material, although conditions for piping were considered to be marginal.

Evaluation of coal refuse embankments

Subsequent to the failure at Buffalo Creek, a United States Department of Interior Task Force (including representatives from the Geological Survey, Bureau of Mines, and Bureau of Reclamation) was assigned the task of evaluating other coal refuse banks in Appalachia (Task Force 1972). In addition to analyzing the Buffalo Creek failure, the Task Force sought to:

1. Identify other potentially hazardous sites.
2. Determine modes of failure and means by which hazardous conditions could be recognized.
3. Develop means for safe disposal and storage of coal mine wastes.

One hundred thirty nine coal refuse disposal sites were investigated.

This study is of particular importance because it established, for the first time, a semi-official method, albeit somewhat generalized, of evaluating coal refuse disposal operations. The sites were rated as: I: Imminent hazard or requiring immediate corrective measures; II: Obvious deficiencies, hazard unknown; III: No obvious deficiencies, probably no serious hazard.

In general, for a site to be rated I the embankment had to show signs of instability such as cracking, slumping caused by burning, previous sliding, excessively steep slopes, piping, or severe surface erosion. In some cases the hydrologic design was inadequate in that adequate spillway or diversion facilities were not available to handle heavy storm runoff. Of particular importance was a high probability of loss of life and property in the event of failure. Ten of the 139 sites investigated met this criteria at some time during the investigation. But most remaining sites required corrective or remedial action of some type, although not necessarily immediate action.

The sites rated I displayed high potential for failure. It was noted that failure can occur in a number of ways, depending on the type of embankment. If the embankment impounded a large volume of water and sludge, a slope failure could cause extensive flooding of property downstream. The failure itself could occur through slope sliding or through erosion caused by overtopping. Even if the embankment did not impound water, sliding or slipping was considered because it could cause damage to nearby structures. In either situation, failure could block main channels resulting in the rapid damming and release of large quantities of water.

The most common shortcomings or deficiencies observed on the coal waste embankments investigated were summarized by the Task Force as follows:

1. Lack of, or improperly designed, spillways:
   a. Many sites had no spillways.
   b. Many sites had pipe spillways that usually discharged onto the downstream face of the dam, were subject to blockage, had inadequate capacity and freeboard, and were set in a loose embankment. These pipe spillways generally were worse than no spillway at all.
   c. A few sites had open-cut spillways across, or partially across, the embankment and were subject to failure by erosion of the embankment.
   d. Some sites had open spillways or diversion channels cut into rock across an abutment. Many of these had inadequate capacity.
2. Improper method of construction:
   a. Many dams were constructed by dumping the refuse from aerial trams. The spilled materials would form an embankment with a slope approximately equal to the angle of repose of the material. Such slopes would have a minimum factor of safety even before the addition of forces of seepage to the already existing loads. The low densities obtained by aerial dumping or other types of dumping without adequate compaction would make a dam particularly susceptible to erosion.
   b. The upstream method of construction was observed to be widely used. As the dam was raised in height, new refuse was placed upstream of the previous crest providing a sludge foundation for the extensive upper portions of the coal refuse dam.
3. Improper method of sludge disposal: Instead of introducing the water and slimes directly behind the upstream face of the dam, it was observed that it is very often released at the upstream ends of impoundments. This practice permits the coarse material to settle immediately at the upstream end of the impoundment and allows the fines and water to accumulate against the dam where they create the greatest hazard, including seepage through the embanking in the case where free water lies against the embankment. Wahler and Associates (1973) concluded that such seepage was prominent prior to the failure at Buffalo Creek.

4. Inadequate freeboard: Many of the dams observed did not provide adequate emergency storage between the pond level and the top of the dam. High storm runoff could cause dam overtopping in such cases.
5. Burning: Burning causes a reduction in volume of the refuse material which, in turn, causes slumping or cracking of the dam surface. Thus, seepage channels can be opened which could cause internal erosion or piping. Saturation of the burning areas can cause explosions that can severely damage the structure.

Included among the conclusions of the Task Force is the fact that, as of July 1972, the technology did not exist to permit accurate determination of the nature and probability of existence of any but the most obvious hazards. Unavailability of acceptable design techniques and adequate materials characteristics data precluded the evaluation or design of coal waste dams constructed of randomly mixed materials. The cost-effectiveness of alternative solutions to the problem had not been considered when the Task Force report was written.

The Task Force concluded, however, that existing technology in civil engineering soil mechanics, and engineering geology—especially as they apply to earthwork and earth-dam engineering—can be of use in analyzing and reducing coal waste embankment problems; but it is necessary to have also a qualified review of plans prior to construction and a continuing inspection of construction as it proceeds.

Safe Disposal Practices

Since the release of the Task Force report in 1972, Wahler and Associates have continued to work on the Task Forces' third objective, the development of means for safe disposal and storage of coal mine wastes. Results are available as Wahler and Associates (1974a). This report defines several characteristics unique to coal mine wastes. Included among them are:

1. The specific gravity of the fine waste (primarily preparation plant waste) is abnormally low, ranging from 1.4 to 1.8, the average for soils being 2.65.
2. The low specific gravity results in low in-place densities (50 to 70 lb/ft^3).
3. The low density makes the material prone to erosion and piping.
4. The low density produces low effective stress and consequently low shear strength unless special engineering techniques are employed to increase the density during emplacement.
5. The coarse material contains platy materials which rapidly weather to clay minerals after emplacement on the waste pile.
6. Coarse material burns and causes a reduction in volume.

On the basis of these unique characteristics and other characteristics, Wahler and Associates concluded that:

1. Coarse coal waste material (non-preparation plant wastes) generally has adequate shear strength, permeability, and compressibility for successful, conventional use in construction of earth and rockfill dams.

2. Conventional earth moving and compaction equipment can be used to construct dams of coarse coal wastes which will meet currently acceptable design standards.
3. Additional research is needed to evaluate the influence of weathering or burning of these materials after emplacement.
4. The fine grained wastes should be excluded from embankment areas unless specialized engineering techniques are employed to maintain their stability against liquefaction and piping.

Other recommendations by Wahler and Associates (1974a) for design and construction of coal refuse embankments are essentially identical to recommended standard procedures for constructing earth dams. These techniques are available from a variety of published sources; consequently, they are not presented here. Wahler and Associates (1974a) are pointedly critical of the practice of constructing impoundment embankments of mine waste materials by tramways, conveyor belts, or trucks without the utilization of demonstrated design and construction techniques, even when only coarse wastes are used in the embankment.

REFERENCES

Blume, J. A. 1973. Probability of Earthquakes and Resultant Ground Motion. In *Tailing Disposal Today* Proceedings, International Tailing Symposium, Tucson, Ariz., ed. C. L. Aplin and George O. Argall, Jr., pp. 325-355. San Francisco: Miller Freeman Publications.

Brawner, C. P., and Campbell, D. B. 1973. The Tailings Structure and Its Characteristics–A Soil Engineer's Viewpoint. In *Tailing Disposal Today*, Proceedings, International Tailing Symposium, Tucson, Ariz., ed. C. L. Aplin and George O. Argall, Jr., pp. 59-95. San Francisco: Miller Freeman Publications.

California State Water Pollution Control Board 1954. Investigation of Leaching of a Sanitary Landfill. Publication No. 10.

Canadian Dept. of Energy, Mines, and Resources, Mines Branch 1972. *Tentative Design Guide for Mine Waste Embankments in Canada.* Technical Bulletin 145, Ottawa, Ontario, Canada.

Cedergren, H. R. 1967. *Seepage, Drainage and Flow Nets* New York: John Wiley and Sons, Inc.

Chow, V. T. 1964. *Handbook of Applied Hydrology.* New York: McGraw-Hill Book Co., pp. 25/16-25/25.

Dahlberg, H. R., and Laird, R. C. 1973. Building Tailings Dams in Arroyos. In *Tailing Disposal Today*, Proceedings, International Tailing Symposium, Tucson, Ariz., ed. C. L. Aplin and George O. Argall, Jr., pp. 317-324. San Francisco: Miller Freeman Publications.

Davies, W. E. 1966. Coal Waste Bank Stability. *Mining Congress Journal*, Vol. 54, No. 7, pp. 19-24.

Dopson, G. W., and McGregor, D. 1973. Magma Copper Company's San Manuel No. 10 Tailings Dam; Design, Construction and Operation. In *Tailing Disposal Today*, Proceedings, International Tailing Symposium, Tucson, Ariz., ed. C. L. Aplin and George O. Argall, Jr., pp. 283-295. San Francisco: Miller Freeman Publications.

Evans, J. B., Ellis, B. V., and Pelletier, C. A. 1973. The Establishment and Implementation of A Monitoring Program for Underwater Tailings Disposal in Rupert Inlet, Vancouver Island, British Columbia. In *Tailing Disposal Today*, Proceedings, International Tailing Symposium, Tucson, Ariz., ed. C. L. Aplin and George O. Argall, Jr., pp. 512-552. San Francisco: Miller Freeman Publications.

Galbraith, J. H., Williams, R. E., and Siems, P. L. 1972. Migration and Leaching of Metals from Old Mine Tailings Deposits. *Ground Water Journal*, Vol. 10, No. 3.

Guerra, F. 1973. Characteristics of Tailings from a Soils Engineer's Viewpoint. In *Tailing Disposal Today*, Proceedings, International Tailing Symposium, Tucson, Ariz., ed. C. L. Aplin and George O. Argall, Jr., pp. 102-136. San Francisco: Miller Freeman Publications.

Hughes, G. M., and Cartwright, K. 1972. Scientific and Administrative Criteria for Shallow Waste Disposal. *Civil Engineering ASCE*, Vol. 42, No. 3, pp. 70-73.

Hughes, G. M., Landon, R. A., and Farvolden, R. N. 1971. *Hydrogeology of Solid Waste Disposal Sites in Northeastern Illinois.* EPA Demonstration Grant Completion Report G06-EC-00006, Report No. SW-12d, Stock No. 5502-0034, Supt. of Documents, U.S. Govt. Printing Office, Washington, D.C.

Kealy, C. D. 1973. Safe Design of Mill Tailings Dams. *Mining Congress Journal*, Vol. 59, No. 1, pp. 51-55.

Kealy, C. D., and Busch, R. A. 1971. "Determining Seepage Characteristics of Mill Tailings by the Finite Element Method." Ph.D. thesis, College of Mines, Univ. of Idaho, Moscow. (Published as U.S. Bureau of Mines Report of Investigations 7477.)

Kealy, C. D., and Williams, R. E. 1971. Flow Through a Tailings Pond Embankment. *Water Resources*, Vol. 7, No. 1, pp. 143-154.

Klohn, E. J. 1972. Design and Construction of Tailings Dams. *Canadian Mining and Metallurgical Bulletin*, Vol. 75, April, pp. 28-44.

Lambie, J. A. 1969. *Development of Construction and Use Criteria for Sanitary Landfills.* U.S. Dept. of HEW, PHS, Consumer Protection and Environmental Health Service, Interim Report to Grant No. Dol-UI-00046, prepared by County of Los Angeles, Dept. of County Engineer.

Miners, K. A. 1973. Alcan's Experience with Disposal of Red Mud Tailings from the Barger Ore Process. In *Tailing Disposal Today*, Proceedings, International Tailing Symposium, Tucson, Ariz., ed. C. L. Aplin and George O. Argall, Jr., pp. 553-562. San Francisco: Miller Freeman Publications.

Mink, L. L., Williams, R. E., and Wallace, A. T. 1972. Effect of Early Day Mining Operations on Present Day Water Quality. *Ground Water Journal*, Vol. 10, No. 1.

Mink, L. L., Williams, R. E., Wallace, A. T., and McNay, L. M. 1973. Renovation of Wastes by Mine Tailings Ponds. *AIME, SME, Mining Engineering*, Vol. 25, No. 7, pp. 81-88.

O'Dell, L. G. 1964. *The Euclid Belt Loader.* Rotational Engineering Training Program Report, Idaho Dept. of Highways, Boise.

Pariseau, W. G., and Dealy, C. D. 1972. *Support Potential of Hydraulic Backfill.* Proceedings, 14th Symposium on Rock Mechanics, Penn. State Univ., Univ. Park, Penn., pp. 501-526.

Pettibone, H. C., and Kealy, C. D. 1971. Engineering Properties of Mine Tailings. *ASCE Journal of Soil Mechanics and Foundations Division*, Vol. 97, No. SM 9, Proceedings Paper 8382, pp. 1207-1225.

Salazar, R. C., and Gonzales, R. I. 1973. Design, Construction and Operations of the Tailings Pipelines and Underwater Tailings Disposal System of Atlas Consolidated Mining and Development Corporation in the Philippines. In *Tailing Disposal Today*,

Proceedings, International Tailing Symposium, Tucson, Ariz., ed. C. L. Aplin and George O. Argall, Jr., pp. 477-511. San Francisco: Miller Freeman Publications.

Sceva, J. E. 1973. *Water Quality Considerations for the Metal Mining Industry in the Pacific Northwest.* EPA, Region 10, Report No. 3.

Sprute, R. H., and Kelsh, D. J. 1973. *Laboratory Experiments in Electrokinetic Densification of Mill Tailings.* U.S. Bureau of Mines Report of Investigations, Dept. of Interior, Washington, D.C.

Stewart, R. M. 1958. Hydraulic Backfilling. *Mining Engineering*, Vol. 10, No. 4, pp. 476-480.

Swaisgood, J. R., and Toland, G. C. 1973. *Control of Water in Tailings Structures.* Dames and Moore Inc., Engineering Bulletin 41, Los Angeles, California, pp. 17-28.

Task Force 1972. *Emergency Investigation of Coal Mine Waste Embankments.* Interim Report, U.S. Dept. of Interior Task Force to Study Coal Waste Hazards, Washington, D.C.

Taylor, R. L., and Brown, C. B. 1967. Darcy's Flow Solution with a Free Surface. *J. Hyd. Div., ASCE*, Vol. 93, No. Hy 2, pp. 25-33.

Toland, G. C. 1971. Gypsum Tailings Pond Design, Kimberly, British Columbia, AIME, SME, preprint of paper presented at SME Fall meeting, Seattle; also available from Dames and Moore Inc., Salt Lake City, Utah.

U.S. Bureau of Reclamation 1965. *Design of Small Dams.* 1st ed., 3rd printing, Dept. of Interior, Washington, D.C.

Wahler, W. A., and Associates 1970. Engineering Evaluation of Mill Tailings Disposal Practices and Potential Dam Stability Problems in Southwestern United States. Technical Proposal to U.S. Bureau of Mines RFP F0110520, Vol. 2, Chapter 7. (Unpublished).

Wahler, W. A., and Associates 1973. *Analysis of Coal Refuse Dam Failure.* Contract Completion Report No. S0122084, U.S. Bureau of Mines, Dept. of Interior, Washington, D.C.

Wahler, W. A., and Associates 1974a. *Summary and Recommendations Relative to Coal Waste Embankment Engineering.* Contract Completion Report No. S0122084, U.S. Bureau of Mines, Dept. of Interior, Washington, D.C.

Wahler, W. A., and Associates 1974b. *Evaluation of Mill Tailings Disposal Practices and Potential Dam Stability Problems in Southwestern United States.* Contract Completion Report No. S011050, U.S. Bureau of Mines, Dept. of Interior, Washington, D.C.

Webb, S. L., and Smith, E. S. 1971. Tailings Dam Sealed by Slimes Slurry. AIME, SME, preprint of paper presented at SME Fall meeting, Seattle.

Weston, R. F. 1971. *Concept Evaluation Report, Taconite Tailings Disposal, Reserve Mining Co., Silver Bay, Minn.* Contract Completion Report, U.S. Environmental Protection Agency, Washington, D.C.

Williams, R. E. 1973. Modernization of Coeur d'Alene Tailings Disposal Practices. In *Tailing Disposal Today*, Proceedings International Tailing Symposium, Tucson, Ariz., ed. C. L. Alpin and George O. Argall, Jr., pp. 634-673. San Francisco: Miller Freeman Publications.

Williams, R. E., Kealy, C. D., and Mink, L. L. 1973. Effects and Prevention of Leakage from Mine Tailings Ponds. *Transactions, AIME, SME*, Vol. 254, pp. 212-216.

Williams, R. E., Wallace, A. T., and Mink, L. L. 1971. Impact of a Well-Managed Tailings Pond System on a Stream. *Mining Congress Journal*, Vol. 57, No. 10, pp. 48-56.

12. *Mined area reclamation*

CLASSIFICATION OF RECLAMATION PROCEDURES

Six categories of procedures are routinely recognized in the field of mined area reclamation, regardless of the mineral mined or the type of mining practiced. These categories are: (1) no reclamation, (2) landscaping, (3) physico-chemico soil stabilization, (4) soil amelioration, (5) revegetation, and (6) reclamation for alternative uses such as the creation of recreation lakes in abandoned open pit mines or behind tailings dams. Clearly these procedures need not be mutually exclusive; however, any single procedure may be applied exclusively in given cases. The selection of the appropriate technique rests on choosing the optimal condition of the abandoned area relative to the environment in which it lies. For example, a permanent waterfowl sanctuary in conjunction with a tailings dam in North Michigan would not be suitable for most of arid Arizona.

Uses toward which the above procedures have been directed include grazing land, park and recreation areas, crop land, sports areas, campgrounds, sanitary landfills, lakes for fish propagation, and home sites. In most instances, areas such as tailings embankments can be reclaimed long before mining ceases (see Ludeke 1973).

The importance of the implementation of reclamation procedures is underscored by the magnitude of mine waste production. About 1.7 billion tons of mineral wastes are discarded annually in the United States. By 1972, total mineral solid wastes accumulated were 25 billion tons covering 2 million acres of land. Mineral wastes comprise 40 percent of the solid wastes produced in the United States annually and are second only to agriculture in this regard.

NO RECLAMATION

Historically, the practice of not renovating mined land was more prevalent in the United States than is the case today. The literature suggests that a similar historical pattern existed in the more developed countries in Europe. However, the trend in the United States is in the direction that many European countries have already gone—toward reclamation and, more ubiquitously, toward revegetation. This trend may be hastened by future legislation and by relatively recent studies which have revealed the consequences of not reclaiming mined lands. The 93rd United States Congress passed legislation to regulate certain types of surface mining (93rd Congress, House Bill H.R. 11500; Senate Bill S425). However, the legislation was vetoed by the President on the grounds that it would be too costly in terms of energy. But the two houses of the 94th Congress have passed similar legislation and at this writing it appears that the proposed act may become law in 1975. Included among the examples of studies that have demonstrated the consequences of no reclamation are those of Corbett (1965), Corbett and Agnew (1968), Agnew (1971), Galbraith, Williams, and Siems (1972), the Truax-Traer Coal Company (1971), U.S. Forest Service (1972), the Appalachian Regional Commission (1969), and other studies discussed in Chapter 4. The literature abounds with additional examples, many of which are tabulated by the Appalachian Regional Commission and by Skelly and Loy (1973). Undesirable consequences in addition to the production of acid runoff include dust and stream siltation caused by rapid erosion of unstable surfaces.

Among the more interesting and somewhat surprising observations reported by Corbett (1965) and Agnew (1971) is the fact that, in southwest Indiana, cast overburden from surface mining of coal produces significant amounts of stream flow during severe drought periods as compared with little or no yield from undisturbed areas. This observation implies that cast overburden induces ground-water recharge because this effect could have been realized only by prolongation of the base flow component of the hydrograph. The Truax-Traer Coal Company supports this observation by citing measurements showing that 46 percent of the rainfall on a coal waste pile infiltrates. In many localities, this effect would be desirable if it could be achieved without a concomitant degradation of water quality. However, Corbett and Agnew (1968) and Lejcher and Kunkle (1972) indicate that this frequently is not the case. Clearly additional research should be directed at increasing good-quality base flow by this method or by preventing ground-water recharge altogether.

The Truax-Traer Coal Company determined that the rate of acid formation on one of its coal piles averaged 198 pounds of acidity (as $CaCO_3$) per acre per day. Acid generation rates of this magnitude cannot be tolerated if stream degradation by runoff from mine wastes is to be precluded and the rights of downstream users protected.

The U.S. Forest Service describes the effects of coal mine wastes on the environment in the Shawnee National Forest in southern Illinois. Table 12.1 presents the characteristics of runoff from the unreclaimed mine waste area. Soil pH of the waste ranges between 2 and 4. The Forest Service reports that the runoff from the unreclaimed mine wastes has caused water pollution problems in three downstream counties. One citizen's suit has been filed against landowners in the area as a result of this runoff.

The Galbraith, Williams, and Siems report is of interest because it demon-

Table 12.1 *Characteristics of Runoff from Coal Mine Wastes in the Shawnee National Forest, Southern Illinois*

Parameters	*Average value in Palzo tract*
Acidity (as $CaCO_3$)	20,000
pH	2.3
Total iron	4,000
Aluminum	2,000
Total manganese	320
Magnesium	890
Copper	5.0
Zinc	20.0
Calcium	490
Chromium (Cr^{+6})	2.00
Total lead	0.25
Total cadmium	0.81
Sulfate	23,700

Source: After U.S. Forest Service 1972.

Note: In mg/l except pH.

strates, as have other studies, that acid lixiviation of mine wastes need not be restricted to coal mine wastes if no reclamation is employed. Acid runoff can be produced by any mine waste containing sufficient pyrite (or other appropriate sulfide) to oxidize and dissolve in precipitation falling on the waste. The mineral sought by the mining operation is in most cases immaterial to the lixiviation process. The critical factors in addition to mineralogy are the topography, hydrogeology, and meteorology in the area where the waste is disposed. Consequently, similar reclamation procedures are applicable to many types of mine wastes, given that mineralogy, local topography, hydrogeology, and meteorology are accounted for. Examples of differences consist of raising or lowering soil pH, depending on initial waste pH; irrigating or not irrigating, depending on meteorological conditions; dewatering or flooding abandoned pits, depending on hydrogeologic conditions; and differences in the extraction-rehabilitation plan caused by differences in the amount of waste produced relative to the volume of the mineral extracted, along with topography.

The fact that the philosophy of no reclamation is hopefully a philosophy of the past is demonstrated by a statement by James D. Reilly, president, Hanna Coal Division, Consolidated Coal Company, in the AIME 1968 volume on the surface mining of coal. According to Mr. Reilly:

> The basic approach to a sound conservation program is the restoration of mined land to its most practical and productive use within the shortest possible time. Mined land should be devoted to the highest and best possible uses compatible with the use of adjoining land. The entire reclamation program should be directed so that the final use of land will not reduce the taxable value below the valuation it carried before mining commenced.

LANDSCAPING

Landscaping is used here in the broad sense of the word. It implies shaping the surface of mine wastes or mined areas to achieve some predetermined objective. In most cases, this objective is control of erosion and leaching; but other objectives such as the shaping of a beach around an area that ultimately will be a lake are also achievable. Surfaces may also be shaped for dust control. Regardless of the objective, landscaping implies that the relocation and re-emplacement of the mine waste are conducted according to a plan. Coates (1973) characterizes this procedure as the "extraction-rehabilitation sequence" and recommends that the two be conducted simultaneously if at all possible.

Coal Mining

An example of landscaping as a part of a reclamation project is described in the U.S. Forest Service (1972) environmental study mentioned previously. In this particular case, the investigators determined that stabilization of coal mine wastes under the topographic and meteorological conditions existing in the Shawnee National Forest in southern Illinois would require reshaping of the surface of the waste to reduce all slopes to less than 15 percent. Since the reshaping had to be done after emplacement instead of according to a plan conceived prior to mining, costs were estimated to be as high as $600 per acre.

The key to success in economical, post-mining, landscape rehabilitation is the formulation and implementation of an extraction-rehabilitation operating plan, as opposed to an extraction plan which simply gets the ore out of the ground and recovers the product from it. Coates (1973) lists the design objectives of such a plan as follows:

1. Creation of final topography which blends with the adjoining undisturbed landscape.
2. Creation of a surface drainage pattern with no areas that promote ponding.
3. Creation of a soil condition capable of supporting plant life equal to that of the regional landscape.
4. Provision for ground cover for erosion control as soon as soil conditions allow.
5. Provision for use at some economic value.
6. Organization of the various steps in the operating sequence for optimum operating efficiency.

Although only the first two of these objectives deal directly with landscaping, they are essential to the last four objectives. It is clear also that in some reclamation projects it is desirable not to landscape for duplication of the surrounding environment. This approach is most applicable to recreation projects in areas where the water table is sufficiently near the surface to facilitate the creation of artificial lakes and to agricultural areas where forested hills are needed to relieve monotonous topography.

Coates continues by recommending that "overburden removal and disposal be designed and managed so that surface materials be replaced as surface materials in relocated positions." However, the author has observed a few cases in which reinstallations of surficial materials are not merited because of poor-quality soils. Strata containing high concentrations of pyrite must also be segregated for reemplacement at greater depths. This technique is particularly applicable to the cut and fill method of flat area strip mining. These methods are discussed in greater detail by Tilman (1968) and Pfeider and Wheaton (1968).

In the case of the contour stripping technique common to the steeper slopes of Appalachia, the two-waste-pile system is difficult to implement because stripped waste tends to move downslope, thereby making retrieval extremely expensive because work must be done against gravity. The downslope movement is also devastating to downslope vegetation and to the quality of down-gradient streams.

Reclamation of such areas through post-mining landscaping and revegetation has been demonstrated to be feasible, but it is expensive. For example, the reclamation costs at Moraine State Park, Pennsylvania, ranged from $420 to $2,700 per acre and averaged $1,455 per acre, although costs in addition to landscaping are included in these figures (Foreman and McLean 1973 and McNay 1970). Landscaping work included backfilling and regrading to terraces, approximation of original contour and special restoration as required. Most of the reclamation included installation of diversion ditches, slope drain flumes, and soil treatment. The strip mine reclamation was conducted in 19 strip pits located throughout the park. Restoration to approximate original contour was begun at, or beyond, the top of the highwall and extended to the toe of the spoil at a maximum angle not exceeding the original contour of the land before mining, with no depressions to accumulate water. Some open pits were backfilled by terracing; the steepest grade of the highwall and top of the spoil was limited to 45 degrees. The terrace was regraded to a maximum gradient of five degrees from the base of the highwall toward the top edge of the toe of the spoil with no depressions to hold water. With one exception, the maximum limits of work were 100 feet above the highwalls and 50 feet below the toe of the spoils. Diversion ditches were excavated above the highwalls as needed. All diversion ditches were designed at a maximum cross-sectional area of 10 square feet and a uniform descending gradient. Slope drain flumes were constructed across the backfill areas as needed. These flumes started at the point of intersection with the diversion ditch above the top of the highwall, extended down across the entire width of the graded backfill area, and discharged below the toe of spoil. The flumes were constructed with bituminized fiber, 1/2-section, 36-inch-diameter pipe (Foreman and McLean 1973).

On the basis of the experience gained at the Moraine State Park project, Foreman and McLean made recommendations with regard to landscaping stripped areas. They recommend terrace-type construction over backfilling to approximate original contour, particularly where final slope grades are steep. If contour slope restoration construction is not done in accordance with strict specifications, serious erosion problems develop and produce extensive wash-out of the seeded soil cover and re-exposure of the refuse to the atmosphere, they state. Terrace-type construction (with auxiliary drainage installation as needed) can eliminate erosion control problems almost completely and prevent the generation of high-turbidity runoff, they point out.

Diversion ditches above the highwall and slope drain flumes across the backfill are utilized in strip mine restoration areas to: (1) reduce surface water infiltration of the buried refuse material, (2) minimize wash-out of the seeded surface, and (3) prevent catastrophic erosion of the surface cover. The greater the drainage area above the highwall, the greater the need for diversion drainage facilities.

Other projects more or less similar to that at Moraine State Park have been conducted. In general, the techniques available for all such projects are similar and the cost per acre reflects the number of adverse conditions and the number of the available techniques employed. Table 12.2 presents costs for several reclamation projects. The costs range from $420 per acre to $2,598 per acre.

Coates suggests that the reclamation achieved at such great expense at Moraine State Park and other projects can be achieved more economically if the extractive process is modified so as to be compatible with his "extractive-rehabilitation sequence." In explaining his concept, the traditional practice of stripping is characterized as the loading of the coal into a hauling unit at the base of the operating face, on the bench or pit floor. The loaded coal hauling unit then moves back across the previously extracted area (bench). This movement pattern utilizes the extracted area as a haul road, thus it requires the overburden to be discarded down the hill or to be hauled a considerable distance to a separate dump area. In a large area stripping operation this does not create much of a rehabilitation problem except for absence of overburden to fill in the last line of extraction cutting. The problem is much more severe with contour stripping on steep slopes in that the overburden is cast downhill, below the bench, and the entire bench remains open as a haul road throughout the life of the operation. Replacement of the spoiled overburden to the original location on the bench can be accomplished after mining has been completed, but this earth-moving job against gravity is the source of much of the expense incurred by reclamation.

Coates suggests that the basic pattern be altered, so that the coal is loaded either at the top or bottom of the operating face and hauled forward across the unextracted areas along the contour of the slope. This would permit overburden to be spoiled directly behind the operating face because there would be no need to maintain the pit floor or bench as an open haul route. This revision in loading and hauling pattern would facilitate an improvement in the handling of overburden for rehabilitation purposes. It would eliminate most of the ecological and visual problems of contour stripping caused by the traditional casting of overburden down the mountainside. The refuse piled behind the stripping machine could be landscaped and covered with topsoil from the unstripped area ahead of the stripping machine—all as mining progressed.

Coates points out that continual operation of the system could be provided by dividing the width of the bench into two segments, a stripping segment and a coal extraction segment. Thus the major equipment item, the stripping shovel, would operate full time.

The cost savings in reduction of environmental damage would be high. The

Table 12.2 *Costs of Selected Reclamation Projects on Abandoned Strip Mines*

Project number	Cost (1967 $)	Total acres	Length of open cut (ft)	Average height of highwall (ft)	Acres of open cut affected	Total ($/acre)	Project cost parameters affected ($/acre)	($/acre/ft highwall)
1	$ 5,455	2.1	2,000	20	2.1	$2,598	$2,598	$104
2	2,680	2.7	500	25	2.7	993	993	50
3	4,110	4.5	800	35	3.7	913	1,142	33
4	7,450	6.0	1,500	50	4.8	1,242	1,552	31
5	4,545	8.1	800	20	3.4	561	1,337	67
6	3,295	8.6	750	12.5	4.1	383	804	64
7	3,875	8.6	1,500	12.5	8.6	451	451	36
8	6,260	10.0	3,000	20	10.0	626	626	31
9	5,600	15.2	2,400	17.5	7.9	368	709	41
10	5,065	15.5	800	15	7.4	327	684	46
11	10,980	17.7	1,000	35	7.4	620	1,484	42
12	13,952	18.0	3,500	30	12.1	775	1,153	38
13	8,365	21.7	3,200	15	15.9	385	526	35
14	12,455	30.0	4,000	20	18.7	415	666	33
15	11,900	32.1	2,600	25	9.2	371	1,293	52
16	23,450	33.1	2,600	35	11.9	708	1,970	56
17	10,155	37.3	6,000	20	21.2	272	479	24
18	13,900	53.3	7,200	20	33.1	261	420	21
19	38,465	57.3	3,500	50	15.0	649	2,564	51
20	28,450	68.9	6,500	30	40.9	413	696	23
21	53,930	86.9	4,800	60	44.3	621	1,217	20
22	88,670	111.0	6,000	50	35.0	799	2,533	51

Source: After Appalachian Regional Commission 1969.

additional operating cost, as compared with the old practice, would be the cost of elevating the coal from the pit floor or bench to the higher haul road level and the additional shovel swing time to move the overburden 180 degrees away from the face instead of 90 degrees.

Allen (1973) describes an experimental coal stripping project directed at the same objective as the technique described by Coates, but with allowance given for a 20-foot (vertical) spoil downslope from the bench. Allen estimated additional costs of $2.60 per ton of coal to achieve the reclamation in the experimental project.

Some states have passed legislation specifying procedures which must be followed by surface coal mine operators to implement Coates' extraction-rehabilitation sequence. Heine and Guckert (1973) set forth the typical method of operation required by law at surface coal mines in areas of less than 14 degrees (25 percent) slope in Pennsylvania (see also Goldberg and Power 1972 and Appendix G). Heine and Guckert list procedures as follows:

1. Brush and trees are cleared for construction of access or haul roads into the area to be mined. Trees of value are harvested to decrease cost. Construction of haul roads is undertaken with care so that fill material does not contain acid-producing minerals. Where banks are created which could erode, quick-growing vegetation is planted. The area where mining is to be initiated and where appurtenant storage, repair and office buildings are to be located is cleared of trees and brush. This is kept to a minimum so that as little area as possible is subject to erosion.
2. Bulldozers and/or scrapers are used to scrape off topsoil and any needed subsoil for segregation and storage. Sometimes this soil is stockpiled above the highwall for easy handling after backfilling is completed. Stockpiled soil is usually not seeded because the practice of concurrent backfilling requires imminent use of this soil before erosion-controlling vegetation could be established on it.
3. The operation of overburden removal is undertaken with the spoil being placed on the "low wall" side of the cut a sufficient distance from the highwall to permit movement of men and machinery in the pit.
4. As the overburden removal operation reaches strata with acid-forming potential, this material is segregated from the clean spoil and stored in a corner of the pit or a prepared area on the low wall where contact with water is minimal. This is one of the most important steps in the entire surface mine operation from the standpoint of prevention of water pollution from acid discharges.
5. After the coal is exposed and removed, reclamation follows as soon as possible. In some large operations, the steps in the entire operation are carried out almost simultaneously, i.e., brush removal, topsoil segregation, spoil removal, acid strata segregation, coal removal, and backfilling. Pennsylvania's regulations do not allow the total length of open cut to exceed 1,500 feet, except where specifically approved by permit when large-size equipment is involved. This requirement has reportedly minimized pollution and has practically eliminated forfeiture of bonds.

6. Reclamation includes layering and compacting of the acid-forming refuse at the bottom of the pit (depending upon ground-water conditions), replacement of spoil material until approximate original contour is achieved, placement of the previously segregated topsoil and finally liming, fertilizing, and planting of the area to establish a quick-growing grass cover.
7. Several design features are included to control water pollution:
   a. Water diversion ditches are usually required along the top of the highwall and at other appropriate locations to minimize erosion and to control the amount of water entering the pit of the mine.
   b. Usually two earthen settling basins are constructed which receive all pumped water from the floor of the pit. This water must be neutralized prior to pumping if it is found to have a pH less than 6.
   c. Under no circumstances may the low wall of the pit be breached to allow a gravity discharge. All water from the pit must be pumped. This is to assure that, after reclamation, no permanent acid seepage will be established at the points where the low wall was breached.
   d. In some extreme cases, particularly when immediate downstream water use must be protected, erosion control basins are constructed which receive runoff from the entire disturbed area of the mine to prevent downstream siltation. In the state of Pennsylvania, newly adopted erosion control regulations will probably result in the use of basins on every operation.
   e. A barrier of undisturbed coal and overburden of at least 25 feet in width is left between the outside wall of the pit and the outcrop of the coal. This helps to promote the re-establishment of the ground-water table above the coal seam after backfilling is completed by retarding ground-water movement through the restored area. The resulting inundation of the seam and the acid-forming refuse practically eliminates acid water formation and subsequent acid seeps.

As indicated previously, these methods are difficult to employ in steep terrain (greater than 14 degrees or 25 percent slopes according to Pennsylvania standards) because of a shortage of space to handle the overburden without allowing it to move downslope under gravity, thereby destroying the entire hillside and making the waste difficult to retrieve, reemplace and landscape. Heine and Guckert propose a modification of the block-cut method of mining as a solution to this problem, with the possible exception of its use with auger mining. The unmodified block-cut method consists of removing overburden and placing it around the periphery of a box-shaped cut. Subsequent to coal extraction the waste is replaced in the cut and its surface landscaped to blend in with its natural surroundings. Drainage is designed to prevent erosion in the standard fashion.

The modified block-cut method discussed by Heine and Guckert adapts this method to steep terrain by backfilling with waste from succeeding blocks rather than from the spoil producing block. The method is similar but more systematic than the method discussed by Coates. Heine and Guckert utilized the diagrams in Figure 12.1 to describe the method as follows:

The overburden is removed from the area noted as Cut 1 (Diagram A) and placed in such a location that it will be retrievable for reclamation or so it can be graded, blended into the topography and vegetated. This initial cut should be carefully selected, preferably in a swale of the hillside or where low areas requiring fill material are nearby. The initial cut is often two or three times larger than subsequent block-cuts, particularly when subsequent blocks will be excavated in both directions around the mountain [as shown in Figure 12.1]. Every block is excavated into the hillside to the maximum highwall height economically feasible. . . . Additional constraints on this distance are existence of deep mineworkings, streams, roads, and power lines. The overburden from Cut 2 is then pushed into Cut 1 as shown in Diagram B. When mining is proceeding in both directions from the first cut, the overburden from Cut 3 is also pushed into Cut 1 as shown in Diagram C. The spoil from succeeding block-cuts is pushed to the previous cut as shown in the subsequent diagrams. In Pennsylvania, an outcrop barrier of 25′ of the undisturbed coal is required as shown in the diagrams.

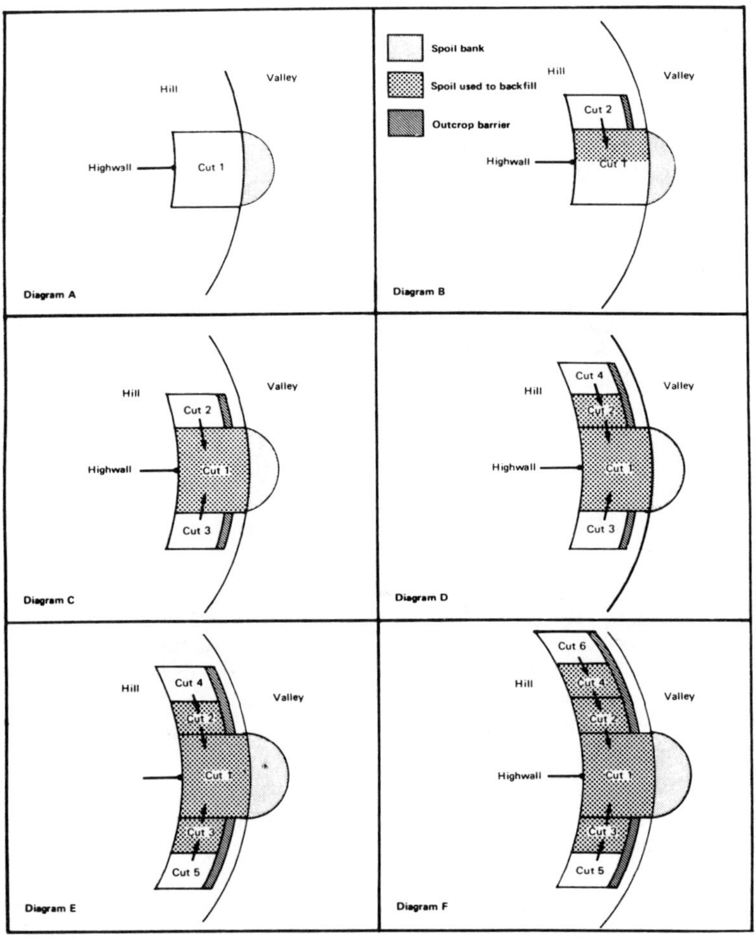

Figure 12.1 *Plan view of block method of strip mining (after Heine and Guckert 1973).*

Care must be taken to assure that the position of acid-forming material (boney) and topsoil are not reversed in the reclaimed areas. Whenever practical, topsoil is removed and stored either above the highwall or downhill from the cut in a place chosen for each operation. Efforts to protect it from erosion include compacting the soil and sloping the adjacent earth to divert surface water.

As the waste is pushed into the preceding block-cut, a depression is left near the highwall of the cut being filled. The boney is deposited in the depression and covered promptly to minimize oxidation and the production of acid water.

Heine and Guckert point out that this method does not disturb the ground surface downhill of the cuts except where the initial cut is taken. They present details of an application of the method in Pennsylvania in which the following advantages were delineated.

1. The bulk of the overburden is to be handled only once, thereby significantly reducing costs.
2. The area disturbed is reduced over conventional methods by keeping most of the spoil inside the mined area.
3. Revegetation costs are reduced.
4. Traditional mining equipment is utilized.
5. If properly planned, the procedure does not interfere with continuous production of coal.
6. Explosives can be used but should be planned to fit procedures involved in this method of mining. The method is, therefore, not limited to operations where the overburden requires no fracturing.
7. The method introduces no new safety hazards and should reduce the accident versus coal production ratio.
8. The concurrent reclamation inherent in the method reduces the amount of open pit and increases the speed of revegetation, thereby reducing the formation of acid mine drainage and siltation.
9. It is applicable to multiple seam mining and still retains the previously noted advantages.

The method known as longwall mining may be the most promising technology of the future with respect to the extraction-rehabilitation sequence in coal mining on steep slopes. The longwall process, mentioned briefly in Chapter 1, has been described by Lusk and Mulhern (1972) and by several other writers. It is now being tested under the sponsorship of the U.S. Environmental Protection Agency as an alternative to surface strip mining. As shown in Figures 12.2 through 12.5, the roof strata above the coal seam is supported by jacks ("chocks") during the process of mining, and after mining it is allowed to cave in a controlled manner. The sequence of events is: (1) the roof is supported so that the overburden load is carried by jacks, (2) the roof in the mined-out area behind is allowed to begin to cave, (3) the roof strata breaks at the jack line and expands to fill the void in the mined-out area, and (4) the mined-out void is filled so that a minimum of the load of the roof is carried by the jacks.

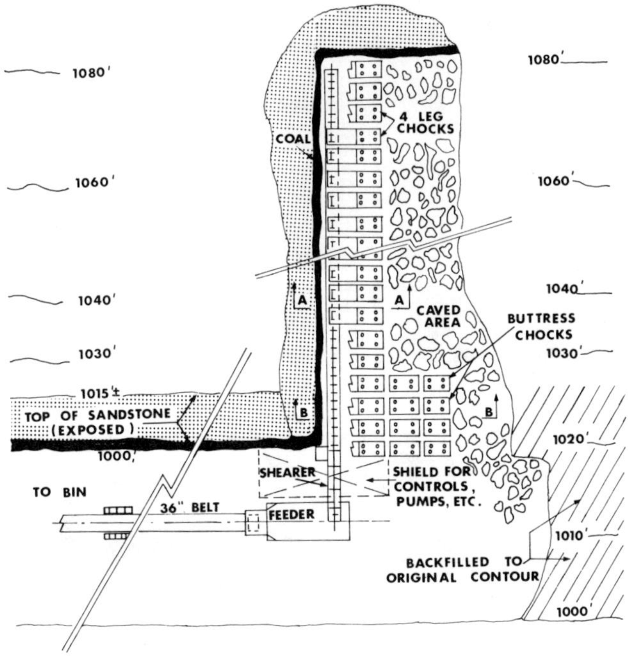

Figure 12.2. *Plan view of a longwall mine (after Lusk and Mulhern 1972).*

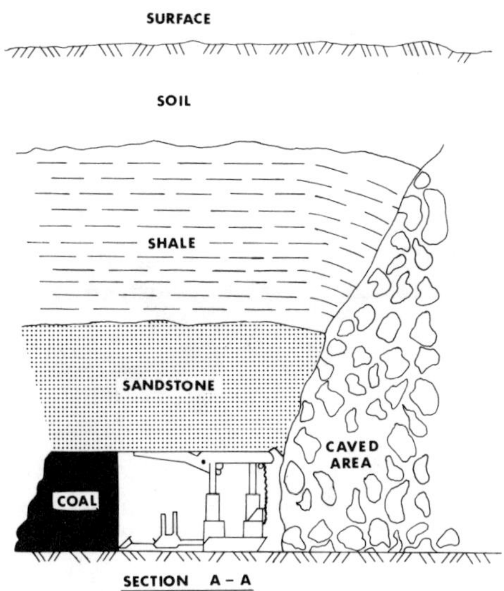

Figure 12.3. *Cross-section view of longwall stripping system (after Lusk and Mulhern 1972).*

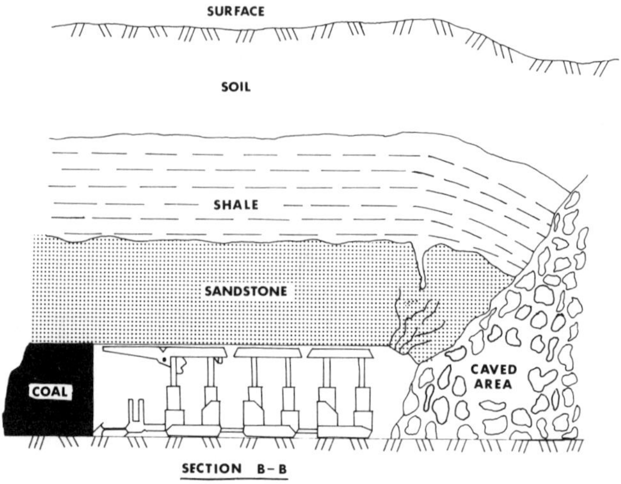

Figure 12.4. *Cross-section of longwall stripping system along highwall (after Lusk and Mulhern 1972).*

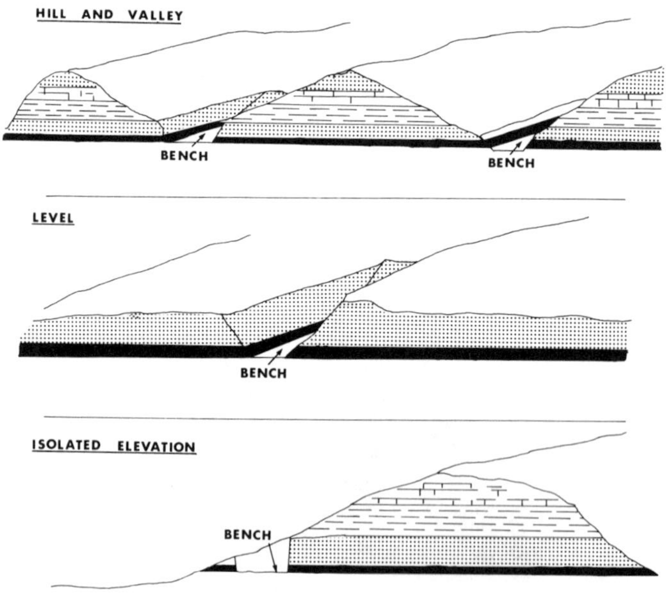

Figure 12.5. *Types of terrain appropriate to longwall stripping system (after Lusk and Mulhern 1972).*

Mining in the EPA-sponsored demonstration project described by Lusk and Mulhern is to be accomplished by:

1. Opening a narrow bench located 10 to 15 feet into the slope from the coal outcrop and running parallel to the coal outcrop.
2. Driving an entry or trench into the coal seam perpendicular to this bench; this perpendicular trench is the longwall face. It will advance along the coal outcrop to recover the coal.
3. Advancing the second bench up or down the valley at a rate permitting continuous progression of the longwall system.

Coal cutting is to be done by a shearer which will travel inward along the face and return to the outcrop end for subsequent cuts.

In the conventional longwall mining operation, mining occurs beneath several hundred feet of overburden so that stress redistribution does not produce caving to the surface. Caving above shallow coal seams, on the other hand, is likely to extend to the surface, particularly near the outcrop of the coal. Horizontal movements in this area may prove troublesome and could cause collapse of the highwall. Considerable effort in the current demonstration project is directed at this problem (see Lusk and Mulhern 1972).

Preliminary estimates indicate that coal can be produced by the longwall method for approximately $6.50 per ton (1972 dollars). One favorable cost item is the fact that the entire coal seam can be recovered.

The anticipated reclamation benefits of the longwall method would include:

1. Little or no landscaping should be required.
2. Production of acid mine drainage should be impeded.
3. Revegetation should be unnecessary because the topsoil is not removed.

Non-Coal Mining

The concept of the extraction rehabilitation sequence for non-coal mining activities is relatively new. Most deep and open pit metal mines and associated mills produce tailings and other wastes which are accumulated in piles allowed to stand at the angle of repose of the coarser fraction. Some are susceptible to erosion and leaching by runoff. Some waste piles are revegetated, but many are revegetated poorly or not at all. In recent years revegetation of tailings has become more prevalent, possibly because optimal management of mine wastes clearly influences the degree to which the public is willing to allow mining to be initiated in new areas.

During the late 1960s and early 1970s, government agencies began to play a significant role in this process because, in the west in particular, much mining occurs on land belonging to the public. As indicated by Paller and Shultz (1973), the process of negotiating reclamation efforts between government and industry is now widespread.

Additional evidence that the non-coal mineral industry is addressing the problems of reclamation more intensely is provided by the papers and discussion at the first International Tailings Symposium in Tucson, Arizona, in 1972.

D'Appolonia, et al. (1973), for example, stated at that symposium that estimates of abandonment costs should be incorporated in the overall economic analysis of every proposed project. Details of calculating the economics of abandonment consist of comparing post abandonment remedial measure costs to costs of preventive measures incorporated into design and construction. Economics are calculated on the assumption that the key factors for abandonment of a tailings disposal facility are ultimate compatibility of the disposal area with its environment and the assurance of a stable structure which will not rapidly erode or be leached. The first factor implies landscaping and revegetation; the second factor implies utilization of sound engineering and site selection practices as discussed in Chapter 11; and the third factor implies a combination of these.

Unfortunately, the literature does not contain sufficient information on the erodibility of tailings under varying conditions of climate to permit generalizations on appropriate landscaping procedures for tailings piles. Optimum embankment slope angles should be a function of erodibility, as well as the engineering factors discussed in Chapter 11, but considerable additional research on erodibility of tailings under different climatic conditions is needed to determine what slope angles are optimal with respect to controlling erosion after abandonment and with respect to maintenance of a vegetative cover. For example, the guideline of 15 percent slopes established by the U.S. Forest Service (1972) for the coal mine wastes in Illinois would probably not be appropriate for tailings because coal mine wastes have some cohesion, whereas tailings are cohesionless.

Improvement of soil properties prior to attempts at vegetation is essential on some mine wastes. The bare rock and salts seen on this embankment are among the problems encountered in preparing for plant growth.

U.S. Bureau of Mines photo

SOIL AMELIORATION

Amelioration of soil properties is essential on some mine wastes prior to attempts at revegetation. Salinity, pH, and nutrient content are among the most important factors which sometimes must be ameliorated. Dean and Havens (1973) cite examples of mine tailings whose salinity is sufficiently high that the concomitant osmotic pressure dehydrates plants to the point of death. Ludeke (1973) and Knudsen, et al. (1971) recommend leaching by irrigation prior to revegetation as the most feasible method of ameliorating such wastes. West Virginia University (1971) presents evidence that presently available native and domestic plants give inadequate ground cover on mine wastes for pollution control unless the plants are rooted partly in materials with pH greater than 4.0. This observation suggests that the upper portion of mine wastes with pH lower than 4.0 must be ameliorated by raising soil pH to 4.0 or above if revegetation is to be completely successful.

Ameliorating agents which have been demonstrated to increase pH are lime, crushed limestone, asbestos mill tailings, fly ash, and sewage (see, for example, Leroy 1973, Dean and Havens 1973, U.S. Forest Service 1972, Ludeke 1973, Peterson and Gschwind 1973, Sutton 1973, Capp and Gilmore 1973, and Skelly and Loy 1973). Ameliorating agents which have been demonstrated to decrease pH include pyrite rich mill tailings, powdered sulfur and any of several acids (see Leroy 1973 for case history utilizing pyrite rich tailings and sulfur). Nutrients may be added by standard fertilizers or along with sewage sludge used for pH control.

Cost of any of the above agents is a function of the distance they must be transported to the waste disposal site, among other things. The somewhat unusual conditions prerequisite to the availability of sewage sludge and fly ash are obvious. Skelly and Loy list the representative average cost of fertilizer as $50 per acre; crushed limestone as $1.10 to $4.50 per ton, depending on location; and fly ash as $0.25 to $1.50 per ton (short tons). Dean and Havens estimate the cost of sewage sludge at $1.00 per ton. Haulage costs are estimated by Skelly and Loy at $0.10 per ton mile, which probably is low.

REVEGETATION

Revegetation is the process of installing a vegetative cover on mineral resource wastes or mined areas. Revegetation is utilized to:

1. Help stabilize erodible slopes in order to minimize stream pollution.
2. Control dust.
3. Maximize the aesthetic value of the area.
4. Maximize evapotranspiration so that a minimum percentage of precipitation enters the runoff cycle.
5. Facilitate the crop-producing potential of some areas for purposes of profit.
6. More rapidly stabilize the oxidation of pyrite and the concomitant production of acid mine drainage by preventing continuous exposure of new pyrite by erosion.

Hydromulching and hydroseeding operations at a tailings dam in Arizona. Note the tailings distribution pipeline where spigotting from the peripheral pipe deposits sand on the top of the dam. *Cyprus Pima Mining Company photo*

According to the Appalachian Regional Commission (1969), practically all states in which there is surface mining for coal require reforestation and/or revegetation to restore mined areas. Table 12.3 presents a summary of the requirements of several states. Similar laws in states with other types of mining are not as prevalent.

Selection of an appropriate vegetative cover depends primarily on the condition of the waste to be covered and on local topographic and climatic conditions. The most significant characteristic of the waste is soil pH; however organics, nutrients and grain-size distribution also play major roles. The most significant climatic factor is the nature (rain or snow) and distribution of precipitation, but mean, maximum, and minimum temperatures (annual and daily) also are significant factors. Heavy metals can affect some plants if their concentrations are sufficiently high. Galbraith (1971) has shown that some plants increase their heavy metal uptake significantly when grown on mill wastes. Others die. Currier (1973) provides additional insight on the significance of each of these variables and instructions for planting seeds so that germination will occur under adverse conditions. Ruffner (1973) provides additional ideas on plant selection along with a classification system for variables affecting plant growth. Ruffner also provides a discussion of recently developed plant material suitable for revegetation of mine wastes, which should be particularly useful on soils with low pH in areas where water is plentiful and

Table 12.3 *Comparison of Strip Mining Requirements*

	Kentucky	*Maryland*	*Ohio*	*Penna.*	*Virginia*	*W. Virginia*	*Tennessee*
Bond required:							
Minimum	$2,000	$1,600	$2,000	$5,000	$2,500	$2,500	None
Add'l per acre	$100-$500	$ 200	$ 300	$ 500	$ 75	$100-$500	$100-$200
Reclamation plan required	Yes	Yes	Yes	Yes	Yes	Yes	Yes
Backfilling/grading required	(1)	(2)	(3)	(4)	(5)	(6)	(7)
Replanting required	Yes	Yes	Yes	Yes	Yes	Yes	Yes
Penalty for failure to reclaim:							
Forfeiture of bond	Yes	Yes	Yes	Yes	Yes	Yes	Yes
Denial of new permit	Yes	Yes	Yes	Yes	Yes	Yes	Yes
State reclamation fund	Yes	Yes	Yes	Yes	Yes	Yes	No

1. Eliminate spoil peaks; grade to contour; backfill highwalls; bury acid-forming materials; impound runoff.
2. Grade spoil banks to approximate original contour; impound water; bury acid-forming materials.
3. Grade peaks and ridges to rolling topography; impound runoff; construct ponds; bury acid-forming materials.
4. Grade peaks and depressions to rolling topography or approximate original contour; bury refuse and debris; impound water for ponds where approved.
5. Grade peaks and ridges to gently rolling topography; impound water where approved.
6. Grade peaks and ridges to a rolling topography; impound runoff water; bury acid-forming materials, construct ponds.
7. Reduce spoil peaks to soft minimum width; cover face of coal seam; bury toxic materials.

Source: After Appalachian Regional Commission 1969.

where winters are mild. The most pertinent portion of Ruffner's discussion of recently developed plant material suitable for mine spoil reclamation follows:

Deertongue (*Panicum clandestinum L.*)

> Deertongue grass is a native grass found throughout the Northeast. It is one of the panicums more frequently found invading low fertility and eroded sites in the Appalachian Region. It commonly is seen along roadbanks where soils are strongly acid, [in] industrial waste areas containing toxic materials, and [in] other critical sites devoid of any other cover. Deertongue grass is persistent once it is established. . . . [The plant] readily adjusts to minor changes in the environment.

This successful growth on a dam in Arizona stabilizes the tailings to prevent wind erosion and, at the same time, enhances the beauty of the area.

Cyprus Pima Mining Company photo

Today, mined out areas are revegetated to take advantage of their crop-producing potential. This former coal strip mine was filled in, rough graded, then seeded and fertilized to turn it into farm land. Amax Journal, *Amax Inc. photo*

Deertongue grass is a warm-season perennial. It grows to a height of 18 to 36 inches (46 to 92 cm) and spreads slowly by rhizomes. Leaves are ½ to 1¼ inches (1.3 to 3.2 cm) wide and 4 to 8 inches (10 to 20 cm) long.

Outstanding characteristics for mine spoil reclamation are: the plant has a large, dense, fibrous root system that binds soil particles together; it tolerates a site low in fertility and acid to a pH of 4.0, although it . . . needs . . . lime and fertilizer applications to establish stands on the more acid sites; it persists at low fertility levels; and it provides good protection from erosion. It is winter hardy in West Virginia at elevations of 4,000 feet (1212 m) and survives the climate in Maine.

A new synthetic variety has been developed at the [United States Soil Conservation Service] SCS Big Flats Plant Materials Center, Big Flats, New York. Twenty strains of deertongue from areas representing the Northeast were bulked to develop a variety having a broad genetic base. It currently is assigned accession number NY-4950. . . .

"Lathco" flatpea (*Lathyrus sylvestris L.*)

"Lathco" flatpea is a viny, rhizomatous, perennial legume with tendril-bearing stems. It normally forms a dense mat, 2 to 3 feet (61 to 91 cm) high, but will climb to 6 feet (183 cm) or more if support is available. It is hardy, drought-tolerant and adapted to a wide variety of soil conditions, and does well on low-fertility sites such as sands, gravels, and shales. It is best suited to well-drained soils. It will grow on sites that are moderately well-drained but is not adapted to poorly drained soils.

Here a vegetative cover on a mined area has produced a lush pasture.

Amax Journal, *Amax Inc. photo*

"Lathco" has been tested as a conservation cover plant on critical areas from Kentucky to Maine. It is suited for cover on haul roads, gravel pits, utility rights-of-way, roadbanks, dams, borrow areas, and mine spoil. It is a valuable wildlife cover plant and is utilized as food by some wildlife species. The primary use of "Lathco" . . . is . . . as a conservation cover plant. It does have some potential as a forage plant, but has not been fully tested, so it is not presently recommended for this purpose.

Flatpea has excellent soil stabilization qualities. It forms a dense mat of herbaceous litter and controls most woody plants invading the site. It is drought tolerant and thrives at a pH of 4.8 to 5.0, which is slightly below site requirements for crownvetch. Its large seed makes it ideal for seeding relatively inaccessible areas with mechanical equipment.

"Arnot" bristly locust (*Robinia fertilis Ashe*)

"Arnot" bristly locust is an attractive open-branched shrub that forms thickets where it is well adapted. It spreads rapidly by root suckers, especially on coarse textured soils. Outstanding with this species is the habit of new growth initiating where rill erosion exposes a root.

It has a broad soil adaptation and is climatically adapted throughout the Northeast [United States]. Hardiness limits have not been determined, but plantings made near Burlington, Vermont, in Zone 5 (−20° to −10°F.) have survived. Successful plantings have been made on acid mine spoil with a pH of 4.0, as well as on alkaline sites with a pH above 7.5. "Arnot" bristly locust is drought tolerant, can thrive and form dense thickets on low-fertility sites.

It grows rapidly, spreads, and thickens, developing a dense canopy and good leaf litter for soil protection. Its nitrogen fixation encourages growth of herbaceous plants which provides additional cover.

It blooms from late May through June with numerous large rose-purple flowers. On fertile sites, mature plants reach a height of 8 to 10 feet, but 5 to 7 feet is more common. Good seed crops are produced annually, and plants often start producing seeds the second year after planting.

"Rem-Red" Amur honeysuckle (*Lonicera mackii maxim*)

Amur honeysuckle, a species introduced from northern China and Korea, is a multi-stemmed, vase-shaped shrub growing to height of 12 and 15 feet (2.7 to 3.6 m). The foliage is similar in appearance to that of Tatarian honeysuckle, which has been widely used as a source of summer-ripening wildlife food. White flowers, which become yellowish with age, are borne during May and June.

The bright red fruit of Amur honeysuckle is about one quarter of an inch in diameter and matures in late September and October. Those fruits not eaten in November or December turn dark brown and become raisin-like. Many remain on the plant throughout much of the winter until used for food. In contrast, fruit of Tatarian honeysuckle is usually completely eaten by songbirds by late August. Plantings made over the past 10 years have shown Amur to be climatically adapted from Massachusetts to South Carolina and westward through Missouri.

Amur honeysuckle grows best on deep, well-drained soils with a sandy, loamy, or moderately fine texture. It is not well adapted to extremely droughty or wet sites. Growth is poor on sites with unusually low fertility. Best growth occurs on mine spoil at acidity levels above pH 5.5. It is not adapted to spoil with pH below 4.8. Response to fertilizer is good at planting time.

Often the growth of new transplants is slow for the first year or two. Once established, transplants reach a height of 6 feet (1.8 m) in 3 or 4 years. Berries may be produced as early as the second year, if the plants are growing on fertile, well-drained sites.

Table 12.4 is an illustrative list of the plants considered by the United States Soil Conservation Service (SCS) to be adaptable to intermediately acid and neutral, loamy mine spoils in Kentucky. Similar data for other areas have been developed by SCS and by the Pennsylvania Research Committee on Coal Mine Spoil Revegetation (1971).

Revegetation may be accomplished by seeding or transplanting. The techniques of seeding are drilling, broadcasting, and hydroseeding. Hydroseeding has been more widely adopted for wastes with appreciable slopes and to which much must be added to enhance germination. Hydroseeding can achieve liming, fertilizing, mulching, and moisturizing in one operation.

Table 12.4 *Vegetative Species Guide for Medium Acid, Slightly Acid, and Neutral (pH 5.5-7.0)*

Slope (%)	*Texture and stoniness*	*Grasses*	*Legumes*	*Shrubs or ground cover*	*Trees*
0-30	Loamy non-stony stony	Annual rye Blackwell switchgrass Orchardgrass Redtop Tall fescue Timothy Weeping lovegrass	Alfalfa Annual lespedeza Crownvetch Flatpea[1] Red clover Sericea lespedeza Sw. clover White clover	Amur honeysuckle[1] Autumn olive Bristly locust[1] Crataegus spp. Fragrant sumac Indigobush[1] Jap. fleeceflower Jap. honeysuckle Kudzu Multiflora rose etc.	Black locust Black walnut Chinese chestnut Cottonwood Europ. bk. alder Green ash Hybrid poplar Loblolly pine N. red oak Pitch pine etc.

Source: After Ruffner 1973.

1. New plant material.

According to Skelly and Loy (1973), costs of broadcast seeding range from $200 to $500 per acre; costs of transplanting seedlings average about $0.06 per plant; and costs of hydroseeding range from $200 to $800 per acre, depending on what materials are applied with the seed. Costs presented by the Appalachian Regional Commission (1969) appear to be slightly higher in some cases and considerably lower in others.

In the case of coal mine wastes, neutralization of soil pH is considered by most investigators to be essential to the establishment of a vegetative cover in a reasonable length of time. Neutralization can be achieved by the well-established use of lime or it can be achieved by more exotic techniques. Lejcher and Kunkle (1972) report successful experiments by the U.S. Forest Service with treated anaerobic sewage sludge on wastes in the Shawnee National Forest, Illinois, with the ultimate objective of using large volumes of sludge from Chicago for this purpose. Additional encouraging results have been supplied by Hunt, Sopper, and Kardos (1971), Peterson and Gschwind (1973) and Sutton (1973).

In the Forest Service experiments described by Lejcher and Kunkle (1972), four 500-square-meter plots were treated with anaerobic sewage sludge in November 1970 at treatment levels of 304, 178, 78, and zero dry metric tons of sludge per hectare. The untreated plot was used as a control. In the spring, the plots were seeded with K-31 tall fescue (*Festuca arundinacea*) and weeping lovegrass (*Eragrostic curvula*) at the rates of 22 and 8 kilograms per hectare, respectively. Within 30 days germination had occurred on all but the control plot. At the end of the first growing season, a 90 percent vegetative cover was reported on the most heavily treated plot; the plots with lower treatments were less than 50 percent covered. In the second growing season only the 304-ton-per-hectare treatment continued to support as much vegetation as in the first growing season.

The maximum level sludge treatment effectively raised the pH of the acid spoil. Figure 12.6 shows the subsurface pH profile of the most heavily treated plot versus the control plot 18 months after treatment (July 1972). The Major difference in pH between the treatment and control plot spoil occurred in the top 20 to 25 centimeters. On the surface, the pH averaged as follows:

Sludge application (dry metric tons/hectare)	*Plot's average soil surface pH**
304	6.2
178	5.2
78	4.7
0	2.3

United States Forest Service (1972) also reports the successful use of redtop, Canada bluegrasss, and annual rye grass on coal mine wastes treated with lime and standard nutrients in the form of fertilizers. Similar results are

* These data represent the average of nine observations per plot taken at various times over the first year following treatment (after Lejcher and Kunkle 1972).

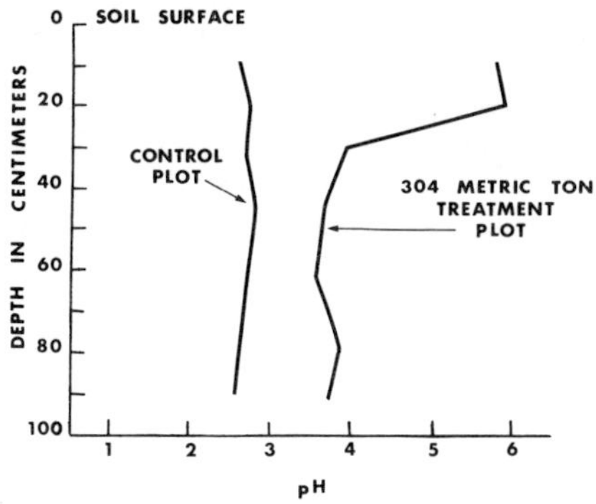

Figure 12.6. *Effect of sludge treatment on pH of mine spoil (after Lejcher and Kunkle 1972).*

reported by the Truax-Traer Coal Company (1971) for orchard grass and lespedeza, especially when topsoil is added with water.

Lejcher and Kunkle (1972) examined the effect of the above treatment on water quality by collecting samples from subsurface drains beneath the plots. These samples were analyzed for chemical and bacterial contamination. The heaviest treatment (304 metric tons of sludge per hectare) reduced acidity as well as aluminum and iron concentrations in the runoff by more than 60 percent and manganese concentrations by more than 50 percent. But unexplained slight increases were observed for chromium, cadmium, and zinc. It should be noted that these reductions may have been due to the application of the sludge rather than the establishment of the vegetative cover.

This U.S. Forest Service project at the Shawnee National Forest in Illinois is continuing on a larger scale with the importation by rail of sewage sludge from Chicago. The continuing project is designed to dispose of the sludge as well as to revegetate the mine waste.*

Revegetation procedures have been demonstrated for mine wastes other than those from the recovery of coal. The work of Dean and others at the U.S. Bureau of Mines' Salt Lake City Metallurgy Research Center is among the more notable. Ludeke (1973) also has achieved impressive results on Pima Mining Company's open pit copper mine near Tucson, Arizona. Leroy (1973) has solved many of the problems of revegetation in harsh environments. Dean and Havens (1972) cite several characteristics common to mill wastes which make revegetation difficult. In summary, these are that tailings usually:

1. Are deficient in plant nutrients.
2. Contain excessive salts and heavy-metal phytotoxicants.
3. Consist of unconsolidated sands that destroy young plants by sandblasting and/or burial during eolian erosion.
4. Lack normal microbial populations.

* Additional reports may be obtained from U.S. Forest Service, Region 9, 633 West Wisconsin Avenue, Milwaukee, Wisconsin.

Other problems are caused by the sloping sides of waste piles which receive varying amounts of solar radiation depending on direction of exposure. Photosynthesis of plants is not continuous and under high-temperature conditions, photosynthesis reportedly may almost stop. The light color of mill tailings may reflect excessive radiation to plant surfaces, thus intensifying physiological stresses. For these reasons, vegetation appropriate on northern and eastern exposures may not be suitable for southern or western exposures. Eolian erosion is considered to be the greatest threat to vegetation longevity. This problem has been overcome during the growth stage of revegetative stabilization by massive sprinkler irrigation, hydroseeding, soil coverings, and chemical-vegetative stabilization procedures. Dean and Havens (1972) have utilized fertilizers on tailings, including nitrogen and phosphorus in the amounts of 30 and 75 pounds per acre as N and P_2O_5, respectively. Some tailings require potassium in amounts up to 40 pounds per acre.

Table 12.5 presents costs which Dean and Havens have derived for various methods of tailings preparation leading ultimately to revegetation. The costs presented assume that extensive chemical treatment is not needed. The investigators point out also that costs for reclaiming sloping dike areas are 25 percent greater than costs for flat areas. The costs presented are estimated for a tailings accumulation consisting of 80 and 20 percent, respectively, of pond and dike areas.

Costs can be greater than those presented in Table 12.5 if excessive salinity or acidity are encountered. Liming and application of sewage sludge have already been mentioned as a means of ameliorating acidity. Soluble salts can be leached downward by excessive irrigation prior to revegetation if water is available. However, the leaching solution is complicated by the fact that tailings may contain both salts and pyrite. In arid environments, the pyrite is likely to be inactive and the initial pH high. Dean and Havens report that the Kennecott Company's Utah Copper Division mill waste is such a material and it is discarded in a semiarid environment. The material has a pH of 7.8 when fresh, along with a salinity equivalent to 2.4 atmospheres osmotic concentration, and approximately 1.3 percent pyrite. The salts in the tailings produce an osmotic gradient that transfers fluid from the plants. Thus vegetation dies of dehydration as irrigation of the plants leaches away the salts; the pyrite oxidizes and the pH may drop from 7.8 to less than 3.0 within a month's time.

Dean and Havens have attempted to overcome this problem with careful use of sewage sludge. They conducted a series of tests in which 2-inch layers of sewage sludge were placed at different depths in 18-inch thicknesses of the Kennecott tailings. In another test series, an additional equivalent of 15 tons per acre of sewage sludge was mixed into the top 3 inches of tailings. Crested wheatgrass, ranger alfalfa, and rye grain seeds were planted for both series of tests with barley used as the fourth seed in one test series, and a small transplanted tomato was used as the fourth plant variety in the other series of tests. The combined germination and survival percentages for wheatgrass, alfalfa, and rye grain plant species, respectively, were 76 to 29, 61 to 7, and 81 to 46. The grain and wheatgrass showed major advantages over the legumes. The overall germination and survival of plants with sewage sludge layers at

Table 12.5 *Cost Comparison of Stabilization Methods*

Type of stabilization	*Effectiveness*	*Maintenance*	*Approximate cost per acre ($)*
Physical			
Water sprinkling	Fair	Continual	–
Slag (9-inch depth)			
By pumping	Good	Moderate	$350- 450
By trucking	Good	Moderate	950-1,050
Straw harrowing	Fair	Moderate	40- 75
Bark covering	Good	Moderate	900-1,000
Country gravel and soil			
4-inch depth	Excellent	Minimal	250- 600
12-inch depth	Excellent	Minimal	700-1,700
Chemical			
Elastomeric polymer	Good	Moderate	300- 750
Lignosulfonate	Good	Moderate	250- 600
Vegetative			
4-inch soil cover and vegetation[1]	Excellent	Minimal	300- 650
12-inch soil cover and vegetation	Excellent	Minimal	750-1,750
Hydroseeding	Excellent	Minimal	200- 450
Matting[2]	Excellent	Minimal	600- 750
Chemical-vegetative	Excellent	Minimal	100- 250

Source: After Dean and Havens 1973.

Note: Based on average tailings, costs could be revised upward for acidic tailings requiring limestone or other neutralizing additives.

1. Generally used on pond area rather than on dikes. Also, not as effective as 12-inch soil cover when tailings are excessively acidic or saline.

2. Based on placing 3-foot-wide matting at 3-foot intervals over the seeded area.

depths of 3, 7, 11, and 15 inches were, respectively, 71 to 22, 75 to 27, 76 to 27, and 70 to 33. On the basis of these data, it was concluded that the sludge at the 3-inch depth gave the poorest results. However, the vegetation appeared healthier in plots with the shallow sludge layers because, once the roots penetrated the sludge layer, the plants became healthier and hardier than plants with roots only in the tailings.

The plants growing through the surficial 3 inches of mixed sewage sludge and tailings were allowed to grow for 10 months, whereupon they were excavated and examined. Practically all roots in plots with the sludge installed at depths of 7, 11, and 15 inches had grown and remained in the upper 3 inches of tailings containing admixed sludge. The main root for the 11-inch buried sludge layer had grown only 3.25 inches downward and had then sent out lateral feeder roots. The main roots in the plots with sludge layers at depths of 7 and 15 inches were much longer but grew in a circular manner, again only in the presence of the organic admixture. Conversely, the main root system for the plot with sludge at 3 inches grew down into and through the sludge layer to within 2 inches of the bottom of the tailings. Much greater plant growth was attained with the sludge layer at shallow levels. Thus it would appear that best results can be achieved with sludge emplaced from the surface down to depths of 6 or 7 inches.

Dean and Havens estimate that the cost of the sewage sludge method of preparing tailings prior to revegetation would be competitive with the costs shown in Table 12.5. The investigators recommend that vegetation indigenous to a locale be allowed to replace the plants listed above as soon as the domestic varieties establish a protective cover for them. Otherwise, maintenance costs will continue to accrue. These observations are corroborated and augmented by the work of Knudsen, Hall, and Horner (1971).

Ludeke (1973) reports that processed sewage sludge (from cattle feed lots) gives much better revegetation results than any other material used alone. He also reports success (and the author has observed it) in hydroseeding with a variety of plants combined with wood fiber and a chemical stabilization agent known as "Soil Seal" (chemical makeup not given). Soil Seal is reported to be an adhesive compound which holds the wood slurry to the tailing slope. The wood fiber slurry serves as the seed bed. Hydroseeding of one acre reportedly requires 1,500 pounds of wood fiber, 40 pounds of fertilizer, 60 gallons of Soil Seal. The soil is first leached with irrigation water to remove salts which would produce deleterious osmotic pressures. Total cost is in the order of $1,000 per acre with the highest cost item being Soil Seal. Ludeke lists 52 plants, both natural and domestic, that have been tested at the Pima operation in Tucson. Their response characteristics are discussed in sufficient detail for the list to be used as a guide to selection of plants for revegetation of mine wastes in semiarid to arid environments.

Leroy (1973) believes that all mill tailings can be "made green." He recognizes the guidelines for revegetation as *amenity* of the product to its surroundings, *unity* in the scheme in order to avoid partial or total failure caused by piecemeal efforts, selection of proper plant "*ingredients*" (plants of specific quality and proven usefulness) and *economics* with respect to most profitable use of the area covered by the tailings. He considers economically competitive, feasible, and alternative uses to be pastureland, cropland, nursery (tree), sod farm, urban development, recreation, and airfields. The vegetation sequence recommended is the installation by mechanical seeding of a fast-growing grass crop to act as a nursery for slower nitrogen fixing legumes—both achieved with the addition of fertilizers (by cyclone), pH adjustment (with crushed limestone), cultivation (with rippers, harrows, and discs), mulching, and artificial membrane spraying. The use of some hard seeds which will lie dormant until the second vegetative cycle is recommended. Leroy cites success of these procedures even with crownvetch in areas where temperatures of −50°F are not uncommon. He recommends against hydroseeding commonly used in the United States because the seedbed is not compacted; thus precipitation is allowed to erode and disrupt the slope and permit losses by predators.

He also recommends against the importation of topsoil because of the unnecessary expense incurred.

Table 12.6 contains the nursery stock that Leroy recommends for northern climates. Table 12.7 presents the site preparation and planting data from one

Table 12.6 *Selected Nursery Stock for Northern Climate*

Evergreens	
Picea alba	white spruce
Picea pungens clauca	Colorado blue spruce
Abies concolor	Colorado fir
Abies balsamea	balsam fir
Thuja occidentalis	white cedar
Larix	tamarack
Taxus canadenses	Canadian yew
Pinus strobus	white pine
Pinus negra austriaca	Austrian pine
Pinus banksiana	jackpine
Pinus montana mughus	mugo pine
Deciduous	
Acer saccarinum	sugar maple
Acer nagundo	Manitoba maple
Acer rubrum	red maple
Acer ginnala	shrub maple
Acer campestre	field maple
Betula papyrifera	paper birch
(will be planted to produce multi-stemmed trees as well as single ones)	
Populus bolleana	silver poplar
Salix	common willow
Fraxinus americana	American ash
Shrubs	
Amelanchier canadenses	
Caragana arborescens	pea bush
Cotone aster	
Chanonelis japonica	
Lonicera vulgaris	honeysuckle
Potentilla fruticosa	
Riebes alpinum	arctic currant
Salix canadenses atropurpurea	
Perennials	
Astilbe	
Heuchera sanguinea	
Parcesandra coccinea	
Vinca minor	
Lupinus	

Source: After Leroy 1973.

Table 12.7 *Preparation and Planting of a Gold Recovery Waste Area*

Preparation:
Grading and shaping of the land, conditioning of the surface
Application of fertilizers:
2 tons of triple superphosphate/acre
1 ton of 15-15-15/acre
2 tons of agricultural limestone/acre
Planting of lining out stock
25,000 evergreens
15,000 deciduous
15,000 shrubs and perennials
Membrane application whenever required
Irrigation system

Source: After Leroy 1973.

case history in which the 30-acre portion of a gold recovery waste area was converted to a nursery. Leroy (1973) presents no figures but he estimates that this scheme will generate a 50 percent return on investment before taxes and the initial capital outlay will have been recovered by the third operating year.

CHEMICAL AND PHYSICAL WASTE STABILIZATION

Physical and chemical stabilization of mine wastes is sometimes used as a substitute for revegetation to minimize air pollution and water pollution. Chemical stabilization is frequently employed simultaneously with revegetation so that the plants are not destroyed by abrasion or erosion before they mature and form a protective cover.

Physical stabilization is usually accomplished with a rock cover composed of erosion-resistant waste rock from the mining operation, if such exists. Gunite, asphalt and concrete are functionable but are expensive. Smelter slag has been used also, and some would consider covers of bark and the importation of topsoil in this category. With the exception of the importation of topsoil, all types of physical stabilization offer the disadvantage of creating a waste pile which seldom blends with its natural surroundings; thus they are not widely utilized. Clay covers are emplaced commonly on sanitary landfills to control leaching by ground water runoff, but the author is not aware of the use of clay on mine wastes.

Skelly and Loy (1973) provide the following information on cost of the physical stabilizers. Clay costs can be expected to range from $1.75 to $6.00 per cubic yard, including installation. Concrete costs approximately $30.00 per cubic yard, and 2-inch- to 3-inch-thick gunite applications cost $15.75 to $18.00 per square yard. Asphalt installation ranges in cost from $2.00 to $5.00 per square yard.

Chemical stabilization under its broadest definition can extend from the use of soil sealants to the application of fertilizers for amelioration of soil for plant growth. However, chemical stabilizers are usually defined as chemical agents which bind waste surfaces to prevent water and eolian erosion. The surfaces are seldom permanent and, even if they were, they would offer the same aesthetic disadvantage as physical stabilizers. The most desirable aspect of chemical stabilizers is the fact that they may prevent the destruction of vegetative covers during early growth stages. Dean and Havens (1973) report testing some 70 chemical stabilizers during their experiments with mill tailings. They have organized the most effective of the 70 reagents tested in the following list according to order of effectiveness based on the cost of the amount of reagent required to stabilize one square yard of waste.

1. Coherex, a resinous adhesive, produced wind-resistant tailings surfaces when applied in quantities costing as little as \$0.01/sq yd, but resistance to water jet testing was not achieved until amounts costing about \$0.10/sq yd were
2. Calcium, ammonium, and sodium lignosulfonates, as well as redwood bark extracts, were all effective surface stabilizers at costs of about \$0.02/sq yd.
3. Cement and milk of lime additives were effective in stabilizing surfaces when applied in amounts costing \$0.03/sq yd.
4. Paracol S 1461 (a blend of wax and resin) and Paracol TC 1842 (a resin emulsion) were effective as stabilizers in quantities ranging in cost from \$0.04 to \$0.10/sq yd on various tailings samples tested.
5. Potassium silicates having SiO_2 to K_2O ratio of 2.5 were effective stabilizers when applied in amounts costing \$0.07 to \$0.15/sq yd.
6. A cationic neoprene emulsion and Rezosol, an organic polymer, effectively stabilized tailings at costs approximating \$0.08/sq yd.
7. Sodium silicate applied in quantities of 4.5 lb/sq yd at a cost of \$0.11/sq yd was an effective stabilizer. Calcium chloride was an effective additive to the sodium silicate, whereas ferrous sulfate was not. The addition of 6 percent by weight of $CaCl_2$ permitted reduction of the cost of sodium silicate used from \$0.11 to \$0.02/sq yd.
8. Peneprime (a bituminous base product), selected amines, dicalcium silicate, and elastomeric polymers (Compound SP-400, Soil Gard, and DCA-70) produced wind- and water-resistant surfaces at costs of \$0.10/sq yd and up.
9. Pyrite treated with sulfuric acid, Aerospray Binder 52 (a synthetic resin), Landscape (a solution of combined sulfur in water-soluble oil), or Water Mate (an organic, nonionic product) were ineffective on the tailings used in the testing.

Dean and Havens do not provide figures on longevity of the chemically stabilized surfaces but they do point out that they are not permanent.

Ludeke (1973) recommends the use of the chemical stabilizer known commercially as Soil Seal for the purpose of stabilizing surfaces during initial stages of revegetation. At an application rate of approximately 80 gallons per acre, the cost is approximately \$300 per acre.

Leroy (1973) also recommends the use of "artificial membranes" which are sprayed onto the waste during initial stages of revegetation. His description of the process is similar to that provided by Ludeke for Soil Seal but Leroy provides no product identification.

Uniroyal (1972) tested the feasibility of using latex as a soil sealant to prevent water infiltration. The method was observed to be functional but the economics proved unattractive (materials alone cost $1,000 per acre). Best results were achieved with a styrene-butadiene (SBR) rubber latex of 2,400 Å average particle size with high mechanical stability. It was applied to the waste by sprinkler irrigation. The latex was observed to deposit progressively in macro openings as it passed through the soil. Most remained in the upper one foot of soil, but the rate of deposition was dependent on soil structure and composition. Addition of excessive anionic or nonionic surfactants improved latex penetration. No deleterious effects on vegetation were observed. However, Parks and Rosene (1971) note that freezing and thawing conditions destroy latex seals.

Water-swellable, cross-linked, high-molecular-weight polymers have been utilized to achieve eolian stability on tailings piles by aircraft spraying (Parks and Rosene 1971). They appear to be effective under ordinary conditions and are compatible with revegetation. But their effectiveness can be destroyed by high winds.

RECLAMATION FOR ALTERNATIVE USES

Reclamation for alternative uses implies that the mined area may be converted to some use other than that prevalent prior to mining. None of the other forms of reclamation are precluded however. Examples include conversion of monotonous, level farmland to aesthetically pleasing rolling, forested hills for recreation projects such as parks, golf courses, and motor sport areas (snowmobile courses and motorcycle courses). Lakes may also be created in terrain devoid of water based recreation (see, for example, Foreman and McLean 1973). Utilization of the extraction-rehabilitation sequence is particularly important for such uses because several reclamation procedures need to be carried out simultaneously if success is to be achieved economically.

LEGAL PROBLEMS OF MINED AREA RECLAMATION

Numerous legal problems are associated with reclamation of active and abandoned mine wastes. The latter are the most serious, especially if the mineral rights and surface rights are owned by different parties. Severed ownership and leasing operating rights for mineral extraction make it difficult to allocate responsibility for the effects of mining on the rights of other resource users, particularly downstream water users and, more recently, on recreational users. In addition some states have no reclamation laws to allocate responsibility regardless of ownership (see Appendix G for summary of U.S. reclamation laws). This tremendous variability among states makes it difficult for mineral resource companies to operate uniformly from one state to another, since operations must be economically competitive on a nationwide basis. The absence of reclamation laws in some states also makes it difficult for management to justify reclamation expenditures to stockholders in all states.

Goldberg and Power (1972) have reviewed the response of legal institutions to these problems in Maryland, Ohio, Pennsylvania, and West Virginia. The following summary presents some of their conclusions, many of which are applicable to other states. The study dealt basically with coal mining.

1. Unlike most industrial side effects, mining's diseconomies often continue after mining terminates, unless abatement or reclamation measures are performed. Due to insufficient legal controls, such measures have not been performed in most operations in the past, and a significant portion of today's mining diseconomies originates in inactive mines.
2. Minimization of side effects both during and after mining is greatest when pre-planned and built into the mining operation. Even then, alleviation of some side effects requires post-operation activity. In situations where side effects cannot be prevented, prohibition of the mining operation is the only suitable legal remedy.
3. Although the owner of land presumptively owns the minerals within, ownership may be severed so that the surface and the minerals belong to different persons. The owner of minerals, whether or not he is the surface owner, may extract them himself or he may give the right to extract them to someone else, usually by means of a lease. The practice of severing ownership was greater in the past than it is today. Severances have produced a significant amount of split ownership.
4. Separate ownership of surface and minerals has complicated procurement of requisite assent to perform reclamation activities. Severed ownership and the practice of leasing operating rights have complicated allocation of responsibility for side effects. In the past, neither the surface owner nor the lessor of the mineral rights was generally held responsible for damage resulting from the mining operation.
5. Private litigation does not afford an effective vehicle through which mine operators can be made to internalize the costs of coal mine diseconomies.
6. All four states have adopted legislation enabling the creation of water quality standards and prohibiting discharges in violation of those standards. The legislation or the regulations thereunder limit acid drainage discharges. The application of these rules to inactive mines has been uncertain. All four states have statutory provisions requiring some form of mine sealing. Except in Pennsylvania, however, these provisions are directed more at diminishing safety hazards than at drainage.
7. All four states have adopted legislation regulating surface mining. This legislation generally requires a license or permit to mine; requires filing of a bond or security deposit; prescribes reclamation standards; provides for revocation of the permit if reclamation requirements are not performed; and provides for creation of a fund made up of fees or contributions by mine operators which has the purpose of financing reclamation of unreclaimed stripped lands. Surface mining legislation generally does not reach reclamation of inactive surface mines. Except for a limited prohibition in West Virginia, no state prohibits strip mining outright, although prohibition may come about through the refusal of a license where grave side effects cannot be prevented.

8. In Pennsylvania and Maryland, bonds have been authorized to finance state reclamation activities on inactive mines.
9. It sometimes is necessary to distinguish active from abandoned mines for regulatory purposes. The definition of "abandonment" developed by the cases in other contexts is often inadequate to meet the needs of effective environmental regulation.
10. In most instances, these four states may limit the discharge of acid drainage from coal operations, without compensating the operator.
11. If application of a ban on surface mining deprives an operator of a fair return on investment, compensation is constitutionally mandated.
12. The states may constitutionally impose the cost of remedying past environmental degradations on the present generation of miners.
13. In Maryland and, perhaps, West Virginia, state constitutional amendments would seem prerequisite to abatement or reclamation programs involving acquisition of land by the state and its subsequent transfer back to the private sector, following restoration.
14. These four states have a variety of constitutionally imposed strictures on the ability of the states to borrow money. These strictures interfere with the public financing of abatement and reclamation projects.

Abandoned Mined Areas

These conclusions indicate that the major legal hangups of mined area reclamation are those associated with wastes that were created and abandoned prior to the passage of reclamation laws (points 6 and 7). It would seem that, under these circumstances, such reclamation would appropriately be authorized and implemented by agencies of government, either state or federal.

The state of Pennsylvania has in fact floated bonds for mine sealing and for mined area reclamation. Maryland has enacted enabling legislation to do so. But Goldberg and Power (1972) point out that there are constitutional limitations on the ability of states to incur debt for such projects. Pennsylvania was required to pass a Constitutional Amendment to create a $500 million conservation and reclamation fund for mined area reclamation. If the states can in fact circumvent the constitutional restrictions on spending money for mined area reclamation, then they must decide whether to acquire title to the land (which adds up to more land than they can handle and which may be against the will of the owner) or try to charge the landowner for the value of the project constructed on his land when the project value may exceed the market value of the land. Clearly their only real choice is property acquisition with the intent of returning it, or project construction by agreement, with no land ownership transfer and no charge. But these alternatives lead to the situation described by Goldberg and Power (1972) as follows:

> Both alternatives present constitutional difficulties. If the state takes property with an intention to later transfer it back to the private sector, constitutional questions arise as to whether the taking was for a "public use"; if the state engages in abatement or reclamation without acquiring the private property interests involved, the questions arise as to propriety of the conference of betterments on private landowners.

With respect to the concept of public use, Goldberg and Power point out that the U.S. Constitution does not contain an express provision which prohibits the states from taking private property for any purpose other than public use, but that, as a matter of federal constitutional law, a state may employ the power of eminent domain only when the taking is to be for a "public use." But what constitutes public use? Goldberg and Power quote the two prominent definitions, the conservative and the liberal. They are respectively:

> . . . "use by the public," that is public service or employment, and that consequently to make a use public a duty must evolve upon the person or corporation seeking to take property by right of eminent domain to furnish the public with the use intended, and the public must be entitled as of right, to use or enjoy the property taken. The term implies the "use of many" or "by the public."

> . . . "public advantage," and that is anything which tends to enlarge the resources, increase the industrial energies, and promote the productive power of any considerable number of the inhabitants of a section of the state, or which leads to the growth of towns and the creation of new resources for the employment of capital and labor, manifestly contributes to the general welfare and the prosperity of the whole community and giving the constitution a broad and comprehensive interpretation, constitutes a public use.

The narrow definition probably would not allow property to be acquired for reclamation then resold to the private sector; but the liberal definition would. Goldberg and Power point out that the U.S. Supreme Court has adopted the liberal definition for the past 75 years.

The technique of spending public funds to reclaim private waste lands also suffers constitutional limitations. Goldberg and Power conclude, for example, that Maryland, Ohio, and Pennsylvania are ostensibly constitutionally forbidden to deal with mined area reclamation problems through incentives, grants, or guaranteee of loans to private operators. However, the courts in some states have upheld such procedures, provided the bonds issued do not constitute a debt to the state. In order to alleviate these problems and to minimize litigation, Goldberg and Power suggest that state funding of mined area reclamation projects be preceded by constitutional amendments wherever appropriate. The existing situation at the state level is described as consisting of anachronistic limitations on legislative powers, coupled with judicially developed means of avoiding the limitations.

Clearly, another alternative is to have legislation and funding handled at the federal level. As indicated in Chapter 1, Public Law 92-500 attempts to do that for the production of acid mine drainage control (including runoff from mine wastes if it constitutes a point source discharge).

Active Mining Operations

A variety of state and federal laws apply in a piecemeal fashion to the production, leaching, erosion, and reclamation of mine wastes (see Appendix G). Historically the trend has been toward progressively greater restriction. For example, the 1937 version of the Pennsylvania Clean Streams Act reads: "The provisions of this article shall not apply to acid mine drainage from coal mines until such time as, in the opinion of the Sanitary Water Board, practical means for the removal of the polluting properties of such drainage shall become known." This legislative exemption for acid mine drainage from coal mines was withdrawn in 1965.

Most states now have some form of water quality control which applies to acid mine water. Such legislation is either a result of the requirements of the Federal Water Pollution Control Act or a result of internal public pressure. In either case, the net effect is increasing control with considerable variability among states. But some of these interstate differences should be eliminated by 1977, when the Water Pollution Control Act Amendments of 1972 become effective. This federal legislation will put all operations on equal ground with respect to the production and required treatment of poor-quality discharges of mining, milling, and metallurgical waste waters.

But mined area reclamation laws still exist essentially at the state level. State legislation ranges from no regulation at all to fairly stringent regulation such as that of Pennsylvania. Thus, pending action by the 94th Congress, mining operations must vary operating procedures from one state to another in order to achieve compliance with different reclamation laws.

Goldberg and Power (1972) have included among their recommendations the following guidelines which might minimize legal inequities at the state level. Their remarks are directed at the mining of coal, but some of their objectives are applicable to other types of mining as well.

1. Authority to regulate side effects of mining should be exercised by a single state agency. This agency should be headed by a single executive and be placed within the state's major natural resource department (e.g., Department of Natural Resources, Department of Environmental Affairs).
2. The agency should have rule-making authority to establish "background" quality standards in mined areas.
3. The agency should have rule-making power to prohibit mine operations which, under the present state of technology, are not subject to effective reclamation or which present a public health or safety hazard.
4. The agency should be empowered to directly engage in projects and to acquire, construct, and operate facilities which deal with side effects of mining. An expedited condemnation power should accompany the powers given the agency.
5. The costs of remedying past adverse side effects resulting from mining operations which were lawful when undertaken should be paid from general (state) revenues.

Recommendation Number 5 may be difficult to accomplish, because political constituents are likely to oppose paying the cleanup bill for the mineral resource industry. Thus, the recommendations of Goldberg and Power would do little to solve the abandoned mined area problem but, if generally adopted, they would constitute the beginnings of a uniform method of reclamation of active mined areas in a fashion that is equitable on an interstate basis.

REFERENCES

Agnew, A. F. 1971. *Coal Mining Hydrology and the Environment, or Give the Devil His Due.* AIME Environmental Quality Conference, preprint volume, paper No. EQC 37, pp. 157-164.

Allen, Natie, Jr. 1973. *Experimental Multiple Seam Mining and Reclamation on Steep Mountain Slopes.* Proceedings, Research and Applied Technology Symposium on Mined Land Reclamation, National Coal Association, Bituminous Coal Research Inc., Monroeville, Penn., pp. 98-104.

Appalachian Regional Commission 1969. *Acid Mine Drainage in Appalachia.* House Document 91-180, Publication No. 35-4320, U.S. Govt. Printing Office, Washington, D.C.

Capp, J. P., and Gilmore, D. W. 1973. *Soil Making Potential of Power Plant Fly Ash in Mined Land Reclamation.* Proceedings, Research and Technology Symposium on Mined Land Reclamation, National Coal Association, Bituminous Coal Research Inc., Monroeville, Penn., pp. 26-41.

Corbett, D. M. 1965. *Water Supplied by Coal Surface Mines, Pike County, Indiana.* Indiana Univ. Water Resources Research Center Report of Investigations No. 1, Bloomington.

Corbett, D. M., and Agnes, A. F. 1968. *Coal Mining Effect on Busseron Creek Watershed, Sullivan County, Indiana.* Indiana Univ. Water Resources Research Center Report of Investigations No. 2, Bloomington.

Currier, W. F. 1973. *Basic Principles of Seeding.* Proceedings, Research and Applied Technology Symposium on Mined Land Reclamation. National Coal Association, Bituminous Coal Research Inc., Monroeville, Penn., pp. 225-232.

D'Appolonia, E., Ellison, R. D., and Gormley, J. T. 1973. Abandonment of Tailings Facilities. In *Tailing Disposal Today*, Proceedings, International Tailing Symposium, Tucson, Ariz., ed. C. L. Aplin and George O. Argall, Jr., pp. 571-605. San Francisco: Miller Freeman Publications.

Dean, K. C., and Havens, R. 1972. Stabilization of Problem Mineral Wastes. Paper presented at Fall meeting, AIME, SME.

Dean, K. C., and Havens, R. 1973. Comparative Costs and Methods for Stabilization of Tailings. In *Tailing Disposal Today*, Proceedings, International Tailing Symposium, Tucson, Ariz., ed. C. L. Aplin and George O. Argall, Jr., pp. 450-474. San Francisco: Miller Freeman Publications.

Foreman, J. W., and McLean, D. C. 1973. *Evaluation of Pollution Abatement Procedures at Moraine State Park.* EPA Document R2-73-140, Supt. of Documents, U.S. Govt. Printing Office, Washington, D.C.

Galbraith, J. 1971. "A Study of Mine Tailings and Associated Plants and Ground Water in the Coeur d'Alene District, Idaho." Unpublished M.S. thesis, Univ. of Idaho, Moscow.

Galbraith, J. H., Williams, R. E., and Siems, P. L. 1971. Migration and Leaching of Metals from Old Mine Tailings Deposits. *Ground Water Journal*, Vol. 10, No. 3, pp. 33-44.

Goldberg, E. F., and Power, G. 1972. *Legal Problems of Coal Mine Reclamation.* EPA Document 14010 FZU 03/72, Supt. of Documents, U.S. Govt. Printing Office, Washington, D.C.

Heine, W. N., and Guckert, W. E. 1973. *A New Method of Surface Coal Mining in Steep Terrain.* Proceedings, Research and Applied Technology Symposium on Mined Land Reclamation, National Coal Association, Bituminous Coal Research Inc., Monroeville, Penn., pp. 105-116.

Hunt, C. F., Sopper, W. E., and Kardos, L. T. 1971. *Renovation of Treated Municipal Sewage Effluent and Digested Liquid Sludge through Irrigation of Bituminous Coal Strip Mine Soil.* Technical paper, Institute for Research on Land and Water Resources, Penn. State Univ., Univ. Park, Penn.

Knudsen, J. F., Hall, K. B., and Horner, J. K. 1971. *Vegetative Stabilization of Tailings Piles at Kennecott Copper Corporation.* AIME Environmental Quality Conference, preprint volume, pp. 181-192.

Lejcher, T. R., and Kunkle, S. H. 1972. *Restoration of Acid Spoil Banks with Treated Sewage Sludge.* Proceedings, Symposium on Recycling Treated Municipal Waste Water and Sludge through Forest and Cropland, Penn. State Univ., Univ. Park, Penn.

Leroy, J. C. 1973. How to Establish and Maintain Growth on Tailings in Canada–Cold Winters and Short Growing Seasons. In *Tailing Disposal Today*, Proceedings, International Tailing Symposium, Tucson, Ariz., ed. C. L. Aplin and George O. Argall, Jr., pp. 411-449. San Francisco: Miller Freeman Publications.

Ludeke, K. L. 1973. Vegetative Stabilization of Copper Mine Tailings Disposal Berms of Pima Mining Co. In *Tailing Disposal Today*, Proceedings, International Tailing Symposium, Tucson, Ariz., ed. C. L. Aplin and George O. Argall, Jr., pp. 377-410. San Francisco: Miller Freeman Publications.

Lusk, B. E., and Mulhern, J. J. 1972. *New Surface Mining Technology to Minimize Environmental Disturbance.* Paper presented at 1972 annual meeting of American Mining Congress; available from American Mining Congress, Washington, D.C.

McNay, L. M. 1970. *Surface Mine Reclamation, Moraine State Park, Pennsylvania.* U.S. Bureau of Mines Information Circular 8456, Supt. of Documents, U.S. Govt. Printing Office, Washington, D.C.

Paller, W., and Schultz, D. A. 1973. *Planning Approaches to Surface Mining in the National Forests.* Proceedings, Research and Applied Technology Symposium on Mined Land Reclamation, National Coal Association, Bituminous Coal Research Inc., Monroeville, Penn., pp. 68-81.

Parks, C. F., and Rosene, R. B. 1971. *An Evaluation of Stabilization of Active Tailing Ponds with Water-Swellable Polymers.* AIME Environmental Quality Conference, preprint volume, pp. 175-180.

Pennsylvania Research Committee on Coal Mine Spoil Revegetation 1971. *A Guide for Revegetating Bituminous Strip Mine Spoils in Pennsylvania.* Dept. of Environmental Resources, Commonwealth of Pennsylvania.

Peterson, J. R., and Gschwind, J. 1973. *Amelioration of Coal Mine Spoils with Digested Sewage Sludge.* Proceedings, Research and Applied Technology Symposium on Mined Land Reclamation, National Coal Association, Bituminous Coal Research Inc., Monroeville, Penn., pp. 187-196.

Pfeider, E. P., and Wheaton, G. F. 1968. Iron Ore Mining. *Surface Mining*, Chapter 13.4, AIME, New York.

Ruffner, J. D. 1973. *Projecting the Use of New Plant Materials for Special Reclamation Projects.* Proceedings, Research and Applied Technology Symposium on Mined Land Reclamation, National Coal Association, Bituminous Coal Research Inc., Monroeville, Penn., pp. 233-242.

Skelly and Loy Inc. 1973. *Processes, Procedures and Methods to Control Pollution from Mining Activities.* EPA Document 430/9-73-011, prepared under contract No. 68-01-1830, U.S. Environmental Protection Agency, Office of Water Program Operations, Washington, D.C.

Sutton, P. 1973. *Establishment of Vegetation on Toxic Mine Spoils.* Proceedings, Research and Applied Technology Symposium on Mined Land Reclamation, National Coal Association, Bituminous Coal Research Inc., Monroeville, Penn., pp. 153-158.

Tilman, I. W. 1968. Brown Coal Mining in West Germany. *Surface Mining*, Chapter 13.6, AIME, New York.

Truax-Traer Coal Company 1971. *Control of Mine Drainage from Coal Mines Mineral Wastes, Hydrology and Related Experiments.* EPA Document 14010 DDH, U.S. Govt. Printing Office, Washington, D.C.

Uniroyal 1972. *Use of Latex as a Soil Sealant to Control Acid Mine Drainage.* EPA Document 14010 EFK 06/72, Supt. of Documents, U.S. Govt. Printing Office, Washington, D.C.

U.S. Forest Service 1972. *Palzo Reclamation Project.* Environmental Impact Study, U.S. Forest Service, Region 9, Milwaukee, Wis.

West Virginia University 1971. *Mine Spoil Potentials for Water Quality and Controlled Erosion.* EPA Document 14010 EJE 12/71, Supt. of Documents, U.S. Govt. Printing Office, Washington, D.C.

Public Law 92-500
92nd Congress, S. 2770
October 18, 1972

An Act

86 STAT. 81

To amend the Federal Water Pollution Control Act.

Be it enacted by the Senate and House of Representatives of the United States of America in Congress assembled, That this Act may be cited as the "Federal Water Pollution Control Act Amendments of 1972".

Federal Wa
Pollution
trol Act A
ments of 1
70 Stat. 4
84 Stat. 9
33 USC 115
note.

SEC. 2. The Federal Water Pollution Control Act is amended to read as follows:

"TITLE I—RESEARCH AND RELATED PROGRAMS

"DECLARATION OF GOALS AND POLICY

"SEC. 101. (a) The objective of this Act is to restore and maintain the chemical, physical, and biological integrity of the Nation's waters. In order to achieve this objective it is hereby declared that, consistent with the provisions of this Act—

"(1) it is the national goal that the discharge of pollutants into the navigable waters be eliminated by 1985;

"(2) it is the national goal that wherever attainable, an interim goal of water quality which provides for the protection and propagation of fish, shellfish, and wildlife and provides for recreation in and on the water be achieved by July 1, 1983;

"(3) it is the national policy that the discharge of toxic pollutants in toxic amounts be prohibited;

"(4) it is the national policy that Federal financial assistance be provided to construct publicly owned waste treatment works;

"(5) it is the national policy that areawide waste treatment management planning processes be developed and implemented to assure adequate control of sources of pollutants in each State; and

"(6) it is the national policy that a major research and demonstration effort be made to develop technology necessary to eliminate the discharge of pollutants into the navigable waters, waters of the contiguous zone, and the oceans.

"(b) It is the policy of the Congress to recognize, preserve, and protect the primary responsibilities and rights of States to prevent, reduce, and eliminate pollution, to plan the development and use (including restoration, preservation, and enhancement) of land and water resources, and to consult with the Administrator in the exercise of his authority under this Act. It is further the policy of the Congress to support and aid research relating to the prevention, reduction, and elimination of pollution, and to provide Federal technical services and financial aid to State and interstate agencies and municipalities in connection with the prevention, reduction, and elimination of pollution.

"(c) It is further the policy of Congress that the President, acting through the Secretary of State and such national and international organizations as he determines appropriate, shall take such action as may be necessary to insure that to the fullest extent possible all foreign countries shall take meaningful action for the prevention, reduc-

Appendix A

PORTIONS OF THE FEDERAL WATER POLLUTION CONTROL ACT AMENDMENTS OF 1972 (PUBLIC LAW 92-500)

As a convenience to the reader, portions of Public Law 92-500, Sections 301, 302, 304, are reproduced here. They concern effluent limitations standards and their enforcement. The U.S. Environmental Protection Agency (EPA) is charged by the legislation with defining currently available practicable control technology in terms of waste water quality criteria. Section 304 of the act contains the charge.

"TITLE III—STANDARDS AND ENFORCEMENT

"EFFLUENT LIMITATIONS

"SEC. 301. (a) Except as in compliance with this section and sections 302, 306, 307, 318, 402, and 404 of this Act, the discharge of any pollutant by any person shall be unlawful.

"(b) In order to carry out the objective of this Act there shall be achieved—

"(1) (A) not later than July 1, 1977, effluent limitations for point sources, other than publicly owned treatment works, (i) which shall require the application of the best practicable control technology currently available as defined by the Administrator pursuant to section 304(b) of this Act, or (ii) in the case of a discharge into a publicly owned treatment works which meets the requirements of subparagraph (B) of this paragraph, which shall require compliance with any applicable pretreatment requirements and any requirements under section 307 of this Act; and

"(B) for publicly owned treatment works in existence on July 1, 1977, or approved pursuant to section 203 of this Act prior to June 30, 1974 (for which construction must be completed within four years of approval), effluent limitations based upon secondary treatment as defined by the Administrator pursuant to section 304(d)(1) of this Act; or,

"(C) not later than July 1, 1977, any more stringent limitation, including those necessary to meet water quality standards, treatment standards, or schedules of compliance, established pursuant to any State law or regulations (under authority preserved by section 510) or any other Federal law or regulation, or required to implement any applicable water quality standard established pursuant to this Act.

"(2) (A) not later than July 1, 1983, effluent limitations for categories and classes of point sources, other than publicly owned treatment works, which (i) shall require application of the best available technology economically achievable for such category or class, which will result in reasonable further progress toward the national goal of eliminating the discharge of all pollutants, as determined in accordance with regulations issued by the Administrator pursuant to section 304(b)(2) of this Act, which such effluent limitations shall require the elimination of discharges of all pollutants if the Administrator finds, on the basis of information available to him (including information developed pursuant to section 315), that such elimination is technologically and economically achievable for a category or class of point sources as determined in accordance with regulations issued by the Administrator pursuant to section 304(b)(2) of this Act,...

"(c) The Administrator may modify the requirements of subsection (b)(2)(A) of this section with respect to any point source for which a permit application is filed after July 1, 1977, upon a showing by the owner or operator of such point source satisfactory to the Administrator that such modified requirements (1) will represent the maximum use of technology within the economic capability of the owner or operator; and (2) will result in reasonable further progress toward the elimination of the discharge of pollutants....

"(e) Effluent limitations established pursuant to this section or section 302 of this Act shall be applied to all point sources of discharge of pollutants in accordance with the provisions of this Act....

"WATER QUALITY RELATED EFFLUENT LIMITATIONS

"SEC. 302. (a) Whenever, in the judgment of the Administrator, discharges of pollutants from a point source or group of point sources, with the application of effluent limitations required under section 301(b)(2) of this Act, would interfere with the attainment or maintenance of that water quality in a specific portion of the navigable waters which shall assure protection of public water supplies, agricultural and industrial uses, and the protection and propagation of a balanced population of shellfish, fish and wildlife, and allow recreational activities in and on the water, effluent limitations (including alternative effluent control strategies) for such point source or sources shall be established which can reasonably be expected to contribute to the attainment or maintenance of such water quality.

Public hearing.

"(b)(1) Prior to establishment of any effluent limitation pursuant to subsection (a) of this section, the Administrator shall issue notice of intent to establish such limitation and within ninety days of such notice hold a public hearing to determine the relationship of the economic and social costs of achieving any such limitation or limitations, including any economic or social dislocation in the affected community or communities, to the social and economic benefits to be obtained (including the attainment of the objective of this Act) and to determine whether or not such effluent limitations can be implemented with available technology or other alternative control strategies....

"INFORMATION AND GUIDELINES

Water quality criteria publication.

"SEC. 304. (a)(1) The Administrator, after consultation with appropriate Federal and State agencies and other interested persons, shall develop and publish, within one year after the date of enactment of this title (and from time to time thereafter revise) criteria for water quality accurately reflecting the latest scientific knowledge (A) on the kind and extent of all identifiable effects on health and welfare including, but not limited to, plankton, fish, shellfish, wildlife, plant life, shorelines, beaches, esthetics, and recreation which may be expected from the presence of pollutants in any body of water, including ground water; (B) on the concentration and dispersal of pollutants, or their byproducts, through biological, physical, and chemical processes; and (C) on the effects of pollutants on biological community diversity, productivity, and stability, including information on the factors affecting rates of eutrophication and rates of organic and inorganic sedimentation for varying types of receiving waters.

"(2) The Administrator, after consultation with appropriate Federal and State agencies and other interested persons, shall develop and publish, within one year after the date of enactment of this title (and from time to time thereafter revise) information (A) on the factors necessary to restore and maintain the chemical, physical, and biological integrity of all navigable waters, ground waters, waters of the contiguous zone, and the oceans; (B) on the factors necessary for the protection and propagation of shellfish, fish, and wildlife for classes and categories of receiving waters and to allow recreational activities in and on the water; and (C) on the measurement and classification of water quality; and (D) for the purpose of section 303, on and the identification of pollutants suitable for maximum daily load measurement correlated with the achievement of water quality objectives. . . .

"(b) For the purpose of adopting or revising effluent limitations under this Act the Administrator shall, after consultation with appropriate Federal and State agencies and other interested persons, publish within one year of enactment of this title, regulations, providing guidelines for effluent limitations, and, at least annually thereafter, revise, if appropriate, such regulations. Such regulations shall—

Effluent limitation guidelines, publication.

"(1)(A) identify, in terms of amounts of constituents and chemical, physical, and biological characteristics of pollutants, the degree of effluent reduction attainable through the application of the best practicable control technology currently available for classes and categories of point sources (other than publicly owned treatment works); and

"(B) specify factors to be taken into account in determining the control measures and practices to be applicable to point sources (other than publicly owned treatment works) within such categories or classes. Factors relating to the assessment of best practicable control technology currently available to comply with subsection (b)(1) of section 301 of this Act shall include consideration of the total cost of application of technology in relation to the effluent reduction benefits to be achieved from such application, and shall also take into account the age of equipment and facilities involved, the process employed, the engineering aspects of the application of various types of control techniques, process changes, non-water quality environmental impact (including energy requirements), and such other factors as the Administrator deems appropriate;

"(2)(A) identify, in terms of amounts of constituents and chemical, physical, and biological characteristics of pollutants, the degree of effluent reduction attainable through the application of the best control measures and practices achievable including treatment techniques, process and procedure innovations, operating methods, and other alternatives for classes and categories of point sources (other than publicly owned treatment works); and

"(B) specify factors to be taken into account in determining the best measures and practices available to comply with subsection (b)(2) of section 301 of this Act to be applicable to any point source (other than publicly owned treatment works) within such categories or classes. Factors relating to the assessment of best available technology shall take into account the age of equipment and facilities involved, the process employed, the engineering aspects of the application of various types of control techniques, process changes, the cost of achieving such effluent reduction, non-water quality environmental impact (including energy requirements), and such other factors as the Administrator deems appropriate; and

"(3) identify control measures and practices available to eliminate the discharge of pollutants from categories and classes of point sources, taking into account the cost of achieving such elimination of the discharge of pollutants. . . .

Appendix B

TEST PROCEDURES FOR THE ANALYSIS OF POLLUTANTS

The following excerpt reproduced from the *Federal Register*, Vol. 38, No. 199, October 16, 1973, includes a list of approved test procedures. References to the "Regional Administrator" are to the U.S. Environmental Protection Agency (EPA) regional administrator.

Identification of test procedures.

Every parameter or pollutant for which an effluent limitation is now specified pursuant to sections 401 and 402 of the Act is named together with test descriptions and references in Table I. The discharge parameter values for which reports are required must be determined by one of the standard analytical methods cited and described in Table I, or under certain circumstances by other methods that may be more advantageous to use when such other methods have been previously approved by the Regional Administrator of the Region in which the discharge will occur, and providing that the Director of the State in which such discharge will occur does not object to the use of such alternate test procedures.

Under certain circumstances the Regional Administrator or the Director in the Region or State where the discharge will occur may determine for a particular discharge that additional parameters or pollutants must be reported. Under such circumstances, additional test procedures for analysis of pollutants may be specified by the Regional Administrator or Director upon the recommendation of the Director of the Methods Development and Quality Assurance Research Laboratory.

Application for alternate test procedures.

(a) Any person may apply to the Regional Administrator in the Region where the discharge occurs for approval of an alternative test procedure.

(b) When the discharge for which an alternative test procedure is proposed occurs within a State having a permit program approved pursuant to section 402 of the Act, the applicant shall submit his application to the Regional Administrator through the Director of the State agency having responsibility for issuance of NPDES permits within such State.

(c) Unless and until printed application forms are made available, an application for an alternate test procedure may be made by letter in triplicate. Any application for an alternate test procedure under this subchapter shall:

(1) Provide the name and address of the responsible person or firm making the discharge (if not the applicant) and the applicable ID number of the existing or pending permit, issuing agency, and type of permit for which the alternate test procedure is requested, and the discharge serial number.

(2) Identify the pollutant or parameter for which approval of an alternate testing procedure is being requested.

(3) Provide justification for using testing procedures other than those specified in Table I.

(4) Provide a detailed description of the proposed alternate test procedure, together with references to published studies of the applicability of the alternate test procedure to the effluents in question.

Parameter and units	Method	References: Standard methods	References: ASTM	References: EPA methods
General analytical methods:				
1. Alkalinity as $CaCO_3$ mg $CaCO^3$/liter.	Titration: electrometric, manual or automated method—methyl orange.	p. 370	p. 143	p. 6. p. 8.
2. B.O.D. five day mg/liter.	Modified winkler or probe method	p. 489		
3. Chemical oxygen demand (C.O.D.) mg/liter.	Dichromate reflux	p. 495	p. 219	p. 17.
4. Total solids mg/liter	Gravimetric 103–105° C	p. 535		p. 280.
5. Total dissolved (filterable) solids mg/liter.	Glass fiber filtration 180° C			p. 275.
6. Total suspended (nonfilterable) solids mg/liter.	Glass fiber filtration 103–105° C	p. 537		p. 278.
7. Total volatile solids mg/liter.	Gravimetric 550° C	p. 536		p. 282.
8. Ammonia (as N) mg/liter.	Distillation—nesslerization or titration automated phenolate.			p. 134. p. 141.
9. Kjeldahl nitrogen (as N) mg/liter.	Digestion + distillation—nesslerization or titration utomated digestion phenolate.	p. 469		p. 149. p. 157.
10. Nitrate (as N) mg/liter.	Cadmium reduction; brucine sulfate; automated cadmium or hydrazine reduction.	p. 458 p. 461	p. 124	p. 170. p. 175. p. 185.
11. Total phosphorus (as P) mg/liter.	Persulfate digestion and single reagent (ascorbic acid), or manual digestion, and automated single reagent or stannous chloride.	p. 526 p. 532	p. 42	p. 235. p. 246. p. 259.
12. Acidity mg $CaCO_3$/liter	Electrometric end point or phenolphthalein end point.		p. 148	
13. Total organic carbon (TOC) mg/liter.	Combustion—infrared method [1]	p. 257	p. 702	p. 221.
14. Hardness—total mg $CaCO_3$/liter.	EDTA titration; automated colorimetric atomic absorption.	p. 179	p. 170	p. 76. p. 78.
15. Nitrite (as N) mg/liter.	Manual or automated colorimetric diazotization.			p. 185. p. 195.
Analytical methods for trace metals:				
16. Aluminum—total [2] mg/liter.	Atomic absorption	p. 210		p. 98.
17. Antimony—total [2] mg/liter.	Atomic absorption [4]			
18. Arsenic—total mg/liter.	Digestion plus silver diethyldithiocarbamate; atomic absorption.[3]	p. 65 p. 62		p. 13.
19. Barium—total [2] mg/liter.	Atomic absorption [4]	p. 210		
20. Beryllium—total [2] mg/liter.	Aluminon; atomic absorption	p. 67 p. 210		
21. Boron—total mg/liter	Curcumin	p. 69		
22. Cadmium—total [2] mg/liter.	Atomic absorption; colorimetric	p. 210 p. 422	p. 692	p. 101.
23. Calcium—total [2] mg/liter.	EDTA titration; atomic absorption	p. 84	p. 692	p. 102.
24. Chromium VI mg/liter.	Extraction and atomic absorption; colorimetric.	p. 429		p. 94.
25. Chromium—total [2] mg/liter.	Atomic absorption; colorimetric	p. 210 p. 426	p. 692 p. 403	p. 104.
26. Cobalt—total [2] mg/liter.	Atomic absorption [4]		p. 692	
27. Copper—total [2] mg/liter.	Atomic absorption; colorimetric	p. 210 p. 430	p. 692 p. 410	p. 106.
28. Iron—total [2] mg/liter	do	p. 210 p. 433	p. 692 p. 152	p. 108.
29. Lead—total [2] mg/liter	do	p. 210 p. 436	p. 692	p. 110.
30. Magnesium—total [2] mg/liter.	Atomic absorption; Gravimetric	p. 210 p. 416 p. 201	p. 692	p. 112.
31. Manganese—total [2] mg/liter.	Atomic absorption	p. 210	p. 692	p. 114.
32. Mercury—total mg/liter.	Flameless atomic absorption [5]			
33. Molybdenum—total [2] mg/liter.	Atomic absorption [4]			
34. Nickel—total [2] mg/liter.	Atomic absorption; colorimetric [4]	p. 443	p. 692	
35. Potassium—total [2] mg/liter.	Atomic absorption; colorimetric; flame photometric.	p. 283 p. 285	p. 326	p. 115.
36. Selenium—total mg/liter.	Atomic absorption [3]			
37. Silver—total [2]	Atomic absorption [4]	p. 210		
38. Sodium—total [2] mg/liter.	Flame photometric; atomic absorption	p. 317	p. 326	p. 118.
39. Thallium—total [2] mg/liter.	Atomic absorption [4]			
40. Tin—total [2] mg/liter	do			
41. Titanium—total mg/liter.	do			
42. Vanadium—total [2] mg/liter.	Atomic Absorption; [4] Colorimetric	p. 157		
43. Zinc—total [2] mg/liter	Atomic Absorption; Colorimetric	p. 210 p. 444	p. 692	p. 120.
Analytical methods for nutrients, anions, and organics:				
44. Organic nitrogen (as N) mg/liter.	Kjeldahl nitrogen minus ammonia nitrogen.	p. 468		p. 149.
45. Ortho-phosphate (as P) mg/liter.	Direct single reagent; automated single reagent or stannous chloride.	p. 532	p. 42	p. 235. p. 246.

Parameter and units	Method	References		
		Standard methods	ASTM	EPA methods
46. Sulfate (as SO_4) mg/liter.	Gravimetric; turbidimetric; automated colorimetric—barium chloranilate.	p. 331 p. 334	p. 51 p. 52	p. 286. p. 288.
47. Sulfide (as S) mg/liter	Titrimetric—iodine	p. 551		p. 294.
48. Sulfite (as SO_3) mg/liter.	Titrimetric; iodide-iodate	p. 337	p. 261	
49. Bromide mg/liter	do		p. 216	
50. Chloride mg/liter	Silver nitrate; mercuric nitrate; automated colorimetric-ferricyanide.	p. 96 p. 97	p. 23 p. 21	p. 29. p. 31.
51. Cyanide—total mg/liter	Distillation—silver nitrate titration or pyridine pyrazolone colorimetric.	p. 397	p. 556	p. 41.
52. Fluoride mg/liter	Distillation—SPADNS	p. 171 p. 174	p. 191	p. 64.
53. Chlorine—total residual mg/liter.	Colorimetric; amperometric titration	p. 382	p. 223	
54. Oil and grease mg/liter	Liquid-Liquid extraction with trichloro-trifluoroethane.	p. 254		
55. Phenols mg/liter	Colorimetric, 4 AAP	p. 502	p. 445	p. 232.
56. Surfactants mg/liter	Methylene blue colorimetric	p. 339	p. 619	p. 131.
57. Algicides mg/liter	Gas chromatography [6]			
58. Benzidine mg/liter	Diazotization—colorimetric [7]			
59. Chlorinated organic compounds (except pesticides) mg/liter.	Gas chromatography [6]			
60. Pesticides mg/liter	Gas chromatography [6]			
Analytical methods for physical and biological parameters:				
61. Color platinum-cobalt units or dominant wave-length, hue, luminance, purity.	Colorimetric; spectrophotometric	p. 160 p. 392		p. 38.
62. Specific conductance mho/cm at 25° C.	Wheatstone bridge	p. 323	p. 163	p. 284.
63. Turbidity jackson units.	Turbidimeter	p. 350	p. 467	p. 308.
64. Fecal streptococci bacteria number/100 ml.	MPN; membrane filter; plate count	p. 689 p. 690 p. 691		
65. Coliform bacteria (fecal) number/100 ml.	MPN: Membrane filter	p. 669 p. 684		
66. Coliform bacteria (total) number/100 ml.	do	p. 664 p. 679		
Radiological parameters:				
67. Alpha—total pCi/liter	Proportional counter; scintillation counter	p. 598	p. 509	
68. Alpha—counting error pCi/liter.	do	p. 598	p. 512	
69. Beta—total pCi/liter	Proportional counter†	p. 598	p. 478	
70. Beta—counting error pCi/liter.	do	p. 598	p. 478	
71. Radium—total pCi/liter.	Proportional counter; scintillation counter	p. 611 p. 617	p. 674	

[1] A number of such systems manufactured by various companies are considered to be comparable in their performance. In addition, another technique, based on Combustion-Methane Detection, is also acceptable.

[2] For the determination of total metals the sample is not filtered before processing. Choose a volume of sample appropriate for the expected level of metals. If much suspended material is present, as little as 50–100 ml of well-mixed sample will most probably be sufficient. (The sample volume required may also vary proportionally with the number of metals to be determined.)

Transfer a representative aliquot of the well-mixed sample to a Griffin beaker and add 3 ml of concentrated distilled HNO_3. Place the beaker on a hotplate and evaporate to dryness making certain that the sample does not boil. Cool the beaker and add another 3 ml portion of distilled concentrated HNO_3. Cover the beaker with a watch glass and return to the hotplate. Increase the temperature of the hotplate so that a gentle reflux action occurs. Continue heating, adding additional acid as necessary until the digestion is complete, generally indicated by a light colored residue. Add (1:1 with distilled water) distilled concentrated HCl in an amount sufficient to dissolve the residue upon warming. Wash down the beaker walls and the watch glass with distilled water and filter the sample to remove silicates and other insoluble material that could clog the atomizer. Adjust the volume to some predetermined value based on the expected metal concentrations. The sample is now ready for analysis. Concentrations so determined shall be reported as "total".

[3] See D. C. Manning, "Technical Notes", Atomic Absorption Newsletter, Vol. 10, No. 6 p. 123, 1971. Available from Perkin-Elmer Corporation, Main Avenue, Norwalk, Connecticut 06852.

[4] Atomic absorption method available from Methods Development and Quality Assurance Research Laboratory, National Environmental Research Center, USEPA, Cincinnati, Ohio 45268.

[5] For updated method, see: Journal of the American Water Works Association 64, No. 1, pp. 20–25 (Jan. 197[illegible]) or ASTM Method D 3223–73, American Society for Testing and Materials Headquarters, 1916 Race St., Philadelphia, Pa. 19103.

[6] Interim procedures for algicides, chlorinated organic compounds, and pesticides can be obtained from the Methods Development and Quality Assurance Research Laboratory, National Environmental Research Center, USEPA, Cincinnati, Ohio 45268.

[7] Benzidine may be estimated by the method of M.A. El-Dib, "Colorimetric Determination of Aniline Derivatives in Natural Waters", El-Dib, M.A., Journal of the Association of Official Analytical Chemists, Vol. 54, No. 6, Nov., 1971, pp. 1383–1387.

†As a prescreening measurement.

Appendix C

METRIC UNITS-ENGLISH UNITS CONVERSION TABLE

This English-Metric unit conversion table is reproduced from a U.S. Environmental Protection Agency development document for effluent limitations guidelines and new source performance standards.

METRIC UNITS

CONVERSION TABLE

MULTIPLY (ENGLISH UNITS)		by	TO OBTAIN (METRIC UNITS)	
ENGLISH UNIT	ABBREVIATION	CONVERSION	ABBREVIATION	METRIC UNIT
acre	ac	0.405	ha	hectares
acre - feet	ac ft	1233.5	cu m	cubic meters
British Thermal Unit	BTU	0.252	kg cal	kilogram-calories
British Thermal Unit/pound	BTU/lb	0.555	kg cal/kg	kilogram calories/ kilogram
cubic feet/minute	cfm	0.028	cu m/min	cubic meters/minute
cubic feet/second	cfs	1.7	cu m/min	cubic meters/minute
cubic feet	cu ft	0.028	cu m	cubic meters
cubic feet	cu ft	28.32	l	liters
cubic inches	cu in	16.39	cu cm	cubic centimeters
degree Fahrenheit	°F	0.555(°F-32)*	°C	degree Centigrade
feet	ft	0.3048	m	meters
gallon	gal	3.785	l	liters
gallon/minute	gpm	0.0631	l/sec	liters/second
horsepower	hp	0.7457	kw	killowatts
inches	in	2.54	cm	centimeters
inches of mercury	in Hg	0.03342	atm	atmospheres
pounds	lb	0.454	kg	kilograms
million gallons/day	mgd	3,785	cu m/day	cubic meters/day
mile	mi	1.609	km	kilometer
pound/square inch (gauge)	psig	(0.06805 psig +1)*	atm	atmospheres (absolute)
square feet	sq ft	0.0929	sq m	square meters
square inches	sq in	6.452	sq cm	square centimeters
tons (short)	ton	0.907	kkg	metric tons (1000 kilograms)
yard	yd	0.9144	m	meters

* Actual conversion, not a multiplier

☆U.S. GOVERNMENT PRINTING OFFICE:1974 546-318/345 1-3

Appendix D

WASTE DISCHARGE PERMIT FOR A MINING AND MILLING OPERATION IN IDAHO'S COEUR d'ALENE DISTRICT

The waste water discharge permit reproduced here was based on the Environmental Protection Agency's interim guidelines (see Chapter 1) and was one of the first issued to the mineral resource industry under the authority of Public Law 92-500. Pertinent data for a mining and milling operation in the Coeur d'Alene District are filled in on the permit, which includes instructions for installing and monitoring four observation wells around a tailings pond to monitor seepage.

The permit was sent to the permittee by the regional manager of EPA's Region X with a letter which said in part:

> I am pleased to announce the issuance of a National Pollution Discharge Elimination System waste water discharge permit to your firm pursuant to section 402 of the Federal Water Pollution Control Act Amendments of 1972. . . .
>
> The terms and conditions of the permit represent the application of the best practicable control technology currently available. Achievement of the conditions of the permit should produce considerable improvement in the water quality of the South Fork Coeur d'Alene basin. Violation of the permit's terms and conditions is subject to stringent civil and criminal penalties. . . .
>
> This permit will become effective on September 30, 1973, unless a request for an adjudicatory hearing is granted pursuant to 40 C.F.R. § 125.34 (c). A copy of the pertinent portions of 40 C.F.R. § 125.34 is enclosed. [See excerpt reproduced from the *Federal Register* for May 22, 1973 at the end of this appendix.]

Application No. : ____________________

Name of permittee: ____________________

Effective date: ____________________

Expiration date: ____________________

NATIONAL POLLUTANT DISCHARGE ELIMINATION SYSTEM

DISCHARGE PERMIT

In reference to the above application for a permit to discharge in compliance with the provisions of the Federal Water Pollution Control

Act, as amended by the Federal Water Pollution Control Act Amendments of 1972, P.L. 92-500, October 18, 1972 (33 U.S.C. §1251 et. seq.) (hereinafter referrred to as "the Act").

____________________ is authorized by Environmental Protection Agency to discharge from

to

in accordance with the following general and special conditions:

General Conditions

G1. All discharges authorized herein shall be consistent with the terms and conditions of this permit; facility expansions, production increases, or process modifications which result in new or increased discharges of pollutants must be reported by submission of a new NPDES application or, if such new or increased discharge does not violate the effluent limitations specified in this permit, by submission to the permit issuing authority of notice of such new or increased discharges of pollutants (in which case the permit may be modified to specify effluent limitations for any pollutants not identified and limited herein); the discharge of any pollutant more frequently than or at a level in excess of that identified and authorized by this permit shall constitute a violation of the terms and conditions of this permit.

G2. After notice and opportunity for a hearing, this permit may be modified, suspended, or revoked in whole or part during its term for cause including, but not limited to, the following:

a. Violation of any terms or conditions of this permit;

b. Obtaining this permit by misrepresentation or failure to disclose fully all relevant facts;

c. A change in any condition that requires either a temporary or permanent reduction or elimination of the authorized discharge.

G3. Notwithstanding (2) above, if a toxic effluent standard or prohibition (including any schedule of compliance specified in such effluent standard or prohibition) is established under section 307(a) of the Act for a toxic pollutant which is present in the discharge authorized herein and such standard or prohibition is more stringent than any limitation upon such pollutant in this permit, this permit shall be revised or modified in accordance with the toxic effluent standard or prohibition and the permittee shall be so notified.

G4. The permittee shall allow the head of the State water pollution control agency, the Regional Administrator, and/or their authorized representatives, upon the presentation of credentials:

a. To enter upon the permittee's premises where an effluent source is located or in which any records are required to be kept under the terms and conditions of this permit;

b. To have access to and copy at reasonable times any records required to be kept under the terms and conditions of this permit;

c. To inspect at reasonable times any monitoring equipment or monitoring method required in this permit; or,

d. To sample at reasonable times any discharge of pollutants.

G5. The permittee shall at all times maintain in good working order and operate as efficiently as possible all treatment or control facilities or systems installed or used by the permittee to achieve compliance with the terms and conditions of this permit.

G6. The issuance of this permit does not convey any property rights in either real or personal property, or any exclusive privileges, nor does it authorize any injury to private property or any invasion of personal rights, nor any infringement of Federal, State or local laws or regulations.

G7. This permit does not authorize or approve the construction of any onshore or offshore physical structures or facilities or the undertaking of any work in any navigable waters.

G8. The specific effluent limitations and other pollution controls applicable to the discharge permitted herein are set forth below in the special conditions. Also set forth below are self-monitoring and reporting requirements. Unless otherwise specified, the permittee shall submit duplicate original copies of all reports to the head of the State water pollution control agency and the Regional Administrator. Except for data determined to be confidential under section 308 of the Act, all such reports shall be available for public inspection at the offices of the head of the State water pollution control agency and the Regional Administrator. Knowingly making any false statement on any such report may result in the imposition of criminal penalties as provided for in section 309 of the Act.

Special Conditions

S1. Initial effluent limitations

During the period beginning on the effective date of this permit and lasting until June 30, 1974, discharges from outfall 001 covered under application __________ **(mine)** and application __________ (mill) shall be limited and monitored by the permittee as specified below:

a. The following shall be limited and monitored by the permittee as specified:

Effluent Characteristic	Discharge Limitation Daily Average	Discharge Limitation Daily Maximum	Monitoring Requirements Measurement Frequency	Monitoring Requirements Sample Type
Flow	10,000 cu m/day (2.64 MGD)	14,800 cu m/day (3.9 MGD)	Daily	-----
Suspended Solids	52 mg/l	75 mg/l	Daily	Grab
Dissolved Copper)	Combined	Combined	Weekly	Grab
Dissolved Cadmium)	total not	total not	Weekly	Grab
Dissolved Mercury)	to exceed	to exceed	Weekly	Grab
Dissolved Zinc)	3.0 mg/l	4.5 mg/l	Weekly	Grab
Dissolved Lead)			Weekly	Grab
Cyanide	0.2 mg/l	0.3 mg/l	Weekly	Grab

For the purposes of this subsection, the daily average discharge is the total discharge by weight during a calendar month divided by the number of days in the month that the production or commercial facility was operating.

For the purposes of this subsection, the daily maximum discharge means the total discharge by weight during any calendar day.

b. The pH shall not be less than 6.0 nor greater than 8.5. The pH shall be monitored as follows: Daily grab samples.

c. A minimum of four observation wells shall be located upstream and downstream of the tailings pond to monitor groundwater. These wells shall be monitored through grab samples taken at least once per month. Constituents to be analyzed shall include dissolved copper, lead, and zinc. The results of this monitoring shall be reported to EPA in the monthly monitoring report as specified in condition S6.

S2. Final effluent limitations

During the period beginning July 1, 1974 and lasting until the date of expiration of this permit, discharges from outfall ________ covered under application __________(mine) and application ____________ (mill) shall be limited and monitored by the permittee as specified below:

a. The following shall be limited and monitored by the permittee as specified:

Effluent Characteristic	Discharge Limitation Daily Average	Discharge Limitation Daily Maximum	Monitoring Requirements Measurement Frequency	Monitoring Requirements Sample Type
Flow	10,000 cu m/day (2.64 MGD)	14,800 cu m/day (3.9 MGD)	Daily	-----
Suspended Solids	20 mg/l	30 mg/l	Daily	Grab
Dissolved Copper)			Weekly	Grab
Dissolved Cadmium)	Combined	Combined	Weekly	Grab
Dissolved Mercury)	total not	total not	Weekly	Grab
Dissolved Zinc)	to exceed	to exceed	Weekly	Grab
Dissolved Lead)	1.0 mg/l	1.5 mg/l	Weekly	Grab
Cyanide	0.2 mg/l	0.3 mg/l	Weekly	Grab

For the purposes of this subsection, the daily average discharge is the total discharge by weight during a calendar month divided by the number of days in the month that the production or commercial facility was operating.

For the purposes of this subsection, the daily maximum discharge means the total discharge by weight during any calendar day.

b. The pH shall not be less than 6.0 nor greater than 8.5. The pH shall be monitored as follows: Daily grab samples

S3. Schedule of compliance for effluent limitations

a. Permittee shall achieve compliance with the effluent limitations specified above for discharges from outfall covered under application ____________(mine) and application _____________ (mill) in accordance with the following schedule:

(1) Completion of final plans for attainment of S2 levels by December 31, 1973.

(2) Attainment of operational level (S2) by July 1, 1974.

b. The permittee shall submit to the permit issuing authority the required report of progress or, where a specific action is required in a. above to be taken by a certain date, a written notice of compliance or noncompliance with each of the above schedule dates, postmarked no later than 14 days following each elapsed date. Each notice of noncompliance shall include the following information:

(1) A short description of the noncompliance;

(2) A description of any actions taken or proposed by the permittee to comply with the elapsed schedule requirement without further delay;

(3) A description of any factors which tend to explain or mitigate the noncompliance; and

(4) An estimate of the date permittee will comply with the elapsed schedule requirement and an assessment of the probability that permittee will meet the next scheduled requirement on time.

S4. Compilation of monitoring data

a. Samples and measurements taken to meet the monitoring requirements specified above shall be representative of the volume and nature of the monitored discharge.

b. Following promulgation of guidelines establishing test procedures for the analysis of pollutants, published pursuant to section 304(g) of the Federal Water Pollution Control Act, as amended, all sampling and analytical methods used to meet the monitoring requirements specified above shall conform to such guidelines. If the section 304(g) guidelines do not specify test procedures for any pollutants required to be monitored by this permit and until such guidelines are promulgated, sampling and analytical methods used to meet the monitoring requirements specified in this permit shall, unless otherwise specified by the Regional Administrator, conform to the latest edition of the following references:

(1) American Public Health Association, Standard Methods for the Examination of Water and Wastewaters.

(2) American Society for Testing and Materials, A.S.T.M. Standards, Part 23, Water, Atmospheric Analysis.

(3) Environmental Protection Agency, Water Quality Office, Analytical Control Laboratory, Methods for Chemical Analysis of Water and Wastes.

The permittee shall submit a description of the sampling and analyses methods it proposes to use to the Regional Administrator within 30 days of the effective date of this permit, specifying their location in a particular reference work. The permittee shall use only such methods in gathering the required data on the quantity and quality of its effluent.

c. Permittee shall take samples and measurements to meet the monitoring requirements specified above in the effluent before its discharge to the receiving water.

S5. Recording of monitoring activities and results

a. The permittee shall make and maintain records of all information resulting from the monitoring activities required by this permit.

b. The permittee shall record for each measurement or sample taken pursuant to the requirements of this permit the following information: (1) The date, exact place, and time of sampling; (2) the dates analyses were performed; (3) who performed the analyses; (4) the analytical techniques or methods used; and, (5) the results of all required analyses.

c. If the permittee monitors any pollutant more frequently than is required by this permit, he shall include the results of such monitoring in the calculation and reporting of the values required in the Discharge Monitoring Report form (EPA Form 3320-1 (10-72)). Such increased frequency shall be indicated on the Discharge Monitoring Report form.

d. The permittee shall retain for a minimum of 3 years all records of monitoring activities and results including all records of calibration and maintenance of instrumentation and original strip chart recordings from continuous monitoring instrumentation. This period of retention shall be extended during the course of any unresolved litigation regarding the discharge of pollutants by the permittee or when requested by the Regional Administrator or the State water pollution control agency.

S6. Reporting of monitoring results

a. Monitoring information required by this permit shall be summarized and reported by submitting a Discharge Monitoring Report form (EPA Form 3320-1 (10-72)), properly filled in and signed, to the Regional Administrator and the State agency at the following addresses:

United States Environmental Protection Agency
Region X
1200 6th Avenue
Seattle, Washington 98101

Attn: Permits Branch M/S 521

Idaho Department of Environmental and
Community Services
State of Idaho
Statehouse
Boise, Idaho 83702

b. Each submitted Discharge Monitoring Report shall be signed as follows:

(1) If submitted by a corporation, by a principal executive officer of at least the level of vice president, or his duly authorized representative, if such representative is responsible for the overall operation of the facility from which the discharge described in the Discharge Monitoring Report originates;

(2) If submitted by a partnership, by a general partner;

(3) If submitted by a sole proprietor, by the proprietor;

(4) If submitted by a municipality, State or Federal agency, or other public entity, by a principal executive officer, ranking elected official, commanding officer, or other duly authorized employee.

c. All information submitted on the Discharge Monitoring Report form shall be based upon measurements and sampling carried out during the previous calendar month. The first Discharge Monitoring Report shall be submitted for the period ending October 31, 1973. Thereafter, reporting periods shall end on the last day of each month. The permittee shall submit a Discharge Monitoring Report postmarked no later than the 28th day of the month following each completed reporting period.

S7. Limitation of discharges of oil and hazardous substances in harmful quantities

The permittee shall not discharge oil into or upon navigable waters or adjoining shorelines in quantities defined as harmful in regulations published at 40 CFR 110, including any amendments or revisions to such regulations effected subsequent to the date of this permit. In addition, the permittee shall not discharge hazardous substances into or upon navigable waters or adjoining shorelines in quantities defined as harmful in regulations promulgated by the Administrator pursuant to section 311(b)(4) of the Federal Water Pollution Control Act, as amended. Nothing in this permit shall be deemed to preclude the institution of any legal action nor relieve the permittee from any responsibilities, liabilities, or penalties to which the permittee is or may be subject under section 311 of the Federal Water Pollution Control Act, as amended, or under any other Federal or State law or regulation.

S8. Limitations of visible floating solids and foam

During the period beginning on the effective date of this permit and lasting until the date of expiration of this permit, the permittee shall not discharge floating solids or visible foam.

S9. Disposal of collected solids

a. Intake water treatment. Solids, sludges, dirt, sand, silt, or other pollutants separated from or resulting from treatment of intake or supply water prior to use by the permittee shall be disposed of in such manner as to prevent any pollutant from such materials from entering navigable waters. Any live fish, shellfish, or other animals collected or trapped as a result of intake water screening or treatment may be returned to their water body habitat.

b. Wastewater treatment. Solids, sludges, filter backwash, or other pollutants removed from or resulting from treatment or control of wastewaters shall be disposed of in such manner as to prevent any pollutant from such materials from entering navigable waters.

S10. Noncompliance with effluent limitations

a. If for any reason the permittee does not comply with or will be unable to comply with any daily maximum effluent limitation specified in this permit, the permittee shall immediately notify the permit issuing authority or his designee by telephone at (206) 442-1200 and provide the permit issuing authority with the following information in writing within five days of such notification:

(1) Cause of noncompliance;

(2) A description of the noncomplying discharge including its impact upon the receiving waters;

(3) Anticipated time the condition of noncompliance is expected to continue, or if such condition has been corrected, the duration of the period of noncompliance;

(4) Steps taken by the permittee to reduce and eliminate the noncomplying discharge; and

(5) Steps to be taken by the permittee to prevent recurrence of the condition of noncompliance.

b. Permittee shall take all reasonable steps to minimize any adverse impact to navigable waters resulting from noncompliance with any effluent limitation specified in this permit, including such accelerated or additional monitoring as necessary to determine the nature and impact of the noncomplying discharge.

c. Nothing in this permit shall be construed to relieve the permittee from civil or criminal penalties for noncompliance, whether or not such noncompliance is due to factors beyond his control, such as equipment breakdown, electric power failure, accident, or natural disaster.

S11. Provision for electric power failure

The permittee shall either:

a. no later than January 1, 1974, certify in writing to the permit issuing authority that the permittee has installed or provided for an alternative electric power source sufficient to operate all facilities utilized by permittee to maintain compliance with the terms and conditions of the permit; or

b. no later than thirty days after the effective date of this permit, certify in writing to the permit issuing authority that upon the reduction, loss, or failure of one or more of the primary sources of electric power to any facilities utilized by the permittee to maintain compliance with the terms and conditions of this permit, the permittee shall halt, reduce, or otherwise control production and/or all discharges in order to maintain compliance with the terms and conditions of this permit.

S12. Prohibition of bypass of treatment facilities

The diversion or bypass of any discharge from facilities utilized by the permittee to maintain compliance with the terms and conditions of this permit is prohibited, except (i) where unavoidable to prevent loss of life or severe property damage, or (ii) where excessive storm drainage or runoff would damage any facilities necessary for compliance with the terms and conditions of this permit. The permittee shall immediately notify the permit issuing authority in writing of each such diversion or bypass in accordance with the procedure specified above for reporting noncompliance.

* * * * * * * * * * * * * * * *

This permit shall become effective on the date of the issuing authority's signature, provided, however, that if the issuing authority is the Regional Administrator, the permit shall become effective on the thirtieth day following the date of the Regional Administrator's signature unless a request for an adjudicatory hearing is filed pursuant to the provisions of section 125.34 of NPDES regulations published at 40 CFR 125.

This permit and the authorization to discharge shall expire on midnight June 30, 1977. Permittee shall not discharge after the above date of expiration. In order to receive authorization to discharge beyond the above date of expiration, the permittee shall submit such information, forms, and fees as are required by the agency authorized to issue NPDES permits no later than 180 days prior to the above date of expiration.

By authority of Regional Administrator, Region X, EPA.

(Date) (Signature)

REGULATIONS GOVERNING REQUESTS FOR ADJUDICATORY HEARINGS PURSUANT TO 40 C.F.R. § 125.34 (1973)

(c) *Adjudicatory hearings.*—(1) Within 30 days following issuance of public notice of a permit application pursuant to § 125.32, or, if a public hearing is held pursuant to § 125.34(b), within 20 days following the issuance of the notice provided in § 125.34(b)(4), any person may submit to the Regional Administrator a request for an adjudicatory hearing to consider the proposed permit and its conditions. If the request for an adjudicatory hearing is granted in accordance with § 125.34(f), any person may submit a request to be a party within 30 days after the date of publication of public notice of an adjudicatory hearing in a newspaper of general circulation as required by § 125.32.

(2) Requests for and adjudicatory hearing and requests to be a party under this paragraph shall:

(i) State the name and address of the person making such request;

(ii) Identify the interest of the requester, and any person represented by issuance or nonissuance of the permit;

(iii) Identify any other persons whom the requester represents;

(iv) Include an agreement by the requester, and any person represented by the requester, to be subject to examination and cross-examination, and in the case of a corporation, to make any employee available for examination and cross-examination at his own expense, upon the request of the presiding officer, on his own motion or on the motion of any party.

(3) In addition to the information required under § 125.34(c)(2), any request for an adjudicatory hearing shall state with particularity the reasons for the request, and the issues proposed to be considered at the hearing.

(4) In addition to the information required under § 125.34(c)(2), any request to be a party shall state the position of the requestor on the issues to be considered at the hearing.

(d) *Filing and service.*—(1) All documents or papers required or authorized to be filed, shall be filed with the regional hearing clerk, except as otherwise herein provided. Except for requests for an adjudicatory hearing or request to be a party, at the same time that a party files documents or papers with the clerk, it shall serve upon all other parties copies thereof, with a certificate of service on each document or paper, including those filed with the regional hearing clerk. Filing shall be deemed timely if received by the regional hearing clerk within the time allowed by this section.

(2) In addition to copies served on all other parties, each party shall file with the regional hearing clerk an original and two copies of all papers filed in connection with an adjudicatory hearing.

(e) *Time.*—In computing any period of time prescribed or allowed by the regulations in this part, except as otherwise provided, the day of the act or event from which the designated period of time begins to run shall not be included. Saturdays, Sundays, and holidays, shall be included in computing the time allowed for the filing of any document or paper, except that when such time expires on a Saturday, Sunday, or legal holiday, such period shall be extended to include the next following business day.

(f) *Notice of hearing.*—Within 5 days following the expiration of the time allowed by § 125.34(c)(1) for submitting a request for an adjudicatory hearing the Regional Administrator shall determine whether such request meets the requirements of § 125.34(c). If any request meets such requirements and sets forth material issues relevant to the question whether a permit should be issued, and what conditions to such permit would be required to carry out the provisions of the Act, the matter shall be assigned promptly for hearing: *Provided,* That if the Regional Administrator holds a public hearing under § 125.34(b), no request for an adjudicatory hearing shall be timely until after the conclusion of such public hearing. The Regional Administrator shall treat all other requests for a hearing as requests to be a party, and shall grant any such request meeting the requirements of § 125.34 (c)(2) and (c)(4). The hearing shall be held in the State in which the discharge or proposed discharge shall occur, or at such other accessible location as is appropriate. The Regional Administrator shall issue public notice of such hearing in the manner specified in § 125.32. The hearing shall take place not less than 30 days after the issuance of public notice of such hearing.

Appendix E

TREATMENT OF A COMBINATION MINING, MILLING, AND METALLURGICAL WASTE WATER*

ABSTRACT

At the Bunker Hill Company mining and metallurgical complex in North Idaho, all metallurgical waste waters that cannot be recycled without treatment are combined with milling effluent and acid mine drainage in one large impoundment pond. A new treatment plant utilizing a lime neutralization process is being installed to treat this discharge in order to minimize discharge of dissolved heavy metals. A bench-scale pilot plant was operated to provide an experimental basis for establishing treatment effectiveness. Based on these tests, a full-scale plant was designed and is now under construction. Capital and operating costs for this plant have been projected.

INTRODUCTION

The Bunker Hill Company operations are conducted at three major subsurface mines and a large metallurgical complex for smelting and refining primary lead and zinc metals. Related by-products, including silver, gold, copper, cadmium, and antimony, are recovered from processed concentrates. Three sulfuric acid plants are operated to recover sulfur dioxide from the smelting processes.

In addition to the mining and metallurgical operations, Bunker Hill, in a joint venture with Stauffer Chemical Company, produces phosphoric acid and ammonium phosphate fertilizers. All operations are in or near Kellogg, Idaho.

SOURCES AND NATURE OF WASTE WATER

Liquid effluents are discharged from the Bunker Hill Mine (acid mine drainage) and all of the surface plants, with the exception of the lead smelter, into a 160-acre central impoundment pond for separation and retention of

* From a paper entitled, "Process Waste Water Treatment at the Bunker Hill Company," presented at the 1973 meeting of the American Mining Congress by Gene M. Baker, Director of Environmental Control Activities and Manager of Plant Engineering, and Andy H. Larson, Manager of Research and Development, Bunker Hill Company, Kellogg, Idaho.

Table E.1 *Characteristics of Waste Water Influents and Central Impoundment Pond Effluent (CIP)*

Dissolved Elements	Influent				Effluent
	Zinc Plant	Phosphoric Acid Plant	Bunker Hill Mine	Mill	CIP (mg/L)
Fe	-	-	X	-	20 - 30
Pb	-	-	X	X	2 - 8
Zn	X	-	X	X	90 - 170
Cd	X	-	-	-	0.5 - 0.9
Hg	X	-	-	-	0.003 - 0.02
F	-	X	-	-	190 - 250
PO_4	-	X	-	-	90 - 120
SO_4	X	-	X	-	1200 - 1500
pH	Low	Low	Low	High	2.5 - 3.0
Suspended Solids (mg/L)	28,000		6,000 - 13,000	130,000	20 - 40

X = present in significant concentrations
- = not present in significant concentrations

most of the suspended solids. The clarified effluent is presently decanted from this area into the South Fork Coeur d'Alene River. The sources and nature of waste water influents and effluents from the central impoundment pond are shown in Table E.1.

At the zinc plant and phosphoric acid plant operations, facilities have been installed to recycle a majority of the water discharged from cooling processes. The remainder of the effluent, which is from metallurgical processing and air pollution control equipment, is not amenable to recycling without treatment. It is transported along with slurried gypsum from the phosphoric acid plant in a 1-mile-long, 12-inch-diameter PVC pipeline and discharged into the central impoundment pond for clarification. This combined discharge is low in pH. It contains dissolved zinc, cadmium, mercury, fluoride, sulfate, phosphate, and includes gypsum as a suspended solid.

Mine drainage is the major volume of waste water. It is acidic in nature and originates from surface and ground waters which seep into the extensive workings of the mine. Combinations of air and water along with complex bacterial action in contact with pyrite create sulfuric acid, lowering the pH of the water and dissolving heavy metals from mineralized areas. The dissolved metals are then carried to the surface in mine drainage water. As shown in Table E.1, this waste stream contains dissolved iron, lead, zinc, sulfate, and considerable suspended solids.

At the mill, waste products from wet grinding and flotation operations are cycloned before discharge into the impoundment pond and the coarse fraction is pumped back into the mine to refill mined-out areas. During this "sand-fill operation," fines drain into the mine drainage system from underground cycloning systems and from the areas being filled with sand slurry. These fines return to the surface as suspended solids contained in the acid mine drainage. The mill effluent is characteristically high in pH and contains dissolved zinc and lead and suspended solids.

The central impoundment pond is equipped with a decant for discharging clarified overflow from the center of the pond. The overflow is low in suspended solids but contains the dissolved elements shown in Table E.1. The concentrations of dissolved heavy metals must be reduced to meet currently proposed state and federal requirements.

CONSIDERATIONS FOR SELECTION OF A TREATMENT SYSTEM

In the selection of a system for waste water treatment, there were four interrelated requirements: acceptable effluent quality, minimum sludge volume, maximum system reliability, and acceptable capital and operating costs. On the basis of previous experience at Bunker Hill and a state-of-the-art technology study by Envirotech* with acid mine water and other heavy-metal-containing waste waters, lime treatment was selected as the only practical system for meeting these criteria.

The initial investigations using lime neutralization were conducted on individual waste water discharges from the zinc plant and mine. These investigations assumed treatment of the discharges at their sources. Although lime additions to a pH of 9.0 to 10.5 produced treated effluents which would meet the proposed regulations for dissolved heavy metals, the large volume of low-density sludge presented impossible storage and disposal problems.

Laboratory tests using lime neutralization of zinc plant waste water produced a sludge containing 6 percent solids by weight (77,000 pounds per day of dry solids). Comparable tests with the mine effluent produced a sludge containing 0.4 percent solids by weight (29,000 pounds per day dry solids). The total volume of sludge produced from the two systems represented 1,135 acre-feet per year: 160 from the zinc plant waste water and 975 from the acid mine drainage.

Chemical analyses of the sludges indicated that zinc plant sludge was predominantly calcium sulfate (due to the high sulfate concentration of the waste water) and zinc hydroxide, while the mine sludge was predominantly hydroxides of zinc and iron.

To reduce the sludge volume, three simultaneous approaches were used:

1. Reduction of the sulfate concentration by recycling within the zinc plant.
2. Combination of the remaining zinc plant effluent with the mine drainage to reduce sulfate concentration to a level at which precipitation of calcium sulfate would not occur during lime neutralization.

* Envirotech Systems, Inc., Menlo Park, California, was retained as a consultant during the initial investigation period.

Table E.2 *Comparison of Sludges from Individually Treated Zinc Plant and Mine Effluents and Treatment of Combined Effluents*

	Zinc Plant	Mine	Combination
Sulfate, ppm	4500 (est)	1000 - 1200	1200 - 1500
Thickener Underflow Solids,* pounds per day	77,000	29,000	35,000
Solids, % by Weight	6	0.4	8
Sludge Volume, acre ft/year	160	975	55

* Suspended solids are not included

3. Utilization of a thickener underflow sludge recycle system whereby the sludge concentration can be increased by including recycled underflow solids with incoming plant feed.*

The potential sludge volume reduction incorporating the approaches listed above is shown in Table E.2.

On the basis of this projected combination advantage, a combined zinc-mine effluent treatment plant was considered mandatory since the problem of disposing of the volume of sludge generated by individual treatment was of major concern in the mountainous terrain of the Kellogg area where space is at a premium. While developing this rationale and considering the location of other effluent sources and previous installations, it became apparent that this combined plant should include treatment for all the waste water presently being discharged from the central impoundment pond. The advantages of such an overall approach are as follows:

1. The total sludge volume is minimized.
2. The central impoundment pond serves as a clarifier for the removal of high concentrations of suspended solids prior to lime treatment, thus removing a heavy burden from the treatment plant—especially important when utilizing a solids recycle system.
3. The high pH mill effluent will help neutralize acid mine drainage.
4. A central plant is more economical than several individual plants from both capital and operating standpoints.
5. The central impoundment pond will provide surge holding capacity during treatment plant upsets or breakdowns, thereby protecting the quality of the river at all times.
6. Processing facilities for possible future recovery of metals from waste sludge will be centralized.

* From a paper by J. Smith III, entitled "The Advantages of A Crowd for Acid Waste Liquors," presented at the 1972 AIME Annual Meeting in San Francisco.

EVALUATION OF A LIME TREATMENT SYSTEM

The dissolved metal constituents of major concern in the central impoundment pond effluent are zinc, iron, lead, cadmium, and mercury. The major dissolved non-metal constituents are sulfate, phosphate, and fluoride. The results of beaker-scale lime treatment of one sample of this effluent water are shown in Figure E.1. With the exception of sulfate, all of the dissolved constituents showed increased precipitation with increased pH levels up to a pH of 9 to 10. It should be mentioned, however, that lead above a pH of 9 and

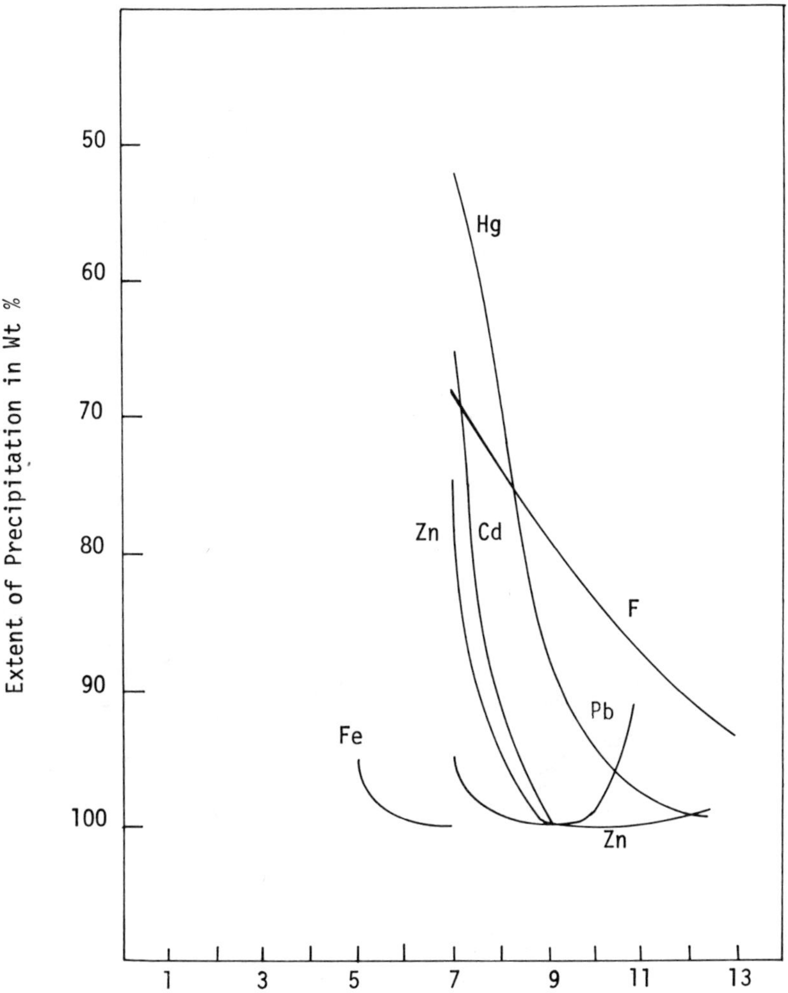

Figure E.1. *Heavy metal precipitation versus pH for central impoundment pond effluent; pH adjustments are by lime additions.*

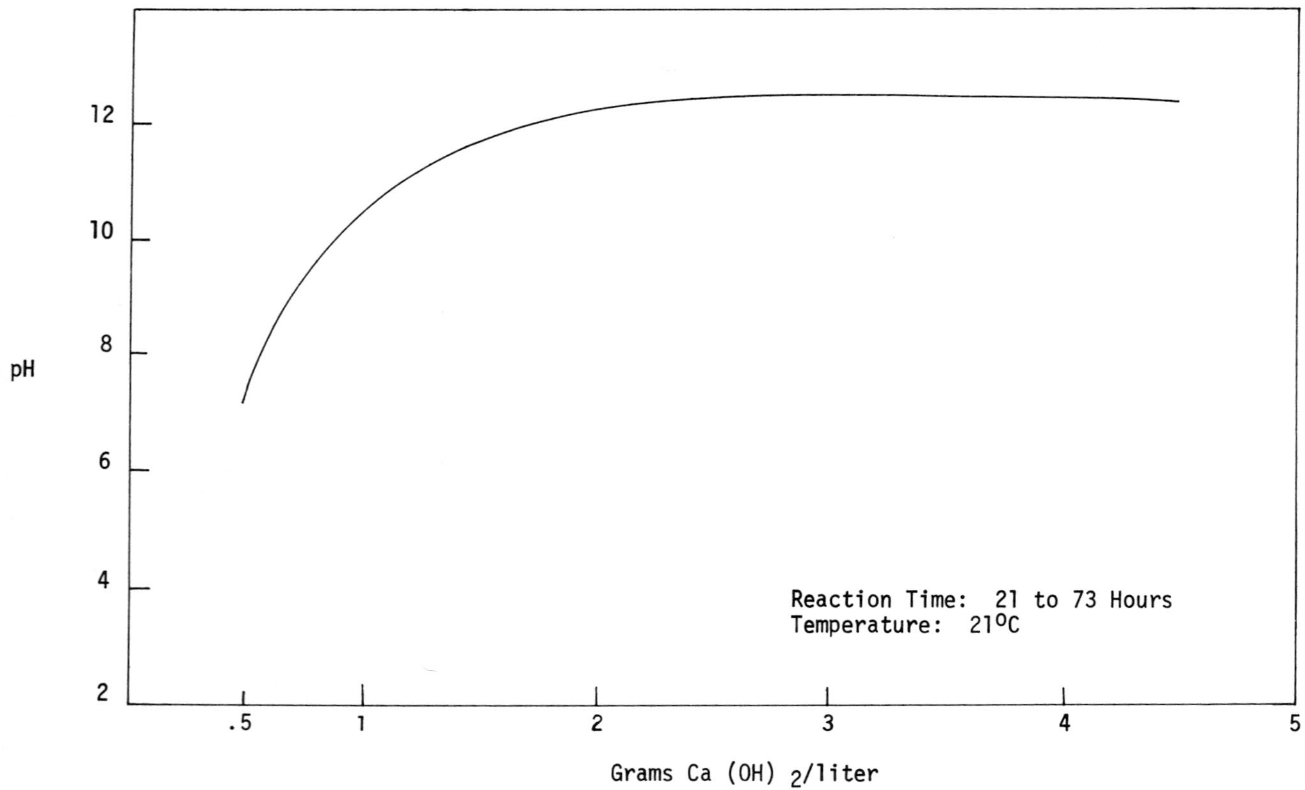

Figure E.2. *Lime consumption versus pH for impoundment pond effluent.*

zinc above a pH of 10.5 exhibited amphoteric behavior; hence, these precipitated metals redissolved. Lime requirements to change the pH are shown in Figure E.2. Over a period of time with different samples the $Ca(OH)_2$ consumption varied from 0.8 to 1.7 mg/l to reach a pH of 10.

In order to evaluate the lime treatment system, a bench-scale pilot plant (miniplant) of the proposed system was set up and operated. This bench-scale circuit is shown schematically in Figure E.3. Since iron is present in the waste water mainly as ferrous iron, it was decided that oxidation to ferric iron by aeration would be necessary to achieve more complete precipitation of iron and better sludge densification. Typical ranges of results obtained by operation

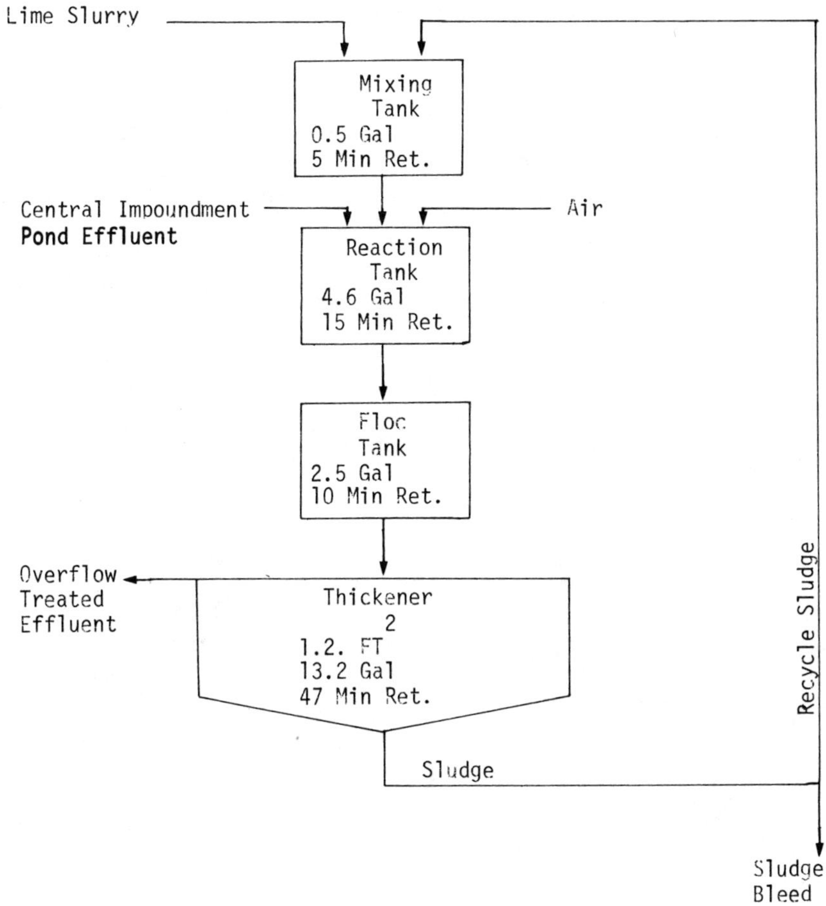

Number of Recycles @ Steady-State Condition = $\frac{\text{Wt Solids in Recycle Sludge}}{\text{Wt Solids Ppt'd from Feed}} = \frac{\text{Wt Solids in Recycle Sludge}}{\text{Wt Solids in Sludge Bleed}}$

Figure E.3. *Laboratory-scale simulation of effluent treatment system for central impoundment pond.*

Table E.3 *Bench-Scale Pilot Plant Results*

Dissolved Element	Thickener Overflow (ppm)	Thickener Underflow* (Wt. %)
Cd	0.04 - 0.06	0.1
Cu	0.02 - 0.13	0.1
Fe	0.05 - 0.6	3 - 6
Hg	0.001 - 0.004	-
Pb	0.3 - 0.6	0.2 - 0.3
Zn	0.05 - 0.6	8 - 14
CaO	700 - 900	17 - 23
SO_4/S	500 - 700	1 - 2
PO_4/P	0.1 - 1.3	1 - 2
F	25 - 35	7
Suspended Solids	15 - 70 ppm (5 - 15 JTU)	-
pH	9.7 - 10.5	-

* Underflow sludge varied from 2 - 8% solids by weight

of the bench-scale pilot plant at feed rates of 10 to 15 gph and sludge recycle rates of 10 to 40 are shown in Table E.3. These tests, as well as laboratory tests by Envirotech on the effect of sludge recycle, proved general applicability of the proposed system for Bunker Hill's use. Although these tests included the waste water from the phosphoric acid plant, it was decided that this plant waste water should be removed from the inflow to the central treatment plant and handled separately by a closed-loop recycle system developed by the Stauffer Chemical Company. In addition to reducing fluoride emissions, this closed-loop system was more economical. The 160-acre impoundment pond has, therefore, been diked into two areas with one area of 43 acres serving as a settling-cooling basin for recycled water to be returned to the phosphoric acid plant process and the remainder serving as a collector for all wastes to be directed to the treatment plant.

DESCRIPTION OF OPERATIONAL TREATMENT PLANT

Treatment facilities are divided into two parts: the central treatment plant and off-site facilities. The central treatment plant is shown schematically in Figure E.4.

Central Treatment Plant

The central treatment plant consists of lime neutralization, aeration, flocculation, sedimentation, thickener underflow solids recycle, and acidification. A

200-ton lime storage bin and 8,000 lb/hr lime slaker unit will provide lime slurry for neutralization. This slurry will be combined with thickener underflow in a 15,000-gallon mixing tank. The decanted effluent from the central impoundment pond will be combined with the output of the mixing tank in a 60x50x10-ft neutralization-aeration basin and controlled at a pH of approximately 10. The overflow from this basin will be conditioned in a 32x32x11-ft flocculation tank. Chemical flocculants will be used as required.

The flow from the flocculation basin will enter a 236-ft-diameter thickener for solids settling and clarification. The majority of the thickener underflow will be recycled through the mixing tank to increase the density of the sludge. The sludge bleed from the thickener will be blended with the mill tailings and distributed around the perimeter of the central impoundment pond.

The clarified thickener overflow will be acidified with sulfuric acid as it flows to the mill water supply reservoir. The mill will recycle part of this water for process requirements. The excess treated effluent will be discharged into the South Fork Coeur d'Alene River.

Off-site Facilities

Off-site facilities include recycle within the zinc plant to reduce sulfate discharge and a pipeline for transporting zinc plant effluent to the central impoundment pond.

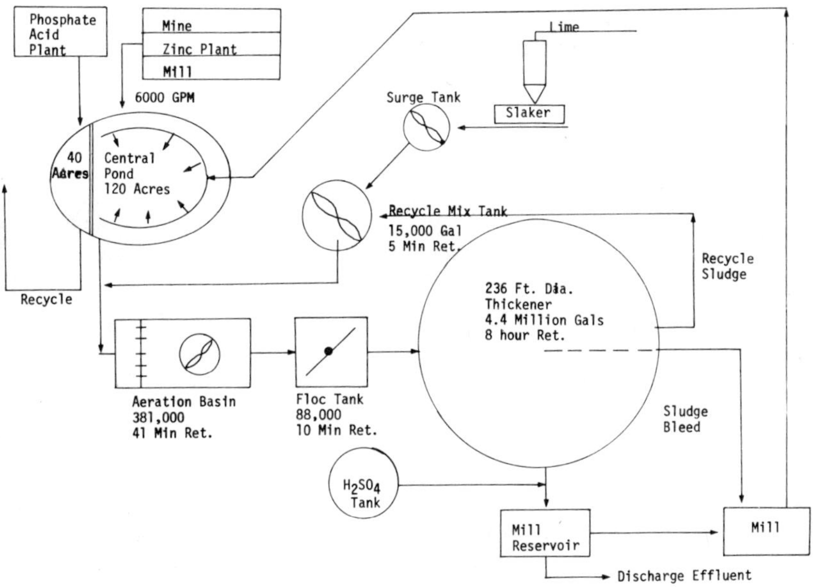

Figure E.4. *Flow sheet for central treatment plant.*

The phosphoric acid plant effluent recycle system is included as part of these facilities, and it will utilize the previously described 43-acre portion of the central impoundment pond for settling and cooling. After settling and cooling, the effluent will be recycled directly back to the phosphoric acid plant. Required makeup water will be supplied from the lead smelter water recycle system. The lead smelter effluent will normally be recycled within the smelter. Provisions will be incorporated to discharge all or any portion of this lead smelter effluent to the central impoundment pond for treatment in the central treatment plant during periods of process upset within the lead smelter.

Projected capital and operating costs for the treatment system are shown in Table E.4.

The central treatment plant is shown diagrammatically in Figure E.5.

Table E.4 *Projected Capital and Operating Costs*

Capital Costs:	
Central Treatment Plant: (6000 gpm Design Flow)	$ 550,000
*Offsite Facilities Costs: (Including pipelines to impounding areas, decants, and various plant effluent recycle systems)	730,000
Total Capital Costs:	$1,280,000
Annual Operating Costs:	
Direct Operating Cost: (at Design Flow)	$ 592,000
Indirect Operating Cost: (including depreciation, taxes, and insurance)	110,000
Total Annual Operating Costs:	$ 702,000

* Not included is $995,000 spent previously for water treatment facilities which were necessary for the proper operation of this facility.

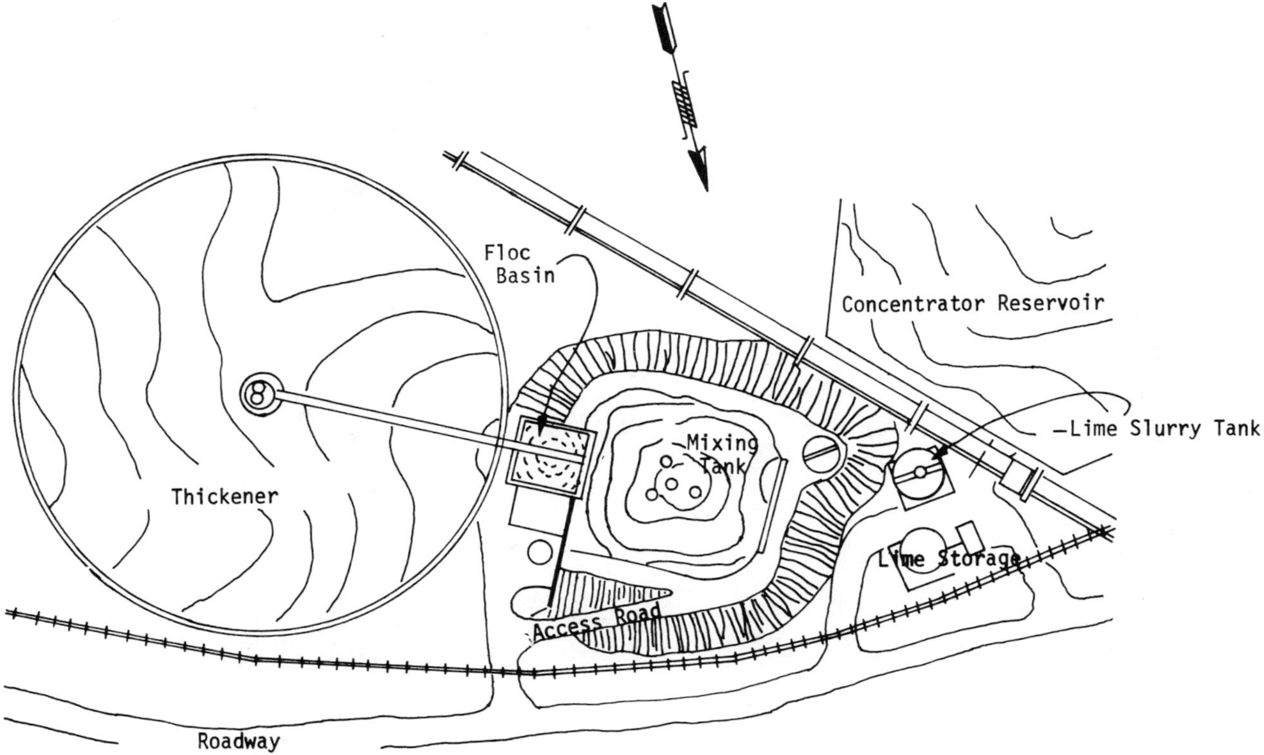

Figure E.5. *Diagram of central treatment plant.*

Appendix F

TREATMENT OF ZINC PLANT WASTES*

SUMMARY

Several sets of effluent samples from the Bunker Hill Company's electrolytic zinc recovery plant were subjected to treatment with lime and coagulant aids in laboratory tests. The primary objectives of the testing were the attainment of an acceptable effluent, low in dissolved heavy metals, and the production of a rapidly settling sludge which would occupy a minimum final volume without additional thickening. Forty-seven proprietary coagulant aids were jar-tested in combination with high-calcium hydrated lime. Of the aids tested, only five showed significant improvement over lime alone. Subsequent steps include a pilot-plant study aimed at producing design data for a treatment system to produce an acceptable effluent and a sludge which may be returnable to the zinc circuit for recovery of its heavy metals values.

INTRODUCTION

The zinc plant area of the Bunker Hill Company consists of several fairly distinct components which process zinc concentrates to produce zinc and cadmium, sulfuric acid, and several types of residues. Wet processing steps within the zinc plant area have been designed for recycling of the aqueous streams where possible. As in all wet processing operations, some discharge of contaminated water occurs; however, most of the effluents from this area have been collected in a single stream which is sent to a settling pond for removal of suspended material. An initial report, based upon studies made by personnel of the Bunker Hill Company, demonstrated the feasibility of lime precipitation of heavy metals in the effluent. However, this study noted problems in settling and concentrating the metal hydroxide sludges. This appendix describes a study supported by a contract between the Bunker Hill Company and the University of Idaho to investigate methods of producing a sludge that would settle rapidly and would also have good compaction characteristics without downstream thickening.

* From a study made for the Bunker Hill Company in 1973 by Dr. A. T. Wallace, professor of Civil Engineering at the University of Idaho, Moscow, Idaho.

Table F.1 *Pond Influent and Effluent Analysis*

	Influent		*Effluent*	
Analysis	*Range*	*Mean value*	*Range*	*Mean value*
pH	1.7-2.3	–	1.7-2.7	–
Temp., °C.	14-40	28	15-35	27
Susp. solids, mg/l	118-977	440	94-379	214
Zn, mg/l	140-824	565	497-985	725
Mn, mg/l	12-1,060	–	23-100	67
Pb, mg/l	0.2-4.4	2.4	1.9-4.7	2.8
Cd, mg/l	1.5-19.6	5.7	1.0-10.3	3.9
Cu, mg/l	0.1-5.7	2.2	0.6-2.3	1.5
Flow, gpm	540-665	593	–	–

WASTE FLOWS AND CHARACTERISTICS

Previous work by the Bunker Hill Company established average values for flow and heavy metals concentrations in both the pond influent and effluent. Table F.1 summarizes the more important characteristics.

Analyses of composites of the first set of samples received for the present testing program were conducted to ascertain whether the waste received was "typical" of zinc plant effluents. The results of these analyses are shown in Table F.2.

Comparison of the data in these tables indicates that the waste used in the laboratory testing program was sufficiently like the "normal" zinc plant waste that a reasonable degree of confidence could be placed in the laboratory results. Composites of three sets of samples received at later dates also could be considered to reflect "average" characteristics.

One important characteristic in which the composite samples differed widely was their titration curves. Figure F.1 shows the titration curves for three composite samples of the waste.

LABORATORY TESTING PROGRAM

The laboratory work consisted of testing the effect of various flocculents and coagulant aids in combination with varying dosages of high-calcium hy-

Table F.2 *Analysis of Zinc Plant Effluent Samples*

Analysis	*Pond influent*	*Pond effluent*
	2.0	2.0
Zn, mg/l	850	707
Mn, mg/l	50	50
Pb, mg/l	2.8	3.1
Cd, mg/l	3.8	3.2
Cu, mg/l	5.0	0.9

drated lime. The standard procedure adopted for these jar tests is as follows:

A Phipps & Bird, six-place gang stirrer was used to mix the waste water and aid being tested in 1,500-ml breakers; 1,000-ml samples were used in all tests. Reagents were added during a rapid-mix (550 rpm) period lasting two minutes. The flocculation period was five minutes at 50 rpm.

When it was desired to check settling rates, the contents of the beakers were transferred, with continuous stirring, to liter graduated cyclinders equipped with slow-speed stirring mechanisms. The subsidence of the solids-liquid interface was timed and recorded. Two settling tests were run for each level of treatment to insure reproducibility.

Observations were made of floc formation, floc settling, final sludge volumes and effluent clarity. A total of 47 proprietary coagulants (listed in Table F.3) were tested over a three-week period.

Each coagulant was added in dosages ranging from 5 mg/l to 30 mg/l and each was tested by adding to the waste solution both before and after lime additions in two series of tests.

In those cases where encouraging preliminary results were obtained, the experiments were repeated using a narrower range of coagulant concentrations. If the results still appeared promising, the contents of the beaker were transferred to 1,000-ml graduated cylinders equipped with slow-speed stirrers (12 to 15 rph) and the settling rates observed. Effluent samples from these settling tests were collected for pH and heavy metals analyses. All heavy metals analyses were made on a Perkin-Elmer atomic absorption spectrophotometer.

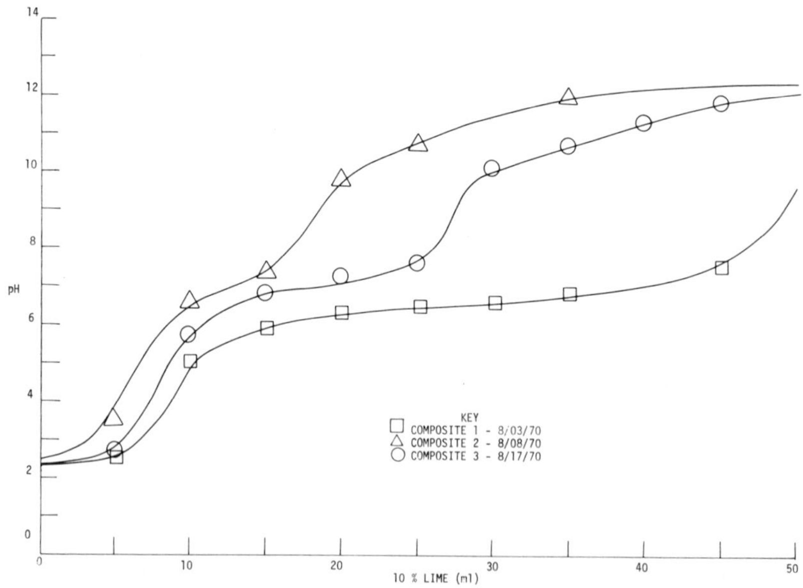

Figure F.1. *Titration curves obtained by titrating 1,000-ml composite samples with 10 percent lime solution.*

Table F.3 *Coagulant Aids Used in Jar Testing Program for Zinc Plant Waste*

Manufacturer	*Product*	*General type*
Allstate Chemical Co.	All-Floc-Ade	sodium aluminate
Baroid Chemical Co.	Baro-Chem 260	unknown
	Surflo A100	anionic
	Surflo 240	cationic
	Surflo 241	unknown
	Surflo 258	unknown
	Surflo 260	unknown
Betz Laboratories	Betz 1100	anionic
	Betz 1110	anionic
	Betz 1120	anionic
	Betz 1130	anionic
	Betz 1150	cationic
	Betz 1160	cationic
	Betz 1170	cationic
Burtonite Co.	Burtonite	gum colloid
Calgon Corp.	Cat-Floc	cationic
	233	nonionic
	243	anionic
	253	anionic
	ST-260	cationic
	ST-266	cationic
	ST-269	anionic
	ST-270	anionic
Dow Chemical Co.	Purifloc C-31	cationic
	A-23	anionic
	N-11	nonionic
	N-12	nonionic
General Aniline & Film	Gafloc A60	anionic
	Gafloc N71	nonionic
	Gafloc C71	cationic
	Gafloc CD181	unknown
General Mills, Chemicals	Gendriv 162	gum derivative
	Gendriv 492	gum derivative
	Gendriv 492-S	gum derivative
	Guartec 574	gum product
	Guartec 1018	gum product
National Chemsearch Corp.	A-Qulex 312	unknown
	A-Qulex 288	unknown
Petrolite Corp.	Tretolite WC 16	unknown
	Tretolite WC 17	unknown
	Tretolite WC 18	anionic
	Tretolite WC 19	unknown
Philadelphia Quartz Co.	N-sol-A	act'd silica
Rohm and Haas Co.	C-7	cationic
Sanfox Co.	Duplix	unknown
Standard Brands, Inc.	Tychem 8011	nonionic
	Tychem 8030	anionic
Union Carbide Co.	Niax X150	cationic

RESULTS

Of the 47 coagulant aids tested in combination with lime, only 5 consistently showed significant improvement over lime alone. All of the 5 were synthetic polyelectrolytes; however, 2 were of the cationic type (Gafloc C-71 and Betz 1150), 2 were of the nonionic type (Dow N-11 and N-12) and 1 was of the anionic type (Betz 1100). The one characteristic the 5 had in common was their extremely high molecular weight. Gafloc C-71 is a very viscous liquid, and although lower in cost than the Dow and Betz products, it proved very difficult to handle; therefore, no further testing was done using this material. The Dow and Betz products are granular free-flowing solids and are relatively easy to handle.

Since slightly superior results were obtained with Betz 1150, an extensive series of tests was made with this polymer in combination with lime to determine whether optimum conditions of heavy metals removal and sludge settling could be obtained. It soon became apparent that raising the pH beyond 8.3 or so did not significantly improve zinc removal but did create sludge which settled more slowly and compacted less than sludge produced at a lower pH. Table F.4 shows the heavy metals concentrations remaining at various pH levels for this series of experiments using a constant polymer dosage of 2 mg/l. The settling curves for this series are shown in Figure F.2.

These data show clearly that an optimum design should result from a pH adjustment in the range from 8.3 to 9.4. A lower pH will not provide for sufficient heavy metals removal and a higher pH will produce a sludge with poor settling characteristics. Table F.5 shows the variation in sludge settling rate, final sludge volume, and sludge solids concentration as a function of pH. Percent removals of zinc and manganese are also given for reference.

CONCLUSIONS

Precipitation of dissolved heavy metal in the Bunker Hill zinc plant area effluent is feasible. A pH adjustment with lime to a range of 8.3 to 9.4, together with a polymer dosage of approximately 2 mg/l, will give satisfactory

Table F.4 *Heavy Metals Remaining in Solution as a Function of pH Adjustment Using Lime and 2 mg/l of Betz 1150*

pH	*Zn, mg/l*	*Mn, mg/l*	*Cd, mg/l*
2.0 (Feed)	850	37	5.7
7.0	350	30	4.0
7.5	40	21	1.3
7.8	20	20	0.8
8.3	1.2	12	0.3
9.4	0.1	3	0.1
9.6	0.1	2	0.1
10.0	0.1	0.2	0.1

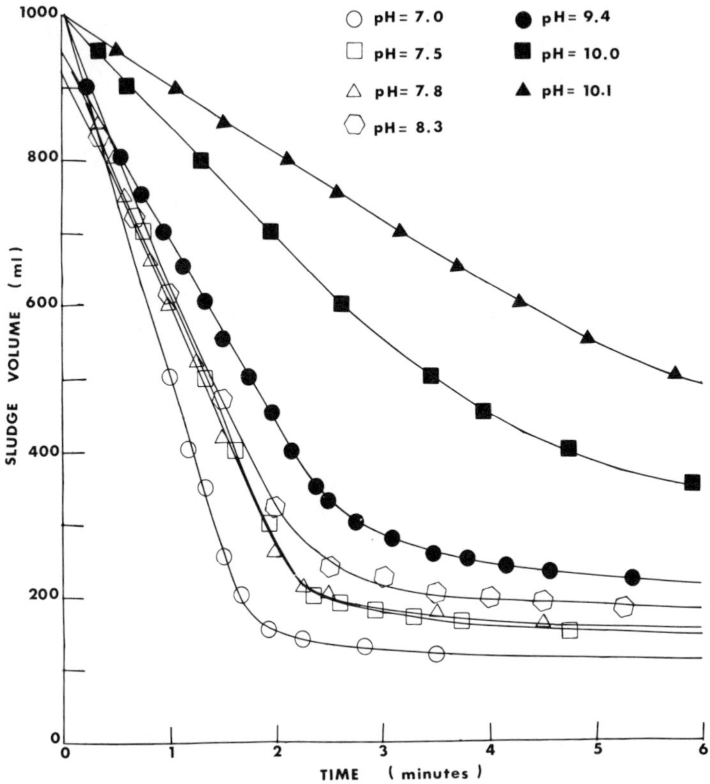

Figure F.2. *Settling rates at various pH levels with 10 percent lime, using a constant polymer dosage of 2 mg/l.*

removal of heavy metals and will produce a sludge which settles rapidly and is relatively easy to handle. Operating the process at the minimum pH consistent with effluent standards will result in the most economical process. It seems possible that the sludge resulting from the treatment might be returned somewhere in the zinc circuit.

Table F.5 *Summary of Sludge Characteristics as Affected by pH Adjustment*

pH	*Settling velocity (ft/min)*	*Final sludge volume (ml/1,000 ml)*	*% Total solids*	*% Removal*	
				Zn	*Mn*
7.75	0.421	140	4.0	97.65	45.3
8.3	0.359	165	3.5	99.86	70.5
9.4	0.254	190	3.0	99.99	93.0
10.0	0.239	210	2.9	99.99	99.4

Appendix G

SUMMARY OF STATE SURFACE MINING AND MINED LAND RECLAMATION LAWS

STATE	TITLE OR CODE CITATION	MINERALS COVERED	LICENSE AND/OR PERMIT REQUIREMENTS: APPLICATION	LICENSE AND/OR PERMIT REQUIREMENTS: FEE	LICENSE AND/OR PERMIT REQUIREMENTS: PENALTY	BOND REQUIREMENTS	RECLAMATION REQUIREMENTS	PENALTY FOR FAILURE TO RECLAIM: FORFEITURE OF BOND	PENALTY FOR FAILURE TO RECLAIM: DENIAL OF NEW PERMIT	REMARKS
ALABAMA	The Alabama Surface Mining Act of 1969. Title 26, Alabama State Code, effective October 1, 1970.	All minerals except limestone, marble and dolemite.	Permit applications must be filed with the Department of Industrial Relations and be accompanied by a plan of reclamation.	Filing fee-$250. $50 fee for amended permit.	Mining without a permit-not less than $500 nor more than $5000 and a requirement that the affected land be reclaimed. Willful misrepresentation of facts on permit application-not less than $100 nor more than $500 for each offense.	$150 for each acre covered by the permit.	Reduce peaks and ridges to a width of 15 feet at the top; cover face of toxic material; divert water to reduce siltation, erosion or damage to streams and natural water courses; plant trees or direct-seed the affected land; revegetate haulage roads and land used to dispose of refuse; and construct fire lanes or access roads in areas to be reforested. Reclamation to be completed within 3 yrs. of expiration of permit period.	Yes	Yes	
ARKANSAS	The Arkansas Open Cut Land Reclamation Act of 1971. Arkansas Statutes Annotated, Title 52, Chapter 9, effective July 1, 1971.	All Minerals	Permit applications must be filed with Arkansas Pollution Control Commission and be accompanied by a reclamation plan.	$25 to $500 depending upon the number of acres to be mined.	Surface Mining without a permit-a fine of not less than $500 nor more than $1000 for each day the violation continues.	$500 for each acre or portion to be affected.	Grade peaks and ridges to a rolling topography; construct earth dams; in areas to be reforested, construct fire lanes or access roads at least 10 feet wide; strike peaks and ridges to a minimum of 20 feet at the top on all land to be seeded for pasture; cover exposed acid forming material; and dispose of refuse so as to control erosion or damage to streams or natural water courses. Reclamation to be completed prior to the expiration of 2 yrs. after termination of permit.	Yes	No	
COLORADO	The Colorado Open Cut Land Reclamation Act of 1969 as amended. Colorado Revised Statutes, Chapter 92, Article 32, effective July 1, 1972.	Coal	Permit applications must be filed with the Land Reclamation Board. A reclamation plan is required.	$50 plus $15 for each acre to be affected.	The Act provides no penalties but contain administrative procedures for dealing with violations.	The bond penalty shall be in such amount as is deemed necessary to insure the operator's performance.	Grade ridges and peaks to a width of 15 ft. at the top; where practial, construct earth dams in final cuts to impound water; cover acid forming material to protect drainage system from pollution; and dispose of all refuse so as to control stream pollution, and divert water to control siltation, erosion, or other damage to streams and natural water courses. The Act further contains specific requirements for reclaiming disturbed areas for various uses including forest, range, agricultural or horticultural crops, homesites, recreational and industrial.	Yes	Yes	
	Chapter 92, Article 32, Colorado Revised Statutes, as amended, effective July 1, 1969.	Minerals other than Coal	----------------	--------	Forfeiture of performance bond.	The Commissioner of Mines may require an operator to post a performance bond conditioned upon the faithful performance of stabilization work.	The Commissioner of Mines is empowered to examine all ore mills, sampling works, smelters, metallurgical plants, rock and stone quarries, clay pits, tunnels, sand and gravel pit excavations and plant and mines, except coal mines, to determine the method of surface stabilization used including vegetation to prevent landslides, floods or erosion. Whenever possible, the type of reclamation to be performed is determined through agreement between the Commissioner and the operator.	Yes	---	
FLORIDA	Chapter 71-105, Florida Statutes. Effective July 1, 1971.	Solid Minerals	------------------	--------	-------------	----------------	The Act imposes a severence tax on the extraction of certain solid minerals. A mine operator may obtain a refund of up to 60 percent of the tax imposed by the Act for developing and instituting a reclamation and restoration program.			Solid minerals which are extracted by the owner of the site of severance for the improvement of such site, or solid minerals upon which a sales tax is paid to the State or sold to governmental agencies in the State, including cities and counties, shall be exempt from the subject tax.

GEORGIA	Georgia Surface Mining Act of 1968. Code of Georgia Annotated, Title 43, Chapter 14. Effective January 1, 1969.	All Minerals	A license must be obtained from the Surface Mined Land Use Board. A mined land use plan is required.	$100 to $500 annually depending upon the number of mining employees employed.	The Act provides Administrative remedies (restraining orders, temporary and permanent injunctions) for violation of its provisions.	Not less than $100 nor more than $500 per acre of land affected.	Grade and backfill peaks, ridges, and valleys to a rolling topography; cover exposed toxic ores or mineral solids with a minimum of 2 feet of soil capable of supporting a permanent plant cover; and establish permanent ground cover on affected lands the first growing season following grading.	Yes	Yes	
IDAHO	The Idaho Surface Mining Act. Idaho Code, Title 47, Chapter 15. Effective May 31, 1971.	All minerals	No permit is required, but persons desiring to conduct exploration and surface mining operations must submit and have approved, by the Board of Land Commissioners, a plan of reclamation.		Any violation of the reclamation plan subjects the operator to a civil penalty, the amount of which is not specified.	Not to exceed $500 for any acre of land affected.	Level ridges of overburden to a minimum of 10 feet at the top; level peaks of overburden to a minimum of 15 feet at the top; prepare overburden piles to control erosion; minimize siltation of lakes and streams as a result of water run-off from affected lands; cross-ditch abandoned roads to avoid erosion gullies; plug exploration drill holes; when possible, top affected land with overburden conducive to erosion control and establishment of vegetative growth; prepare tailings ponds so as not to constitute a hazard to human or animal life; and complete reclamation within 1 year after surface mining operations permanently cease or are abandoned.	Yes	Yes	
	The Idaho Dredge and Placer Mining Protection Act. Idaho Code, Title 47, Chapter 13. Effective May 31,1971.	Minerals recovered with the use of dredge boat, sluice washing or other method capable of removing more than two (2) cubic yards of material per hour.	Permit application must be filed with the Board of Land Commissioners.	$50 for each 10 acres involved in the application.	Termination of permit, enjoinment from operating dredge or mine placer.	$10,000 for the 10-acre tract or less than 10 acres.	Every person conducting dredging or placer mining operations in the State is required to level and smooth the affected area comparable with the natural contour of the ground prior to the disturbance, and to a condition conducive to the growth of verdure. Whenever such operations result in substantive removal of topsoil, the operator shall be required to restore the land to its original condition by adding topsoil and planting grass, trees, and other vegetation.	Yes	Yes	
ILLINOIS	The Illinois Surface-Mined Land Conservation Act. Illinois Annotated Statutes, Chapter 93-201, Mines and Miners. Effective July 1, 1975.	All Minerals	Applications for permits must be filed with the Department of Mines and Minerals for all operations exceeding 10 feet in depth or affecting more than 10 acres during the permit year. A reclamation plan is required.	$50 plus $25 for every acre to be affected.	Surface mining without a permit-not less than $50 nor more than $1000. Each day's violation is deemed a separate offense.	$600 to $5000 for each acre to be affected including slurry and gob disposal areas.	Grade affected land to a rolling topography with slopes having no more than a 15% grade, except land reclaimed for forest plantation, recreational or wildlife, the final cut spoil, the box-cut spoil, and the outside slopes of all overburden deposition areas, the grade shall not exceed 30%; return land to be used for row crop to approximate original grade and, when available, segregate and replace at least 18 inches of topsoil; impound runoff water to reduce soil erosion, damage to unmined lands, and pollution of streams and waters; cover exposed acid forming material with not less than 4-6 feet of water or other materials capable of supporting plant and animal life; confine slurry in depressed or mined areas; remove and grade all haulage roads and drainage ditches; and plant trees, shrubs, grasses and legumes. All reclamation except slurry and gob areas in active use shall be completed prior to the expiration of 3 years after termination of the permit year.	Yes	Yes	
INDIANA*	Indiana Code 1971, 13-4, 6. Effective February 2, 1974.	Coal, clay and shale	Applications for permits must be filed with the Department of Natural Resources. A reclamation plan is required.	$50 plus $30 for each acre to be affected.	Not less than $1000 nor more than $5000.	The greater of $5000 or $600 multiplied by the number of acres for which the permit is issued.	Grading to reduce peaks and ridges to a rolling, sloping or terraced topography; construct earth dams in final cuts to impound water; bury all metal, lumber, or other debris or refuse resulting from mining; and revegetate affected areas as soon as practicable after initiation of mining operations.	Yes	Yes	
IOWA	Iowa Surface Mining Law, Title V, Chapter 83A. Effective January 1, 1968, amended August 15, 1973.	All Minerals	Permit applications must be filed with the Department of Soil Conservation.	License-$50, $10 renewal.	$50 to $500 or imprisonment not to exceed 30 days or both.	An amount equal to the estimated cost of rehabilitating each site affected.	Grade spoil banks to slopes having a maximum of 1-foot vertical rise for each 3-feet horizontal distance, except where the original topography exceeds these stipulations, and spoil bank shall be graded to blend with surrounding terrain; construct an earth dam where a lake or pond may be formed to properly control the drainage of acid water from the site; and cover acid forming material with at least 2 feet of earth or spoil material. Operators shall rehabilitate affected areas within 24 months after mining is completed.	Yew	Yes	

KANSAS	The Kansas Mined-Land Conservation and Reclamation Act. Kansas Statutes Annotated, Article 4, Chapter 49-409. Effective January 1, 1975.	Coal	Permit applications must be filed with the Mined Land Conservation and Reclamation Board. A reclamation plan is required.	$50 plus $25 per acre.	Not to exceed $250. Each day violation continues constitutes a separate offense.	Not less than $200 nor more than $500 per acre with a $2000 minimum.	Grade each pit to a flat surface with a width equal to at least 60% of the original pit; cover the face of coal or other minerals with non-acid bearing and non-toxic materials to a distance of at least 2 feet above the seam being mined, or by a permanent water impoundment; control flow of all runoff water to reduce soil erosion, damage to agricultural lands, and pollution of streams and waters; and grade overburden to provide suitable vegetative cover. Reclamation must be pursued as soon as possible after mining begins and completed within 12 months after the permit has expired.	Yes	Yes	
KENTUCKY*	Kentucky Strip Mining Law. Kentucky Revised Statutes, Title 28, Chapter 350 as amended by Chapter 3, Laws of 1972. Effective January 1, 1973.	All Minerals	Permit applications must be filed with the Division of Reclamation. A reclamation Plan is required.	$150 plus $35 per acre.	$100 to $1000 for each violation plus an additional $100 to $1000 for each day violation continues. Willful violation-not less than $500 nor more than $5000 for each day violation continues.	Operations in existence before June 23, 1974-$500 per acre or $5000, whichever is greater. Operations started on or after June 23, 1974-$1000 per acre or $5000, whichever is greater.	Complete backfilling not to exceed the original contour with no depressions to accumulate water is required of all land affected by area mining. All highwalls resulting from contour strip mining shall be reduced or backfilled, the steepest slope of the reduced or backfilled highwall and the outer slope of the fill bench being no greater than 45 degrees from the horizontal. The table portion to be terraced with a slope not greater than 10 degrees. The restored area to have a minimum depth of 4 feet of fill over the pit floor. Revegetation shall include planting trees, shrubs, grasses legumes. Reclamation to begin as soon as possible after strip mining begins and completed within 12 months after the permit has expired.	Yes	Yes	
MAINE	Conservation and Rehabilitation of Land. Maine revised Statutes, Annotated, Part 5A, Chapter 451. Effective June 1, 1971.	All Minerals except sand, gravel and borrow operations.	Permission to conduct surface mining is contingent upon approval of the operators mining plan.	$50 plus $25 for each acre to be affected not to exceed a total of $500.	Not more than $100 for each day a violation continues.	An amount to be determined by the Mining Commission of not less than $100 nor more than $1500 for each acre to be affected.	Varied-depending on planned future use of reclaimed land. The intent of the Commission is to insure that an approved permanent vegetative cover is established where possible on affected land, and that the condition in which the land is left is not conducive to erosion or pollution.	Yes	No	
MARYLAND*	Maryland Strip Mining Law. Annotated Code of Maryland, Natural Resources Article, Title 7, Subtitle 5, Strip Mining. Effective July 1, 1971.	Coal	A license and permit must be obtained from the Bureau of Mines. A reclamation plan is required.	License-$100 plus $10 for each renewal.	Failure to obtain a license-not less than $5000 nor more than $10,000 or imprisonment not to exceed 6 months, or both. Failure to obtain a permit-not less than $500 nor more than $5000. Failure to backfill prospected areas-not less than $200 nor more than $500.	$400 per acre with a $3000 minimum. A special reclamation fee of $30 per acre of land affected and a revegetation bond of not less than $50 nor more than $125 per acre are also required.	Grade spoil banks to reduce depressions between peaks of spoil to a surface which restores the terrain to a condition prescribed by the Director, Bureau of Mines; if overburden deposits are composed of materials which are suitable for supporting vegetative growth, it shall be graded so as to cover the final pit; and seal-off, with a fill, underground mining operations at the base of the final cut.	Yes	Yes	
MICHIGAN	Reclamation of Mining Lands. Michigan Statutes Annotated, Act No. 92 of the Public Acts of 1970, as amended by Act No. 123 of the Public Acts of 1972. Effective March 29, 1973.	All Minerals except Clay, gravel, marl, peat or sand.	--------------	-------	---------------	If there is doubt as to the operator's financial ability to comply with the rules of the Act, he may be required to post a performanance bond or other security.	The Act authorizes the Chief of the Geological Survey to conduct a comprehensive study and survey to determine the type of regulation needed to protect the public interest. Upon completion of the survey, rules may be promulgated governing: Sloping, terracing or treatment of stockpiles and tailings to prevent damage to fish and wildlife, pollution of waters or injury to persons or property; vegetation or treatment of tailings basins and stockpiles where natural vegetation is not expected within 5 years and where research reveals vegetation can be accomplished within practical limitations; and stabilization of the surface overburden banks of open pits in rocks and the entire bank of open pits in unconsolidated material.	---	---	

MINNESOTA	Reclamation of Lands. Minnesota Statutes Annotated, Title 8, Chapter 94, effective August 1, 1973.	Metallic Minerals	A permit to mine must be obtained from the Commissioner of Natural Resources. A reclamation plan is required.	-------	Failure to comply with the provisions of the Act-not more than $1000 for each day such failure continues.	The Commissioner determines whether or not a bond may be required.	The Commissioner of Natural Resources shall conduct a comprehensive study and survey to determine the extent to which regulation is needed to protect the public interest giving due consideration to the environment, future land utilization, protection of other natural resources and the future economic effects of such regulations on mine operators and landowners, the surrounding communities and the State of Minnesota.	---	---	Public liability insurance in an adequate amount to provide personal injury and property damage protection is also required.
MISSOURI	Reclamation of Mining Lands. Vernons Annotated Missouri Statutes, Vol. 23, Section 444.500, effective September 28, 1971.	Coal and barite.	Permit applications must be filed with the Land Reclamation Commission. A reclamation plan is required.	$50 plus $17.50 for each acre to be affected.	Mining without a permit-$1000 per day for each day the violation continues.	Not less than $300 for coal and $200 for barite nor more than $700 for coal and $500 for barite for each acre of land affected, with a $2000 minimum.	Grade peaks and ridges of overburden, except where lakes are to be formed, to a rolling topography traversable by farm machinery. The slopes need not be reduced to less than the original grade prior to mining, and the slope of overburden ridge resulting from a box cut need not be reduced to less than 25 degrees from the horizontal. Dispose of all debris, material or substance removed from the surface prior to mining.	Yes	Yes	
	The Land Reclamation Act. Vernons Annotated Missouri Statutes, Vol. 23, Section 444.760, effective September 28, 1971.	Clay, limestone, sand and Gravel.	Permit applications must be filed with the Land Reclamation Commission. A reclamation plan is required.	$50 plus $17.50 for each acre to be affected.	Mining without a permit-not less than $50 nor more than $1000. Each day violation continues is deemed a separate offense.	$500 for each acre to be affected.	Grade peaks and ridges to a rolling topography traversable by machines; construct fire lanes or access roads through areas to be reforested; strike peaks and ridges of overburden to a minimum of 25 feet at the top on all land to be reforested; on land to be used for crops, grade peaks and ridges of overburden so that the area can be traversed by farm machinery; construct lakes from mined pits and dams in final cuts; cover exposed face of mineral seam with not less than 4 feet of earth that will support plant life; and sow, set out or plant upon the affected land plants, cuttings of trees, shrubs, grasses or legumes appropriate to the designated type of reclamation.	Yes	Yes	
MONTANA	The Montana Strip Mining and Reclamation Act. Revised Code of Montana, Replacement Volume 3, Part 2, Title 50, Chapter 10, effective March 16, 1973.	Coal and uranium.	Permit applications must be filed with the Department of State Lands. A reclamation plan is required.	$50 for mining permit. $100 for prospecting permit.	Violation of provisions-fine of not less than $100 nor more than $1000. Willful violation-not less than $500 nor more than $5000. Each day violation occurs constitutes a separate offense.	Not less than $200 nor more than $2500 per acre with a $2000 minimum.	Bury under adequate fill all toxic materials; seal off breakthrough of water creating a hazard; impound, drain or treat runoff water so as to reduce soil erosion, damage to grazing and agricultural lands, and pollution of surface and subsurface waters; and remove and bury all refuse resulting from the operation. Segregate, preserve, and replace topsoil. All highwalls must be reduced, the steepest slope of which shall be no greater than 20 degrees from the horizontal. Backfilled, graded and topsoiled areas shall be prepared and planted with legumes, grasses, shrubs, and trees. Reclamation to begin as soon as possible after beginning strip mining.	Yes	Yes	
	Reclamation of Mining Lands. Chapter 12, effective September 15, 1971.	Any ore, rock or substance other than oil, gas, bentonite, clay, coal, sand, gravel, phosphate rock or uranium.	Exploration license and Development permit must be obtained from the Department of State Lands. A reclamation plan is required.	Exploration License-$5.00. Development permit-$25.00.	Not less than $100 nor more than $1000 with an additional $100 to $1000 for each day violation continues.	Not less than $200 nor more than $2500.	Reclamation of the affected land must be performed in accordance with the approved reclamation plan which contains measures for: surface gradient restoration suitable for proposed land use; revegetation or other surface treatment; public health and safety; disposal of mining debris; diverting water to prevent pollutin or erosion; reclamation of stream channels and banks to control erosion, siltation, and pollution.	Yes	Yes	
	Strip Mined Coal Conservation Act, Chapter 14, effective March 8, 1973.	Coal	-------	-------	Mining without an approved strip mining plan-not less than $100 nor more than $1000 and an additional $100 to $1000 for each day violation continues.	-------	The Act provides for the conservation of strippable and marketable coal by requiring each operator wishing to conduct strip coal mining within the State to submit, for approval of the Department of State Lands, a strip mining plan which outlines the planned course of conduct of a strip mining operation including plans for the removal and utilization of strippable and marketable coal located within the area planned to be mined.	---	---	
	The Open Cut Mining Act. Chapter 15, effective March 16, 1973.	Bentonite, clay, scoria, phosphate rock, sand and gravel.	Applications for contracts must be made to the Board of Land Commissioners if the planned operation involves removing 10,000 cubic yards or more of product or overburden. A reclamation plan is required.	$50	Mining without a contract-not less than $500 nor more than $1000. Each day's violation is considered a separate offense.	Not less than $200 nor more than $1000 per acre.	Reclamation must be carried out in accordance with the approved reclamation plan which requires that the land be reclaimed for specified uses including forest, pasture, orchards, cropland, residence, recreation, industry, or wildlife habitat. Reclamation requirements include: establishment of vegetative cover; control water drainage; grading; removal or burial of metal or waste; and revegetation of affected area.	Yes	Yes	

MONTANA (cont.)	The Strip Mine Siting Act, Chapter 16, effective January 1, 1974.	Coal and uranium.	A minesite location permit must be obtained from the Department of State Lands before preparatory work (construction of buildings, railroads, storage and train load-out facilities, transmission lines, etc.) begins.	$50	Not less than $100 nor more than $1000 for the violation, plus an additional $100 to $1000 for each day violation continues. Willful violation-$500 to $5000. Each day's violation constitutes a separate offense.	Not less than $200 nor more than $10,000 per acre with a $5000 minimum.	The type of reclamation is determined by future suitable use of the land including removing or burying facilities, backfilling, grading, and topsoiling.	Yes	Yes	
NEW MEXICO	Coal Surface Mining Act. New Mexico Statutes Annotated, Replacement Vol. 9, Part 1, Article 34, effective February 29, 1972.	Coal	Application for permit must be filed with the Coal Surface Mining Commission. Mining plans must accompany permit applications.	$50 application fee. $10 initial acreage fee. Annual fee of $20 per acre for each acre affected during the preceding year.	$1000 for each day violation continues.	The Surface Coal Mining Commission may require an operator to file a bond in an amount sufficient to insure compliance.	Grade to produce a gently undulating topography or such other topography as is consistent with planned end use of the land. Grading shall be done in such a manner as to control erosion and siltation of the affected area and surrounding property and water courses. Revegetation of the affected area must be accomplished in accordance with the previously approved mining plan.	---	---	
NEW YORK	The New York State Mined Land Reclamation Law, McKinney's Consolidated Laws of New York, Environmental Conservation Law, Title 27, effective April 1, 1975.	All Minerals	A permit is required for any operator mining 1000 tons or more in 12 successive months. A reclamation plan is required.	$100 annually or $200 for 3 years.	Suspension or revocation of permit.	The amount of bond is to be determined by the administering agency.	Reclamation to be accomplished in accordance with the approved reclamation plan which shall indicate specifics covering revegetation, disposal of debris, refuse, tailings, waste, and spoil grading. Reclamation to be completed within a 2-year period after mining ceases.	Yes	Yes	
NORTH CAROLINA*	The Mining Act of 1971. General Statutes of North Carolina, Chapter 74, Article 7, Amended 1973, effective July 1, 1974.	All Minerals	Application for a permit must be filed with the Department of Conservation and Development. A reclamation plan is required.	No fee is required. Permit will be granted if the reclamation plan is approved.	Willful violation $100 to $1000 fine. Each day constitutes a separate violation.	$2,500 to $25,000 depending upon the number of acres to be affected.	Reclamation must be performed in accordance with approved reclamation plan which must meet the following standards: The final slopes in all excavations in soil, sand, gravel, and other unconsolidated materials shall be at such an angle as to minimize the possibility of slides and be consistent with the future use of the land. Provisions for safety to persons and to adjoining property must be provided in all excavations in rock. In open cast mining operations, all overburden and spoil shall be in a configuration which is in accordance with accepted conservation practices and which is suitable for the proposed subsequent use of the land. Suitable drainage ditches or conduits shall be constructed to prevent collection of small pools of water that are noxious, odious, or foul. The type of vegetative cover and method of its establishment shall conform to accepted agronomic and reforestation practices.	Yes	Yes	
NORTH DAKOTA	Reclamation of Strip Mined Lands, Title 38, Chapter 38-14, effective July 1, 1973.	All Minerals	Applications for permits must be filed with the Public Service Commission for all planned operations exceeding 10 feet in depth. A reclamation plan is required.	Up to ten acres-$25 plus $10 times the number of acres to be affected between two and ten; eleven to fifty acres-$100 plus $10 times the number of acres between eleven and fifty. More than fifty acres-$275 plus $10 times the number of acres in excess of fifty acres.	Mining without a permit-fine of not less than $50 nor more than $1000. Each day violation continues constitutes a separate offense.	$500 for each acre to be affected.	Regrade affected area to approximate original contour, or rolling topography or topography for higher end use; spread topsoil or other suitable soil material over the regraded area to a depth of two feet; impound or treat runoff water to reduce soil erosion, damage to agricultural lands and pollution of streams; back-slope final cuts and end walls to an angle not to exceed 35 degrees from the horizontal (operator may propose alternative to backfilling if consistent with the Act); remove or bury all debris; and sow, set-out, or plant cuttings or trees, shrubs, grasses, or legumes. All reclamation shall be carried to completion prior to the expiration of three years after termination of the permit term.	Yes	Yes	

OHIO	Ohio Reclamation Law. Ohio Revised Code Annotated, Title 15, Chapter 1513, effective April 10, 1972.	Coal	Applications for licenses must be filed with the Division of Reclamation. A reclamation plan is required.	$100 plus $30 for each acre to be mined.	Mining without a permit-$5000 plus $1000 per acre of land affected. Exceed limits of license-$1000 per acre of land affected that is not under license. Willful misrepresentation-$100 to $1000 or 6 months. Violation of any other provision-$100 to $5000 or 6 months in prison, or both.	Sufficient to cover the cost of reclamation but not less than $5000.	Cover all acid producing materials with nontoxic material; construct and maintain access roads; prevent the pollution of waters, erosion, landslides, flooding and the accumulation or discharge of acid water; contour the affected area unless the mining and reclamation plan provides for terracing or other uses; and replace segregated topsoil and grow vegetative covering.	Yes	Yes	The Act levies a severance tax on the following minerals for the purpose of reclaiming land affected by strip mining: (a) four cents per ton of coal; (b) four cents per ton of salt; (c) one cent per ton of limestone and dolomite; (d) one cent per ton of sand and gravel; (e) three cents per barrel of oil; (f) one cent per thousand cubic feet of natural gas.
	(Ohio Revised Code Annotated, Title 15, Chapter 1514, effective September 27, 1974.)	All minerals except coal.	Application for permit must be filed with the Division of Reclamation. A reclamation plan is required.	$150 plus $30 for each acre to be affected but not to exceed $1000 per	Mining without a permit-$5000 plus $1000 per acre of land affected. Exceeding limits of permit-$1000 per acre of land affected. Willful misrepresentation-$100 to $1000 or 6 months, or both. Violation of any other provision-$100 to $5000 or 6 months, or both.	$500 per acre of land to be affected.	Grade, contour, or terrace final slope; resoil affected land with topsoil or suitable subsoil; establish diverse vegetative cover of grass, legumes, or trees; remove or bury refuse resulting from mining; insure prevention of contaminating underground water supply; and control drainage to prevent floods or landslides.	Yes	Yes	Public Liability Insurance: $100,000 to $200,000 for bodily injury and $100,000 to $300,000 for property damage.
OKLAHOMA*	The Mining Lands Reclamation Act. Oklahoma Statutes Annotated, Title 45, Chapter 8A, effective June 12, 1971, amended April 17, 1972.	All Minerals	Application for permits must be filed with the Department of Mines and Mining. A reclamation plan is required.	$50	Mining without a permit-not less than $50 nor more $1000. Each day constitutes a separate offense.	Not less than $350 nor more than $650 for each acre to be affected. For coal and copper mining the minimum bond shall be $5000. For all other mining the minimum bond shall be $1000. Sand and gravel operators who sell less than $1000 per year may be exempt.	Grade peaks and ridges of overburden to a rolling topography, but the slopes need not be reduced to less than the original grade prior to mining, and the slope of ridge resulting from the box cut need not be reduced to less than 25 degrees from the horizontal; construct earth dams to form lakes in pits resulting from surface mining operations; cover exposed faces of mineral seams with not less than 3 feet of earth to support plant life or with a permanent water impoundment; and revegetate affected land, except that which is to be covered with water or used for homesites or industrial pruposes, by planting trees, shrubs or other plantings appropriate to future use of the land.	Yes	Yes	
OREGON	An Act Relating to mining. Oregon Revised Statutes, Title 43, Mines and Minerals, effective July 1, 1972.	All Minerals	Permits must be obtained for all operation exceeding 10,000 cubic yards of material extracted or at least 2 acres of land affected within a period of 12 consecutive calendar months. A reclamation plan is required.	Basic fee-$150. Annual renewal fee-$50.	Mining without a permit-a fine not exceeding $1000. Violation of any rules or regulations is punishable by a fine of not less than $25 nor more than $250, or imprisonment for not more than 60 days, or both.	Not to exceed $300 per acre to be surface mined.	Reclamation of the affected land must be performed in accordance with the approved reclamation plan which must contain: measures to be undertaken by the operator in protecting the natural resources of adjacent lands; measures for the rehabilitation of the surface-mined lands and the procedures to be applied; procedures to be applied in the surface mining operation to control the discharge of contaminants and the disposal of surface mining refuse; procedures to be applied in the rehabilitation of affected stream channels and stream banks to a condition minimizing erosion, sedimentation and other factors of pollution; such maps and other documents as may be requested by the Department of Geology and Mineral Industries; and a proposed time schedule for the completion of reclamation operations.	Yes	Yes	

PENNSYLVANIA*	Surface Mining Conservation and Reclamation Act. Pennsylvania Statutes Annotated, Title 52, Chapter 1396, approved May 31, 1945, amended by Public Acts of 1963 and 1968; Public Act No. 147, November 30, 1971, effective January 1, 1972; Public Act 355, approved December 28, 1972; Public Act No. 941, approved October 18, 1974.	All Minerals	Application for permits must be filed with the Department of Environmental Resources. A reclamation plan is required.	$50 for persons mining 2000 tons or less of marketable minerals other than coal per year, and $500 for mining coal or more than 2000 tons of other marketable minerals per year. Annual renewal-$50 for mining 2000 tons or less of marketable minerals other than coal and $300 in the case of all other minerals.	Mining without a permit-$5000 or an amount of not less than the total profits derived from unlawful activities, together with the cost of restoring the land to its original condition or 1 year imprisonment, or both.	An amount sufficient to insure completion of the reclamation plan, but not less than $5000, except in the case of minerals other than anthracite and bituminous coal where it is determined that the amount of marketable minerals to be extracted does not exceed 2000 tons, no bond shall be required. Liability under the bond shall be for the duration of the operation and for 5 years thereafter.	Backfill all pits within 6 months after completion of mining. Such backfilling shall be terraced or sloped to an angle not to exceed the original contour. Plant grasses and trees or grasses and shrubs upon affected land within 1 year after backfilling.	Yes	Yes	Operators mining minerals other than anthracite and bituminous coal, the amount of which does not exceed 2000 tons, shall be exempt from obtaining the required $100,000 certificate of public liability insurance and posting the required bond.
SOUTH CAROLINA*	The South Carolina Mining Act. Code of Laws of South Carolina, Title 63, Chapter 5, effective July 1, 1974.	All Minerals	Applications for permits must be filed with the Land Resources Conservation Commission. A reclamation plan is required.	--------	Willful violation of the Act or misrepresentation of facts or giving false information on permit applications-not less than $100 nor more than $1000 fine for each day the violation continues.	$2,500 to $25,000 or a greater amount depending upon the number of acres to be affected.	Reclamation to be performed in accordance with the approved reclamation plan which must meet the following standards: The final slopes in all excavations shall be at such an angle so as to minimize the possibility of slides; provide safety of persons and of adjoining property; in open cut mining, overburden and spoil shall be left in a configuration suitable for subsequent use of the land; and construct suitable drainage to prevent the collection of small pools of water that are noxious or likely to become noxious, odious, or foul. The type of vegetative cover and method of its establishment shall conform to accepted recommended agronomic and reforestation practices. The plan must further provide that reclamation activities be completed within 2 years after completion or termination of mining on each segment of the area for which a permit is issued unless a longer period is specifically authorized.	Yes	Yes	
SOUTH DAKOTA	Surface Mining Land Reclamation Act. South Dakota Compiled Laws of 1967, Title 45, Chapter 45-6A, effective July 1, 1971.	All Minerals	Permit applications must be filed with the State Conservation Commission. A reclamation plan is required.	$50 - $25 for each renewal.	Violation of the Act's provisions-a fine of not less than $1000 for each day the violation continues.	An amount sufficient to cover the cost of reclamation.	Isolate all toxic or other material that have a damaging effect upon ground and surface waters, fish and wildlife, public health and the environment; reclaim surface mined areas to control erosion, provide vegetation, and eliminate safety hazards; replace topsoil evenly over reclaimed area; revegetate in accordance with agronomic and forestry recommendations; and upon completion of operations, remove all structures, machinery, equipment, tools and materials from the site of operation.	Yes	Yes	
TENNESSEE[a]	The Tennessee Surface Mining Law. Tennessee Code Annotated, Chapter 15, effective March 23, 1972, amended by House Bill 630, approved March 20, 1974.	All minerals except limestone, marble, and dimension stone.	Applications for permits must be filed with the Commissioner, Department of Conservation. A reclamation plan is required.	$250 plus $25 for each acre to be mined. The total amount not to exceed $2,500.	Violation of the Act-fine of not less than $100 nor more $5,000 for each day violation continues. Willfull violation-not less than $1000 nor more than $5000 or imprisonment not to exceed 1 year, or both.	Not less than $400 for minerals other than coal and not less than $600 for coal for each estimated acre to be affected.	Coal: cover all acid producing material; seal off any breakthrough in mine or pit walls which creates a hazard; control drainage to prevent damage to adjacent lands, soil erosion and pollution of streams and waters; remove all refuse except vegetation resulting from the operation; provide adequate access roads to remote areas; on steep slopes, regrade area to approximate original contour or rolling topography and eliminate highwalls, spoil piles and water-collecting depressions. (Grading and other soil preparation to accomodate vegetation shall be completed within 6 months following initiation of soil disturbance.) Revegetate the affected area with grasses or legumes to prevent soil erosion. Minerals other than coal: regrade the area to approximately the original or rolling topography, and eliminate all highwalls, spoil piles, and water collecting depressions; control drainage to prevent soil erosion, damage to adjacent lands, and pollution of streams and other waters; and revegetate with trees, grasses, or legumes.	Yes	Yes	

VIRGINIA	Chapter 17, Title 45.1, Code of Virginia (1950), as amended. Effective April 10, 1972.	Coal	Permit applications must be filed with the Department of Conservation and Economic Development. A reclamation plan is required.	Prospecting permit-$10 per acre. Surface mining permit-$12 per acre. Annual fee-$6 per acre.	Violation of the Act-Fine of not more than $1000 or imprisonment for not more than 1 year or both. Each day violation continues constitutes a separate offense.	Prospecting-$300 per acre. Surface mining bond-not less than $200 or more than $1000 per acre to be mined. Minimum bond-$2500, except when the operation involves less than 5 acres, the bond shall not be less than $1000.	Remove all debris resulting from mining operations; regrade the area in a manner established by rules and regulations; grade overburden to reduce peaks and depressions between peaks to produce a gently rolling topography; preserve existent access roads; and plant trees, shrubs, grasses or other vegetation upon areas where revegetation is practicable.	Yes	Yes	
	Title 45.1, Chapter 16, Code of Virginia, 1950 as amended. Effective June 27, 1966.	Other minerals	Permit applications must be filed with the Department of Conservation and Economic Development. A reclamation plan is required.	$6 for each acre to be affected, not to exceed a total of $150.	Violation of the Act-maximum fine of $1000 or 1 year in jail, or both.	$50 per acre based upon the number of acres to be disturbed. The minimum amount of bond furnished shall be $1000.	Same as for coal, except that in the case of dimensional stone and quarry operations, special consideration is given to the peculiar nature of the excavated cavity.	Yes	Yes	
WASHINGTON	Surface-Mined Land Reclamation Act, Revised Code of Washington Annotated, Title 78, Mines and Minerals, Chapter 78-44, effective January 1, 1971.	All Minerals	Permit applications must be filed with the Department of Natural Resources. A reclamation plan is required.	$25 per permit year for each location plus $5 per acre for all acreage exceeding 10 acres which was disturbed during the previous permit year.	Mining without a permit-misdemeanor. An operator can be enjoined or otherwise stopped. Each day's violation constitutes a separate offense.	Not less than $100 nor more than $1000 per acre.	In reclaiming excavations for use as lakes, all banks shall be sloped to 2 feet below the groundwater line at a slope no steeper than 1½ feet horizontal to 1 foot vertical. In all other excavations, the side slopes shall be no steeper than 1½ feet horizontal to 1 foot vertical for their entire length. All strip pits and open pits shall be no steeper than 1 foot horizontal to 1 foot vertical. The slopes of quarry walls shall have no prescribed slopes, except where a hazardous condition is created the quarry shall be graded or backfilled to a slope of 1 foot horizontal to 1 foot vertical. In strip mining, peaks and depressions of spoil banks shall be constructed to a gently rolling topography. Suitable drainage shall be constructed to prevent the collection of stagnant water. All grading and backfilling shall be made with non-noxious, non-flammable, noncombustible solids. All acid-forming materials shall be covered with at least 2 feet of clean fill. Vegetative cover shall be required and all surface mining that disturbs streams must comply with State fisheries laws.	Yes	Yes	
WEST VIRGINIA*	West Virginia Surface Mining Act, West Virginia Code, Vol. 8, 1970 Replacement Volume, Chapter 20, effective March 13, 1971.	All Minerals	Applications for permits must be filed with the Department of Natural Resources.	Prospecting-$300. Surface mining-$500. Annual renewal-$100. Personal injury and property damage insurance of $100,000 and $300,000 respectively are also required.	Violation of the law's provisions-$100 to $1000 fine or 6 months imprisonment, or both. Deliberate violation-$1000 to $10,000 fine or 6 months imprisonment, or both.	Not less than $600 per acre nor more than $1000 per acre with a $10,000 minimum. A special reclamation tax of $60 per acre is also required.	Cover the face of coal and disturbed area with material suitable to support vegetative cover; bury acid-forming materials, toxic material, or materials constituting fire hazard; impound water. Bury all debris. The law also contains requirements for regrading surface mined areas where benches result specifying the maximum bench width allowed. On land where benches do not result complete backfilling is required but shall not exceed the original contour of the land. The backfilling shall eliminate all highwalls and spoil peaks. Planting is required.	Yes	Yes	An operator who causes damage to property of others shall be liable to them in an amount not in excess of three times the provable amount of such damage. A $10,000 certificate of insurance must be filed to cover such damage.
WYOMING	The Wyoming Environmental Quality Act, Wyoming Statutes, Title 35, Article 4, Land Quality, effective July 1, 1973.	All Minerals	Applications for permits must be filed with the Administrator, Division of Land Quality. A reclamation plan is required.	Surface mining fee-$100 plus $10 for each acre to be affected with a $2000 maximum. Amended permit-$200 plus $10 per acre with a $2000 maximum. License fee for mineral exploration-$25.	The Act imposes fines ranging from $10,000 to $50,000 per day depending upon the violation involved. Criminal penalties are also prescribed for certain violations ranging from 6 months to 2 years imprisonment.	Not less than $10,000 except for scoria or jade and sand and gravel, in which case the bond shall not be less than $200 per acre.	Protect the removed and segregated topsoil from wind and water erosion and from acid or toxic materials; cover, bury, impound or otherwise contain radioactive material; conduct contouring operation to achieve planned use; backfill, grade, and replace topsoil or approved subsoil; replace vegetation; prevent pollution of surface and subsurface waters; and reclaim affected land as mining progresses in conformity with the approved reclamation plan.	Yes	Yes	

*-MEMBER OF THE INTERSTATE MINING COMPACT

Source: Division of Environment, U.S. Bureau of Mines, Department of the Interior, Washington, D.C. 1975

Index